THE OXFORD COMPANION TO AUSTRALIAN FOLKLORE

THE OXFORD COMPANION TO AUSTRALIAN FOLKLORE

EDITED BY

GWENDA BEED DAVEY
and
GRAHAM SEAL

Melbourne
OXFORD UNIVERSITY PRESS
Oxford Auckland New York

OXFORD UNIVERSITY PRESS AUSTRALIA
Oxford New York Toronto
Delhi Bombay Calcutta Madras Karachi
Kuala Lumpur Singapore Hong Kong Tokyo
Nairobi Dar es Salaam Cape Town
Melbourne Auckland Madrid
and associated companies in
Berlin Ibadan

First published 1993

National Library of Australia
Cataloguing-in-Publication data:

Seal, Graham, 1950– .
The Oxford Companion to Australian Folklore.

Includes index.
ISBN 0 19 553057 8.

1. Folklore — Australia. I. Davey, Gwenda. II. Title.
398.20994

Edited by Sarah Brenan
Typeset by Bookset Pty Ltd, Victoria
Printed by Impact Printing Pty Ltd, Victoria
Published by Oxford University Press,
253 Normanby Road, South Melbourne, Australia

PREFACE

One of Australia's many unique features has been, until very recently, an almost total disregard for the collection, preservation, study and general recognition of its folk heritage and its contemporary folklore. Unlike those of most other nations, Australian governments have never found it necessary to provide institutions for the study of folklore; no universities provide degrees or other formal accreditation for folklore studies; and apart from a lamentably restricted repertoire of mostly 'bush' songs, Australian folklore is unknown in primary and secondary schools. Almost all the work of collection, scholarship and preservation of folklore that has been carried out here has been by private individuals and organizations. While collections of bush ballads were made earlier in the nineteenth century by interested amateurs, A. B. Paterson has the honour of being the first to attempt a serious collection of one aspect of Australian folklore — the 'old bush songs', as he termed them in his anthology first published in 1905. Like virtually all those who succeeded him, Paterson carried out his work with no institutional support or funding. His only chance of recouping expenses and perhaps of obtaining some financial reward for his efforts was through publishing — and that only after the work had been completed.

Here is an intriguing problem. Since the late nineteenth century there has been a flourishing Australiana publishing industry presenting Australian folklore, amongst other things, to an avid public. The continuing existence of this readership indicates a very high level of popular interest in aspects of Australian folklore — songs, music, yarns, verse, speech, crafts and other traditions. Why then, given this interest and activity, at a time when Britain, the United States and many European countries were developing impressive bodies of folkloric knowledge and scholarly facilities for analysing that knowledge, did Australia not do likewise? Why, given the close connection between national aspirations and folklore studies, did Australia not investigate its own traditions at a time when nationalism was fervent, as the century began? Perhaps another manifestation of the pervasive 'cultural cringe'? Such questions are of considerable interest, and though they require far more discussion than can be given here, the answers to them may well reveal a good deal about our unspoken assumptions as a nation state. These unanswered questions lie behind the necessity for a book such as this, the first Oxford Companion to Folklore in the world; an innovation occasioned in Australia by a paucity rather than a surfeit of knowledge, by an absence of co-ordinated collection and study of national traditions and folklore. It is an Australian 'first' because few other countries need such a work. While it is not the purpose of this *Companion* to provide even a representative sampling of the diversity of Australian folklore — if such were possible — it does attempt to draw together the various threads of what is known and to present these to the reader in an informed and accessible manner. As well, it tracks the history of the development of folklore studies in this country, identifying significant individuals, institutions, events and movements that have both shaped and been shaped by involvement —

whether scholarly, performing, conserving, or simply consuming — with the broad field of Australian folklore.

While there has been so little official activity in Australian folklore, the efforts of numerous individuals and community groups since the 1950s have increased our knowledge, and to some extent our understanding, of the processes and the people involved in Australian folklore, past and present. The increasing recognition of Australia's polycultural society in the last two decades has been a great stimulus to such efforts, as has the maturing awareness of the value of Australia's cultural distinctiveness. While those aspects of folk heritage revolving around the bush and the 'Australian Legend' remain important to many Australians, it is also clear that more recently imported folk traditions have taken root here, in many cases interacting with earlier lore, and that these have become an integral part of Australian folklore. As well, the ever-changing nature of modern life and technology has brought many new folklore forms into existence and led to the adaptation of many others. *The Oxford Companion to Australian Folklore* attempts to notice all these diverse and sometimes conflicting trends and themes. While it has not been possible in the space available to do more than briefly summarize even major aspects of folklore, we have gone to considerable lengths to be as comprehensive as possible. Readers will find at least some information on most areas of interest in this *Companion*, together with some indication of sources for further investigation.

One of the enjoyable and encouraging things about compiling a work such as this has been the discovery of a number of people who have been working in related areas of study. While these individuals may not use the term 'folklore' in describing their work, it is clear from their willingness to contribute to this volume that they find something of value in the term and the notions involved in it. Precisely what the term should encompass has long been a matter of debate; the ins and outs of that debate are canvassed in some detail in the Introduction. Suffice to say here that any reasonably comprehensive attempt at definition will include the face-to-face elements of much folklore; its informality and unofficial nature; the variation of content it typically displays; its group orientation and its shared or communal character. The term 'communal', by the way, should not be taken to mean that all folklore is necessarily pleasant or positive. Folklore can be extremely offensive, articulating sexist, racist, sectarian or otherwise bigoted points of view, as discussed in some of the entries herein. This is one important reason why we need to know more about our folklore — to find ways of combating the prejudices, ignorance and misconceptions that often underlie certain kinds of folk expression. Folklore is now recognized as an important social and cultural process that is in many ways distinct from other aspects of culture, often labelled 'high' and 'popular', yet is also in constant interaction with them. The *Companion* adopts this broad contemporary understanding of folklore and folklife.

Few limitations have been placed upon contributors, all of whom have written in accordance with their own perspectives on the subject. Articles have been sought where expertise is known to exist; most of our sins of omission arise from a lack of suitable contributors. Our aim has been to provide authoritative information about the state of knowledge of our folklore up to the present. The result is a rich and varied documentation of a largely unrecognized and ignored facet of Australian heritage and contemporary culture — the expressions and the actions of the Australian people in all their ethnic, religious, regional, occupational and other diversities.

While attempting to be as comprehensive as possible, the *Companion* cannot hope to discuss or even mention all of the multitude of characters, events, beliefs, customs and other details of Australian folk expression and practice. One thing that has become clear to us in preparing this book is that there is a great need for a number of other, cognate works on Australian folklore. One such work would be an authoritative dictionary noting and explaining the minutiae of Australian folklore. Another would be a comprehensive anthology of Australian folklore, or rather Australian folklore*s*, as the need to encompass and comprehend non-Anglo-Celtic traditions as discrete yet interacting elements of Australian folklife is now urgent. If the present work can stimulate the production of such essential references, the efforts of the many who contributed to it will be more than justified.

As this book is published near the start of the last decade of the century, nearing the first century of federation, the long neglect of Australian folklore may be coming to an end. Most of the nation's major cultural retrieval and preservation institutions are actively, if sometimes tentatively, involved in folklore research of various kinds. Folklore courses are now being taught in a few academic institutions. An Australian Folklife Centre has been established, currently housed in the National Museum of Australia. The National Library of Australia, the National Film and Sound Archive, and the Australian National Gallery all have active programmes of various kinds related to folklore and folklife. The Western Australian Folklore Archive is situated in the Centre for Australian Studies at Curtin University of Technology, while the earliest archive to be established devoted specifically to folklore, the Australian Children's Folklore Collection, is now a part of the Australian Centre at the University of Melbourne. *The Oxford Companion to Australian Folklore* has received contributions from all these institutions, as well as from many others, from some of the country's most distinguished academics and from private individuals and organizations across the country who continue to provide the essential information necessary for the development of folklore studies in Australia. We are greatly in their debt.

The *Companion* follows the model of its numerous sister volumes published by Oxford University Press. Entries are arranged in strict alphabetical order, with extensive cross-references to other relevant topics. Within the constraints of space, we have attempted to indicate further readings and relevant sources.

The editors would like to acknowledge the support or advice of the following institutions and individuals: School of Communication and Cultural Studies, Curtin University of Technology, particularly Rae Kelly; Centre for Australian Studies, Curtin University, particularly Don Grant and Thelma Buktenica; Oxford University Press, particularly Sandra McComb, Yvonne White, Peter Rose and Sarah Brenan; National Library of Australia; Bert Davey; June Factor; Keith McKenry; Maureen Seal; Richard Walsh and Australian

Consolidated Press for permission to reproduce sections of Monika Allan's article on Country Music, first published in *Australian Almanac and Book of Facts 1988* (Angus & Robertson), and Susan Jarvis, Kevin Knapp, Richard Keams for assisting with the revision of that article.

The following major contributors are most sincerely thanked for their work: Monika Allan, Sydney; Shirley Andrews, Melbourne; Alex Barlow, Canberra; Philip Batty, Central Australian Aboriginal Media Association, Alice Springs; Kevin Bradley, National Library of Australia, Canberra; Colleen Burke, Sydney; Dr Philip Butterss, University of Adelaide; Frank Campbell, Deakin University, Geelong; Anna Chatzinikolaou, University of Melbourne; Professor Michael Clyne, Monash University, Melbourne; Peter Ellis, Bendigo; Warren Fahey, Sydney; Feng Wei, the Chinese Folk Literature and Art Society, Beijing and Guangzhou (Canton); Dr Eve Fesl, Koorie Research Centre, Monash University, Melbourne; Dr Stathis Gauntlett, University of Melbourne; Dr Hazel Hall, Canberra; Peter Hamilton, The Channon; Dr Ian Harmstorf, University of Adelaide; David Hults, Perth; Nancy Keesing, Sydney; Lin Zeshen, the Chinese Folk Literature and Art Society, Beijing and Guangzhou (Canton); Dr Keith McKenry, Canberra; John McPhee, National Gallery of Victoria, Melbourne; Nita Murray-Smith, Mt Eliza; Dr Joan Newman, Curtin University of Technology, Perth; Helen O'Shea, Melbourne; Dr Bill Ramson, Australian National Dictionary Centre, Australian National University, Canberra; Professor Michael Roe, University of Tasmania, Hobart; Margaret Rolfe, Canberra; Dr Graeme Smith, Melbourne; Dr Bob Reece, Murdoch University, Perth; Dr Jill Stubington, University of New South Wales, Sydney; Dr Edgar Waters, Canberra; Professor Jerzy Zubrzycki, Canberra.

Gwenda Beed Davey and Graham Seal

INTRODUCTION: WHAT IS FOLKLORE?

THERE IS little disagreement amongst folklorists about the notion that folklore is a distinct level of culture and that folk culture is different in many ways from either high/elite culture or commercial popular culture. For example, in his book *The Study of American Folklore* (1978) Jan Brunvand from the University of Utah categorized four levels of culture, defined as elite (academic, progressive, etc.), normative (popular, mass, mainstream, etc.), folk (conservative, traditional, peasant, etc.), and primitive (nonliterate, uncivilized etc.). According to the Australian government's Committee of Inquiry into Folklife in Australia (1987), the distinction between high, popular and folk culture is in 'common usage in Australia'. In *The Hidden Culture: Folklore in Australian Society* Graham Seal wrote that folklore differs from 'popular culture', which is generally understood to refer to the products and effects of the mass communications industries, and is also clearly distinct from anything that might be labelled 'high' or 'elite' culture, such as the opera or the ballet. Folklore, however, may both influence these forms and be influenced by them (see **Popular culture and folklore**).

In Australia at least, these distinctions acquired considerable political importance at the time of the Australian Parliament's Inquiry into Commonwealth Assistance to the Arts. Its report, *Patronage, Power and the Muse* (1986) was produced by a sub-committee of the House of Representatives Standing Committee on Expenditure, commonly known after the sub-committee's chairman Mr L. B. McLeay MP as 'the McLeay Report'. The Inquiry produced bitter argument about the allocation of funds by the Australian government's arts funding body, the Australia Council, in particular about the three 'flagship' companies, the Elizabethan Theatre Trust, the Australian Opera and the Australian Ballet. These were 'the three largest clients of the Australia Council, accounting together for almost 26% of its arts support budget in 1984–85' (McLeay Report, p. 133). The heavy funding of 'high culture' was opposed by a wide range of persons from community arts, the rock music industry, and by folklorists.

The different and sometimes oppositional relationship between the levels of culture is widely accepted in folklore, anthropology and many other disciplines. For example, the anthropologist Redfield distinguished between 'Great Tradition' and 'Little Tradition', the historian Lawrence Levine between 'Highbrow' and 'Lowbrow' and the historian and folklorist Richard Dorson referred to 'folk culture [as] in one sense opposed to the elite culture'. Yet there is much less agreement about the precise nature of folk culture, and many attempts have been made to define it. The English antiquary William Thoms is usually credited with the introduction of the term 'folklore' into nineteenth-century and subsequent scholarship. Using the name of Ambrose Merton, he wrote to the London magazine the *Athenaeum* on 22 August 1846, in a letter which almost incidentally proposed the use of a word which was a 'good Saxon compound, Folklore'. Thoms sought to enlist the aid of the magazine in gathering 'the few ears . . . remaining' of 'the manners, customs, observances, superstitions, ballads, proverbs, etc., of the olden time'. Yet antiquarian scholarship such as that of Thoms is much older than the mid-nineteenth-century. From 1812 the brothers Jacob and Wilhelm Grimm published many volumes of folk tales and mythology, collected from oral tradition. The Grimms had in turn been preceded by sixteenth- and seventeenth-century folk tale collections in France and Italy, such as those of Perrault, Basile and Straparola. Although none of these collections was produced for children, it is interesting in view of the contemporary 'appropriation to the nursery' of much traditional verbal folklore such as folk tales and nursery rhymes to note that one of the earliest extant collections of folklore was in fact produced for a child. *El Conde Lucanor o Libro de Patronio* (the Count of Lucanor or the Book of Patronio) was produced for his son by the Infante Don Juan Manuel in 1335. Three of these stories are reprinted in Carmen Bravo-Villasante's *Antologia de la Literatura Infantil en Lengua Espanola*, Vol. I (1973). The collection is of particular interest because it appears to contain the earliest known written version of 'The Emperor's New Clothes', later used by Hans Christian Andersen, and also 'The Man Who Married a Shrewish Woman', the story used by William Shakespeare for *The Taming of the Shrew*.

Folk tales were not the only items of interest to earlier collectors and recorders of 'popular antiquities'. In his Foreword to *Nordic Folklore: Recent Studies* (Kvideland & Sehmsdorf [eds], 1989, pp.viii–ix), Dan Ben-Amos reproduces the full text of an extraordinary document signed by King Gustavus Adolphus of Sweden on 20 May 1630, on the eve of his departure for the Thirty Years' War. The proclamation established a Council on Antiquities and outlined a brief 'which those who are appointed to be the antiquarians and historians of the Kingdom should follow'. The objects to be collected should include old runic inscriptions, calendars, almanacs, ancient law books, chronicles and histories,

> immemorial tales and poems about dragons, dwarfs, and giants, as well as tales about famous persons, old monasteries, castles, royal residences, and towns . . . the music of old warriors and runic songs.

Old letters should be examined and copied, and old coins and currencies collected. 'Thorough inquiries' should be made about the topographical features of the country, about household records 'and all kinds of economic matters', and all kinds of land measurements should be listed. The names of field tools and ship tools should be recorded, and deposits of ore should be sought.

> [They should] record the different cures and search for medical books, noting down the names of herbs and trees. [They should find out about] the signs for forecasting the weather . . . [and] describe the old costumes and weapons that are used in each province and [write] about the drinking vessels, kosor, horns and other objects etcetera.

It would be some time before any one document again provided so much in the way of description or definition of a nation's folk culture. In 1873 Edward Burnett Tylor, the pioneer of modern anthropology, described many aspects of folk culture to support his 'survivalist' theories of culture. In Chapter III of *The Origins of Culture* (1873, 1970), Tylor discusses superstition, building techniques, children's games, fire-lighting techniques, games of chance, traditional sayings, nursery poems, proverbs, riddles, divination and fortune-telling. In his view, many of these activities derived from archaic times, when they performed a radically different function.

In 1888 the American Folklore Society was founded by William Wells Newell in Cambridge, Massachusetts. In her book entitled *American Folklore Scholarship: A Dialogue of Dissent* (1988), Rosemary Levy Zumwalt describes the first volume of the *Journal of American Folklore* (1888) which opened with a statement by Newell indicating that the journal would provide

> 1. For the collection of the fast-vanishing remains of Folk-Lore in America, namely:
>   (a) Relics of Old English Folk-Lore (ballads, tales, superstitions, dialect, etc.).
>   (b) Lore of Negroes in the Southern States of the Union.
>   (c) Lore of the Indian Tribes of North America (myths, tales, etc.)
>   (d) Lore of French Canada, Mexico, etc.
> 2. For the study of the general subject, and publication of the results of special studies in this department.

There is a clear emphasis on verbal folklore here, and such was the emphasis in American folkloristics for some time. Material culture was not mentioned. Over eighty years later, Stith Thompson also omitted material culture from folklore in his definition in *Funk and Wagnalls' Standard Dictionary of Folklore, Mythology and Legend* (1972). Thompson's definition was one of twenty-one separate definitions of folklore included in Funk and Wagnalls' encyclopaedia, a clear indication not only of the complexity of the field but also of the lack of consensus among scholars about the nature of folklore.

In an earlier attempt (1954) to reach agreement at least about one genre of folklore, namely music, the International Folk Music Council defined folk music thus:

> Folk music is the product of a musical tradition that has been evolved through the process of oral transmission. The factors that shape the tradition are:
> 1. continuity, which links the present with the past;
> 2. variation, which springs from the creative impulse of the individual or group;
> 3. selection by the community, which determines the form or forms in which the music survives.
>
> The term 'folk music' can be applied to music that has been evolved from rudimentary beginnings by a community uninfluenced by popular or art music and it can likewise be applied to music which has originated with a known composer and has subsequently been absorbed into the unwritten living tradition of a community.
>
> The term does not cover popular composed music that has been taken over ready-made by a community and remains unchanged, for it is the re-fashioning and re-creation of the music by the community that gives it its folk character. (Scholes, *The Oxford Companion to Music*, 10th edn, 1970)

In 1989 the General Conference of UNESCO adopted a recommendation on the Safeguarding of Traditional Culture and Folklore, and defined folklore as follows:

> Folklore (or traditional and popular culture) is the totality of tradition-based creations of a cultural community, expressed by a group or individuals and recognized as reflecting the expectations of a community in so far as they reflect its cultural and social identity; its standards and values are transmitted orally, by imitation or by other means. Its forms are, among others, language, literature, music, dance, games, mythology, rituals, customs, handicrafts, architecture and other arts.

The UNESCO recommendation was the result of an extraordinary twelve-year series of meetings of government experts initiated in 1973 in order to reach international agreement on the definition of 'folklore'. By the time agreement was reached in 1985, the UNESCO definition had long been superseded by a much more influential document, the *American Folklife Preservation Act* (1976). For its bicentennial year, the American Government enacted measures to establish in the Library of Congress an American Folklife Center 'to preserve and present American folklife'. By the 1990s the terms 'folklore' and 'folklife' had become more or less interchangeable, despite several years of debate about the two terms. In 1972 the doyen of American folk scholarship, Professor Richard Dorson (then from the University of Indiana) edited *Folklore and Folklife: An Introduction*; he wrote that the term 'folklore' had been of value in defining 'a new area of knowledge and subject of inquiry', but that the term had also created some problems and confusion:

> To the layman and to the academic man too, folklore suggests falsity, wrongness, fantasy, and distortion. Or it may conjure up pictures of granny women spinning traditional tales in mountain cabins or gaily costumed peasants performing seasonal dances.

(In the case of Australia, one might substitute for the 'granny women', images of swagmen with corks hanging from their hats.) He also challenged the notion that folklore belongs to the past:

> folklore reflects the ethos of its own day, not of an era long past. Legends in the seventeenth century dealt with witches; in the twentieth century they deal with automobiles.

It was in order to overcome these delusions and to indicate that folklore is still alive and living in cities as well as in remote communities that the Scandinavian term 'folklife' (*folkliv*) was increasingly adopted, culminating in the 1976 Act. The American Folklife Preservation Act defined folklife thus:

> The term 'American folklife' means the traditional expressive culture shared within the various groups in the United States: familial, ethnic, occupational, religious, regional; expressive culture includes a wide range of creative and symbolic forms such as custom, belief, technical skill, language, literature, art, architecture, music, play, dance, drama, ritual, pageantry, handicraft; these expressions are mainly learned orally, by imitation, or in performance, and are generally maintained without benefit of formal instruction or institutional direction.

Further support for the term 'folklife' had come from the powerful Smithsonian Institution, which in 1967 mounted its first Festival of American Folklife in the Mall in Washington DC. A description of the 1988 Festival makes it clear that *living* culture is involved:

> The twenty-second annual festival on the Mall drew an estimated 1.5 million people to a vibrant 'living museum', giving traditional artists needed national visibility and exposing the public to the skills, knowledge, and creativity of diverse peoples. At this year's festival . . . the focus was the folklife of Massachusetts, the music of the Soviet Union, the culture of immigrants to metropolitan Washington D.C., and the work of the American Folklore Society. (*Smithsonian Year 1988*, p. 154)

In 1976, the same year in which the American Folklife Center was established at the Library of Congress, the Smithsonian Institution created its Office of Folklife Programs.

Although folklore scholarship was well established in Europe and North America during the nineteenth century (e.g. the Folklore Archives of the Finnish Literature Society, founded 1831), such was not the case in Australia. In this respect Australia has paralleled England, where folklore has only a slender foothold in academia, despite a considerable amount of research from scholars such as Cecil Sharp (e.g. *English Folk Song: Some Conclusions*, 1907). Folklore was almost an unknown term in Australia until the 1950s, although A. B. Paterson (q.v.) had published his collection of *Old Bush Songs* in 1905 and the English itinerant (later folklorist) A. L. Lloyd (q.v.) had noted many bushworkers' songs in Australia in the 1920s and 1930s. Folklore as such was scarcely discussed in Australia until the formation in the 1950s of organizations such as the Sydney Bush Music Club (q.v.) and the Folk Lore Society of Victoria.

Only a few efforts at definition were made in the pioneer scholarly publications of the 1950s. Russel Ward's article in the third issue of *Meanjin* in 1954 is entitled 'Australian Folk-Ballads and Singers' and deals with the origins of folksongs without attempting to define the genre. Hugh Anderson's *Colonial Ballads* (1955, p. 84) discusses several contentious aspects of definition of folksong, including oral transmission, communal origin, authorship and comparisons with British folksong and American frontier songs. The Bush Music Club (NSW) produced its first issue of its magazine *Singabout* in summer 1956, and in that issue the words 'folk', 'folk songs' and 'folk music' are used widely, but there are no attempts at definition. The nearest is a comment by E. Lancaster in an article (p. 14) on 'Singing folk songs', that 'our folk songs are the heritage of all our people', and that singers should emulate the singing styles of 'traditional singers and players', such as singing unaccompanied and without harmony. The most important definition is in the editorial in the third issue of *Singabout* (winter 1956) which discusses 'the printed folk song' (p. 2) and notes that 'an inherent characteristic of folk song is the change and modification that the words and music undergo in the process of oral transmission'. This important statement was presumably made by John Meredith (q.v.), who was the editor of the journal at that time.

The Victorian Bush (later Folk) Music Club began its newssheet in August 1960, which shortly after this date took the title of *Gumsuckers' Gazette*. In February 1964 it took more substantial form as the *Australian Tradition* magazine. These publications include considerable discussion about aspects of folklore, such as Joy Durst's article in *Tradition* in March 1964 which considered the changing characteristics of folk songs. Nevertheless there were no substantial efforts at definition in these early publications, despite a particular preoccupation in Victoria as to whether contemporary songs by known writers could be considered 'folk' songs.

In 1961 the American academic Dr John Greenway from the University of Colorado published an article 'Folklore Scholarship in Australia' in the *Journal of American Folklore*, based on his visit to Australia during the 1950s. Greenway's article was the first international review of folklore in Australia. Greenway divided his review of folklore publications in Australia into two sections, I The Aborigines and II The Settlers. Greenway's 'settlers' were all of Anglo-Celtic origin, not surprising perhaps in a period in which Anglo-Celtic dominance in Australia was undisputed, even by the new folk clubs and organizations of the 1950s. Greenway's only 'definition' of a people's folklore was a passing reference to 'the vestiges of their time of unsophistication and innocence' — a very nineteenth-century 'survivalist' view. The specific forms of Australian folklore which Greenway discussed were as follows:

I The Aborigines

Mythical beasts (the bunyip)
White beliefs about Aborigines' telepathic abilities
Games
Mythology, legend
Religion
Folk tales
Music
Rock carvings and paintings
Ritual

II The Settlers

Folksongs
Bushrangers
Ballads
The Australian Legend (Ward)

Folk Songs of Australia and the Men and Women Who Sang Them was co-authored by John Meredith and Hugh Anderson and published in 1967. Their preface includes Meredith's definition that 'a folk song is one composed to describe some happening or some aspect of the life of the singer, or of someone near to him, and "written purely for the purpose of self-expression or commemoration"'. In 1969 Ian Turner (q.v.) included a lengthy discussion about the meaning of folklore in his publication on Australian children's play rhymes, *Cinderella Dressed in Yella*. Turner wrote about the nineteenth century's 'great flowering of the collection and study of folklore — the music, songs, dances, stories, rhymes, sayings, language, belief, superstitions and customs of the folk, the common people' and goes on to discuss particularly the documentation and significance of children's games and rhymes.

In 1981 the Department of Home Affairs and Environment in Canberra published the *Report of the Working Party on the Protection of Aboriginal Folklore*, a report which grew out of 'some well-publicised instances of the unauthorised use of Aboriginal paintings and designs', for example, the use of sacred designs on tourist items such as tea-towels. Like those put forward by the Folklife Inquiry of 1986–87, the recommendations of the Working Party were never to be implemented. Nevertheless the Working Party Report is important as the first Australian government document dealing with folklore, and it was one which addressed questions of definition. The Report asked whether the use of the term 'folklore' was appropriate in an Aboriginal context, and stated that the term was being used 'partly because of its international legal connotations' and partly for other reasons:

> Use of the term 'folklore' recognises that traditions, customs and beliefs underlie forms of artistic expression, since Aboriginal arts are tightly integrated within the totality of Aboriginal culture. In this sense folklore is the expression in a variety of art forms of a body of custom and tradition built up by a community or ethnic group and evolving continuously. In the words of a UNESCO study 'it constitutes an artistic manifestation of a people . . . the flowering of the popular consciousness'.

In 1985 the Department of Arts, Heritage and Environment published a 'guide to Commonwealth activities and resources' entitled *Folklife and the Australian Government.* The Introduction stated that

> Australia has a rich heritage of cultural traditions, the diverse elements of which may be brought together under the general heading of Australian 'folklife'.
>
> This heritage draws upon not only at least forty thousand years of Aboriginal history but also upon the cultural backgrounds of Australians whose forbears came to this country during the last two hundred years, or who have come here during the course of their own lifetimes . . .
>
> Australian folklife embraces not only the various forms of oral folklore — sayings, songs, superstitions, stories, folk poetry etc — and the yarn-spinning, singing and recitation styles associated with their transmission, but also the Australian languages, dialects and accents themselves. It also includes folk architecture and social folk customs, such as festivals and ceremonies, games and dances, and the skills and knowledge associated with the practice of folk crafts.

In 1987 the Committee of Inquiry into Folklife in Australia drew on the American Folklife Preservation Act (1976) and the UNESCO definition of folklore (1985) as well as the extensive public submissions to the Inquiry to produce its own definition of folklife:

> Folklife is tradition-based and/or contemporary expressive culture repeated and shared within a community, and accepted by it as an adequate reflection of its cultural and social identity. It embraces a wide range of creative and symbolic forms such as custom, belief, mythology, legend, ritual, pageantry, language, literature, technical skill, play, music, dance, song, drama, narrative, architecture, craft. Its expressions are mainly learned orally, by imitation or in performance, and are generally maintained without benefit of formal instruction or institutional direction.

The Committee considered both the terms 'folklore' and 'folklife' and stated a preference for folklife since 'It is generally perceived as a more comprehensive term than "folklore", and as relating more directly to living culture'. It also noted, however, that the terms have 'sufficiently close meaning to be used interchangeably', and such continues to be the practice in Australia as elsewhere.

Graham Seal's publication of *The Hidden Culture: Folklore in Australian Society* (1989) was the first attempt to provide a textbook in Australian folklore, and the first full-length analytical work on Australian folklore as a whole. June Factor's earlier work *Captain Cook Chased a Chook* (1988) had dealt extensively and analytically, but exclusively, with children's folklore in Australia. Graham Seal's discussion of the definitions includes his own useful statement that 'folklore is best understood as a historical continuum, a continuing and infinitely flexible tradition in which the forms may change but the process remains the same'. He includes some modern forms of folklore such as jokes, graffiti and 'new forms generated and transmitted by the photocopier and the computer'. Seal sees folklore as 'a network of informal communication' with the following characteristics: folklore is informal, group-oriented and universal, and shows variation of form and content across time and space.

Most of the above definitions have involved simple enumerations of different types of folklore and folklife. There have, however, been a number of attempts to systematize the field of folklore studies. In 1978, in *The Study of American Folklore*, Jan Brunvand described folklore as the 'traditional, unofficial, non-institutional part of culture' and suggested three major categories of 'traditional materials (or products), namely oral, customary and material':

1 Oral folklore
Folk speech
Folk proverbs and proverbial sayings
Folk riddles
Folk rhymes and other traditional poetry
Folk narratives
Folk songs
Folk ballads with their music

2 Customary folklore
Folk beliefs and superstitions
Folk customs and festivals
Folk dances and dramas
Gestures
Folk games

3 Material folk traditions
Folk architecture
Folk craft
Folk arts
Folk costumes
Folk foods

Brunvand made the important statement that folklorists must be aware of the total context surrounding any individual item. For example, in the case of folktales,

> folklorists should not overlook collecting and studying such nonverbal elements as facial expressions, gestures, and audience reactions while tales are being told. The ideal study is one that takes into account the entire traditional event, which is what we are referring to as a 'performance' (p. 5).

Dorson (*Folklore and Folklife*, 1972) categorized the field into oral literature, material culture, social folk custom and the performing folk arts. He acknowledged, however, that

> the four divisions sketched here are not all-inclusive or mutually exclusive. Where do we place traditional gestures, for instance? They might be bracketed with 'folk speech' . . . even though such placement leads to the paradox of subsuming gestures under oral folklore.

The Report of the Committee of Inquiry into Folklife in Australia (1987) used the following classification for its chapter on 'The Nature and Diversity of our Folklife':

Making Things
Folk crafts
Vernacular buildings

Oral Folklore
Australian folk speech
Languages other than English
Yarns and other tales
Jokes
Recitation, folk poetry and popular verse

Music and Dance
Music
Dance

Customs, Pastimes and Beliefs
- Calendar customs
- Sport
- Rites of passage
- Death
- Foodways
- Gardening
- Folk medicine
- Beliefs

Occupational Folklore
- Some types of occupational folklife
- Endangered occupations

Children's Folklore
- Children's own folklore
- Folklore for children
- Folklore about children
- Children as apprentices in folklife

As can be seen from the above discussion, there have been numerous attempts to define and systematize folklore and the whole field of folklore studies. This is an ongoing process of debate and discovery that indicates a healthy engagement with the informal facets of culture. It is the editors' belief that the Australian contributions included in this volume will considerably enrich this important undertaking.

Gwenda Beed Davey and Graham Seal

A

Aboriginal folklife in central Australia One can talk about folklife among Aborigines in the sense of traditional (but not secret-sacred) dance, music, foodways, and so on, in communities where these traditions are still alive; in central Australia there are a number of such communities, as there are in the north and west of Australia. One can also look at the contribution of Aborigines to Australian folklore in general. Racist jokes denigrating Aborigines have probably been with us for a very long time, but in most other areas of Australian life Koories have been ignored until very recently. Things are changing, however. In the Northern Territory today there exists an Aboriginal-controlled and -operated radio network that broadcasts daily in six Aboriginal languages; it is delivered via satellite as far north as Bathurst Island off the Arnhem Land coast, and as far south as the Great Australian Bight, covering a third of the continent. The former Director of the National Art Gallery, James Mollison, said that Aboriginal art is 'the most interesting art Australia has ever produced'. Aboriginal food is on the menu at a number of exclusive city restaurants. Aboriginal writers are attracting a large readership; Sally Morgan's book, *My Place*, was on the best-seller lists for many weeks following its release. An appreciative audience for Aboriginal music, both traditional and contemporary, is also developing. In many fields of cultural endeavour, the power and uniqueness of Aboriginal life is at last having a positive impact on the Australian consciousness, and changing the way Australians think about themselves.

Old ways, new ways
The following story was told by Napaltjarri, a Pintupi woman, in 1978. It is about the experience of meeting white people for the first time.

Early one winter's morning Napaltjarri was preparing to leave her family's camp to gather food. Her younger sister had burnt herself on the fire the night before and was unable to join her. Her husband was still asleep, and in any case would not usually go gathering food with his wife. Her two sons were awake and rekindling the fire, but they would not go with her either. They knew that their father was planning to find a mulga tree from which he would make a shield. They would wait and go with him when he awoke. This would be preferable to tagging along with their mother looking for lizards and dried-up berries.

Napaltjarri left the camp alone. Balanced on her head was a large *piti* or wooden carrying dish. She walked south towards a distant range. She soon found a small tell-tale hole in the ground indicating the presence of a *lugata*, or blue-tongue lizard. Using her digging stick she quickly located the lizard, broke its neck and tied it around her waist with a piece of hair string.

Near the foothills of the range she discovered a thicket of *ngamanpurru* bushes laden with sweet black berries. She picked and ate one, then gathered a few handfuls, placing them in the *piti*. By midday she had caught two more lizards and collected enough grass seed to make a large cake. Satisfied with her efforts, Napaltjarri began the long walk home.

On returning to the camp she was surprised to find her sister missing. Her husband and sons were probably out hunting, or perhaps looking for a suitable tree to make the shield. Her sister, however, should have been resting by the windbreak, not roaming around in the bush. Napaltjarri took the *piti* from her head and placed it next to the grinding stone. As a way of establishing the whereabouts of her family, she made a close inspection of the tracks that covered the ground around the camp. She noticed that all the footprints led in one direction, due east. Perhaps they had all gone to find the mulga tree together.

Looking up at the afternoon sun, Napaltjarri realized that if the grass-seed cake was to be cooked before sunset she would have to start grinding the seed as soon as possible. Putting aside her fears about the family, she sat down by the heavy grinding stone, sprinkled some seeds into its narrow groove and began the long and arduous task of turning them into flour, confident that her family would return before nightfall.

The sun disappeared over the western horizon. The moon rose in the east. Napaltjarri removed a crusty cake from the ashes of the fire. No one had returned. Something must have happened. Her husband was always home before sunset. She put the uneaten cake on top of the windbreak and began to make a torch from acacia twigs and spinifex grass. She would find her family by following their footprints.

Beneath the flickering light of the burning torch their footprints seemed strangely regular, one following the other in straight lines. Napaltjarri's sister was obviously finding it difficult to walk. Her left foot made a heavy impression in the soft sand while the right foot barely left a mark. Napaltjarri's two young sons were usually difficult to track as they liked to jump and skip, but here their footprints were easy to see. She continued to follow the tracks for what seemed a lifetime, when the footprints came to a sudden end. Had her family vanished into the night? Had they grown wings and flown away? Napaltjarri quickly lifted the torch above her head, spreading the light over a wider area. Next to where the footprints ended there were two snake tracks, both the size of a man's leg. They ran across the ground, parallel to one another, disappearing into the darkness.

Barely conscious of what she was doing, Napaltjarri ran screaming in the direction of the camp, swinging the flaming torch around her head. She tripped over and sent a shower of sparks into a dry clump of spinifex, setting it ablaze. Bruised and covered in scratches, she eventually reached the camp and threw herself down beside the windbreak, convinced that her family had been eaten by serpents.

Napaltjarri was still lying by her windbreak, three days later, when the truck that had taken her family away arrived to pick her up.

It was only weeks later, after being taken to a distant government settlement, that she was reunited with her family. It was many more weeks before she began to comprehend what had happened. She had never seen white people nor the snake-like tracks of trucks before.

The Dreaming continues

The 'Dreaming' still exerts a powerful influence over the daily lives of the Aboriginal people of central Australia. Underpinning the Dreaming is a vast body of myth that continues to be passed on from one generation to the next. Gaining a full knowledge of these mythologies requires a lifetime of patient learning at numerous ritual ceremonies. A knowledge of traditional myth is significant both in a religious and practical sense. For example, by acquiring mythological knowledge a person will come to understand his or her social obligations under traditional law.

Perhaps the most important myths are those that describe the journeys of ancient ancestors as they travelled across the earth during the Dreamtime, creating present-day landforms and all living things. Each one of these ancestors is associated with a 'dreaming line' or track that may extend for hundreds of miles. There are literally thousands of these 'dreaming-lines' criss-crossing central Australia. Dreaming lines are marked out by mountains, caves, creeks, swamps, waterholes or other natural landforms. An understanding of a particular legend will provide a person with a mental map of waterholes, creeks, food sources, and thus the ability to safely navigate vast distances across the desert.

The song describing the entire journey

of an ancestor as it travelled along a particular dreaming line may take many hours or even days to complete. There are also dances, ground paintings, body decorations and sacred objects associated with every particular mythological ancestor.

The legends and stories associated with the 'dreaming tracks' are still taught in ritual ceremonies throughout central Australia. Aboriginal people do not travel on foot any more, but still need to travel over the land to attend ceremonies, see relatives and meet other social obligations.

Tribal areas were (and still are) delineated to a degree by the dreaming lines. In times past trade between coastal and inland tribes was conducted via the dreaming tracks and even now navigation through the desert is made possible by a knowledge of the lines, as in the following account:

> A trip was organized with seven traditional Aboriginal custodians of a large area of land about 480 km west of Alice Springs, near Lake Mackay. The seven Aboriginal men had not been back to this, their traditional country, for many years. There are no roads or any settlements in this area at all. Maps of the area were deliberately left behind. Throughout the whole trip, the traditional system of finding one's way around the land was used. Once the party was well away from the nearest roads, one of the old men would occasionally stop the convoy, climb up on a nearby sandhill and sing. This singing would continue for perhaps ten minutes. The old man would then point the rest of the convoy in the right direction. What in fact he was doing was going through a particular mythological song associated with the land through which the party was travelling. He would look for natural landmarks that were mentioned in the song, and thereby get his bearings. The trip continued for more than two weeks in this manner, through an extremely remote part of Australia, and not once did we get lost.

The following story also illustrates the continuing strength of traditional beliefs among some Aboriginal groups.

In 1983 the Northern Territory government wanted to build a recreation lake north of Alice Springs. A section of the Todd River (running through the centre of the town) would be dammed and a large area flooded. However, the local Arrernte people protested strenuously, claiming that this would submerge one of the last remaining sacred sites within the Alice Springs area. The sacred site was associated with the 'Seven Sisters Dreaming' which describes the creation of the stars in the Southern Cross. It is particularly significant to women, and is known as *walyatjathere*, meaning 'two breasts' in Arrernte. The dreaming line of the Seven Sisters starts in South Australia and travels into the Northern Territory, up the MacDonnell Ranges and into Arrernte tribal lands.

The Northern Territory has legislation protecting Aboriginal sacred sites. In order to have a site registered, and therefore protected, the traditional custodians of the site must prove that they still have traditional knowledge of the site and a full understanding of its significance. There were only a handful of very old Arrernte men and women who had complete knowledge of the songs, dances and rituals associated with the threatened sacred site. While a political battle began to rage between the NT government and the old Arrernte custodians, the most senior woman in the traditional custodian camp suggested that the Pitjantjatjara be called in to help win the fight. The Pitjantjatjara people live on their traditional lands in the extreme north of South Australia. The old Arrernte woman knew that the Seven Sisters dreaming started in South Australia in Pitjantjatjara lands, and that the Pitjantjatjara women would know the complete song including the section running through Arrernte country. She also knew that they would have a good knowledge of the sacred site now under threat of destruction.

Following some hasty calls via the radio telephone network (there were no regular telephone services to the Pitjantjatjara lands at that time) and a brief discussion on intertribal protocol, the Pitjantjatjara women decided to help their Arrernte sisters. Eventually, a convoy of women and their families, well over six hundred people in all, arrived in Alice Springs. After lengthy discussions with the Arrernte custodians, the Pitjantjatjara women decided to perform part of the Seven Sisters ceremony, at the threatened site. This would help 'regenerate' the land around the site and give the Arrernte women added strength to carry on the fight.

The ceremony of the Pitjantjatjara women, complete with decorated ritual poles, elaborate headresses and symbolic body painting, began in the evening and went on into the early hours of the morning. The Arrernte people were indeed 'energized' by this powerful performance, and maintained their opposition to the government's plan for another three months. In the end, the traditional custodians won the fight and another area was found for the recreation lake.

In central Australia the Dreaming continues.

Foodways

The traditional diet of Aborigines in central Australia is derived from an abundant assortment of native plants and animals. There are also a variety of ways in which this food can be gathered and cooked. The following represent a very small selection of the many traditional Aboriginal foods still gathered and eaten in central Australia.

Traditional bread or mayi

There are several native grasses that can provide seed for traditional bread. The Panicum species in particular provides a delicious flour. One can find ample supplies of this grass in open plains country anywhere in central Australia. After about ten bundles of grass (a bundle being about as much grass as can be stuffed into one pillow case) have been gathered, a space is cleared on the ground, the grass piled into a heap and beaten with a stick for twenty minutes. The grass is cleared away, leaving the seeds on the ground. The seeds are gathered up and placed in a large billy. The seed is poured from one coolamon (wooden carrying dish) to another until all sand and other extraneous matter is separated from the seed.

The next step is to place the grass seed on a traditional grinding stone, mix in a little water and slowly grind the seed to a paste using a clean, smooth stone. A coolamon at the edge of the grinding stone catches the paste as it is produced. All the seeds are ground in the same manner until the coolamon is full. A shallow hole is dug in the ground, and a fire lit in the hole; when it has burnt down to a suitable collection of hot coals, the paste, or dough, is poured on to the coals and allowed to cook for five to ten minutes. Some smouldering wood is balanced over the top of the loaf (not touching it) to cook the upper surface of the bread. It is removed from coals after a further twenty minutes.

Bush honey or wama

There are two main sources of native honey in central Australia; honey ants and native bees. Since honey ants are far more abundant and easier to find than native bees, the following is a guide to finding the ants and how to eat them.

Although honey ants are relatively easy to find, extracting them from their nests can be hard work, especially on a hot Centralian day.

The honey ant usually builds its nest beneath clumps of mulga trees. To determine the whereabouts of the nest, one must look for a fine white powdery substance on the ground around the trunks of the mulga trees, then dig a hole big enough to climb into. Sometimes it is necessary to dig down to about 1.8 metres before finding the horizontal shafts containing the ants. Traditionally, Aboriginal people used sharpened digging sticks to extract the ants; these days a crowbar or shovel is preferred.

One nest usually contains enough ants to fill two medium-sized billy cans. To eat them, one simply detaches the abdomen (containing the honey) from the rest of the ant and sucks out the contents. The smoothness and rich taste are surprising.

Goanna or nintaka

Traditionally, the Aboriginal people of central Australia caught and ate a variety of goannas and lizards. Now, however, there is no need to supplement a mainly plant-based diet with any kind of edible lizard.

These days, most Aboriginal people are only interested in the large Western Desert goanna or *nintaka*.

The goannas are usually found in shallow holes, just beneath the surface of the ground in most sandy plains areas. On occasion they can be spotted above ground level, sometimes on roadways and in small trees. It is much easier to catch them underground in their burrows as they tend to move very fast out in the open. The entrance to their burrows is in the form of a small hole, about 5 cm in width. If the hole is still in use there will be fresh tracks of the beast or some freshly disturbed sand. The hunter plunges a digging stick or crowbar into the ground, about one metre from the hole, continuing this process until a large cavity is found. He slowly digs into this cavity until moisture is evident (this is the urine of the goanna) and, using his hand, pushes away the sand and catches the goanna, preferably by the tail.

The underbelly is cut so as to gut the goanna. It is important to thoroughly clean out all extraneous matter from inside the goanna as this material can at times be poisonous. Once cleaned, the goanna is placed on a bed of hot coals and turned over occasionally until cooked. The taste is similar to chicken, although the meat is rather more sinewy and oily.

Kangaroo or marlu

It is illegal to shoot wildlife on traditional Aboriginal lands unless you are a traditional owner of the land, and there is a great deal of religious significance (for Aboriginal people) associated with the killing of a kangaroo for food. It is therefore advisable, as a matter of respect, to acquire a kangaroo through the services of an Aboriginal hunter. Aboriginal people rarely hunt with spears these days. A rifle, a few dozen bullets, a reliable four-wheel-drive vehicle and a keen eye are the basic requirements of today's hunter. The best time to catch a kangaroo is a few hours before sunset or early in the morning. This is particularly the case in the hot summer months. It may take a few hours of driving through the bush to find a kangaroo — they are not as plentiful in the desert as in other parts of Australia — then the engine must be immediately turned off and no one must make a sound. The hunter will leave the vehicle and circle, downwind, around the prey, until he is in a position to shoot the animal at close range. Once wounded, the kangaroo is usually finished off with a quick blow to the back of the neck. The carcass needs to be disembowelled as soon as possible. This is accomplished by making a small cut in the side of the beast, just below the ribcage, through which the intestines are removed. The incision is then closed up using a stick and a piece of the small intestines. The back legs are dislocated and the lower parts broken off. This will produce two stumps, making the bones visible. The tail is usually cut off at this stage. The kangaroo is now ready for cooking.

The first thing to do is to prepare the cooking pit. An oval hole is dug into the ground, about half a metre in depth. A large fire is made in the pit to ensure a good supply of hot coals. The carcass is thrown onto the fire for a few minutes to singe off the fur, then removed and scraped clean. Once the fire has died down, the carcass is carefully placed back in the ashes then covered with earth with only the bones of the back legs protruding. The kangaroo should be left to cook for about two hours or so.

According to traditional law, the hunter must retain the responsibility for cutting up the meat and also for distributing the pieces to the correct people. He is obliged to give certain parts of the animal to certain relatives, depending on their relationship. However, there is usually plenty of meat to go around, even after the family has eaten. Kangaroo meat is a light pink colour and tastes a little like veal with a strong gamey flavour; it is also fairly sinewy.

Desert yams, bush tomatoes and wild onions or yala, pura *and* yalka

The desert yam (*Ipomoea costata*), or *yala*, can be found most abundantly along the banks of dry creekbeds or around natural springs. Aborigines look for dark-green leaves or dormant stems on the surface of the ground and dig down directly below the runners, about 70 cm or so, to get the tuber, between 10 and 15 cm long. Some plants bear a dozen tubers or more. The bush tomato (*Solanum petrophilium*), or *pura*, can be found in almost any desert location, although it tends to grow more profusely in the foothills of ranges. Each tomato is about 4 cm in diameter and a yellow-green colour. They are generally only available during the hot summer months. Before they are eaten they must first be cut open to clean away all the seeds and other internal matter, as the seeds, if eaten in quantity, can make one very ill. The bush onion (*Cyperus bulbosus*), or *yalka*, can be found in most locations, especially after rains. They have small green stems, something like the shoots on domestic onions. Pulling up the stems reveals a small bulb, about the size of a peanut, attached to each plant.

All these foods can be eaten raw, although the yams and bush onions are sometimes better eaten after their outer skins have been removed.

The emu or tjakapirri

Emus are not in large supply in central Australia. They are obviously fast on their legs and therefore difficult to catch using the traditional hunting methods. To overcome this problem the Aboriginal people of the desert look for a regular watering hole for emus and poison it with a naturally-occurring narcotic (*Duboisia myoporoides*). The hunter hides behind a windbreak, especially built for the purpose, and waits until an emu comes and drinks. Within minutes of consuming the poison the emu becomes stupefied and unable to run. A quick blow to the head finishes the bird off.

There are numerous styles used by various Aboriginal groups in central Australia to prepare and cook an emu. The Pitjantjatjara method is to first pluck out all the feathers and then remove the skin by making a circular cut around both legs. An experienced hand can remove the skin in one complete movement — like taking a rubber glove off. To butcher the emu, both legs are cut off and the sinews pulled out (they are exceptionally strong and used to bind spears and other implements). Next, the abdomen is cut open and all the intestines removed. The carcass is now filled with feathers and grass to ensure that the meat is evenly cooked. A long cut is made along the emu's neck then the head is pushed through the opening, making a loop. The emu is cooked in the same way as the kangaroo.

Traditional medicine

Many naturally available plants are still used by Aboriginal people to cure or alleviate a range of medical conditions. The 'medicine man' or *nankarri* has the ability to use such remedies, but he also possesses special healing powers of a spiritual nature. In *Nomads of the Australian Desert* C. P. Mountford provides a stunning illustration of the work of the *nankarri*:

> I was present on another occasion when Namienja [the medicine man] treated an Aboriginal girl who, suffering from a violent headache, had become unconscious when placed in the smoke of a fire to ease the pain. The medicine man was summoned immediately to treat the patient.
>
> However, Namienja did not hurry but, guided by the commotion around the girl's camp, strolled nonchalantly across. Instead of even looking toward the unconscious girl, he searched among the leaves of the wind-break until he had apparently captured something in his cupped hands. He then turned the patient on her back and pressed his cupped hands on the pit of her stomach, telling her at the same time that as the missing kuran (spirit) had now been returned to her body she would get well. The girl sat up straight away.
>
> The medicine man explained later that the 'smoking' of the patient had so scared the kuran of the girl that it had left her body and sought refuge among the leaves of the windbreak. Later the girl assured me that the moment the kuran was restored to her body her headache had disappeared and she felt much better. The next morning I again saw Namienja

> at the camp of the same girl, who was complaining that her head was opening and shutting (an excellent description of a splitting headache). The medicine man massaged the aching head of the patient for a while, pressing it together. He continued to massage the head when, turning around suddenly, he thrust his hand in the sandy creek bed, announcing he had just extracted from her head the 'stick' that had been causing all the pain. The girl was asleep in a few minutes. Later the writer inspected the girl, whose pulse and temperature appeared to be normal.

While the work of the *nankarri* is generally left to men, the experts on herbal medicine are usually women, as the following story illustrates:

> A few years ago I was visiting a remote community in Walpiri country when I came down with an extremely severe case of diarrhoea (which is not uncommon in central Australia). I was given some drugs by the local nurse, but to no avail. I spent two or three days in a state of complete exhaustion, weak from lack of food and water (I couldn't keep anything in my stomach for more than ten minutes). It looked as though I would have to be evacuated to Alice Springs and hospitalized. At this point, an old Aboriginal woman by the name of Emily Nakamarra offered to help out. She went away into the bush and a few hours later came back with a small billy of warm liquid. It was a deep red colour. I was told to drink the entire contents of the billy, slowly, bit by bit. It had a disgusting taste, but I was prepared to try anything and swallowed every drop. Almost immediately the stomach cramps abated. Within a few hours I was able to eat a piece of bread and keep it in my stomach. By the end of the day I felt almost well again. What was this amazing cure I had been given? I went off and found old Emily, the maker of miraculous potions. After thanking her, I asked her how she made such powerful medicine. She pointed into the bush and explained that it came from an 'orange tree'. She would show me. We walked for an hour or so, away from the community, until we came to a small tree growing next to a dry creek bed. Emily scraped some bark from the trunk and said that to make the medicine you boiled the bark from the tree in a billy until it turned red, then drank it, simple as that. She said that her mother had shown her how to make it, although in the past they would put the bark on the fire then suck it. There were no billies in her mother's day. At a later stage I identified the tree as *Capparis umbonata*, a relative of the bush orange tree.

Over the past few years the West has looked to the East for alternative solutions to health problems; witness the growth in acupuncture and natural health centres and the keen interest in herbal medicines. This interest in taking a less 'scientific' approach to the art of healing has, if belatedly, generated an awareness of traditional Aboriginal medicine within the broader Australian community. At least two recent publications list an extraordinary number of traditional Aboriginal herbal medicines used to cure almost anything including laryngitis, diarrhoea, rheumatism, toothache, fevers, colic, ulcers, congestion, and a host of other ailments. It is unfortunate, but not surprising, that white Australia has taken so long to recognize the value and wisdom of Aboriginal approaches to health.

'New traditional' music

The present-day music of the central Australian Aboriginal people ranges from traditional ceremonial chants through to rock music performed in Aboriginal languages.

There are many traditional musical forms still performed and still passed on to the younger generation. While some of this music is secret-sacred and can only be performed during particular ceremonies by particular people, much traditional music is open to the public. For instance, there is a large repertoire of love songs that are sung by both men and women, and many other songs relating to traditional lands. T. G. H. Strehlow's *The Songs of the Aranda* documents the vast range of traditional music of just one of the tribal groups of central Australia. In the following quotation, taken from *Aranda Traditions*, Strehlow describes the setting of a traditional song and provides sections of the song in translation:

> A group of honey-ant men leaves its old home at Ljaba and goes on a long journey west to Popanja, another honey-ant totemic centre, situated in Kukatja territory. For a time they march along the edge of the foothills fringing the Western MacDonnells. Finally they have come to a curve in the lowflanking range, and they know that Ljaba will hereafter be hidden from their view. So they halt for a moment, and glance back towards Ljaba. They see the pale purple peaks which surround it gleaming dim through the fast gathering haze. Like a pall of smoke the haze enfolds the hills; and the men grieve for the home which none of them expects to see again. Tears fall from their eyes, and their last chant goes up:
>
> *Enfolded by plains lies Ljaba;*
> *Beyond the far horizon lies Ljaba;*
> *Enfolded by plains lies Ljaba;*
> *Dimmed by the enveloping mists.*
>
> Longing for home — this is the motive which leads most of the weary ancestors of legend back to the place whence they originated. The euro brothers, Ntjikantja and Kwaneraka, return to their old home at Kaput' Urbula when their toil of instructing men in the use of the spear had been ended. The head of Ankotarinja rolls back to the little sandy rill at Ankota where he had first emerged from his long sleep. The death-sick Ulamba chief slowly plods on over the many weary miles that separate him from his mountain home; he must journey thither to lay his head down for his final sleep:
>
> *The fatal combat is done; he wishes to return:*
> *High in the heavens shines the afternoon sun:*
> *His heart is filled with yearning to turn home.*
>
> His body gleams in the waning sun; he is painted with the device of the blood-avengers. He sees his old home in the distance:
>
> *My own home, my dear home*
> *O Ulamba, rugged, chasm-cleft.*
>
> He hears the familiar twittering of birds in the vicinity of the mountain:
>
> *The birds are speaking with many voices*
> *At Ulamba, chasm-cleft Ulamba.*
>
> He sees his old camping ground. He sees the hollow where he used to sleep covered with the marks of birds' feet; their scratching feet have disfigured it; no one had tended it in his absence:
>
> *My own home, my dear home —*
> *Whose feet have disfigured it?*
>
> *The mulga parrots have disfigured it;*
> *Their feet have scratched the deserted hollow.*
> *He sinks down; and sleep shuts his eyelids for ever.*

Apart from such rich traditional songs (which are still sung on many occasions), there are also new musical forms emerging. Indeed, over the last ten years, there has been a virtual revolution in the development of contemporary Aboriginal music. In the early 1990s there were at least fifty different Aboriginal bands with albums on the market. This modern sound draws influences from a variety of sources including country and western, reggae, rock, blues, gospel, church music and traditional forms of Aboriginal music. While most of these influences come from external cultures, the resulting musical mix has a unique quality. This is particularly the case with regard to the lyrics. Young Aboriginal people, perhaps unable to find in traditional music the means to express the dilemmas and frustrations of contemporary Aboriginal life, have used popular musical forms to talk about their experiences in an often perplexing cross-cultural world. For instance, the members of the Warumpi band from the settlement of Papunya write their songs in their own language, Luritja, and set them to a rock-and-roll beat. The following is an example with both the English and Luritja lyrics:

Kintorelakutu

Irrirti arnangu waltja tjurta
Wilurarranguru ngalyanu
Ngalyankulaya kakarrara nyinangu
Nganarna rawa nyinangu
Ngurra waltja nyakunyt jawiya
Nganarna rawa nyinangi

Arnangu tjurta kakarrara nyinapayi
Tjurratja tjikira nganarna pikarripayi
Nganarna kakerrara wiyarrinyi
Nganarna watjilarrinyi

Arralaka wilurarra Kintorelakutu
Tjamuku ngurrangka palya nyinant jaku
Arralaka wilurarra Kintorelakutu
Tjanuku ngurrangka playa nyinant jaku

Tjamu tjana marlu wakaragalkupaye
Kaparli tjana pura mantjirangalupayi
Ngurra panys Walangurrunya
Wati, minma, tjitji tjurtangku
Kurwarrikutju ngangu Kintorenya

Arralaka wilurarra Kintorelakutu!
Tjanuku ngurrangka palya nyinant jaku
Arralaka wilurarra Kintorelakutu!
Tjanamuku ngurrangka palya nginant jaku

Arnangutjurta! Arnangutjurta!
Irrirtitja nyinapayi ngurra panya Walangurrunya
Kuwarrilatjju nyinanyilpi Kurarrilatju nyinanyilpi
Tjanuku ngurrangka ngurra panya Kintorela
Arralaka! Wilurararra Kintorelakutu

Towards Kintore

In the olden times all the families
came from the west
and sat down in the east
We stayed a long time
without seeing our country
we were sitting a long time

A lot of people always stay in east
after drinking grog we always start fighting
In the east we are becoming nothing
We are yearning for our own country

We must go west towards Kintore!
We'll be better in our Grandfather's Country
We must go west towards Kintore!
We'll be better in our own Grandfather's Country

Those Grandfathers always speared and ate kangaroo
Those Grandfathers always gathered and ate bush apple
at the same place — Kintore
A lot of young men, women and children
have only now seen Kintore

We must go west towards Kintore!
We'll be right in our Grandfather's country
We must go west towards Kintore!
We'll be right in our Grandfather's country
Mobs of people! Mobs of people
The olden time ones always lived at Kintore
Finally now we are sitting! Finally now sitting!
at our Grandfather's camp in the same home at Kintore
We must go! West towards Kintore!

The notion that Aboriginal culture is static and unchanging is entirely erroneous. People make culture, and people change. The view that an Aboriginal person will cease being Aboriginal if he or she absorbs influences from other cultures is false. Contemporary Aboriginal musicians have shown that they can absorb modern influences yet produce a form of music that rests on traditional musical forms.

Philip Batty

Aboriginal folklore The question of whether the term 'folklore' should be used in relation to any aspects of Aboriginal culture is a contentious one, both within Aboriginal and non-Aboriginal populations. Should the fieldwork of anthropologists, professional and amateur, among Aboriginal communities since the nineteenth century be treated as 'folklore'? Certainly, some collectors did think of the material they collected as such, in accordance with intellectual notions of their time. However, since the term 'folklore' has, in English-speaking cultures, come to have negative connotations, especially since the Nazi period in Germany, such materials would not today be considered folklore, either by Aboriginal people, by anthropologists or by folklorists. Indeed, to call such significant, often sacred, cultural items by a term that is now understood to refer to informal, sometimes subversive, practices and expressions is seriously to demean core aspects of Aboriginal culture. The Report of the Committee of Inquiry into Folklife in Australia (1987) stated that it would not report on traditional Aboriginal ceremony, lore and belief, although it would address 'other aspects of Aboriginal folklife, most particularly Aboriginal craft and the contemporary folklife of both urban and rural Aboriginal communities'. The

Committee considered that contemporary Aboriginal folklife 'is in major aspects fundamentally inter-twined with the folklife of other communities within Australian society'. The Committee received a major submission from the Central Australian Aboriginal Media Association (CAAMA) in Alice Springs which discussed personal oral histories, traditional legends, stories and myths, traditional foods and cooking techniques, traditional medicine, traditional body and ground paintings, traditional dancing, traditional tool production, traditional navigation techniques, traditional iconography and sand drawing and traditional and adapted music. The report also discussed CAAMA's involvement in the documentation of Aboriginal folklife.

In 1981 the then Department of Home Affairs and Environment published the Report of the Working Party on the Protection of Aboriginal Folklore. The Working Party was established because of the public attention directed to the unauthorized use of Aboriginal paintings and designs for commercial and tourist purposes. The Report asked whether the use of the word 'folklore' was appropriate in the Aboriginal context, and stated that the term had been used because of its international legal connotations, including ongoing work by UNESCO (See Introduction). One focus of the Working Party concerned copyright, and it also examined wider issues of protection in other countries 'where there is both a large body of traditional art and where traditional society is still a living entity'. The Report recommended that there be an Act of Parliament to protect Aboriginal folklore by measures including prohibitions on non-traditional and debasing, mutilating or destructive uses of secret-sacred materials, payments to traditional owners for use of songs, dances, art motifs, etc., and the establishment of an Aboriginal Folklore Board. The recommendations of the Working Party have, however, not been implemented by successive Australian governments.

See also: **Aboriginal folklife in central Australia**; **Australian Institute of Aboriginal and Torres Strait Islander Studies**; **Children's folklore**; **Craft**; **Festivals**; **Koorie languages and folk speech**; **Narrative literature and folklore**.

ADAM-SMITH, PATSY (1924–), born in Melbourne, has made a lifetime study of folklore and oral history in Australia, Ireland, the United States, England and elsewhere. She has published twenty-eight books and more than 300 feature articles and in 1980 was awarded the OBE for her services to literature and oral history. One of her particular interests, which derives from her childhood, is in Australian railways and railway employees, and her books on this topic include an autobiography, *Hear the Train Blow* (1964) and *Folklore of the Australian Railwaymen* (1969).

Between 1954 and 1960 Patsy Adam-Smith worked on ships in Bass Strait, and is the first woman in Australia to have been articled on a trading vessel. Her sailing experiences and knowledge of Tasmania and the Bass Strait islands are described in books such as *There Was a Ship* (1967) and *Moonbird People* (1965).

During the 1970s Patsy Adam-Smith created the position of manuscripts field officer within the La Trobe Library at the State Library of Victoria. She was president of the Victorian Fellowship of Australian Writers in 1973 and federal president in 1974. In 1976 Patsy Adam-Smith tape-recorded the memories of First World War survivors for the State Library of Victoria, and studied in the Imperial War Museum in England in 1977. Her book *The Anzacs* shared the 1978 *Age* Book of the Year Award, and in 1982 she produced a companion volume, *The Shearers*. She has worked extensively in the television industry and was the creator of all characters and story consultant to the major TV series *Anzacs* shown in the 1980s. All her writing is based on intensive field work, and between 1954 and 1989 she spent most winters when in Australia north of the twenty-sixth parallel, often camping out. For three years she worked with stockmen, some white but mostly Aboriginal, and published *Outback Heroes* in 1981. She then published *Australian Women at War* in 1984. In 1986 she published *Heart of Exile*, based on research in Dublin, Australia and the USA into the letters and diaries of seven Irish rebels.

AFFLEY, DECLAN JAMES (1939–1985), one of Australia's best-known exponents of Irish and Australian traditional music, was born in Cardiff, Wales, but many of his ancestors were from County Cork, Ireland, and Declan inherited a deep feeling for Irish music from his parents. Some of his earliest memories were of his father playing Irish music on the Scottish bagpipes at St Patrick's Day marches and attending *ceilidhs* at St David's parish hall in Cardiff. His formal musical education started at the age of five and he took up the clarinet when he was eight years old, performing with various bands and orchestras. Declan was a student at the Royal Welsh College of Music until he joined the British Merchant Navy at the age of sixteen. At sea he bought himself a guitar and taught himself to play it. Declan's sensitive and melodic guitar playing was based on his appreciation and understanding of the Irish harp. On accompaniment to traditional songs Declan wrote:

> A fretted instrument tuned to the diatonic scale cannot have the range of an untrammelled voice or a fretless instrument — this is why many traditional fiddlers sound out of tune with a piano or a guitar. Remember too, little is better than too much ... Learning a song for singing is not just a matter of getting (a) the words and (b) the melody and then putting them together. The song is a complete unit, created and polished as such over many years, by skilful performers-creators, and each note and word must be understood and matched correctly. Forget all your half-learned ideas of time and rely on the natural rhythm of the words and the feeling they express. Infinite subtle variations are available.

In 1960 Declan decided to live in Australia and frequented 'pubs and low dives' in both Sydney and Melbourne where he came into contact with performers such as Brian Mooney, the late Don Ayrton, Martin Wyndham-Reed and Paul Marks. He continued to learn and sing Irish traditional and rebel songs and later became interested in Australian folksongs. Eventually giving up the sea, he became a regular performer at the Troubadour coffee lounge in Edgecliff, Sydney. Declan participated in Irish dances and sessions around Sydney, including the lively sessions at the Wentworth Park hotel run by the late Tommy and Joan Doyle.

At the age of twenty-seven Declan taught himself the fiddle, mainly from the tapes of the Irish fiddle player and storyteller Johnny Docherty. He also played tin whistle and banjo. He was a regular performer at Traynor's folk club and the Dan O'Connell hotel when he lived in Melbourne for two years. In 1972 he toured Ireland with the production *The Restless Years*, a dramatic presentation of Australian history illustrated by stories, poetry and folksongs. He was also one of the singers in the award-winning ABC television documentary upon which the stage presentation was based.

As well as radio and television programmes, Declan was also involved in several films including Tony Richardson's *Ned Kelly*, Peter Weir's *The Last Wave*, and Richard Lowenstein's *Strikebound*, on which he was also employed as musical director. He performed in many traditional music groups and in 1969 started the Wild Colonial Boys, which was one of the first Australian groups to combine Australian and Irish music. A regular participant and guest at Australian national and local folk festivals, he also toured New Zealand several times and performed in Papua New Guinea and Indonesia.

In 1970 the Sydney Irish community lent Declan a set of Uileann pipes, made in Cork at the end of the nineteenth century, which had belonged to the late Sydney-based piper Frank Heffernan. This changed Declan's life and musical focus. The Uileann pipes became his love and passion. As he was virtually the only Irish piper in Australia, he spent time in Dublin with the Irish piper Dan Dowd having the set repaired and learning playing and reed-making tips. Two of his favourite pipers were Seamus Ennis and Willie Clancy, and he was fortunate enough to meet Willie Clancy in 1972.

In 1984 Declan was employed as a teacher of fiddle and banjo at the Eora Centre, an Aboriginal culture centre in Redfern, Sydney.

Declan had one of the best singing voices in Australia and the rare quality of being able to interpret a song and make it his own, as he did so well with the Irish traditional song 'Carrickfergus' and the Henry Lawson song 'Do You Think I Do Not Know', arranged and set to music by Chris Kempster. He also sang songs by contemporary Australian songwriters such as Harry Robertson, John Dengate and the late Don Henderson. For his performance of political/satirical songs he was often subjected to warnings and acts of censorship; this never deterred Declan, but served to reinforce his commitment to singing these songs in public. He would never have turned his back on such a challenge. Describing himself as a socialist, Declan was outspoken on many issues, in particular traditional music and song and Irish, international and working-class politics.

Declan's sudden death on 27 June 1985, at the age of forty-five, left a great gap in the Australian/Irish traditional music scene as well as in the Australian folk scene generally. He is survived by his two children, Bridie and Patrick, and his wife, Colleen Burke.

Declan Affley contributed to a number of record albums, but left only three solo recordings. These are *Rake and Rambling Man* (1967), *The Day the Pub Burned Down* (undated, probably 1973–75) and *Declan Affley* (1987), produced posthumously by Mark Gregory, Colleen Burke and Peter Parkhill.

In many ways Declan Affley is representative of the great contribution which expatriate Irish, Welsh, English and Scottish singers and musicians have made to folk music in Australia. After his untimely death the Australian Folk Trust (q.v.) inaugurated an annual Declan Affley Memorial Songwriting Award which is announced at the National Folk Festival.

Colleen Burke

AH FOO is a central figure of Chinese descent in a cycle of humorous tales told particularly in the north of Australia.

See also: **EDWARDS, RON**.

ANDERSON, HUGH McDONALD (1927–), is a widely published writer in fields as diverse as history, literary criticism, school texts and biography, and has also made significant contributions to the study of Australian folklore, most particularly in the field of folksong. For his work in Australian history, he was elected a Fellow of the Royal Historical Society of Victoria in 1974. Since 1954, beginning with the Black Bull Chapbooks series published by the Rams Skull Press, Anderson has written, edited and co-edited works central to folklore studies in this country. These include *Colonial Ballads* (1955, and subsequent revised and enlarged editions, including that of 1970 under the title *The Story of Australian Folksong*); *The Goldrush Songster* (1958); *Farewell to Old England: A Broadside History of Early Australia* (1964); *Folk Songs of Australia* (with John Meredith) (1967, and subsequent editions); *Time Out of Mind: Simon McDonald of Creswick* (1974, 1987); and Charles Thatcher's *Gold Diggers' Songbook* (1980). More recently, Anderson has compiled and published *George Loyau, the Man Who Wrote Bush Ballads* (1991), and *'On the Track' with Bill Bowyang* (six parts, 1991–92). He also contributes articles and reviews on folklore to literary and folklore journals both in Australia and internationally, including *Overland*, *Folklife* (UK) and *Folk Music Journal* (UK). Anderson is a member of the editorial board of *Australian Folklore* and review editor of that journal.

By invitation of the Chinese Writers' Association and the Chinese Folklore Society, Hugh Anderson has spent much time studying folklore in China, and was the leader of several tours by Australians in the 1980s.

As a publisher, Anderson has been instrumental in producing a number of folkloric studies and collections through the Red Rooster Press (q.v.), which he founded as a family partnership in 1979, and of which he has been managing editor since. Its folklore titles include *Frank the Poet: The Life and Works of Francis MacNamara* by John Meredith and Rex Whalan (1979); *The Wild Colonial Boy* (1982) and *Duke of the Outback* (1983), both by John Meredith, as well as several of Anderson's own works listed above.

In 1986 Hugh Anderson was appointed Chairman of the Australian Government's Committee of Inquiry into Folklife in Australia (q.v.) which involved considerable travel gathering material throughout Australia and in central and eastern Europe. This led to the publication of the Report of the Inquiry as *Folklife: Our Living Heritage* (1987) under the name of Anderson and his fellow committee members, Gwenda Davey and Keith McKenry. His personal view of Australian folklore and its study is set down in *Taking Aim Against the Sun: The Making of Traditions in Australia* (forthcoming).

ANDREWS, SHIRLEY, received her early training in folk dance in Melbourne with the Borovansky Ballet and the Unity Dance Group. Later, in the early days of the folk revival, she commenced research into Australian traditional dance. She has written *Take Your Partners: Traditional Dancing in Australia* (1979) and (with Peter Ellis, q.v.) *Two Hundred Dancing Years* (1988). Shirley Andrews is a life member of the Victorian Folk Music Club and the Folk Song and Dance Society of Victoria. She is now a committee member of the Traditional Social Dance Association of Victoria after some years as president and vice-president. She was a member of the organizing committee for the first and second National Folk Festivals in 1967 and 1968 and the first National Folklore Conference in 1984.

Anzac Day At dawn on 25 April 1915, British and allied troops were landed at various points along the Gallipoli Peninsula (*Canakkale* in Turkish) near the Dardanelles to attack waiting Turkish forces. Australian and New Zealand troops (hence the acronym ANZAC, from Australian New Zealand Army Corps) were among those landed, this being the first significant experience of Imperial battle for both countries. Since 1916 Anzac Day has been celebrated on 25 April in both Australia and New Zealand as a commemoration of this military event. As with many other formal observances, such as Christmas and Easter, Anzac Day has developed a number of folk traditions. These include the swapping of tales, jokes, songs and verse by reuniting ex-service personnel at convivial occasions throughout the country, the often excessive intake of food and alcohol within these groups and the playing of the gambling game 'two-up' or 'swy', an illegal activity that is usually tolerated by the police on this 'one day of the year'.

See also: **Wartime folklore**.

Apodimi Compania is a Melbourne Greek ensemble specializing in rebetika (q.v.) and Greek-Australian folksong (q.v.). Revivalist bands known as *kombanies* have been a feature of the nostalgic cult of rebetika in Greece since the mid-1970s. Comprising for the most part young enthusiasts of both sexes, these bands characteristically focus their repertoire on historic or little-known recordings of the pre-war era and seek to reproduce their supposedly authentic, pre-commercial style on acoustic traditional instruments. The results have aptly been compared with 'old 78s without the scratches' and are sometimes open to criticism as contrivedly primitive, hypercorrect and unhistorical.

Melbourne's Apodimi Compania (Expatriate Ensemble) has evolved beyond the typical mould of metropolitan Greek revivalist bands. The band was formed by a group of university students in Melbourne in 1984 out of the Parikiaki Compania (Parochial Ensemble) which performed at the Tsitsanis memorial concerts of that year, and initially comprised bouzouki, baglamas, guitar, accordion and a vocalist.

For some years thereafter the nucleus of the band consisted of the Galiatsos brothers, for whom music is a part-time activity pursued in parallel with jobs ranging from teaching Greek to community arts work, but no less ardently for that. They are avid collectors of old recordings of every kind of traditional Greek music, and Manolis Galiatsos manufactures traditional Greek and Middle Eastern stringed instruments. In defiance of the conventional wisdom that Greek music in Melbourne revolves primarily round the cosmopolitan bouzouki and its tourist appeal, the band in 1992 recorded a compact disc of rebetika and Greek folk songs using over a dozen different folk instruments.

Their first engagements included performances at Greek community functions, student concerts, and at the Dan O'Connell hotel, Carlton. In the mid-1980s the band began appearing at multicultural folk festivals all over Australia, including the National Folk Festivals of 1986, 1988 and 1990. In 1988 members of Apodimi Compania accompanied the Greek-Australian composer Costas Tsikaderis to Greece to participate in the Greek government-sponsored *Apodimia* Festival, and in 1989 moved into a regular Sunday-night spot at the Retreat hotel in Brunswick. They have also appeared on SBS TV ('Vox Populi' 1990) and ABC Radio ('Music Deli').

Out of their contact at festivals with exponents of other traditions arose a particularly close co-operation with Celtic musicians, and the Irish bouzouki player Andy Irvine sang an Irish ballad on the band's 1989 LP *Patris*, titled after the ship which brought their parents to Australia. The album also featured a diverse selection of Greek folk songs, one rebetiko, and Balkan instrumentals, representing a considerable broadening of their repertoire since their first LP, *Rebetika* (1987, reissued in 1990), and their contributions to the composite albums *The Music of Migration* (1986) and the 'Music Deli' LP (1988). For special occasions the band's repertoire has also expanded to include non-traditional songs, such as the poetry of Ritsos, Hikmet and Livaditis set to music, and compositions of Loizos and Xarhakos.

The continuing evolution of Apodimi Compania is shaped by a confluence of multicultural musical forces, which appears unique, but probably has parallels all over the Greek diaspora, particularly in the USA where Greek folk songs and rebetika were first recorded in the same studios as Negro blues in the 1920s, and often on the same instruments.

At least one other Greek *kombania* of this kind is currently active in Australia, To Meraki of Adelaide, headed by the ethnomusicology student Demeter Tsouni.

REFERENCES: S. Gauntlett, 'Orpheus in the Criminal Underworld. Myth in and About Rebetika', *Mandatoforos* 34, 1991; G. Holst-Warhaft, 'The Rebetika Revival', *Aegean Review* Fall/Winter, 1988; G. Michelakakis, 'On Greek Song in Melbourne' [in Greek], *Chroniko Melbourne* 3, 1980–1.

Stathis Gauntlett

Architecture, vernacular: see **Australian house, The**; **Building and folk architecture**.

Art: see **Folk and popular art**.

Audio preservation Sound is invisible, and it is precisely this intangible aspect that is, to collectors of folklore, so appealing. The fact that, in the majority of cases, information about things folkloric are passed on aurally means that recording of the information on to a sound carrier is appropriate. To store this information in a transcribed and printed form is to store only part of the data as the recorded word carries levels of meaning and nuance that are not conveyed in the printed form. Preservation of sound is therefore vital to the preservation of folklore materials.

To date, no medium for the storage of sound has been produced which might be called permanent. All recording mediums carry within them the seeds of their own self-destruction, so that either by use or by time, the item will eventually become unusable. However, a copy of the sound, made accurately and without added noise, is just as valuable as the original recording.

The majority of Australia's field recordings have been made since the late 1940s on magnetic tape of one form or another, or in a very few cases have been copied on to magnetic tape for preservation from early wax cylinders or lacquer discs. Though the recording industry is undergoing great change in the face of digital technology, analogue tapes are still the prime medium for storing Australia's intangible folk heritage.

Tapes and their composition

Magnetic tapes are composed in layers: the layer which carries the information is known as the *oxide layer*, as traditionally the magnetic information has been stored on ferric (iron) oxide. The oxides are suspended in a *binder*, and this layer is spread on to the *substrate*, a film layer whose purpose is to carry the oxide information. In professional tapes, a rough coat is applied to the rear of the tape to aid in winding. This is known as backcoating.

Substrates

From the advent of tape until 1963 when compact cassettes came on the market the most common substrate was cellulose acetate, a high-strength clear film plastic. Cellulose acetate tapes are considered archivally unstable when compared to modern tape stock. The most common problem in sound archives with these tapes is the loss of adhesion of the oxide layer to the tape base, resulting in some cases in loss of portions of the signal. The problem is exacerbated by the dimensional instability of cellulose triacetate which expands as temperatures rise. A 7″ reel of standard play tape would increase or decrease its length by nearly 60 mm for every 1°C change in temperature. Some later tapes, particularly those made by the German company BASF during and for some time after the Second World War, used specially prepared polyvinyl chloride (PVC) as substrate. PVC is even more sensitive to temperature changes: a 7″ reel on PVC would lengthen or shorten by nearly 100 mm per °C.

As temperature rises the increased tape length must be accommodated on an already wound reel. This will most often result in the physical deformation of the tape, producing a corrugated appearance. If the tape retains the corrugation then it can no longer intimately contact the heads of the replay machine. A significant amount of information is lost, particularly with shorter wavelength, with even a small movement away from the heads. Corrugation or movement which can be seen with the naked eye would represent a very significant loss of audible information.

A reduction in temperature results in a reduction in the circumference of the reel, and a resultant increase in the pressure on the hub on which the tape is wound. If exposed to a cycling variation in temperature (or humidity, which can operate in a similar way), the tape expands and contracts, and any single point on the tape will not necessarily return to exactly the same point in relation to the next layer of tape. Variations develop in the packing tension at various points along the length of a reel, resulting in an uneven wind. The narrower the diameter of the wind, the greater the effect.

In a modern professional tape the substrate is most likely a polyethylene terapthalate (PET) commonly known as polyester or mylar. Polyester tape substrate is less susceptible to variations in climate than its predecessors, but changes in the storage conditions can still adversely affect a polyester tape and need to be controlled if physical damage is to be minimized.

The stresses developed in a reel of tape can be relieved by the periodic rewinding and consequent replaying of the tape for storage. A polyester tape stored under stable conditions at 18°C would need to be respooled at intervals of four years; at a temperature of 28°C the tape would need to be respooled every four months. Variation of the temperature of storage would increase the requirement for respooling.

Binders

The binder is usually a polyester urethane, though sometimes it is a PVC. Magnetic particles are suspended in the binder, and this is spread evenly on the substrate. Included in the binder are a number of other substances which enhance the performance of the tape, including plasticizers and lubricants to improve the physical playing of the tape. The lubricants are often fatty acids, and these affect the stability of the binder.

The most likely mode of failure or instability in a polyester urethane tape is the hydrolysis of the tape binder. Under conditions of higher than normal temperature and humidity, a sticky deposit forms which causes a characteristic squeal. If a fatty acid is present then the hydrolysis will probably be permanent. In storage conditions of under 18°C and 40 per cent humidity, hydrolysis is less likely.

When hydrolysis occurs, the signal quality is impaired in the case of analogue recordings, and a significantly higher error rate occurs in the case of digital recordings. Though the hydrolysis is not completely reversible, trials have shown that it is possible to play a tape, under conditions of very low humidity and, to speed the process, higher temperatures.

All binder formulations, even those which are not polyester-based, may exhibit the characteristics of a hydrolysed tape. This is brought about by the loss of lubricant from the surface of the tape. This can be treated in a similar way to the problem of hydrolysis, but the temperature is the important factor in the process, rather than the humidity. In the case of tapes which do not respond to heating, the topical application of a suitable lubricant may improve the playability of the tape, with only small losses generated by the loss of head contact.

Acetate-based tapes may also deteriorate, producing acetic acid in the process. Acetic acid has a fairly distinctive smell and this process of degradation is known as the 'vinegar syndrome'. The smell of acetic acid is detectable in very small concentrations, certainly before any major damage is done. At the first sign of vinegar syndrome, the information on a tape should be immediately transferred to a stable archival medium. Long-term breakdown of acetate tape eventually renders it unplayable.

Oxide layer

There are several problems that can occur with the oxide layer of a tape — print-through, distortion produced by external magnetic fields, and self-demagnetization.

When a tape is wound on to a reel, the layers of the tape are in close contact with the adjacent layer of tape, and under these conditions the field associated with a recording of a particular signal may imprint itself on the adjacent layers. This is known as *print-through*, and is only apparent on analogue recordings. Print-through is typified by a pre- or post-echo of a high-level signal. It occurs from the moment of contact in a wound tape, and continues as long as the layers remain in contact; however the greatest proportion of print-through occurs within the first forty-eight hours after contact. Changes in temperature can alter the print-through ratio, as can the frequency of the recorded item, the thickness of the tape, and the speed of the tape.

When a tape is stored on a reel, the signal prints to both the layer before and after it; however, the signal is printed more to the outside layer than the inside layer. If the tape is stored *tail out*, then the louder signal will be an echo rather than a preview, and will be less obvious. There are also other advantages to storing tail out. In print-through only the easily magnetized low-coercivity particles are affected. These particles can be mechanically realigned, which is achieved by rewinding; a tape which is stored tail out must be rewound to be played. Print-through which is not effectively reduced by this method may be ameliorated by the application of a weak bias signal, which will realign low coercivity particles. Some machines have the facility as an option for archival work.

If a tape is duplicated without the print-through being eliminated first, the new tape will be recorded using the same bias field as the wanted part of the signal, the print-through then cannot be removed by any of the described methods, it has become an unalterable part of the signal.

Magnetic fields, AC or DC, have detrimental effects on a recorded tape, and magnetic fields are legion in any modern installation where electrical equipment is used. Low-level DC fields will cause an increase in noise, medium levels a loss in playback level, and high DC levels an increase in distortion. It is therefore important that a tape does not come in contact with a DC magnetic field, any stronger than 25 oersteds. DC fields greater than this can be found in speakers, headphones, microphones, most electrical equipment; even the metal shelves commonly used to store material in libraries and archives retain a significant DC field, a by-product of their method of construction.

AC fields develop as a result of the operation of almost any mains-powered piece of equipment. Low levels of AC field can cause partial erasure of the signal, and the recording of unwanted noise as a result of the mains signal. It can also increase print-through and make it permanent. Very high levels can erase a tape. Tapes should not be exposed to an AC field greater than 5 oersteds. AC fields occur in fluorescent light fittings at various levels according to the type of ballast, in all mains-operated motors, in equipment like vacuum cleaners, polishers and so on, none of which should be operated in close proximity to stored tapes.

Magnetic field strengths are proportional to the inverse of the square of the distance, which means that they drop rapidly with increasing distance from the source. However, as the fields associated with so many pieces of equipment are so large, considerable caution should be exercised in the use of any magnetic equipment in the vicinity of a recorded tape.

Self-demagnetization (self-erasure) occurs when the magnetic field of one part of a tape realigns the field in another, effectively losing that piece of information. It can only occur with low tape speeds, high frequencies, and tapes of high 'coercivities' — a metal formulation cassette will have a greater tendency than standard oxide cassettes to self-erasure, both from adjacent magnetic fields, and from adjacent layers. A digital tape recording using rotating head technology, such as modified video machines or DAT recorders are very likely to exhibit this fault. The rotating head and slow tape speeds mean that the wavelength on the tape is very short, the recorded domains being very small and close together. The loss of this information would in most cases be hidden by the error correction facility in the replay equipment; however, it may present a long-term archival problem.

Maintenance and restoration

Storage

Every facet of a tape's performance is affected by the environment in which it is stored. The chemical, magnetic, dimensional, cleanliness and electro-acoustic condition of a tape is in one way or another determined by the way a tape is stored. For example, tapes with a polyester-based binder stored below 40 per cent relative humidity are less likely to suffer from hydrolysis than at higher humidities. The dimensional instability of polyester also means that humidity should be very stable, within a total range of 5 per cent. Care should be taken in the setting of the minimum also, as taking the humidity very low may cause problems with the older cellulose acetate tapes, which can become more brittle.

A practical temperature for tape storage is 18°C. Lower temperatures can increase the life of a tape by reducing the speed of any chemical degradation, particularly with PVC binders; however, fluctuations in a tape's environmental conditions can engender greater problems than the low temperatures would save. Changes in temperature ideally should be no more than 1°C. If a choice has to be made, then a stable temperature of 18°C is better than a lower but unstable temperature. However, if a duplicate of a tape exists which is not expected to be used frequently, then that tape will benefit from being stored at a temperature around 8°C.

Air cleanliness is an important factor in the storage of tape. Small airborne particles will degrade the performance of an analogue tape and displace the information in a digital recording, whether optically or magnetically read. All storage areas should have positive air pressure, that is the air coming into a room should be forced in at slightly above ambient air pressure. This excludes dust as the only air input then is via the air-conditioning input, which should be suitably filtered.

Gases and other pollutants may cause increases in the chemical degradation of tapes and other sound carriers, quite apart from their effect on the replay of the tape. Breakdown of certain tape binders may generate components that are detrimental to the stability of other types of binders; it is advisable for example, to store PVC-bound tapes separately from polyester-bound tapes.

The pros and cons of using plastic bags for storage have been much discussed. The polyethylene bag that is usually supplied with a reel of tape will protect it from dust and accidental contact with liquid contaminant, providing this is cleaned off and dried quickly. The bag has very little effect on temperature and humidity. On balance it would seem valuable to store tapes in bags with the ends folded rather than sealed, though further research may shed more light on this.

Cleaning and demagnetization

Tapes may be damaged by being replayed on a badly maintained machine. All surfaces in the tape path should be routinely cleaned to remove build-up of oxides, dirt and loose binder materials. All worn surfaces should be replaced, as they can damage the smooth surface of a tape. All ferrous surfaces on a tape path should be demagnetized at regular intervals as the residual magnetic fields in close contact with a tape will be detrimental to its condition. Surfaces become magnetized by the playing and winding of tapes.

Moulds and fungi

Under conditions of high humidity and temperature, moulds and fungi can grow. Though they are most likely to occur on the glues and papers used to make the boxes that store the tapes rather than the tapes themselves, these growths can transfer themselves to the tape. Growths can also occur on debris deposited through careless playing or handling of a tape. The hands can transfer oily substances which may attract dust and airborne spores. When playing, repairing or storing a tape, care should be taken to avoid, as much as possible, touching the oxide surface of the tape.

Most modern tapes contain fungicides in the binder which prohibit growths, so any moulds are likely to be surface growths. If a tape so affected is exposed to a cool dry environment, the growth will most likely die, and the tape can be easily cleaned. The material should be brushed off and vacuumed (taking due care to avoid the magnetic parts of the cleaner coming close to the tape), while the tape is still wound. The tape can then be cleaned while being spooled, and consequently played and duplicated.

Repair

Even under well-managed storage conditions, tapes may suffer damage for a variety of reasons. If repair is necessary, it is generally a sign that there is either fault in the tape itself, the replay equipment or the conditions of storage; to avoid further damage it is necessary to rectify the problem.

One kind of damage is caused by *pack deformation*. For example, a tape that has been fast-wound will often exhibit an uneven tape pack where some layers of tape will protrude above or below the normal height of the tape. This is known as 'leafing'. Tapes that are handled while in this condition will generally suffer edge damage, where small parts of the tape are damaged and that amount of information lost. If the tape is stored for a long time with a poor pack the substrate will eventually 'remember' the form of the pack and the tape will wander in relation to the tape path, and not make the necessary intimate and accurate contact with the tape head. A similar fault can occur if a tape is not respooled regularly. The deformation in the tape pack can be reversed to some extent by respooling slowly and with gradually increasing packing tension. The tape should then be allowed to acclimatize to the environment before any changes in storage, which takes about eight hours for polyester tape, and a good deal longer — maybe days — for an acetate or PVC tape. The tape will eventually take on the shape of the new smooth pack and retain its shape.

Tape breakage can be repaired using splicing tape in the same way as tapes are cut-edited in a recording studio. The splicing tape should be non-bleeding, as any migration of the adhesive on to adjacent layers causes the tape to stick and this degrades the consistent replay speed of the tape. The tacky substance will also get between the heads and the tape and degrade the sonic performance of the recording. This can be cleaned off using methanol or some such substance, which if applied sparingly will remove the adhesive and not the binder of the tape. When cleaning in this fashion, the tape should pass a swab slowly, the alcohol being allowed to evaporate, before it is wound on to the next layer. The tape should then be copied. All solvents degrade the performance of a tape by either directly attacking the components of a tape, or by importing or dispersing impurities over the surface.

Acetate tapes become brittle with age but tend to break without elongation, making repairs fairly simple. Once repaired, the tape should be copied on to a stable tape. Polyester tapes will stretch before breaking, and the area surrounding the tape will be unplayable. In this situation it is often necessary to cut or remove the damaged portion of the tape, sacrificing that part of the information, to make the tape playable. A demagnetized cutting instrument and splicing block must be used, as any magnetized object very close to a tape will add noise to the signal.

Most digital tapes cannot be seamlessly cut and spliced. With the exception of DASH tapes, the information on a digital tape is interleaved in a complex pattern, and cutting and splicing two segments of a tape will produce a non-continuous data stream that the tape player will be unable to read. The error-correction circuitry will most likely blank out the offending section. If a tape is repaired, the tape player, again with the exception of DASH, will probably have difficulty in playing the segment of the tape as the thickness of the tape is critical in the case of rotating-head replay machines. Also, the handling of the tape will increase the errors produced in that area. Any tape that has been repaired should be transferred to a non-destructive electronic editing system, the faults remedied as far as possible, and the signal transferred to a new storage system.

When a cassette is repaired, it should be remembered that the tape is packed with the oxide facing out, unlike most open-reel tapes. Care should be taken to ensure that the splicing tape is on the back coat, not the oxide side of the tape.

All edits and repairs are impermanent, and an archival tape should be stored without any splices throughout its length.

Sonic restoration

The signal quality of a tape may be enhanced, the sonic balance restored, or unintended noise removed, with varying degrees of success, using electronic filtering equipment. However, a filtered tape should never be stored as an archival master; it should be a second copy, as the processes involved in the manipulation of sound are not accurately reversible. The storing of a 'flat' or unfiltered copy means that future generations, who may be able to carry out sonic restoration to a much higher standard than is possible today, will have all the original information available to them, information that is lost by present-day filtering systems.

The compact cassette

The ubiquitous compact cassette is to be found recording information all over the world. It is small and easy to use, produces a reasonable level of reproduction, and is cheap and easily distributed. It is not, however, suitable for an archival master. The cassette is designed to be compact and to play for lengthy periods, making very thin tape a necessity. The thinner tape exacerbates some of the main problems associated with archiving sound, such as print-through, mechanical instability and liability to damage. The tape, by virtue of the very narrow track width, has a higher inherent noise level, and, as tape is recorded in both directions, there is no 'tail out' position on the tape.

For these same reasons the cassette is not an appropriate choice for original recording either. When an analogue signal is transferred from one tape to another, a certain amount of the signal is lost, and faults in the original recording, such as speed variations, print-through and extraneous noise, become an intrinsic part of the preservation duplicate. Moreover, the cost of preserving a cassette by making a preservation duplicate is greater than the cost of storing a preservation original.

The shell that encases the cassette affects the quality of the signal, and in time most cassettes develop problems related to the shell. The shell will influence the accuracy with which a tape is applied to the head, or 'azimuth', and small changes here can cause large changes in signal quality. The shell can also degrade the short-term speed variation performance, causing 'wow' and 'flutter', and this is a basically an irreversible problem once the tape is recorded. Most portable cassette recorders produce a lot of wow and flutter.

As well as being less robust than the thicker reel tapes, cassettes are also more difficult to repair because of the shell's construction. Welded, as opposed to screwed, cassette cases are a great problem and should never be used, as it is necessary to split the case to gain access to the tape inside. The low speeds and narrow track widths mean that damage to a small portion of the tape leads to impairment of a comparatively large segment of the recorded signal.

Digital recording

All sound archiving is built on the premise that as a sound nears the end of its carrier the signal is transferred to another carrier. Digital offers great benefits under these circumstances, but the carrier may not offer the archival life necessary to take advantage of them.

When a signal is recorded in a conventional form, an analogue of the signal is recorded. Upon replay the noise components and signal degradation brought about by the medium and the transfer become a part of the signal. When a signal is recorded digitally, it is encoded and the coded information is stored on the recording medium. When it is replayed, any noise introduced to the coded signal is ignored by the decoding device and the signal is replayed with the same quality as it was encoded. A digital recording can be transferred to another digital recording without ever being decoded, and this transfer method is very accurate, resulting in almost no variation between the original and the next-generation copy. In the very long term, a signal archived in analogue form will show gross degradation due to continual duplication, and a digital recording will show little or no difference to the original. The hidden disadvantage is the recopying period. A digital recording on the type of carrier in use today will require recopying between two and five times as often as an equivalent analogue recording.

Digital recordings are at present made on a number of different formats, and according to a number of different standards. These incorporate such things as domestic and professional video cassettes, open-reel DASH, DAT recordings and a now-increasing number of disc-style formats, closely aligned with the computer industry. Archivally all are untested and it remains to be determined which format will eventually provide the archival life and professional support required of a long-term archival storage device. Until such time as there is a clear choice, all digital material to be archived should have a suitable analogue copy made as a backup.

The digital original should be retained, as it can be used to produce a lossless copy in the digital domain, providing that it is replayable at the time it is required.

At the time of writing, the manufacturers of domestic equipment are poised to launch two competing digital recording devices, the digital Compact Cassette, and the Minidisk. These devices use data compression, a system by which a portion of the information about a signal is stored and later used as a representation of the whole. It is important, for archival recordings, to store as complete a recording as is technically possible, rather than modified data.

Until the almost certain development of a suitable digital archival format, analogue tape will continue to be the main storage medium for archival recording in Australia.

Recording standards

When sound is recorded on to a tape it is done according to a standard, so that tapes and equipment can be exchanged between users. For analogue tapes Australia is theoretically a signatory to the European standard, IEC, however, many commercial users prefer the American standard, NAB. Most libraries and archives in Australia use IEC as the standard for storing and recording analogue tapes. A digital standard for archiving does not yet exist.

It is crucial that any archival standard be exchangeable with the wider recording community. The archival industry is a very small part of the recording industry as a whole, and technologies for storage are only those adapted from their original role in recording. If a collection is to be useful, the equipment to replay it must be available in the future as well as the item itself. This will only be possible for technology that has the support of the industry as a whole. For these reasons archives must standardize in consultation with the industry.

Brief recommendations

1 Record for storage only on appropriate archival medium (at the time of writing, analogue reel).
2 Record at the appropriate standard for archiving (in Australia, IEC).
3 Only play original or archival tapes for the purpose of making a reference copy.
4 Make a copy of all tapes in accordance with the preservation specification, preferably on a tape of a different makeup or manufacturing process.
5 Wind off tape in play mode at or close to the storage temperature; store tail out.
6 Avoid extremes or variations in temperature. Respool and acclimatize before storing.
7 Regularly maintain recording and replay equipment.
8 Ensure storage area meets required conditions of temperature, humidity, and cleanliness and is free of magnetic fields.
9 Respool regularly.

The archiving of recorded collections of folklore requires the management of collections to ensure the longevity of the material as a whole. The value of field recordings is in the information they carry rather than the tape themselves (rather than a painting, for example, which may have more value as an object than a bearer of information). The sound archivist's job is to ensure that the aural information will be useful and intelligible in the future, but which may involve the discarding of present formats that store the information. Under these circumstances it is vitally important that the steps taken to record, copy and preserve the recordings are as accurate and predictable as the technology will allow. If this does not happen then the documentation of the folk heritage of Australia will be degraded and the important work of the collectors of this material will be lost.

REFERENCES: *Phonographic Bulletin:* Journal of the International Association of Sound Archives (all issues); N. Bertram and A. Eshel, *Recording Media Archival Attributes* (Magnetic), The Rome Air Development Centre, Air Force Systems Command, Griffiss Air Force Base, New York, April 1980; E. F. Cuddihy, 'Ageing of Magnetic Recording Tape', *IEEE Transaction on Magnetics*, Vol. MAG-16, No. 4, July 1980.

Kevin Bradley

'Australia All Over': see **Radio and television programmes.**

Australian Children's Folklore Collection, The, possibly the world's largest collection of children's folklore, was established by June Factor and Gwenda Beed Davey in 1979 at the Institute of Early Childhood Development (now School of Early Childhood Studies, University of Melbourne). The collection consists of over 10 000 children's playground rhymes and games, together with photographs, audio and video tapes, monographs and artefacts. It is now managed by Dr June Factor from the Australian Centre at Melbourne University and is housed in the university archives. Since 1979 the Collection has published two copies per year of the *Australian Children's Folklore Newsletter* (q.v.).

As well as its extensive collection of playground rhymes and games collected since 1971, the Collection includes adult recollections of childhood games and pastimes, the Dorothy Howard (q.v.) Collection from the 1950s, and the original field recordings and photographs from two major projects, the Multicultural Cassette Series (1976–77) and the Multicultural Playground Project (1984). As well as children's own playground lore, the Collection includes folklore addressed to children by adults (e.g. family sayings and aphorisms) and folklore about children (such as old wives' tales) (see **Women and folklore**). In addition to the English language, about thirty other languages spoken in Australia are represented in the Collection. All the material is indexed and available to researchers.

The Australian Children's Folklore Collection has substantially assisted major exhibitions at the Museum of Victoria and the Royal Children's Hospital in Melbourne. It publishes monographs and recordings as Australian Children's Folklore Publications (q.v.).

See also: **Children's folklore.**

Australian Children's Folklore Newsletter, was founded in 1981 by June Factor (q.v.) and Gwenda Beed Davey (q.v.). It has been published twice yearly under the continuing editorship of its founders and in close association with the Australian Children's Folklore Collection (q.v.). Devoted to all aspects of Australian children's lore and play, the Newsletter publishes news and reviews, as well as substantial research articles.

Australian Children's Folklore Publications is the publication arm of the Australian Children's Folklore Collection (q.v.) at the Australian Centre, University of Melbourne. It publishes the *Australian Children's Folklore Newsletter* (q.v.) twice yearly and publishes and distributes a number of audio tapes, video tapes and monographs. Materials are in English and several other languages and include *Children's Folklore in Australia: An Annotated Bibliography* (1986), compiled by June Factor. A publications list is available on request.

Australian Folk Trust Inc., The, is the national co-ordinating body promoting the performance, development and practice of folk arts and the collection and preservation of folklife heritage, and is funded by the Australia Council. It originated from the establishment of a National Folk Festival Trust in 1966 which was formed to promote and help finance an annual festival on an Australia-wide scale. The National Folk Festival has subsequently expanded into one of the major events of the Australian arts calendar, while the AFT, incorporated in 1977, is now a firmly established national body with professional staff and an office located in Canberra. Management is by a Board of Trustees with representatives from each state and territory. The member organizations of the AFT are: NSW Folk Federation; Queensland Folk Federation; WA

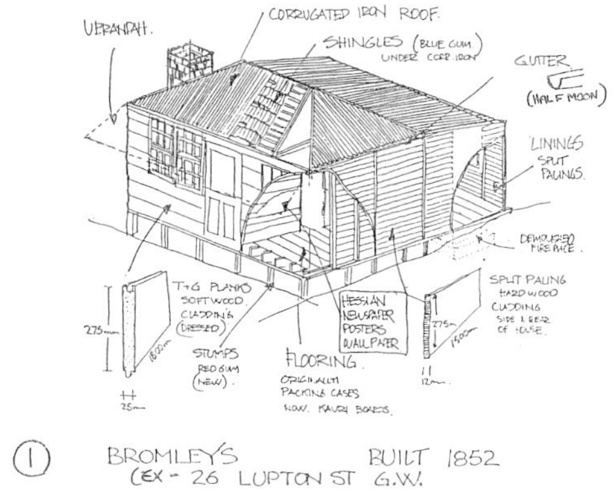

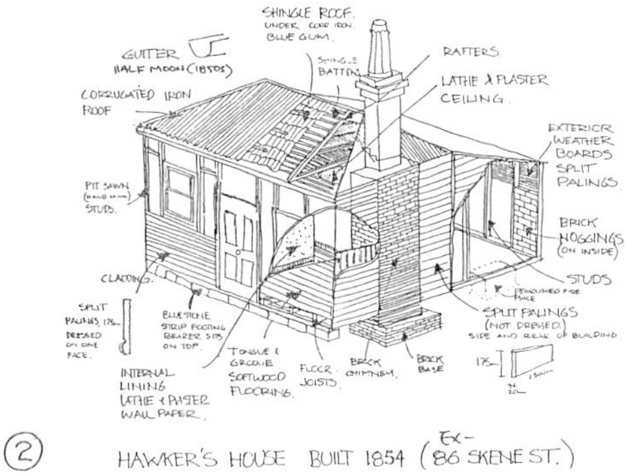

Figures 1 & 2 Bromley's and Hawker's houses, drawn by John Graham. (Campbell and Fox, *The Australian House Museum: A Guide to the Project*, Deakin University, 1986.)

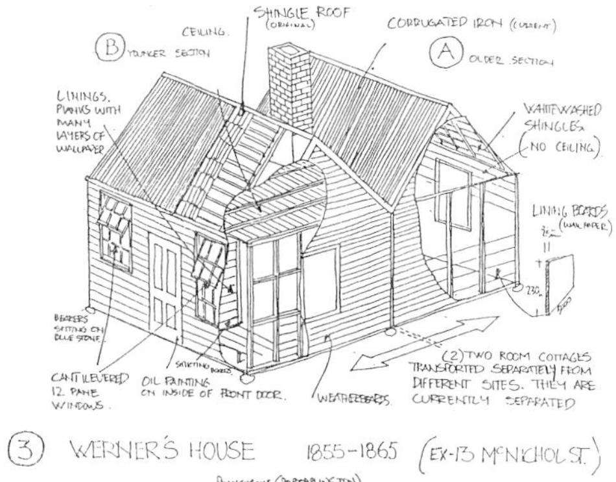

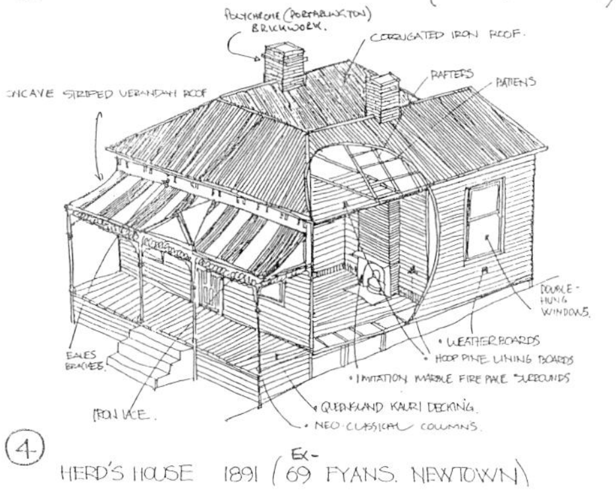

Figures 3 & 4 Werner's and Herd's houses, drawn by John Graham. (Campbell and Fox, *The Australian House Museum: A Guide to the Project*, Deakin University, 1986.)

Figure 5 Prefabricated house (before restoration) built in 1853. Formerly at Geelong West, Vic., now at the Australian House Museum, Deakin University, Geelong. (Frank Campbell)

Figure 6 Elizabeth Farm, Parramatta, NSW, built in 1793. (Frank Campbell)

Figure 7 Middle-class Australian Queen Anne, Broken Hill, NSW, 1984. (Frank Campbell)

Figure 8 Early Penang houses, late eighteenth century. (Lee Kip Lin, *The Singapore House*, Times Editions, Singapore, 1988.)

Figure 9 'Aleena', Wagga Wagga, NSW, 1989. California bungalow obedient to Georgian style. (Frank Campbell)

Figure 10 Prize-winning historicism: revival or continuity? Geelong, Vic., 1988. (Frank Campbell)

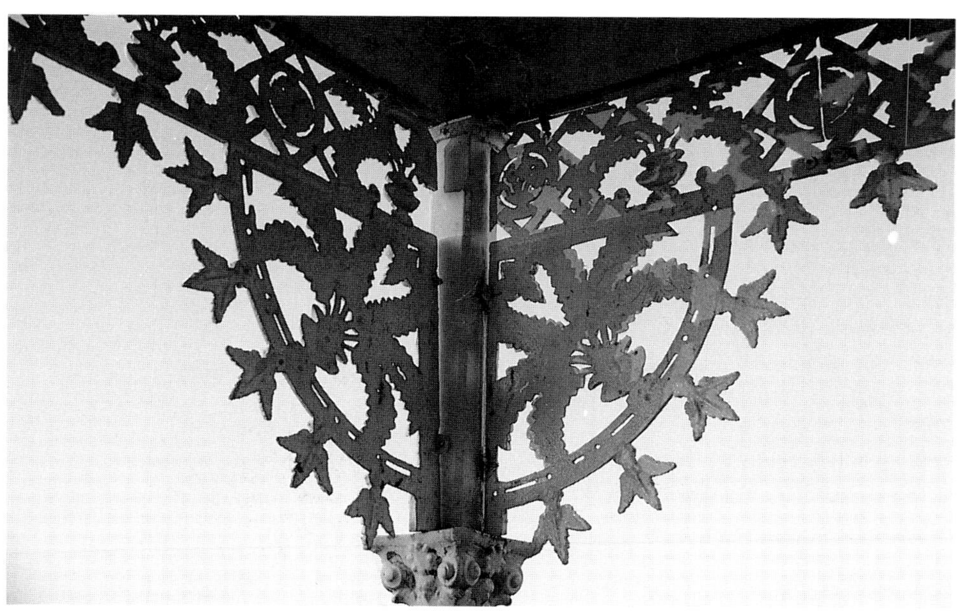

Figure 11 Iron lace: No lace please, we're British. Geelong house, 1904. (Frank Campbell)

Folk Federation; Tasmanian Folk Federation; Top End Folk Federation; SA Folk Federation; Folk Song and Dance Society of Victoria; and the Australian Folklore Association. Board meetings are held twice a year.

The AFT has three broad aims: to assist with community development in all aspects of traditional and contemporary folklore, including music, song, dance, craft, lore, verse, customs, beliefs and traditional lifestyles, speech, games and food practices; to promote, maintain and preserve all facets of folklore/folklife; and to co-ordinate and assist member bodies and all interested parties in the attainment of these general aims. The AFT advances its aims through a varied programme which includes the devolved grants scheme, National Folk Festival; National Folklore Conference; Declan Affley Memorial Songwriting Award; publications such as the *Australian Folk Directory*, *Australian Folk Resources*, *Music for Australian Folk Dancing*, proceedings of National Folklore Conferences and the AFT monthly journal; dissemination of information to members, government bodies and the general public; and promotion of folk arts/folklife through specific projects and advocacy.

Australian Folklife Centre, The, was established on 3 December 1990 by the Australian Folk Trust (q.v.) to preserve and present the traditional culture and folklore of multicultural Australia. Its goals are:

to promote and facilitate the collection of Australian folklife of all types and from all the various peoples of Australia, and make it accessible to scholars and the general public;

to co-ordinate information about folklife collections elsewhere throughout Australia;

to initiate fieldwork;

to help researchers and disseminate the results of their work;

to run training programmes in collection, documentation, conservation and preservation;

to provide public education by such means as concerts, conferences, workshops, lectures and exhibitions.

The Australian Folklife Centre has a programme advisory committee consisting of representatives of the National Library of Australia, the National Film and Sound Archive, the National Museum of Australia and the Australian Folk Trust. It seeks independent status through funding from a variety of sources in both the public and private sectors. Its first project grant was from the Office of Multicultural Affairs (Department of Prime Minister and Cabinet) for a demonstration project documenting multicultural music and dance in the Monaro region. The Centre is currently housed in the National Museum of Australia at its Mitchell offices in Canberra.

Australian Folklore, established in 1987 by Graham Seal and David S. Hults (qq.v.), provides for the publication of scholarly articles, reviews, correspondence, news and reports on various fieldwork projects. Until 1991 *Australian Folklore* was published through the Centre for Australian Studies at Curtin University, Perth, and jointly edited by its founders. Since then the journal has been published by the Australian Folklore Association (q.v.), under the editorship of Professor John Ryan (q.v.).

Australian Folklore Association, The, (Incorporated) was formed in November 1988 following discussions at the forum described in the report: *Collecting Folk Music in Australia, Report of a Forum at the University of New South Wales, 4–6 December, 1987*. The aims of the Association are to promote the collection, preservation and study of folklore in Australia; to foster the discussion and dissemination of information about folklore in Australia, and to promote understanding and appreciation of the important social and cultural role of folklore in Australian society. Communications of the Association are published in the journal *Australian Folklore* twice-yearly, and in a newsletter. Currently the AFA is developing a code of ethics for Australian folklorists and a programme of regional folklore surveys. The organization operates with a national executive of elected representatives.

Australian Folklore Society, The, was formed in 1979 to represent the views of collectors of Australian folklore. At the time all the existing folklore societies were concerned mainly with performance and performers, and their interest was almost entirely in songs. By the nature of its charter the Australian Folklore Society is limited to collectors and so is, and must remain, a numerically small society. However, it has worked well as a lobby group with the Australian Folk Trust, and in recent years the Trust has been paying much more heed to the suggestions that have come from the collectors. This has resulted in a greater number of grants being approved for the collecting of folklore in Australia.

The Society began publishing a journal in September 1984. The latest issue, the fifteenth, was 49 pages long and was produced by Ron Edwards (q.v.). It is available only to members.

Australian house, The, is as much an anthropological artefact as the boomerang. It is not simply a repository of people and objects, but a cultural entity. To the English, Australians live not in houses but in 'bungalows'; to the Americans, a bungalow is a specific form of early twentieth-century house, the 'California' bungalow. To Australians, a bungalow is that detached fibro flatlet in the back garden. To a Malaysian, a bungalow is a massive-roofed, detached, two-storeyed colonial dwelling surrounded by verandas. The word is said to derive from Bengali, as is perhaps the idea itself. In all these countries there would also be different understandings of the word 'house'. The focus here will be on the 'ordinary' Australian house — a free-standing, one-storeyed building on a private allotment — not on flats or mansions, or on pioneer desperation and improvisation (see **Building and folk architecture**), although these other dwellings show by no means separate evolutionary lines.

So, is the ordinary Australian house a folk creation? It is true that many of the slab, wattle-and-daub, turf and paling dwellings of pioneer Australians were built by their inhabitants, and 'home-made' houses did proliferate during periods of austerity — the 1890s, early 1920s, 1930s and early 1950s. None the less, as time and the industrial revolution progressed, house components were increasingly made in factories in Australia and abroad. Pitsawing and handsplitting scarcely survived the nineteenth century, except deep in the backblocks. Mechanization and standardization became mandatory and building the preserve of artisans and other professionals. This is not to say that Australian houses were not products of a folk culture. The essence of the argument here is that the Australian house is culturally distinctive and consistent, although it was originally a borrowed form. Indeed, the history of the Australian house is to a remarkable extent the story of waves of foreign and local high-culture influence breaking over the stubborn rock of the vernacular dwelling. For example, in the densely packed urban centres of the later nineteenth century, the terrace house eliminated the standard Australian house. A product of the British industrial revolution, it was rarely the preferred model, however. With privatized and improved transport, the Australian vernacular type reasserted itself.

Australia was a *tabula rasa* in the eyes of the British in the late eighteenth century. There was no vast and seductive culture to become entangled with as in India. Hence, the overwhelming dominance of Georgian style in housing was natural, as that was the ruling passion 'at home'. Once established, Georgian style became fixed as *the* Australian style. For sixty years after 1788 the Australian colonies were weak and distant offshoots of Britain and remained infused with a dependent mentality, controlled by the military in many respects. Early Australia was not fertile ground for new high-style ideas, nor folk-style innovation. Georgianism was unchallenged until the surge of wealth and free immigrants in the 1850s.

In America a prolonged and complex struggle took place between the new eighteenth-century high style of Georgianism and various deeply entrenched vernacular styles. The same was true in Britain and much of Europe, while in a colony such as Singapore, British Georgian had immediately to contend with Chinese and Malay influences. In Australia the official vernacular could proceed unhindered. Consequently Australia developed the purest and most thorough example of vernacular Georgian architecture in the world.

What then defines a house? — the form? the façade? the structure or materials? the floor plan? In vernacular conversation, Australian houses are defined by . . . walling material! 'That is a brick house.' Walling material has been a key status differentiator in this culture, but form, period and scale must also be taken into account. The following typology was developed in the 1980s for a survey in a rural city, Geelong, in Victoria, but smaller surveys elsewhere in Victoria and in South Australia and New South Wales have used similar schema. There were eight major criteria: scale (including volume as well as height and area), form and style, floor plan, materials, structure, class, stylism, and decoration.

Scale

The Geelong survey and a series of related studies showed that there were only two house types as defined by scale between c. 1840 and 1940. In Australian idiom all were 'houses', but some were 'small' and most were 'ordinary'. The figures illustrating Australian house types appear on pp. 23–6. Bromley's house (Fig. 1), built in 1852, and Hawker's (Fig. 2), 1854, are now considered to be very small houses indeed. Bromley's is 18′ × 18′, consisting of four rooms, with ceilings about 6′ high. It is the house of a convict, at that time employed as a carter. At the time of construction it was a respectable working man's house, not at all the meanest available. Hawker's by contrast is a lower-middle-class dwelling, somewhat larger, with plastered ceilings and walls. Herd's house (Fig. 4), built at the height of the boom in 1891, is a full-scale ordinary Australian house. It was also a house of the lower middle class. It has seven times the volume of Bromley's and four times that of Hawker's. Occupants per room numbered three for Bromley's (they were a family of twelve in a four-roomed house) compared with just over one for Herd's.

These changes took place over four decades of increasing colonial wealth. It was as if the inhabitants had a model of the ideal house in their heads, and it expanded to that scale when economic conditions permitted. The surviving small houses very quickly descended in the social order as the century progressed. The cut-off point between a 'small' and 'ordinary' house by scale was approximately 30′ × 40′ and a ceiling height of about 9′. 'Small' became culturally defined as 'sub-standard'. The working and middle classes finally lived in houses of very similar scale. (Note that the Geelong survey stops at the middle of the middle class: no assumptions are made about upper-middle-class or upper-class houses here.)

It seems likely that the detached English and Celtic cottage forms provided one direct template for Australia, but that this small form gave way to the comparatively massive oblong box derived from the British Empire bungalow. Figs 1–3 represent the British cottage scale and form, whereas the photograph of the prefabricated house (Fig. 5), erected in Geelong West in 1853 and probably built in Singapore, shows the British Empire form. A famous example of a common late eighteenth-century version is Elizabeth Farm, Parramatta, built in 1793 (Fig. 6). The Geelong West survey suggests a gradual phasing out of the British Georgian 'cottage', although it was still appearing for the bottom of the market at the turn of the century.

The Geelong house, one-storeyed and detached, is in terms of scale remarkably homogeneous. Most of the population was housed in the nineteenth century, as in the twentieth century, in one type of house. To the extent that Australia does have egalitarian traditions, commonality of size of residence is one of them.

Form/style

In the Geelong survey there is only one basic style, or culturally derived form, before the advent of the Queen Anne style later in the century. That style is vernacular Georgian. The elements which define 'Georgian' in this colonial context are:

1 horizontal line (as opposed to the vertical line of Gothic for example);
2 parallel orientation to the street frontage (given narrow urban allotments, this is no mean feat; it is often achieved by artifice, such as the masking of valley roofs by transverse pitched roofs, or occasionally by false parapets -- see Fig. 4 for example);
3 religiously symmetrical façade elements. This may include matching single and double chimneys for instance, but the essentials include doors and windows balanced, veranda posts likewise, and the positioning of the façade in relation to the allotment boundaries. Gardens usually followed suit.

The crucial façade elements of Australian Georgian retain the proportions and often the details of Renaissance classicism. Exceptions occur in the smaller, early structures in which the prefabricated windows are often far too large for the façades. In some cases, windows have been shortened later to harmonize the whole according to vernacular Georgian taste.

No element of the Australian house has been so commented upon as the veranda. Baglin wrote: 'The verandah is perhaps the outstanding contribution to the distinctive architecture of Australia in the nineteenth century' (D. Baglin, *The Australian Verandah*, 1976), and many others have written in similar vein. The origins of this splendid icon are variously given in the literature as Australia or British India. The word itself probably derives from *verano*, Spanish for 'summer', but verandas were commonplace in the ancient Mediterranean world, and in other cultures as well. Each of the European empires used them in their tropical and temperate possessions. They were an integral part of the British Empire house before Australia became a part of the Empire. In a sense they probably did 'come from India' in that the civil servants and military men of early Australia were ex-India, or ex-Penang, or ex-Singapore, or ex-Jamaica . . .

Floor plan

Symmetry penetrated into the floor plan. The basic pattern is two rooms deep symmetrically arranged on either side of a central passageway running the length of the house. Other rooms were often added at the rear, usually less felicitously. Early small houses, such as Bromley's and Werner's (c. 1855, see Fig. 3), often lack a central passage — in fact both of these were built originally as ceilingless single rooms, albeit partitioned very soon after construction, after which the central front door typically opens into the larger of two rooms — but halls were sometimes added later by means of more partition walls. The point here is that the cultural model for an acceptable house in this culture demanded a hall. In the early, often hastily built houses, this requirement is not met, which in turn could affect the disposition of elements of the façade, i.e. the front door and two windows. Windows were occasionally moved later to achieve symmetry and the verandas, which were almost all added later, brought such houses into conformity by symmetrical placement of posts and the virtual obscuring of offending windows.

Materials

The materials were immaterial, or at least they remained subordinate to style and form. The stone houses of tree-scarce South Australia are the same in most details as their iron cousins in Broken Hill or weatherboards in Melbourne. The rapid capitulation of even the ethnocentric South Australian Germans can be seen in towns such as Hahndorf. Amid the Bavarian fakery in what was originally an austere Silesian enclave, a few fine *fachwerk* (half-timbered) North German vernacular houses survive. After about 1870, however, the Australian vernacular predominates. Nor is the famous Queensland house a distinctive house type: it is the universal late nineteenth-century vernacular Australian British Empire bungalow with a few details (stilts, latticework) altered (see Peter Bell, *Timber and*

Iron, 1984). Nor does Tasmania have conspicuously different traditions. The marked regional variation in Australian domestic architecture depends on how much survives of early housing. Thus, one finds whole suburbs of Australian Queen Anne or California bungalow houses; stagnant Tasmania now capitalizes on its unfeaturist Georgian townscapes.

Structure

The commonest structure of the Australian house is the light timber (stud) frame. The poverty in Australian architectural history is crudely revealed by the debate over the origin of this wooden frame, which of course lies behind both weatherboards and 'brick veneer'. J. M. Freeland (*Architecture in Australia: A History*, 1968) repeats the now-discredited American claim that the light stud frame was 'invented' in Chicago in 1833. He states that it was brought out from California by gold-seekers in the early 1850s. The main deficiency of this theory is that the light stud frame was common in Australia before this date. Later writers, having rejected the USA, turned inevitably to Britain as the source. Possibly something similar was known in one or two south-eastern counties, but what of British India and the Malay world? Here, from Bengal eastward, existed a bungaloid house form, made of light timbers and placed upon a platform atop heavy posts. The British knew little or nothing of the platform and light timber frame house and to this day look with contempt upon one-storeyed houses and houses clad or framed in light wood. But the Malays, Thais, Burmese and others were there to be learned from in the Imperial way: borrow ideas freely, but launder them in the Metropolis. Early (1786) British colonial houses in Penang (see Fig. 8), show the form, the style, the veranda, the structure and the materials which were to dominate Australian domestic architecture. The heavy mediaeval house frame was lightening for centuries for many reasons, but the assemblage techniques still common in Australia today are essentially south-east Asian.

Class

The general class-status hierarchy in Australia is expressed with precision in the built environment. The following conceptual oppositions usually divide folk architecture from its self-conscious relatives:

High-status criteria	*Low-status criteria*
public	private
large	small
grand	ordinary
decorated	undecorated
exotic	vernacular
stone/brick	weatherboard
architect	no architect
non-domestic	domestic

(Buildings in the left-hand column are usually given a style label: those on the right, almost never.)

Status hierarchies are of course much finer instruments of discrimination than these simple oppositions. Taking walling materials alone, the most potent status marker in Australian housing, a hierarchy such as this is suggested:

stone
rendered brick
'solid brick'
'brick veneer'
weatherboard
fake brick
fibro-cement sheet
galvanized corrugated 'iron'

Prices vary accordingly.

Stylism

Stylism is closely akin to 'featurism' as described by Robin Boyd in *The Australian Ugliness* (1968). Such was the ubiquitous influence of the Georgian style that the vernacular could, as Robin Boyd so eloquently describes, wear a new stylist hat with abandon. Thus, what passed for Gothic, Italianate, Baroque, Art Nouveau, and so on, could be tacked on to the basic model. 'Features' such as porticos, columns and pediments commonly found in high-culture Georgian buildings are rare in Australia. (This is not so in buildings of the same class and period in, for example, the United States. Intriguingly, the stylistic basics of Georgian architecture as noted above are very often abused in the American vernacular, but there is no end to the triumphalist features.) Features from other styles are often applied superficially, and usually inappropriately, to the basic form. Several such stylisms may coexist in (on) the one house. Thus, one sees many Australian Georgian houses with a Gothic flourish or a California bungalow façade. In the latter part of the nineteenth century many houses, labelled 'asymmetrical' by architectural historians, acquired a gable on one side. This did not, in fact, affect the Georgian basics: one could still have an 'Italianate' bay-window gable projection, or a Gothic gable with finial, and still retain the horizontal line and symmetrical balance around the central hallway.

Only once in Australia's first century did the established folk style and plan suffer a telling blow from the ever-more frequent stylisms that blew in from metropolitan centres overseas. The jumbled and imitative houses of the rich had much earlier succumbed to the temptations of Gothickry and Caesarian pomposity, but it was the Queen Anne revival that seemed to penetrate deep into the vernacular Georgian of even the Australian middle class. At the turn of the twentieth century, the simple oblong box with its horizontal symmetry was giving way to the 'Federation House', a busy collocation of gables, angles and turrets. It was asymmetrical in every eclectic aspect. Stylism had apparently triumphed. In the lower-middle- and working-class versions, however, the change was stylist and slight; and after scarcely a decade, the Georgian Empire bungalow reappeared as if nothing had happened (Fig. 7). Perhaps wartime austerity provided a good excuse for a return to folk vernacular basics. In any case, the rise of the California bungalow in the 1920s soon tested the folk idiom to its limits. The Geelong West survey included hundreds of timber-frame weatherboard houses of this description. The amazing thing is that only one 'California bungalow' was genuine. All the rest were stylist. The central hallway, disposition of rooms and façade elements, the overall symmetry of the Georgian form were generally left intact, while the essential aspects of the real California bungalow, asymmetry and very low roof pitches for example, were evaded. Gable entry, which is a North American and northern European tradition, is contrary to deep folk tradition in Australia. The new stylism insisted on wide gables covering the whole façade, but often these are simply tacked on over the Australoid box (see Fig. 9). This tension was later relieved in the 1930s with the reversion to folk basics. 'Modern', 'streamline moderne', and 'functionalist' influences appear as stylisms in succeeding decades. Historicism becomes intensely popular, particularly in the 1970s and 1980s, with attempts at replication of nineteenth-century colonial style and stylisms. Entire new suburbs look like their counterparts of the 1860s and 1880s (see Fig. 10). The basic structure remains the light timber frame, though increasingly steel stud frames are appearing. There are important changes in materials and floor plan, but the continuities are profound. And in the underlying preferences for detachment, in scale and in style, the vernacular traditions hold fast.

Decoration

By world standards, the façades of Australian houses are often very copiously decorated. As the nineteenth century unfolded into greater and greater wealth, and the austere detailing of the Georgian style faded in a way that the style itself did not, a veritable cornucopia of decoration encrusted Australian houses. Florid timber fretwork, lavish iron lace, convoluted Art Nouveau leadlight, intensively embossed metal ceilings, pillars, pilasters and posts proliferated (Fig. 11). The façade of Herd's house (Fig. 4, 1891) is typical of southern Australia, with heavy iron lace and cast-iron columns with mock Corinthian capitals, framed by toothsome dentilation and a geometric timber frieze. To this day there has been no study of regional variation in exterior or interior decoration, yet there are marked differences, some of which may not relate to period but to cultural variation. The origin of much decoration is as complex as that of structure and style. The repetitive lacework owes much to Mogul India, as does the dentilation; the columns are of course Greco-Roman; Belgian Art Nouveau, nationalist gumnuttery, classical urns and ferns each have obvious if devious ethnicity. Some of these elements may be seen in the United States, for instance, but the mix is never quite the same. The magic of the folk type is that the alchemical ingredients are often the same or nearly so, but the special amalgam is unique.

It is just possible to discover a perfect Australian 'Federation' House in Mexico, contrary to every Mexican folk tradition. A freak, no doubt, but what of the streets of early Australian timber-frame houses complete with iron lace (in wood) to be found near the Laotian border in Thailand, or uncannily familiar Malay kampongs? On the other hand, Australia's ethnic clone, New Zealand, shows marked differences in vernacular architecture tradition. The exploration of the New World, in terms of folk architecture, has scarcely begun.

REFERENCES: R. Boyd, *Australia's Home*, 1978 (1951); F. J. Campbell, *Culture, Architecture and the Australian House*, 1983; F. J. Campbell and I. Fox, *The Australian House Museum: A Guide to the Project*, 1986; H. Fraser and R. Joyce, *The Federation House: Australia's Own Style*, 1986; H. Glassie, *Folk Housing in Middle Virginia*, 1975; R. Irving (ed.), *The History and Design of the Australian House*, 1985; H. Joseland, 'Domestic Architecture in Australia', in *The Centennial Magazine* 3, 1890, pp. 5–11; A. D. King, *The Bungalow*, 1984; Lee Kip Lin, *The Singapore House*, 1988; M. Lewis, *Victorian Primitive*, 1977; M. Lewis, 'The Corrugated Iron Aesthetic', in *Historic Environment* 2, 1982; B. S. Saini, *The Australian House: Houses of the Tropical North*, 1982.

Frank Campbell

Australian Institute of Aboriginal and Torres Strait Islander Studies, The (AIATSIS), Canberra, was first established as the Australian Institute of Aboriginal Studies by an Act of parliament in 1964 and continued under its present name by a further Act in 1989. The major function of the Institute is to increase specialist knowledge and general community understanding of Aboriginal and Torres Strait Islander societies by encouraging research (training Aboriginal and Islander people themselves as research workers is a priority), by publishing the results, and by building up a cultural resource collection open to all.

AIATSIS is the major source of research funds in Aboriginal studies. Grants may be given in any discipline, including social anthropology, prehistory, history (including oral history), linguistics, education and ethnomusicology. The Institute also funds research Fellowships on topics considered of high priority, for example recently a Research Fellow completed preparation of a biographical database of Aboriginal artists to assist the development of the Aboriginal art industry. The Research Section also organizes a weekly seminar series and a workshop series, and edits the Institute's twice-yearly journal, *Australian Aboriginal Studies*.

The results of research are published by the Institute's press, Aboriginal Studies Press. It publishes about twelve books per year, a report series and the journal, as already mentioned, as well as audio-visual material (films/videos and records/cassettes), making it the main publisher of Aboriginal material. Publications cover all relevant disciplines; of particular interest are audio-visual recordings of traditional ceremonies and song series, such as the award-winning film *Waiting for Harry* (a mortuary ceremony from Arnhem Land, Northern Territory) and the record/cassette *Aboriginal Sound Instruments*, which includes examples of Aboriginal musical instruments from across the continent.

The resource collection consists of both print and non-print material and is by far the largest collection of Aboriginal cultural and historical materials anywhere. The library's print collection includes books, theses, pamphlets, posters and published and unpublished papers. Of particular note are: (1) the Australian Languages Collection, which includes material prepared for bilingual education programmes and materials published by Aboriginal language/ cultural resource centres around the country; and (2) the collection of Aboriginal community newsletters and publications by other Aboriginal organizations. The pictorial collection includes several hundred thousand photographs (slides and prints), the Film Archive is a major historical repository of films made since the turn of the century and its video collection is expanding rapidly, and the Sound Archive, which includes speech and music, has over 20 000 field tapes. The library publishes an annual bibliography, a microfiche publication of selected news clippings, *Aborigines in the News*, and a series of regional filmographies. The library is open to the public five days a week and it offers a specialized reference service to researchers and members of the public.

In recognition that Aboriginal and Torres Strait Islander culture belongs to Aboriginal and Islander people and that the Institute holds a major record of that culture, the Community Access Programme (CAP) and the Return of Material to Aboriginal Communities programme (ROMTAC) were established to assist Aboriginal and Islander people to use the Institute's holdings and to return copies of material to Aboriginal and Islander communities and organizations. Under CAP the Institute holds regional meetings of representatives of Aboriginal and Islander organizations to inform them of the resources and services available and how to use them, to be informed about their views on research priorities, and to assist in planning local resource centres and 'keeping places' (the Koorie term for museum). Under ROMTAC, video copies of the Institute's films, copies of slides and audio tapes from the Institute's archives, and copies of books published by the Institute are provided free of charge to Aboriginal and Islander community libraries, keeping places and resource centres.

Australian Legend, The, is a phrase most often used in reference to Russel Ward's (q.v.) book of that title, first published in 1958 and ever since a touchstone in discussions about Australian nationalism and national identity. The book's opening paragraphs declared:

> According to the myth the 'typical Australian' is a practical man, rough and ready in his manners and quick to decry any appearance of affectation in others. He is a great improviser, ever willing 'to have a go' at anything, but willing too to be content with a task that is 'near enough'. Though capable of great exertion in an emergency, he normally feels no impulse to work hard without good cause. He swears hard and consistently, gambles heavily and often, and drinks deeply on occasion. Though he is 'the world's best confidence man', he is usually taciturn rather than talkative, one who endures stoically rather than one who acts busily. He is a 'hard case', sceptical about the value of religion and of intellectual and cultural pursuits generally. He believes that Jack is not only as good as his master but, at least in principle, probably a good deal better, and so he is a great 'knocker' of eminent people unless, as in the case of his sporting heroes, they are distinguished by physical prowess. He is a fiercely independent person who hates officiousness and authority, especially when these qualities are embodied in military officers and policemen. Yet he is very hospitable and, above all, will stick to his mates through thick and thin, even if he thinks they may be in the wrong. No epithet in his vocabulary is more completely damning than 'scab', unless it be 'pimp' used in its peculiarly Australian slang meaning of 'informer'. He tends to be a rolling stone, highly suspect if he should chance to gather much moss.
>
> In the following pages we shall find that all these characteristics were widely attributed to the bushmen of the last century, not, primarily, to Australians in general or even to country people in general, so much as to the outback employees, the semi-nomadic drovers, shepherds, shearers, bullock-drivers, stockmen, boundary-

riders, station hands and others of the pastoral industry.

This was so partly because the material conditions of outback life were such as to evoke these qualities in pastoral workers, but partly too because the first and most influential bush-workers were convicts or ex-convicts, the conditions of whose lives were such that they brought with them to the bush the same, or very similar attitudes.

While Ward linked the Legend to particular groups, he claimed its widespread diffusion, at least among the ordinary folk of Australia. They, by the very reason of their modest socio-economic status, were the readier to adopt a 'local' patriotism, he has insisted; whereas those higher in the social scale inclined more to an 'imperial' patriotism — beneficial enough in its way, but increasingly lacking relevance and power as the twentieth century progressed (Ward, 'Two Kinds of Australian Patriotism', *Victorian Historical Magazine*, 1970). The 'local' patriotism of the masses and admiration for the bushman's ethic, runs the broad argument, helped elevate it into 'the Australian Legend'. Nor was this admiration merely abstract; much Australian behaviour has accorded with Legendary precepts.

Ward's spirited and stylish book pursued these themes. He argued that in the early decades of colonial history native-born British Australians, and especially those of Irish descent, were quick in responding to the Legend. Its essential form was well established by mid-century and so, from Ward's perspective, the gold rushes were less significant than in the established historiographical canon: the gold-seekers accepted Australian images rather than created them. In either half of the century, Ward continued, popular idolizing of bushrangers — above all, Ned Kelly (q.v.) — sprang from and in turn intensified salient features of the Legend. The 1880s were vital in the Legend's history. Then developed powerful trade unions, especially among shearers and miners, whose principles of solidarity and activism embodied and diffused the Legend's tenets. Social radicalism found other expressions in Australia in this decade, often intermeshed with assertive nationalism, itself having both literary and political forms. Several journals, above all the Sydney *Bulletin*, pursued such ideas, and their contributors often presented the bushman in terms appropriate to Ward's thesis. There yet remained scope for the Legend's expansion. Ward recognized that ideas similar to his own had been expounded by C. E. W. Bean, most famously in presenting the Australian soldier ('digger') of the First World War as being influenced by bush values. In more recent years this conflation of bushman and digger legends has intensified.

Appropriately to his theme, Ward drew upon a wider range of sources than was conventional among academic historians. Popular materials, including song and anecdote, were among them. Earlier books which offered such data were especially valuable for him — supreme examples being Alexander Harris, *Settlers and Convicts* (1847) and Charles Macalister, *Old Pioneering Days in the Sunny South* (1907). The Legend gained all the more vitality from such innovation.

Withal, Ward built upon an established tradition of Australian commentary. Writers of the *Bulletin* type — most pertinently Joseph Furphy, T. A. Browne, and A. B. Paterson — received his due appreciation. Most remarkable of all Ward's precursors in that generation was F. W. L. Adams (1862–93), an Englishman sensitive to the day's cultural modernism. During his Australian residence, 1885–90, Adams lived mainly in Queensland, writing much. In *The Australians* (1893) he contrasted the inland worker with the urbanite, celebrating the former as the country's 'powerful and unique national type'. Bean developed the same point several times over in pre-1914 journalism, setting the context for his digger legend. Then too the young Vance Palmer developed his faith in 'the real Australia' as being rooted in a masculine, outback world. Palmer's fiction often embellished this notion and in 1954 he developed it further in his book, *The Legend of the Nineties*. Palmer helped shape Ward's thought, while his name and ideas appeared several times in a chapter entitled 'The Australian Legend' in R. M. Crawford's short history, *Australia* (1952). Crawford long headed the history department in the University of Melbourne, and Australianists trained within that department were to foster ideas akin to the Legend: probably the most important exemplar is Geoffrey Serle's *From Deserts the Prophets Come* (1973). Manning Clark's attitude to the Legend has varied, with some rejection in the 1950s but substantial endorsement thereafter. Some literary criticism likewise developed and upheld the Legend: particularly relevant were A. A. Phillips, *The Australian Tradition* (1958) and T. Inglis Moore, *Social Patterns in Australian Literature* (1971).

All this notwithstanding, Ward's thesis is open to dispute. As earlier suggested, much debate as to the nature of Australian identity has centred on his ideas, and some of the argument thrusts against him. To separate the strands of this critique is a necessary task, however difficult.

An especially crucial and complex matter is the objective truth of various claims in the original argument. Have Australians at large, and/or bushworkers more particularly, behaved and believed in accord with the Legend? From time to time, Ward has suggested that such questions are irrelevant. 'The average frontiersman was of course the actual, "average" bush-worker who, at any given time differed not so graphically from other Australians or indeed from Englishmen', he characteristically wrote in 1978, drawing a contrast with the '*typical*' bushworker; this latter was he 'who differed most graphically from the real or *average* bushworker, and differed at any given time in the most "Australian" or un-English direction' ('The Australian Legend Revisited', *Historical Studies* 18, 1978). Yet right from the opening sentences of *The Australian Legend* Ward has always insisted on the substantial degree to which 'a more solid sub-stratum of fact' has underpinned the Legend. Likewise, while accepting that Australia is notably urbanized, and that urban forces have shaped much of Australian history, he insists upon bush associations being important for many Australians in their lived experience as well as in imagery creation.

The true nature of the outback worker remains a matter for debate. As Ward emphasized, pastoral employment in the early days fell largely to convicts — labour had to be coercible and coerced to undertake such work. No doubt some developed an affinity for the task but others became eccentric mad hatters — rejects and rejectors of society rather than its value-setters. This lunatic–eccentric tradition persisted at least to the end of the nineteenth-century (S. Garton, *Medicine and Madness*, 1988, pp. 118–20). A variant upon the theme occurred after both world wars when migrants were compelled to do outback work as a condition of being subsidized to come to Australia.

Similar points may be developed as to that particular sub-group of Legendary heroes, the bushrangers. Many of them were ruthless and greedy, securing support from surrounding communities through terrorist threats rather than popularity. Recent research indicates that Ned Kelly himself was a police informer at one stage, as was his father (transported in 1842 to Van Diemen's Land) in Ireland before him (J. Clifton, 'A New Kelly Letter', *Royal Historical Society of Victoria Journal* 57, 1; R. Reece, 'Ned Kelly and the Irish Connection', in C. Bridge, *New Perspectives in Australian History*, 1990). Granted that Kelly and some earlier bushrangers do have a place in Australian popular lore and affection, the argument might go, they cannot seriously be accepted as exemplars of the Wardian ethic.

Also problematic is the role of convictism at large. As indicated, one of Ward's striking innovations was to give convicts so positive a place within Australian traditions. Most earlier Australians had detested convictism. This was especially true of groups whom Ward linked with convicts in the Legendary tradition — the gold-seekers of the 1850s and the *Bulletin* nationalists of a generation later. On the goldfields fierce animosity prevailed between (ex-) convict and free migrant, and Victoria's politicians of the 1850s flouted constitutional law and civil liberty by denying entry from Van Diemen's Land to persons holding conditional pardons from the Crown. The *Bulletin* saw convicts not as proto-nationalist heroes but as an integral part of that British imperial system which every true Australian should abhor.

Since 1958 there has been a great deal of research on convicts, with particular reference to the British background. Ward took no hard-line stance on the convicts' criminality or morality, so much of the consequent discussion is irrelevant to the Legend. However, the essay of M. B. and C. B. Schedvin ('The Nomadic Tribes of Urban Britain', *Historical Studies* 18, 71) is germane. The Schedvins agree with Ward that traits developed in Britain were carried into Australian popular behaviour. But they see these characteristics as expressing a 'psycho-social dimension' having scant appeal. The convicts lived in a world of suspicion, hate and inferiority. They had but little to offer each other, and nothing to society at large. The Schedvins' presentation negated any likelihood of seeing the convicts as founders of Australian nationhood.

An even more fundamental claim against Ward's thesis may be that basic Australian values are not, nor have they ever have been, such as the Legend upholds. Here arise enormous, perhaps insoluble, issues — but if the Legend is to be judged, then they must receive some account. Several historians — often developing their arguments specifically against Ward — have argued that dominant Australian values accord only tangentially with the Legend, being in essence conformist and bourgeois. Thus, Michael Roe argued in *Quest for Authority in Eastern Australia* (1965) that such notions as temperance and self-help had established a firm hold on the Australian consciousness even by 1850. More pungent and comprehensive development of such themes is apparent in such other studies as Donald Horne's *Australian People* (1972) and *The Lucky Country* (1964). From the perspective of New Marxism came the not-so-different critique of Humphrey McQueen's *A New Britannia* (1970). 'While rejecting Ward's conclusions and argument concerning the mythical Australian, I believe there is some merit in considering yet another version of the frontier thesis', wrote McQueen. Australia, he insisted, was a frontier of white capitalism in Asia. That caused the strongest theme in Australian nationalism to be an obsessive conviction that British power should shield Australia from a hostile world. Thence developed also an anti-Asian racism which eroded all human values. Consonant with this, a generation of Australian types — including convicts, gold-diggers, bushrangers and unionists — had made the acquisition of personal property their supreme goal: ownership and display of a piano symbolized the aspirations of mainstream Australia. In McQueen's eyes, Ward and other backers of the Legend were — in effect if not in intent — reinforcing conservative hegemony in Australia. McQueen, by contrast, presented his interpretation of Australian history as a tool for revolution.

Permeating the criticism from such sources is the idea that the Legend's validity is diminished because of Australia's level of urbanization — even more remarkably high, by way of international comparison, in the nineteenth century than since. If the masses live in towns, in some significant degree because of apparent preference, can a rural-oriented set of values have real truth and meaning for them? Against such argument Ward and others not only insist upon the continuing ties of many Australians with the countryside, but see the demographic statistics as having little ultimate relevance. The very intensity of urbanization, they contend, could generate a kind of counter-attachment to the rural world.

Other critics of Ward themselves argued from a rural perspective. In one of the few attempts to diagnose 'The Smallholders' Place in the Australian Tradition' (*Tasmanian Historical Research Association Papers and Proceedings* 10, 2, 1962), Douglas Pike disparaged Ward's 'hypothetical nomad tribe':

> Where the nomad was escapist, the smallholder stayed to face domestic crises, often with the courage and understanding that marked his greater emotional stability . . . At home and work he gathered more gear and needed more ingenuity than any roving pastoral worker . . . In his endless struggle with nature the small holder could be beaten but his resolution always revived . . . Quiet pride in self, family, and the property he hoped some day to pay off might yield small dividends, but it was a comforting quality easily transmitted through children and assimilated into the familar patterns that rule suburbia today.

One influence on Pike appears to have been D. W. Meinig's study of agriculture in the Yorke Peninsula, South Australia (*On the Margins of the Good Earth*, 1962), while further contributors to such an image of the small farmer include G. L. Buxton (*The Riverina 1861–1891*) and Charles Fahey ('The Wealth of Farmers', *Historical Studies* 21, 82). There is continuity also between Pike's essay and J. B. Hirst's analysis of 'The Pioneer Legend' (*Historical Studies* 18, 71, 1978). 'Courage, enterprise, hard work, and perseverance' were the qualities expounded in this tradition. While not in flat opposition to the Wardian ethos, Hirst remarked, the consequent value-set was far more conservative and celebratory; landowners — big or small, but rarely their employees — were its heroes.

So has proceeded questioning of the truth of Ward's argument. A further query becomes more cogent: if the Legend is not founded on truth, whence did it derive? The theme of several answers is that the Legend has served to uphold the socio-ideological values and fashions favoured by its creators and supporters; thus, the Legend appears as an intellectual construct rather than a transcript of reality.

Just as Ward anticipated criticism as to the Legend's objective truth, likewise he affirmed that it belonged to a broad context of ideas. His book's final chapter, even better written than the rest, surveys the history of frontier romanticism. The American historian F. J. Turner commanded most of Ward's attention, but his compass ranged from Tacitus to T. S. Eliot. Especially notable was Ward's recognition that the crystallizing of the Legend in the 1880 and 1890s cohered with widespread Anglo-Saxon triumphalism, manifest not only in Turner, but still more remarkably in such figures as Rudyard Kipling and Theodore Roosevelt. The critics now to be mentioned develop such insights, rather than discover them.

A seminal essay was by Graeme Davison, 'Sydney and the Bush' (*Historical Studies* 18, 71, 1978). It affirmed that

> A fundamental weakness of folk-history — of which Ward's book is a superior example — is its assumption that popular values may be abstracted from literature without direct reference to the ideas and situation of those who created it. We are required to look beyond the mediating author to divine the *conscience collective*. The hazards of this method are obvious even with an unselfconscious, traditional culture but they are greatly multiplied when we apply it to a post-industrial, culturally-derivative society like nineteenth-century Australia. In such a culture, I suggest, we do better to begin, as we would any other exercise in the history of ideas, with the collective experience and ideas of the poets and story-writers themselves.

Davison referred especially to the *Bulletin* writers, so important in shaping the Legend. He presented them as 'an emerging urban intelligentsia', often suffering a marginal life in the half-poverty of Bohemia. Particularly with the advent of economic depression in the 1890s, such people rejected the society around them and developed an alternative set of values. These suffused their literary work. The apotheosis of the Legend, argued Davison, 'was not the transmission to the city of values nurtured on the bush frontier, so much as the projection on to the outback of values revered by an alienated urban intelligentsia'. Davison's work has had perhaps its most notable development in Leigh Astbury, *City Bushmen* (1985), which applies a similar analysis to the painters of the Heidelberg school, creators of Australian images in Legend-like style.

Davison's theme had been anticipated by others apart from Ward. David Walker, in *Dream and Disillusion* (1975), studied Vance Palmer and his associates from a similar perspective. As young men, Walker showed, Palmer's group had been much influenced by radicalism generated in early twentieth-century Britain — the Irish Renaissance in some degree, and more importantly the mix of post-Nietzschean ideas developed by A. R. Orage. Hence, rather than from Australian reality, derived the inspiration behind Palmer and company.

Inventing Australia (1981) by Richard White expanded such analysis. Sub-titled 'Images and Identity 1688–1980', this work applied Foucaultian insights with wit and skill. Modern national identities had been constructed within the whole framework of Western ideas, argued White. Appropriately important in the Australian process was an intelligentsia, which in turn, and despite some dissidence, worked within the orbit of socio-economic power groups. 'The most influential images are those which serve the interest of a broader ruling class, on whose patronage the intelligentsia rely.' White applied this analysis to that end-of-century group vital in the Legend's formation. Influenced by worldwide modernist ideas, these people were also sensitive to their own livelihood and prestige. The Australian national hero they fashioned had among his roles the defence of Britishness in an increasingly perilous world. Bean's digger legend complemented this part of the story: White himself, and various others, have presented Bean as very much a myth-maker, inspired by ideals of Anglo-Austral hegemony.

An analogous critique might be applied to Ward himself. His ideas chimed with a radical–egalitarian style which has been strong among Australia's intelligentsia and academics. Ward was a member of the Communist Party through the 1940s and beyond, as were several of his closest colleagues. While remote from schematic Marxism, the Legend as presented by Ward could seem attractive and appropriate to people of such disposition. With Cold War acerbity, James McAuley in 1962 argued that 'the Communist apparatus' had sought to impose monolithic control over 'the Australian Tradition', thus attempting 'a promotional device in the interests of particular individuals, or a politico-ideological manoeuvre, or both' (P. Coleman, *Australian Civilization*, 1963).

Interweaving with criticisms of Ward and the Legend, but conceptually separate, runs a motif that whatever the origins or truth of the Legend, its influence has been malign. Thus, Peter Coleman, an associate of James McAuley, was vehement in alleging its fostering of 'naive humanism and nihilism', these leading *inter alia* 'to the cult of the bacchanal, the cult of sport, the singing of vindictive folk songs, or the preaching of bullying racialism' (*Australian Civilization*, p. 3). If more temperately, others also alleged a link between the Legend and scorn for intellect and contemplation. Humphrey McQueen, from the opposite ideological extreme to that of McAuley and Coleman, also saw the Legend as suffocating creativity in that it had inhibited the growth of a proletariat infused with a truly revolutionary consciousness.

In an analogous way, Miriam Dixson voiced a critique expressive of early radical feminism in Australia. 'Their implications are mysognist to the core', she wrote of the Legendary traits of national character (*The Real Matilda*, 1976). They had set a pattern whereby women suffered harsher subordination in Australia than in other Western societies. While subsequent feminists have often deprecated Dixson's picture of Australian womanhood as crippled and inert, there has been widespread recognition of the ultra-masculinity of the Legend. David Walker made the point against Palmer, and Richard White against the *Bulletin* Bohemians. Kay Schaffer has returned close to Dixson's position in arguing that Ward, Palmer and others have sought to create an Australian identity which sustains hegemonies of race, sex and class: 'it allows the native white speaker to control the threat of sexual, moral and physical corruption which he fears in himself by locating its cause in the other' (*Women and the Bush*, 1988).

Ward himself always recognized and abhorred the racist theme within the Legendary ethic. In 1958 he described this much more in terms of anti-Chinese feeling on the goldfields, and subsequently, than by reference to European–Aboriginal relations. In accord with the greater sensitivity to such matters which has developed worldwide in the past generation, not only have such commentators as McQueen and Schaffer stressed the Legend's racism, but Ward himself has become more vigorous in reprobation. His general writings on early Australian history recognize the Aboriginal experience in full and sympathetic terms.

There is much other evidence of the Legend's continuing flexibility and strength. Beyond doubt many Australians — everyday folk and image-makers alike — cherish some or other of its qualities. *The Australian Legend* has sold far beyond the academic average, over many years. In successive editions Ward has affirmed his original argument and arraigned its critics. A special number of the journal *Historical Studies* celebrated the twentieth anniversary of *The Australian Legend*'s first publication. Ward has written general histories of Australia and (with John Robertson) edited documentary collections, all running congruent with the themes of the Legend.

A particularly notable item among Ward's *oeuvre* is his autobiography, *A Radical Life* (1988). As a schoolteacher's son, young Russel lived in Queensland and Western Australia before the family resettled in Adelaide where Ward senior long headed the prestigious St Peter's College. Russel studied History and English at the University of Adelaide, and then himself became a schoolteacher in the pre-war period at Geelong Grammar School. The 1940s he passed largely in Sydney. There he joined the Communist Party and moved among the radical intelligentsia. The reader learns that *The Australian Legend* had its genesis when one wartime night Ward heard four Britons singing items from Paterson's *Old Bush Songs*. 'Like most of my compatriots at that time I had never heard of an Australian folksong. Yet here were a pair of Scotch immigrants and Pommy visitors singing them, inevitably in bizarrely unsuitable accents, to a born and bred Australian couple'. Vance Palmer and Nancy Keesing encouraged the interest thus conceived. It bore fruit during Ward's tenure (1952–55) of a doctoral scholarship at the Australian National University — a time and place which he well described. *Radical Life*'s final chapter, 'Folksong and Fury', described Ward's continuing interest in exploring and proclaiming this aspect of popular culture. It is a key source as to folklore enthusiasm, especially in the 1950s.

A Radical Life affirms its author's warmth and power of personality. True to this, he exercised great influence through and beyond his years as a teacher of history at the University of New England (1956–79). One early associate there was his co-editor, John Robertson, whose own final book, *Anzac and Empire* (1990), echoed the Wardian heritage in arguing that the Anzac digger was essentially as Bean had portrayed him. Earlier one of Ward's graduate students, A. M. Grocott, had documented the disdain for religion among the convicts of early New South Wales. J. B. O'Hara has likewise followed his mentor in detailing Australian delight in sport and chance.

The burgeoning of an Australianist school around 1980 entailed fresh honour for Ward. Thus John Docker (*In a Critical Condition*, 1984), while advancing some criticisms of the radical–nationalist tradition, saw it as more vigorous and fruitful than the metaphysical academicism of its opponents. Stephen Alomes (*A Nation at Last?*, 1988), similarly trained his guns on the upholders of 'a modern international version of dominion culture' which he saw as imposing sterile conservatism on the Australia of the 1950s; Ward, Palmer and Phillips had stood against that tide in gallant optimism. Noel McLachlan's *Waiting for the Revolution* (1989) is probably the major single contribution to the nationalist debate since 1958; although it makes little direct reference to Ward, the book's prevailing assumptions are similar to his. For Alomes, McLachlan and many other Australianists the great need is for sensitivity to local specialness and to reject those influences (derived from England in the past, more widely in recent times) which would rank Australia as derivative and third-rate. By such criteria, Ward could stand well.

Despite Humphrey McQueen, Ward's ideas have had some backing from the latter-day New Left. Larrikin–mobsters were the sole urbanites to win much of his favour, their behaviour impressing him as supportive of Legendary traits. There has been little academic study of the streetboys but R. W. Connell and T. H. Irving offer some pregnant ideas in their *Class Structure in Australian History* (1980). They recognized affinity between the intelligentsia's treatment of larrikins on one hand and bushworkers on the other. Underneath that surface lay 'a distinctive working-class youth culture'; and so larrikins 'were still culturally the least tractable section of the working class' in Australia's big cities before the First World War. Ward would have had no problem in endorsing this view. Eighty years on, the thorough-going larrikin had receded, yet his tradition continues in the 'ocker', the Everyman of Australia: crude, cheerful, noisy, hedonist. Probably most ockers would like to see themselves as heirs of Ward's archetype.

The Australian Legend thus remains a vital and vitalizing element in Australian historiography. Is it based on a delusion, and does it foster delusion — even in an ugly and obscurantist way? Or does the Legend focus upon the best of human experience in European Australia and help inspire national achievement? The argument must continue.

Michael Roe

Australian National Gallery, The The Gallery's commitment to collecting examples of Australian folk and popular art dates from the late 1970s and the acquisition of an important collection of Queensland ceramics. This collection, which had been assembled by the artist Robert MacPherson (born 1937), included a range of objects, both utilitarian and decorative, from bread plates and mixing bowls to more sculptural pieces such as money boxes and chimney-piece ornaments. What unified this diverse group of objects was Robert MacPherson's taste, which ensured that each item was not only an example of local pottery-making but also representative of the potter's skills. Many of the sculptural pieces in the collection are outstanding examples of Australian folk art and were the first works of their kind to enter the collection of the Australian National Gallery. Their acquisition opened up the possibility of assembling a small but select collection, which would complement the existing displays reflecting the history and development of Australian art. In collecting and displaying Australian folk and popular art, the Australian National Gallery has been the first major gallery or museum in this country to acknowledge a tradition that in many instances is anonymous and exists outside the accepted canons of fine art.

The Gallery's small collection of Australian folk and popular art has been acquired over the past decade, with many objects being received as gifts from interested and generous donors. The collection includes work in a range of media, produced by people of different cultural backgrounds and circumstances who have worked at different times, under a variety of conditions, and with widely varying ideas of the nature of the purpose of their art. It is the best collection of its kind in an Australian art museum.

See also: **Folk and popular art**.

REFERENCE: John McPhee, *Australian folk and popular arts in the Australian National Gallery*, 1988.

Australian Plaiters' and Whipmakers' Society, The Plaiting with kangaroo leather is a traditional Australian craft. Because the material is regarded as the strongest for its thickness in the world, craftspeople in this country have been able to develop very advanced techniques, with the result that most experts in the field consider that the best whips in the world are made in Australia. The Australian Plaiters' and Whipmakers' Society was established in October 1985. It has a strong and active membership of both amateur and professional leatherworkers, and most of the professional whipmakers in Australia are both members and contributors to its journal, published about three times a year. Membership is limited to those who are working in the medium. Present membership is about sixty.

Australian Stockman's Hall of Fame and Outback Heritage Centre, The, was opened at Longreach, Queensland, in April 1988. Apart from the initial State and Commonwealth government loans to establish the Centre, which cost approximately $12.5 million to build and outfit, the operation is self-funding, depending on public attendance fees, merchandise sales, memberships, sponsorships and donations. The Hall of Fame has won several tourism industry awards.

The aims of the Hall of Fame and Outback Heritage Centre are to retrieve, preserve and disseminate, in a variety of ways, the social history, traditions and culture of the Australian outback. In addition to the displays at Longreach, the Hall of Fame conducts a variety of activities, including Australian and international touring exhibitions, a rural oral history programme, the 'Unsung Heroes' videodisc programme, a library and general curatorial activities and exhibitions.

Australian Tradition Evolving from the *Gumsucker's Gazette*, the publication of the Victorian Folk Music Club (q.v.), *Australian Tradition* (usually known simply as *Tradition*), was published in co-operation with the Folklore Society of Victoria from 1963 to 1975. Edited by Wendy Lowenstein (q.v.) and featuring the contributions of many people, *Australian Tradition* provided for all the varied interests involved with the folk revival of the period. It included reports on the fieldwork of collectors and on contemporary artistic creations in the folk idiom (mainly music, song and dance), research articles, polemic, correspondence and news. The pages of *Australian Tradition* are an invaluable resource for folklorists, though little use has been made of these to date, possibly because very few libraries or even private collections possess the full series.

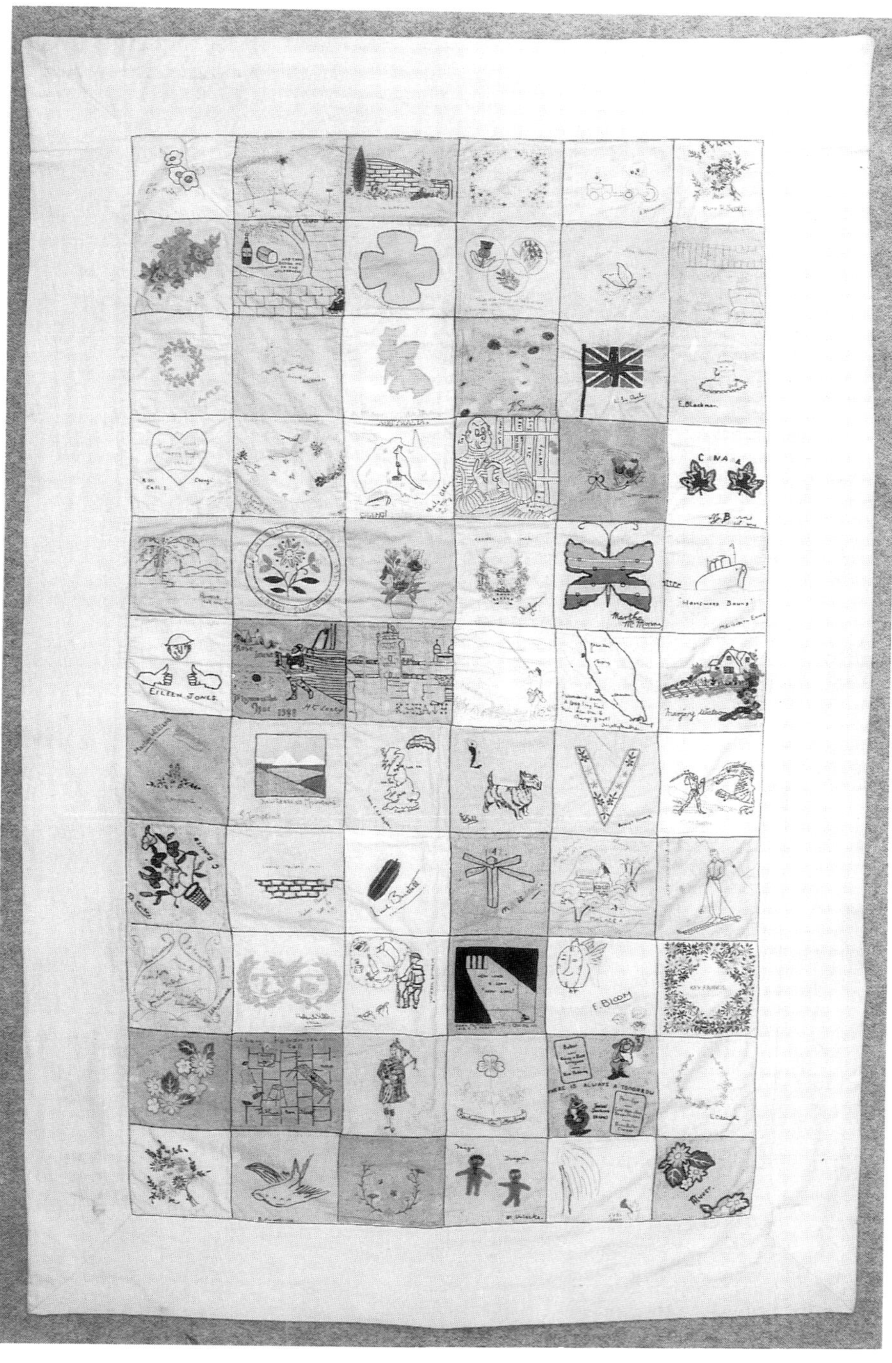

One of the Changi quilts made in 1942 by some of the 400 women interned by the Japanese in Changi Prison, Singapore. Although most of the makers of this quilt were British rather than Australian, it still occupies a powerful place in Australian history and folk memory. (Australian War Memorial)

Crochet panel designed by Mary Card, *c.* 1916, Tas. (Australian National Gallery)

This linen tablecloth, with fine drawn-thread work and crochet edging, is believed to have been made in the early twentieth century by Miss Amy Watkin, great-grandmother of the present owner. (Photograph, Australian Foreign Affairs and Trade Department)

A traditional Ukrainian pattern found on cushions, clothing and china.

Quiltmakers displaying their skills at Mudgee, NSW, 1988. (Photograph, Keith McKenry)

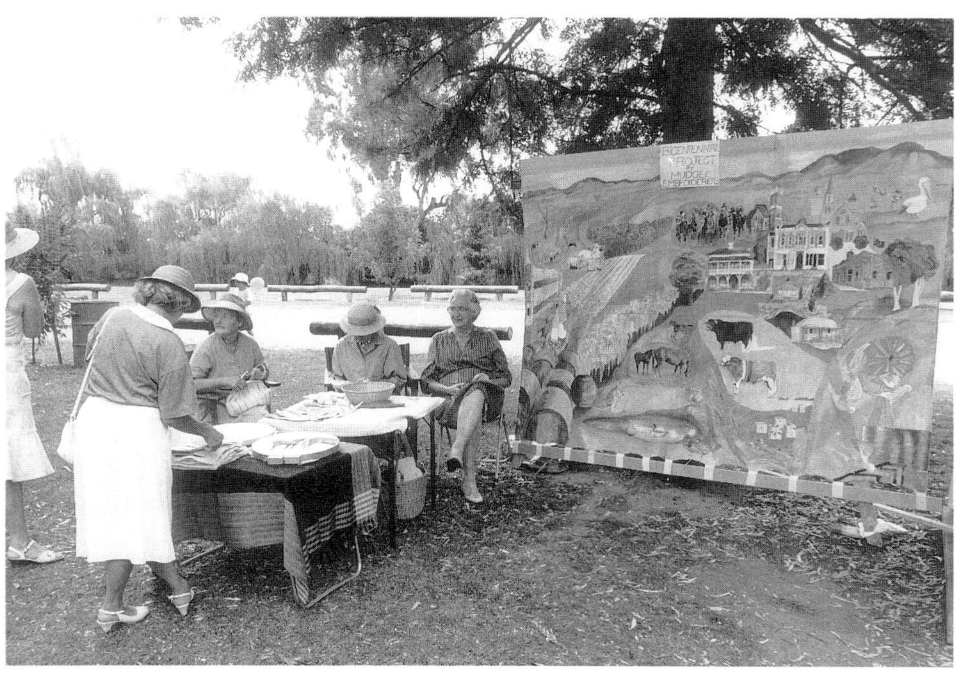

A bicentennial quilt made by women in Mudgee, NSW. (Photograph, Keith McKenry)

Audrey Kennett, of the Canberra Lacemakers' Group, with original lacework and bobbins. (Photograph, Australian Foreign Affairs and Trade Department)

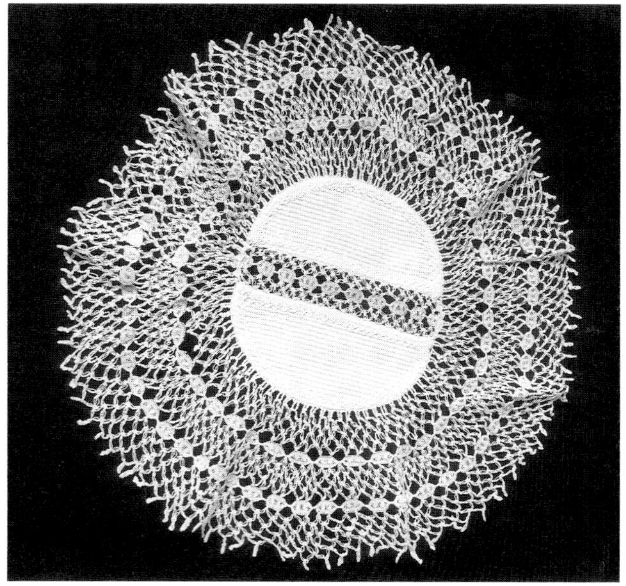

Handcrafted doiley believed to have been made in Australia in the early twentieth century and passed down through four generations to the present owners. (Photograph, Australian Foreign Affairs and Trade Department)

B

BAKER, SIDNEY JAMES (1912–76), is widely regarded as an outstanding figure in the documentation of Australian English, including folk speech (q.v.). In his classic work *The Australian Language* (1945), Baker made a useful distinction between three types of slang used in Australia, namely slang words which Australia had imported from Britain without alteration, like 'all Sir Garnet', 'keep it dark', 'rattletrap' and 'Jesus wept!', slang and dialect words which Australia had imported from Britain and given new meanings to, like 'cobber', 'barrack', 'dinkum' and 'smoodge', and slang words which Australia had invented, like 'Buckley's chance', 'googly', 'pozzie' and 'whacko!'. In this book, Baker declined to include as a part of Australian English the slang which had been imported from Britain without alteration, but in his later work *Australia Speaks: A supplement to 'The Australian Language'* (1953) he had a chapter entitled 'Overseas Influences' where he included incorporations into Australian English from English and American slang. Sidney Baker was born and educated in New Zealand, and his first publication was *New Zealand Slang* (1940), closely followed by *A Popular Dictionary of Australian Slang* (1941).

Bandicoot Ballads, a series of publications of mostly traditional folksongs edited by John Manifold (q.v.) and Ron Edwards (q.v.), was published between 1951 and 1955. These early presentations of Australian folk song were particularly influential in the collection, study, and performance of such material, and in promoting general public awareness of it.

See also: **Folk song**.

Bawdry Obscenity, scatology and plain bad taste are essential elements of all folklores and the oft-observed preference in Australian speech for the vernacular, usually of the cruder variety, means that it is particularly characteristic of some aspects of Australian folklore. In many cases, its *raison d'être* is to deal with matters generally considered unwholesome or otherwise unacceptable in everyday discourse and interaction. Generally this material is in the form of songs, recitations, jokes, anecdotes and, in modern times, reprographic (q.v.) cartoons or texts. The usual topics are sexual activities (often bizarrely improbable), gender prejudice and racial and ethnic prejudice; sometimes these are combined. Other aspects of human life, including politics, occupational relations and topical issues of various kinds, may also be treated in a bawdy manner.

Bawdry in Australian folklore is of both the internationally distributed type, including poems such as 'The Bloody Great Wheel' and 'Eskimo Nell', and the decidedly indigenous, including such items as 'The Bastard From the Bush' and 'The Broken-Mouthed Shepherd', and 'The Man from Kaomagma' parody of 'The Man from Snowy River', to name only a very few. While bawdry is generally considered to be typical of all-male groupings, such as sailors, shearers, soldiers and so on, it is important to note that adult males do not have a monopoly of such material. Women have folk traditions of bawdy expressions and practices (although there has been no major investigation of it to date), as do children. The bawdry of childhood has been well documented in Turner, Factor and Lowenstein's *Cinderella Dressed in Yella* (1969, 1978) and recently in Factor's *Captain Cook Chased a Chook* (1988) and elsewhere. Male bawdry is also well-served, with Hogbottel & Ffuckes [Stephen Murray-Smith and Ian Turner] *Snatches and Lays* (1962; expurgated 1973), Brad Tate's *The Bastard from the Bush* (1982) and Ron Edwards'

Australian Bawdy Ballads (1973, 1986), Don Laycock's *The Best Bawdry* (1982), rather tamely in Bill Wannan's *Robust, Ribald and Rude Verse in Australia* (1972) and in numerous journal articles and other, less specialized collections of folklore. Anonymous, pseudonymous and otherwise clandestine or semi-clandestine collections of bawdry are, and no doubt have always been, continually circulated through Australian society.

BEATTY, BILL (1912–79), known as the 'Ripley of Australia', in the early 1940s presented a weekly radio show called 'Australoddities' on the ABC. The growth of radio and popular magazines and newspapers in Australia during the 1920s and 1930s created a niche for writers and broadcasters about certain kinds of Australian folklore. Some of their columns and programmes attracted audience response, and compilations were published, under the name 'Believe' Bill Beatty, such as *This Australia: Strange and Amazing Facts!* (1941) and *Australia the Amazing* (1944). Beatty's 1944 compilation included numerous so-called Aboriginal legends such as 'How the Kookaburra Began to Laugh' and the origin of the dance of the brolga as well as a great deal about curious animals and natural phenomena and historical tidbits. Bill Beatty produced *Come A-Waltzing Matilda* in 1954 and *A Treasury of Australian Folk Tales and Traditions* in 1960.

Blessing of the Fleet: see **Festivals**.

Bloke, as in 'There was this bloke ...', 'Have you heard about the bloke who ...', is a term used to begin many yarns. 'Bloke' is a standard item of male address in Australian folk speech (q.v.).

Body language: see **Non-verbal communication**.

BOGLE, ERIC: see **Folk performers**

***Boîte*, The**, was established in 1979 as a non-profit, community-based organization located in Fitzroy, Melbourne. Supported by the Victorian Ministry for the Arts and the Australia Council, the *Boîte* presents and encourages ethnic performing arts of all kinds, particularly music and dance, providing performance venues and appropriate training. The World Music Café is held every Friday night, complementing a programme of concerts, festivals, a weekly radio show on 3PBS and the publishing of five newsletters each year, all organized by a mainly volunteer workforce.

Broadside ballads The term 'broadside' is used to describe a wide variety of popular street literature printed cheaply on single sheets of paper; it includes handbills, proclamations, advertisements and ballads. These narrative songs had been printed in Britain since the early sixteenth century, and in Ireland from the seventeenth century, but the first fifty or so years of European settlement in Australia coincided with a significant revival of the industry. In Britain and Ireland broadside ballads often passed quickly from printed sheet to oral circulation, and many ballads learnt from broadsides must have come to Australia in the memories and song repertoires of the colonists. Occasionally the sheets themselves were brought out to Australia — for example Rev. John McGarvie's collection of Scottish broadsides, including almost forty ballads from the early 1820s, which is now held in the Mitchell Library. By the middle of the nineteenth century the broadside's popularity had undergone a rapid decline in Britain and its impact on the songs composed in Australia in the second half of the century diminished correspondingly, although the form remained popular in Ireland for several decades. Throughout the nineteenth century tunes which had originally been associated with broadside ballads were often used with words composed in Australia.

In the past folklorists and enthusiasts revered the traditional ballads or narrative songs which F. J. Child collected in his five-volume *The English and Scottish Popular Ballads* (1882–98), and also the folk songs gathered from an oral tradition by collectors such as Cecil Sharp; but they looked down their noses at the broadsides, which were regarded as the work of hack writers, tainted both by the haste with which they were composed and by their commercial nature. In many ways this was a class-based judgement made by the generally well-educated and often wealthy collectors and connoisseurs. Certainly the people who *sang* traditional ballads, folk songs and broadside ballads made no such distinctions. Indeed, it is impossible to make any strict distinction between these different groups: many traditional ballads and folk songs were printed as broadsides and, conversely, many ballads which originated as broadsides were taken into oral tradition. Towards the middle of the nineteenth century it was common for songs composed by music-hall entertainers to be produced on broadsides.

Among the vast outpourings of the British and Irish broadside industries in the first half of the nineteenth century were numerous ballads dealing with transportation to Australia. A number of these have been collected in Australia since the 1950s, but it is clear that they were not composed in the southern hemisphere and that their presence is the result of convicts and emigrants bringing their song repertoires with them. Elaborate arguments have been constructed to suggest that some of these ballads may have been composed in Australia, for example John Manifold claims that 'Van Diemen's Land' may have originated in the penal colony of that name and then been taken back to the northern hemisphere by returned convicts (*Who Wrote the Ballads?*, 1964). But it is almost certain that this and the other broadsides about transportation were originally composed in Britain and Ireland for the broadside press.

The equipment unloaded from the First Fleet at Sydney Cove in 1788 included a screw-press capable of producing broadside ballads, but it was never used for this purpose. It was not until 1795 that the authorities were able to find someone who could operate it. The oldest surviving Australian broadsides are government notices setting out the duties of constables and watchmen which were produced on this press in late 1796. The authorities in the fledgling penal colony kept a tight rein on what was printed and this control, along with the small number of potential buyers, made a commercial broadside ballad industry impossible.

In 1803 George Howe began printing the *Sydney Gazette*, the colony's first newspaper, and very soon was including poetry composed in New South Wales. Some folklorists have argued that the verses contributed to newspapers in early colonial Australia were a continuation of the verse broadside tradition in Britain and Ireland, but the verse published in the early colonial press very rarely bore a resemblance to broadside balladry. The verse in early colonial newspapers was generally high cultural in style and strictly controlled. As R. B. Walker has pointed out, the *Sydney Gazette* was 'kept at Government house, and printed on a government press with government ink on government paper', and it was censored until 1824 (*The Newspaper Press in New South Wales*, 1976). Other newspapers such as the *Australian* and the *Sydney Monitor* could, at times, sharply criticize the way the colony was governed, but the verse which the papers printed was almost always directed towards an educated few. One such poem was the earliest extant ballad produced in Australia, which appeared in the *Sydney Gazette* on 8 April 1804. Although it was inspired by the most significant convict uprising, the rebellion of several hundred convicts at Castle Hill, it was far from a celebration of that event. 'The Volunteers', subtitled 'A Ballad', was lavish in its praise for those ready to jump to the defence of the existing order:

When menac'd with civil commotion and
  noise,
Shall Britons inactively slumber?
Then away to the field, the bright musket to
  poise,
With courage, regardless of number.

With patriot firmness the laws we'll maintain;
With spirit and vigor we'll brave the campaign;
Our women and children relinquish their fears,
And trust to the prowess of bold Volunteers.

Very occasionally the newspapers would publish a poem which was modelled on a broadside ballad. One such piece which appeared in the *Sydney Gazette* on 26 May 1829, was 'The Exile of Erin, On the Plains of Emu', long seen as an Irish convict's wistful thoughts of home, although it was actually written by the Rev. John McGarvie, a Scottish Presbyterian minister with literary aspirations. He had based the poem on his compatriot Thomas Campbell's 'The Exile of Erin' which had been widely circulated in Britain and Ireland, at times on broadsides. Campbell's version is said to have been sung in Australia in 1800 by James Harold, one of those involved in the 1798 rebellion in Ireland, and Harold is supposed to have found an appreciative audience for the song (Patrick O'Farrell, *The Irish in Australia*, 1986). Even if Campbell's ballad gained some local circulation, and there is nothing to suggest this, McGarvie would have been familiar with it before he left Scotland as he was both widely read and had an interest in broadsides. There is no evidence that his version of 'The Exile of Erin' was performed as either a song or a recitation until the 1950s when Nancy Keesing and Douglas Stewart rediscovered it while they were working on their *Old Bush Songs* (1957). The tune to Campbell's original was added to McGarvie's words and the end product came to be regarded by some of those involved in the folk revival as an Australian 'folksong'.

As is to be expected, few of the songs composed and sung by convicts and their fellows have survived. The earliest reference to the composition and performance of popular ballads in Australia is in the journal of Peter Cunningham, a surgeon on some of the ships which transported convicts to New South Wales in the 1820s. Cunningham records that during this period songs about the exploits of bushrangers were 'often being made . . . by their sympathizing brethren' (*Two Years in New South Wales*, 1827, p. 282). None of the pieces Cunningham refers to is now extant, but a couple of bushranger ballads probably dating from the early 1830s have survived, each showing strong links with British and Irish traditions. One of these is 'Johnny Troy', a ballad celebrating the deeds of a bushranger who was hanged in 1832, texts of which have only been found in America in a clearly corrupt form.

The other ballad probably composed about this time is 'Bold Jack Donahoe', which tells of its hero's resistance to the crown and his defiant death in 1830. This song's conections with broadside balladry are numerous. Some collected versions specifically link Donahoe with the outlaws 'Freincy, Grant, bold Robin Hood, and Brennan and O'Hare'. Some at least of these were the subjects of popular Irish broadsides. The close connections between 'Bold Jack Donahoe' and the Irish broadside tradition are also evidenced by its complex textual history. Given Cunningham's report of the popularity of bushranger ballads in New South Wales, it seems likely enough that 'Bold Jack Donahoe' was composed in the colony, but the earliest surviving texts are British and Irish, suggesting that the ballad was taken back to the northern hemisphere to become part of the broadside tradition in which it had its roots. An Irish broadside version with significant alterations later found its way back to Australia in the 1890s and was collected by Alan and Gay Scott in the late 1950s (*A Collector's Songbook*, 1970).

Composers of convict ballads and verse also used other types of broadside ballads as models. The most popular British ballads about transportation were those which warned against crime. Although broadsides such as 'Van Diemen's Land', 'The London Apprentice Boy' and 'Jamie Raeburn' (each of which has been collected in Australia) could express some sympathy towards the criminal, their overall impact was probably to stress the power of the legal apparatus, suggesting that crime had inevitable and often brutal consequences. In the penal colonies the narrative structure of this type of ballad was appropriated for radical purposes, as it had been on a few occasions in Britain. This is clearest in 'Jim Jones at Botany Bay' which first appeared in Charles MacAlister's book of reminiscences, *Old Pioneering Days in the Sunny South* (1907). Cliff Hanna has recently suggested that the ballad may be wholly or largely MacAlister's own work (Hergenhan et al. [eds], *Penguin New Literary History of Australia*, 1989). This is by no means certain, as MacAlister attributed a number of the compositions in his book to himself, whereas he describes 'Jim Jones at Botany Bay' as a 'typical song of the convict days'. In either case the ballad clearly derives most of its structure from the British broadsides, deviating significantly from their narrative pattern only in the last six lines. As was common in the British warning ballads, 'Jim Jones at Botany Bay' describes the trial in England, the judge's harsh words, the rough and dangerous voyage to the colonies and the hardships of life in chains. But it finishes with the expression of a rebellious ideology very different from that of its British predecessors:

But bye-and-bye I'll break my chains: into the bush I'll go,
And join th' brave bushrangers there — Jack Donohoo and Co.;
And some dark night when everything is silent in the town
I'll kill the tyrants, one and all; and shoot th' floggers down;
I'll give th' Law a little shock: remember what I say,
They'll yet regret they sent Jim Jones in chains to Botany Bay.

Another ballad which has links with the British warning broadsides is 'The Seizure of the Cyprus Brig in Recherche Bay, Aug, 1829'. Its first verse has some echoes of that tradition and one version has been collected which begins with a verse directly from 'Van Diemen's Land'. Like 'Jim Jones at Botany Bay', the ballad does not follow the structure of its British predecessors, and instead provides its audience with a powerful celebration of convicts who take up arms against their guards.

The convict ballad 'Moreton Bay' also shows traces of a broadside ancestry. It is primarily a strong critique of the violence meted out to convicts in places of secondary punishment, and some of that violence is graphically depicted. Yet what is surprising about 'Moreton Bay' is that it concludes passively, particularly in early versions, such as the one which appears in Jack Bradshaw's *The Quirindi Bank Robbery* (*c*. 1899). In that text the ballad finishes with an apparent endorsement of the hegemonic values of patience and resignation — convicts are reminded that upon the expiry of their sentences they will gradually forget their previous hardships when they return across the ocean, as the speaker is about to do. This passive ending can be explained by the ballad's use of the narrative structure of a group of Irish love songs which tell a tale centring on the separation of two lovers by transportation. It is common in these broadsides for the woman's father to arrange for his daughter's lover to be transported to Australia, but the ballads conclude favourably for the young man, usually with his return to Ireland. 'Moreton Bay' derives its tune from an Irish broadside, 'Youghal Harbour', with which it has some slight textual links, but its narrative shape is derived from the transportation love songs as the beginning of the Bradshaw text indicates:

I am a native of the land of Erin, and lately banished from that lovely shore,
I left behind me my aged parents, and the girl I adore.

Henry Mayhew's descriptions of street life in *London Labour and the London Poor* (1861–62) indicate that the production of broadside ballads was well past its heyday by the middle of the nineteenth century. In London the street ballads lost their dominant position in the realm of popular song to the compositions of the music hall, which was undergoing a rapid rise. In Australia among the early examples of the music-hall genre were the hundreds of songs produced by professional entertainers on the goldfields. In his song 'London Cries' Charles Thatcher (q.v.), the most popular of the goldfield composers, records the English decline of interest in broadside ballads and also suggests that the same was true of Australia by the late 1850s. The song lists positive aspects of life in the colonies, one of which is the absence of broadside sellers:

Here no dreadful ballad singing,
Slumber from our eyelids keeps,
And no bells do we hear ringing,
But in peace a fellow sleeps.

As the popularity of broadsides gradually waned in the northern hemisphere, their impact on the verse and song produced in Australia also faded. In the second half of the nineteenth century, paradoxically, songs were occasionally published on broadsides in Australia, but they were very much the exception. Some of these were clearly in the tradition of their British and Irish predecessors, such as 'The Fight on George's River Ground' which related the details of a boxing match between Larry Foley and Sandy Ross on 28 March 1871. M. J. Conlon wrote the ballad immediately after the fight and quickly published it as a broadside which was then sold on the streets of Sydney for threepence each, with the profits going to Foley (Russel Ward, *Three Street Ballads*, 1957). During the gradual decline of the broadside tradition in Britain and Ireland, some songs which had their origin in the music halls found their way on to broadsides, and this also occurred on a small scale in Australia. A handful of Thatcher's songs survive in this form, and seven of these are held in the La Trobe Library in Melbourne, including 'Where's Your Licence?', one of the few goldfield songs which has also been collected from oral sources. The *Argus* for 7 April 1854 records that Mr M'Donogh, a lawyer who heard Thatcher sing in Ballarat, arranged for 'some dozen of the best' of the entertainer's pieces to be printed in Melbourne as broadsides. But in the second half of the nineteenth century the most common form of publication for popular song was in 'songsters' which were cheaply produced booklets, generally about thirty pages in length. During the 1850s and 1860s dozens of songsters were printed, containing hundreds of popular pieces from Australian and English concert halls. Production of these flimsy booklets continued throughout the rest of the century, and their contents suggest very strongly that it was songs in the music-hall tradition which were marketable, not those whose origins lay in the broadside era.

Nevertheless, British and Irish street songs continued to be sung throughout the nineteenth century, and are still sometimes collected. While most of those which were popular in Australia were performed in something like their original form, many became localized to varying degrees. For example one of the several British broadsides called 'Botany Bay' has been collected by John Meredith under the title 'Moreton's Bay' (*Folk Songs of Australia*, 1967). Essentially the ballad's action remains the same, but it has been resituated so that instead of describing an English felon about to be transported to New South Wales, it tells of a criminal in Sydney who is sentenced to transportation to 'Moreton's Bay'. Other ballads could acquire local characters such as the version of the Irish street song 'There's Whiskey in the Jar' which features Sir Frederick Pottinger, the leader of the police hunt for the bushranger Ben Hall (*Folk Songs of Australia*, 1967). More thoroughly localized is the song known variously as 'Banks of the Condamine' and 'Banks of the Riverine'. These Australian versions record a conversation in which a horse-breaker or shearer sets forth the reasons why he must depart into the bush alone, while a woman argues that she should accompany him. In the earlier street song called 'Banks of the Nile', known in both Britain and Ireland, a similar discussion takes place between a soldier and the woman who wishes to accompany him to war. Just as the texts of the broadsides could alter over time, the tunes too could take on a new character in their new context. Thérèse Radic writes that the tune used by Simon McDonald for his version of 'Moreton Bay' is derived from the Irish broadside 'Youghal Harbour' but it 'elaborat[es] on the original melodic line and rhythm, though retaining its structure, so that it becomes something more than a variant' (*Songs of Australian Working Life*, 1989, p. 171).

Although it is clear that the performance of British and Irish broadside ballads must have been common enough up to the 1850s, much more information remains about their currency in the latter part of the nineteenth century, when their influence on local song production was in decline. In March 1888 the *Bulletin* published an article titled

'Bush Songs' in which 'Gambasino' reminisced about songs he had heard as far back as 1865. What is perhaps most interesting is the diversity of song types which he mentions: local songs such as 'Bold Jack Donahoe', local authors such as 'Frank the Poet' (q.v.), British broadsides such as 'Van Diemen's Land', Irish broadsides such as 'Gallant Michael Hayes', as well as songs of American origin, sea shanties, and music-hall pieces. A similar range exists in Henry Lawson's sketch 'The Songs They Used to Sing', which he wrote in 1899. In their famous verse debate about the advantages and disadvantages of the bush, Lawson and A. B. Paterson both refer to the broadside 'Willy Reilly' as being popular among shearers, and Paterson included 'The Wild Rover' in his *Old Bush Songs* (1905), evidently mistaking it for a local production. Nor was it only in the bush that British and Irish broadside ballads were performed. In his *Old Books, Old Friends, Old Sydney* (1952), James Tyrell records that in the latter part of the nineteenth century the street character 'Blind Billy Huntingdon' used to sit on a city corner in Sydney, playing a concertina and singing 'Ten Thousand Miles Away', a sea shanty which also had a life as a broadside. Since the folk revival which began in Australia in the early 1950s, many broadside ballads have been collected from oral sources. Just how popular British and Irish street ballads had once been is evident from the considerable number which appear in the repertoires of two well-known performers, Sally Sloane (q.v.) and Simon McDonald (q.v.).

In the mid-twentieth century there was a profound resurgence of interest in broadside ballads. Geoffrey Ingleton's influential *True Patriots All: Or News from Early Australia as told in a Collection of Broadsides* (1952) put Australian newspaper poems, popular verse and traditional songs alongside British and Irish broadsides, sometimes printing them in a form which suggested that they had originally been broadsides. Partially modelled on Ingleton's book was Hugh Anderson's (q.v.) *Farewell to Old England: A Broadside History of Early Australia* (1964), which also used a variety of sources including a number of Australian-composed poems and songs. In his *Colonial Ballads* (1962) Anderson had earlier made a strong statement as to the continuity between the broadsides and early colonial newspaper poetry: 'we find such choice pieces in the newspapers of the day as . . . 'The Exile of Erin' [and] 'Botany Bay Courtship' . . . In every way except in the manner of publication, such songs are identical with broadsides'. While these books stressed the links between nineteenth-century Australian material and the broadsides, other revivalist publications suggested links between their own role as song publishers and the broadside press. In Sydney, the Bush Music Club produced roneoed song-sheets titled Bushwhacker Broadsides, and in Melbourne Rams Skull Press published the Black Bull Chapbooks — a series 'devoted to little known aspects of Australiana, and in particular the study of early colonial ballads'. These were printed on 'Tudor Antique' paper, in imitation of northern hemisphere chapbooks, a form which derived from broadsheets.

Stress on the continuity of 'the ballad tradition' was a prominent feature of the early years of the revival, perhaps partly arising from many of the revivalists' desire to establish their own relationship to the culture or cultures which they memorialized. In fact the discontinuities between different ballad groups are more striking than the similarities, and this is particularly true of the differences between the northern-hemisphere broadsides and the bulk of the the songs and verse composed in Australia.

Philip Butterss

Brunswick Recordings was established in 1985 as part of the City of Brunswick's Community Arts Programme to produce, promote and market recordings featuring the work of local ethnic, folk and traditional musicians. The label seeks to produce high-standard recordings which reflect the character and concerns of the Brunswick community in Melbourne, and to feature music which falls outside the narrow parameters of the mainstream recording industry. Brunswick Recordings is possibly the world's only municipal recording label, and has opened up opportunities for local musicians to bring their cultural traditions to the wider attention of both the local and the Australian community. The company also strives to utilize and develop local technical, production and administrative skills.

The catalogue contains recordings from the Turkish, Arabic, Greek (for example Apodimia Compania, q.v.), Latin American, Celtic and English traditions through to the contemporary urban sound of the Whirling Furphies and the children's playground songs and stories of *Toodaloo Kangaroo*.

Building and folk architecture Vernacular or folk architecture has also been called 'layman's architecture' and might be defined in terms of non-academic building styles and practices which are characteristic of a particular locality. J. M. Freeland, in *Architecture in Australia* (1968), described the 'layman's architecture' which characterized European Australia's first buildings after the arrival of the First Fleet as 'the plain hip-roofed box still to be found through the rural areas of the Grampians, the Midlands or the Cornish countryside of home. Only the nature of the materials used gives it any individuality'. There is, however, evidence that Australian vernacular architecture or folk building, including the Australian veranda, drew not only on British practices but also substantially on influences from Asia (see **Australian house, The**).

The Australian house veranda, particularly as expressed in its most elaborate form in the Queensland elevated timber house, is frequently regarded as one of Australia's most distinctive architectural features. In a letter written from Sydney on 19 April 1839, Dr James Satchell, an Englishman visiting Australia, expressed his surprise on seeing the many buildings with verandas, and commented that 'the appearance of the roofs made to reach beyond the walls on every side and supported on pillars is very strange'.

Whereas the Australian veranda may have had strong Asian influences, other vernacular building techniques such as the use of weatherboards rather than brick for much domestic building reflects regional British building methods such as those of East Anglia. According to David Denholm in *The Colonial Australians* (1980), early weatherboarding was fraught with difficulties due to the use of unsuitable materials such as red gum in the Australian climate, and the true weatherboard house was 'introduced to Australia a second time' by American immigrants to the goldfields of the early 1850s. This method only slowly replaced Australia's more common methods of vernacular building, namely rammed earth or pisé, and wattle and daub. The latter technique combined both earth and timber, and the common use of the native Australian acacia tree for 'wattling' or making a lattice framework gave our national flower its popular name of 'wattle'.

David Denholm argues that Australians should not romanticize the proverbial 'old bark hut' as our major national form of folk building, since 'there are parts of Colonial Australia which hardly knew what a bark hut was'. The hut is certainly celebrated (though scarcely romanticized) in one of Australia's best-known folk ballads, 'The Old Bark Hut':

I've had the rain come pouring in just like a
 perfect flood,
Especially through the great big hole where
 once the table stood;
It leaves me not a single spot where I can lay
 me nut
But the rain is sure to find me in the old
 bark hut.

The Australian bark hut's greatest claim is probably that it is a truly indigenous building style, with the bark roof — 'one of the most successful of colonial innovations', according to Miles Lewis — replacing less effective thatch made of reeds or straw. Lewis also pointed out that 'bark building is an unequivocally local response to the presence of suitable barking trees'. The methods of construction, however, were based on overseas precedents, as the use of horizontal and vertical poles to hold the bark in place was the same system as was used with thatched roofs in Serbia, Russia, north Germany and Canada. A question which should be answered concerns whether the early settlers in Australia learnt about the use of bark for building purposes from Aborigines, as seems likely. Lewis argues for international origins for the vernacular buildings which he characterized in the title of his book as *Victorian Primitive* (1977). Mud brick may be from East Anglia, France or California or Mexico, and wattle and daub from Leicestershire, Cheshire, East Anglia and 'the North' (originally from the Romans). Other methods such as horizontal slab construction and half-timbering appear not only in Australia but also in Canada, England, France and Germany.

Gwenda Beed Davey

Bunyip This is the name, taken from Aboriginal Wergaia dialect of the Wembawemba of western Victoria, of a fabulous large, black, amphibious monster supposed to inhabit waterways in various inland locations, especially in lakes, swamps, pools and rivers in south-eastern mainland Australia and Tasmania. Its lore has been further coloured by two forces: Irish memories of the *poukha* (a similar threatening and mysterious night monster) and, at least in the earlier nineteenth century, natural scientists like G. F. Angas comparing it to the much larger Maori *Tanniwha* of New Zealand, which was capable of devouring 'men, women, children and all weapons of war'.

In the minutes of the Philosophical Society of Australasia, for a meeting held in Sydney in December 1821, it is recorded that the explorer H. Hume had noted the existence of a 'hippopotamus'-like creature in Lake Bathurst, and that the Society had desired to obtain a specimen. In 1823 one E. S. Hall, resident of this alleged Bunyip region, reported his own various sightings there of a monster 'resembling a porpoise' and with 'a bulldog's head', which 'the natives of the lake and its neighbourhood ... describe as having formerly taken their children'.

In early 1847 W. H. Hovell, another southern explorer, wrote to the press that he had much 'recent' corroboration of a 'huge animal desporting' in inland waters like the Lachlan, Murrumbidgee and Murray Rivers, and recording its name as Katenpai, Kyenprate or Tunatabah, adding that the natives near Great Corangamite Lake called it the Bunyip. About that time such a Murrumbidgee monster's skull was retrieved, examined and pronounced to be that of a calf. Over the next six years numerous similar reports were published, one calling a like menacing creature on the Hunter River a 'Yaa-hoo'. Others again reported a baboon-like water creature on Phillip Island, a bunyip on the Hawkesbury River, and a strange monster in the bulrushes near Mount Gambier. A report of December 1853 deemed such a creature to be horse-like and 'its entire length from fifteen to eighteen feet'. Over the next twenty years more caution or clearer daylight viewings suggested the creatures to be seal-like or resembling the sea-elephant, a view also advanced in 1857 by the naturalist, Stocqueler, of various water monsters seen in the Murray.

The best early clue as to the nature of this strange amphibian is afforded by the long-time wild white man, William Buckley, in his reported *Adventures*, published in 1852. He refers to the fact that the monster did much to hurt the Aboriginal people, especially women who strayed from their camps, coming suddenly on them and making terrible groanings and bellowings, but that he himself had only seen their backs, and then recording a Barwon River belief that they had 'some supernatural power over human beings, so as to occasion death, sickness, disease, and such like misfortune ... the natives considering the animal something supernatural'. This report and others have led to speculation that the stories had their origin in cautionary tales told to keep children from wandering away from camp and women from approaching certain places at times of initiation, the averred location of the creatures always being somewhat removed from the zones for regular social activity or encampment. There is also evidence that 'bunyip', the word, is linguistically connected to Bunjil, a mythic 'Great Man' who had made the mountains and the rivers and man and all the animals, as was explained by Aldo Massola, anthropologist and archaeologist, in his important study, *Bunjil's Cave: Myths, Legends and Superstitions of the Aborigines of South-East Australia* (1968). These localized beliefs about water monsters in Victoria were widely popularized in the mid-nineteenth century and probably led to the recording of similar tales among Tasmania's lakes and Jordan River, along the Cox River valley west of Sydney, and as far inland as Conargo, New South Wales, more than 800 miles upstream from the sea.

An extension of the work is its occasional use, since 1875, in a sense meaning 'mature and unbranded', applied to full-size yet wild (swamp) cattle. Another is the use of 'bunyip' to refer to an Australian marsh bird of the genus *Botaurus*, the male of which species has a booming note in the breeding season, the full folk term for it being 'the bunyip bird'. Another adaptation was noted in 1852: 'A new and strong word was adopted into the Australian vocabulary: Bunyip became, and remains, a Sydney synonym for *imposter*, *pretender*, *humbug* and the like'. The following year Daniel Denihy opposed the proposed Australian local 'Bunyip peerage', a democratic attitude which firmed from this point on, so that in 1956 in his *Overlander Songbook* Ron Edwards could observe that 'The very large stations . . . fostered the class feeling that was summed up with the title of "Bunyip Aristocracy".'

In the last third of the twentieth century the term 'bunyip' — as noun or adjective — is used figuratively to refer to anything chimerical or well-nigh impossible of attainment, be this personal as with marital happiness, or more publicly, the quest for national identity, it being wryly styled 'our century-long pursuit of the bunyip'.

After more than two centuries of white settlement it can be said that a (once Victorian) mythic word for a fabulous monster of swamps and lagoons, like the western European word 'dragon' for 'the Devil', has become the most engaging and widespread of Aboriginal folklore concepts for an unseen monster, having largely displaced its many regional competitors such as the more ape-like *wowie* or *yowie*. While some raconteurs may insist that the 'bunyip' comes down the stream of time as an Aboriginal folk memory of the great wombat-like Diprotodon which once roamed the continent, it is not contentious that this native monster has a secure niche in the affection of storytellers and illustrators, as well as for all those non-urban folk young enough in heart to require an intriguing mystery to explain strange noises out there in the night. Whether or not we choose to link it with 'quinkins' (as Percy Trezise did in *The Quinkins* in 1980), the affectionate lore of the bunyip is strong and prolific, be it in poems like John Manifold's 'The Bunyip and the Whistling Kettle', or Kath Walker's (Oodgeroo Noonuccal) 'Biami and Bunyip', or 'Tia-Garn', or in modern fantasy tales, often splendidly illustrated, for reading aloud, such as Jenny Wagner's *The Bunyip of Berkeley's Creek* (1973); Bill Hornadge's *The Boy Who Talked with Bunyips* (1980); Edel Wignell's *A Boggle of Bunyips* (1981); or Graham Jenkin's Mount Gambier story and already ubiquitous folk poem, *The Ballad of the Blue Lake Bunyip* (1982).

While, in the 1980s and 1990s, the somewhat cynical have been able to describe the quest for national identity as the impossible 'pursuit of the bunyip', the word itself has already entered the hearts of the caring young as the spirit of mystery, loneliness, exile and of painful estrangement from human society. While few would go so far as to compare the Bunyip closely with the Germanic world's Grendel (from the epic, *Beowulf*), each monster is as old as humanity and can generate potentially eloquent and emotional images, capable in the right hands of affording the listener deep psychological satisfaction. Even more significantly the Bunyip comes from the timeless Australian landscape, reconciles the present with the past and, by its socially corrective role for the trespasser or intruder, underscores the need for personal harmony and social order in both the smaller environment and the general human community.

See also: **Folk Tales.**

J. S. Ryan

Bush dance The name of this popular style of dancing is misleading, as it has not come from the Australian bush tradition of social dance. Instead it started in the cities with the folk revival of the 1950s. At that time the Bush Music Club (q.v.) in Sydney and the Victorian Bush Music (later Folk Music) Club in Melbourne included some folk dancing at their regular functions which were known as 'singabouts'. The name 'Bush Music Club' was chosen by John Meredith and other founders of this club because it was set up to introduce the music that was being collected, mainly from bush musicians, to other people (see *Stringybark & Greenhide* 6, 2).

Only a few folk dances were brought out to Australia and danced here in the early days of settlement (see **Dance**). These were soon replaced by the new dance fashions of the nineteenth century which quickly travelled to Australia and were adopted by all classes of society. It was these later dances (the quadrilles and couples dances) which survived into this century in the bush, not the folk dances. The folk dances were danced by small specialist groups and as entertainment and competition items. Certainly some of the folk dances first performed at the bush music clubs were those known to have been danced in the early days in Australia, but the people who taught them had themselves learnt them at special folk dance classes, not in the bush. They did also try to introduce a few of the old-time dances still done in country areas such as the Varsoviana, Waltz Cotillon and the Alberts. The major part of the music collected from traditional bush musicians had been for these types of dances. However, it proved to be easier to persuade the people who came to singabouts to try the simpler folk dances rather than the real bush dances that needed some expertise in waltzing, etc.

This preference for easy dances led to the folk dances being more strongly featured in clubs' activities, although the research into our dance history and into dance music soon made it clear that few of the folk dances survived on the social dance scene after the 1850s. Some of these early dances, such as the Soldier's Joy, were revived from printed instructions and added to the repertoire. Others were borrowed from the British folk movements which had been active in reviving their own folk dancing this century.

This revival style spread out from the bush music clubs, at first through the national folk festivals, and was taken up by other folk clubs, and from the 1970s many bush bands started up to cater for its growing popularity. These bands then spread its popularity further, and out into the general community. In this period, the folk enthusiasts were strongly influenced by the British folk revival movements. Most bush bands played mainly jigs, reels and hornpipes acquired from these revival movements, ignoring the dance music that had been collected from our bush musicians.

The typical programme of a 'bush dance' includes a mixture of folk dances, mainly of British origin with a few European ones, which are danced to the lively music of a bush band with the dances led by a caller. This caller is all too often a member of the band with no special dance expertise, which has led to a rather pedestrian standard of dancing. The dances have tended to lose their characteristic styles and to be danced with simplified steps. The Colonial Dancers in Melbourne introduced a simplified step–close–step which has been adopted as the basic travelling step. Generally, this bush dancing is marked by energy and enthusiasm, but is lacking in the co-ordination and stylishness for which the dancing of our colonial ancestors was renowned.

The few collected dances usually included in the bush dance repertoire have had rather rough treatment. The Brown Jug Polka is now danced in the English folk style rather than as collected by John Meredith (see *Singabout* 1, 2). The revival bush dancers' Manchester Galop is very different from its original version, and those dances with polka stepping are rarely done properly.

The sociable nature of bush dancing and its casual style have ensured its continuing popularity among some sections of the population. New dances are made up to suit this style of dancing. This has been encouraged by competitions organized by the Bush Music Club and the Traditional

Social Dance Association of Victoria. At the same time, the pioneers of this revival dance, the Sydney Bush Music Club and the Victorian Folk Music Club (q.v.), as well as many specialist dance groups, now have an enlarged dance repertoire which includes the quadrilles and couples dances which were the mainstream of our colonial dance tradition. Some of their functions feature only these latter dances, others feature the revival bush dances, and some feature a mixture of colonial and revival styles. Only time will determine what may develop into a modern folk dance that will be popular in our present multicultural society.

See also: **Dance; Colonial or heritage dance**.

Shirley Andrews

Bush Music Club, The, was founded in Sydney by a group of folksong and dance enthusiasts in 1952, and is one of the oldest of the various Australian folk organizations still in existence. Emphasizing the collection, research and performance of traditional bush music, song, dance, yarns and recitations, the Bush Music Club has extensive archives of such material held by various members. The club has published many works on aspects of bush traditions, regularly publishes *Mulga Wire* (q.v.), originally *Singabout*, and organizes festivals, concerts, dances and social occasions of all kinds. Many of the country's most noted collectors of bush music, song and dance are members of the club.

Bushrangers Bushranging, particularly the folklore associated with it, is important in Australian history, both because of the very real threat that it often posed to social order and because one particular bushranger has attained the status of a national hero. Only a select few of the thousands of bushrangers in Australia's past have become folk heroes. Most were mere criminals, driven to their acts by hardship, cruelty or plain stupidity. Those who did become heroes, however, sometimes despite themselves, were usually seen by those who celebrated them as representative of a legitimate grievance or grievances against the government, the police and the economic powers of the day. This generally translated into the trooper police versus escaped convicts or, later, squatters versus free selectors. Bushrangers like Jack Donohoe, Ben Hall, Ned Kelly and others celebrated in folklore almost all were the product of conflict between various social groups in colonial Australia. The bushranger who was seen as a hero by substantial segments of the population was a very real threat to the forces of law and order in the very thin 'civilization' of nineteenth-century Australia.

Those bushrangers treated as heroes took and/or were given, the mantles of bush Robin Hoods, friends of the poor and oppressed. In this respect the bushrangers were a continuation of a long tradition of noble outlaws that goes back at least to the mythical Robin Hood of England, a tradition also seen in the heroization of some figures from the American west, such as Jesse James and Billy the Kid.

The essentials of the outlaw hero tradition were that the outlaw hero was driven to his activities by injustice or oppression, acted bravely and chivalrously, was a friend of the poor, was often betrayed and ultimately 'died game', either in a last glorious battle with the forces of oppression or with some style and defiance at the end of a rope. Posthumous legends of various kinds have grown up about bushrangers, as about other folk heroes. Most important, the outlaw had the sympathy and usually active support of his own social group. The heroic folklore surrounding the bushranger is a direct continuation and adaptation of the traditions that developed around British highwaymen such as Dick Turpin, William Nevison and William Brennan, among others.

A 'bushranger' was literally someone who ranged the bush, and the term first appears in a diary entry of 1801, though it had no doubt been in use for some time. The first bushrangers were escaped convicts or 'bolters' who fled the ironed gangs and eked out a miserable survival in the bush by robbing or 'bailing up' travellers. Bolters were often aided by other convicts, particularly those working away from authority, such as shepherds or other rural workers. Noted bolters who became bushranger folk heroes were Martin Cash, Mike Howe and Mathew Brady in Van Diemen's Land and Jack Donohoe in New South Wales. Donohoe, a transported Irish convict, took to the bush in 1826. He soon collected a gang of bolters and quickly became notorious for his robberies. Donohoe was a hero to the convict population of the colony and a number of ballads about him were composed and circulated. The one that has lasted best in folk tradition is 'Bold Jack Donohoe', which fits its hero neatly into the traditional mould of the gallant outlaw defying the forces of authority and dying gamely (1830). In one version of the song Donohoe is explicitly and favourably compared with Robin Hood, William Brennan and other highwaymen; in another version he 'did maintain his rights, my boys, and that right manfully'. In most versions of this song Donohoe 'scorns to live in slavery or humble to the Crown', a recurring theme in bushranger balladry exemplifying the strong Irish influence on our nineteenth-century folklore.

Undoubtedly the best-known figure in bushranging folklore never existed. The song and its hero, 'The Wild Colonial Boy', seem to have been modelled on the exploits of Jack Donohoe, though 'The Wild Colonial Boy' ballad did not arise until the 1860s. Since then, in one version or another, it has spread throughout the English-speaking world and, in its waltz-time version, the song's tune has become a widely-known signature of Australia, rivalling that of the Cowan tune for 'Waltzing Matilda' (q.v.). Like all bushranger heroes, the Wild Colonial Boy is disdainful of danger, authority and death, and like Donohoe, also usually 'scorns to live in slavery . . .'.

The 1860s was a period in which most of the real figures famous in bushranging folklore flourished gloriously — if briefly. In the New England area of New South Wales (mainly), between 1863 and 1870, Frederick Ward or 'Thunderbolt' (sometimes 'Captain Thunderbolt') became a hero of folk balladry:

> My name is Frederick Ward, I'm a native of this Isle.
> I rob the rich to feed the poor and make the children smile.

Daniel 'Mad Dog' Morgan terrorized both sides of the New South Wales/Victoria border between 1863 and 1865, and according to one ballad about him was 'the travellers' friend' and 'the squatters' foe'. Francis Christie alias Gardiner alias Clarke, usually known as Frank 'Darkie' Gardiner, led a gang of mostly youthful bushrangers around the Bathurst–Forbes area from 1861 to 1864, during which time he organized the famous Eugowra Rocks gold escort robbery of 1862. Some of our most folklorically famous bushrangers were associated with Gardiner, including John Gilbert, Johnny Dunn, Michael Burke, John Vane and John O'Meally, all of whom are celebrated in various ballads and folk tales.

The most famous of Gardiner's accomplices was Ben Hall, whose bushranging career lasted from 1862 to 1865. Small farmer Hall appeared to be a classic example of the upstanding citizen hounded by the unjust forces of oppression into outlawry. While the historical records indicate that Hall was hardly as innocent as his folkloric image implies, they also indicate that he was subject to particularly harsh treatment by the trooper police. Ben Hall folklore, however, unreservedly portrays Hall as the friend of the poor and oppressed, a hero in the mould of the highwaymen of old, and as a man betrayed to death by an unfaithful accomplice. These elements are contained in all the Ben Hall ballads and are also integral to the considerable body of narrative traditions about the bushranger. Ben Hall ballads include 'The Death of Ben Hall', 'The Streets of Forbes', 'Bold Ben Hall' and 'Ben Hall'. As well, Hall features in other bushranging songs, such as 'Dunn, Gilbert and Ben Hall', which includes the lines:

Ben Hall he was a squatter who owned six
  hundred head,
A peaceful man he was until arrested by
  Sir Fred
His home burned down, his wife cleared out,
  his cattle perished all:
'They'll not take me a second time', says valiant
  Ben Hall.

Between 1878 and 1880 Australia's most famous outlaw hero and our central national hero ranged the bush in Victoria's north-east. Edward (Ned) Kelly (q.v.), his younger brother Daniel and two accomplices, Steve Byrne and Joe Hart, carried out the best-planned and most threatening 'social banditry' in white Australia's brief history. The exploits of the Kellys generated a very considerable body of folksong and narrative that placed the gang, particularly Ned, firmly in the tradition of the outlaw hero. According to folklore, the small-selector, Irish Catholic Kellys were oppressed by just about everyone, including the police, the squatters, the government and their non-Roman Catholic neighbours. While such claims are exaggerated, there is no doubt that the family was on the receiving end of continuing coercive treatment, as were the sympathizers and supporters who helped the Kellys dodge the police and the army for almost two years. The extent to which the Kellys were supported by local small selectors will be debated for decades, but in folklore there is no doubt about Ned Kelly's status as friend of the poor and avenger of the oppressed: 'The Governor of Victoria was an enemy of this man' as 'Kelly Was Their Captain', one of the numerous extant Kelly ballads, puts it.

One recurring element of outlaw and other hero traditions is the supposed survival of the hero after death. While Ned Kelly was definitely hanged in Melbourne in 1880 (thus generating the folk belief that he uttered the immortal last words 'Such is life'. He probably said 'I suppose it had to come to this'), the death of his younger brother, Dan, was much more ambiguous, involving the shootout and fire at the Glenrowan Hotel, followed by the family's spiriting away of the charred body. It seems that the folk belief in the survival of the hero was transferred to the figure of Dan Kelly, who was often reported to be 'alive' in Australia and elsewhere well into the 1920s.

What has survived Ned Kelly's death has been the image of the bushranger as a defining national symbol. This has been discussed in detail elsewhere (G. Seal, *Ned Kelly in Popular Tradition*, 1980), and there is only space here to mention the persistence of Ned Kelly in folk speech ('game as Ned Kelly') and in narrative traditions, where it is difficult to find an Australian of Anglo-Celtic background without a forbear who *didn't* meet, see, hear or know one or more members of the Kelly gang, open or close railway gates for them, or perform some other minor service. As well, Ned Kelly has a continuing appeal as a character in popular song, film, theatre, art, fiction, non-fiction, tourism and books about Australian folklore. 'St Ned', as the title of one of the innumerable books about Kelly has it, is with us to stay.

Apart from the bushranger folk heroes maintained above, there are a number of less well-known characters who appear in folklore. Some of these are apparently fictional, such as Jim Jones, Johnny Troy (whose exploits are now known only in American versions of his ballad), Jack Lefroy, and the bushranger 'Jack' in the song 'Taking His Chance'. Others certainly existed, including the West Australian convict escaper extraordinaire, 'Moondyne Joe' or Joseph Johns, whose career spanned the 1860s, and later; and Harry ('Jack') Power, a Victorian bushranger who influenced and instructed the young Ned Kelly in 1870. As well, there are undoubtedly other local figures around whom similar, if less developed, traditions have cohered.

Much has been written, and will continue to be written, about the history and folklore of bushranging, and the bushranger will be an enduring figure of Australian folklore and popular culture. It is unfortunate that much of this writing will continue to suppress, ignore or downplay the extent to which the many real and fictional bushranger folk heroes were articulators — if often unwittingly — of genuine forms of popular dissent and protest against perceived injustice.

REFERENCES: J. McQuilton, *The Kelly Outbreak, 1878–1880: The Geographical Dimension of Social Banditry*, 1979; G. Seal, *Ned Kelly in Popular Tradition*, 1980; Bill Wannan, *Tell 'em I Died Game: The Stark Story of Australian Bushranging*, 1963 and numerous subsequent editions.

Graham Seal

C

Chain letters Arriving anonymously in the mail, traditional chain letters have been circulating the world's postal systems probably since such systems were introduced and certainly since early this century. The most common traditional chain letter is one that usually begins 'This paper has been sent to you for good luck ...' and goes on to promise the arrival of great riches if the receiver will only send the letter on — 'do not break the chain'. If the chain is broken by the receiver, says the letter, bad luck, possibly including death, will befall him or her. While many people simply smile and throw such letters away, others take them seriously and may well be distressed at receiving such communications.

Other traditional chain letters include the one aimed specifically at women, usually headed 'To the women friends in my life who know how to dream and create their own reality'. This one asks for money to be sent on, with the promise of receiving a vast return on this initial investment. This letter, while of folk origin, may be the inspiration for (or may have been inspired by) the definitely un-folkloric chain letters usually termed 'pyramid selling schemes'. Such letters and their senders have an unscrupulous vested interest in the chain and some Australian states have made these selling schemes illegal.

A relatively recent and definitely folkloric chain letter is the spoof 'Chain Letter for Women', in which husbands and/or male lovers are 'mailed' on to other women.

Children's folklore can be discussed under four headings: folklore *of* children, folklore *for* children, folklore *about* children and children as apprentices in learning adult folkways.

Folklore of children

Children's own playground folklore has been described in several well-known books; the first and second editions of *Cinderella Dressed in Yella* compiled in 1969 by Ian Turner (q.v.) and in 1978 by Turner et al., and June Factor's (q.v.) several compilations such as *Far Out, Brussel Sprout!* (1983). Factor's analytical study *Captain Cook Chased a Chook* (1988) won the American Folklore Society's Opie Prize for the year's best publication in children's folklore. All the above books have been best-sellers in Australia, indicating the strong interest and appreciation which not only children but also adults (the buyers of the books) have for the humorous, inventive and irreverent traditions of the playground:

> Captain Cook chased a chook
> All round Australia;
> He lost his pants
> In the middle of France
> And ended up in Tasmania.

The above rhyme is only one of the many variations of Captain Cook, and Turner et al. (1978) list nearly twenty. It is also only one of several types of playground lore including games, rhymes, chants, parodies, jokes, tricks, counting-out routines, autograph album entries, nicknames, taunts and fortune-telling. One favourite autograph album entry was as popular in the nineteenth century as it is today:

> True friends are like diamonds,
> Precious and rare:
> False friends are like stones,
> Found everywhere.

Children's own playground folklore is mainly passed down from one generation of primary school children to another, by observation, imitation, trial and error and sometimes direct instruction. Sometimes friendly adults will teach children an item remembered from their own childhood, but most of the transmission is from child to child.

Children's games and pastimes are one of our oldest traditions. The ancient Romans played knucklebones, and one game which is sometimes call 'Buck Buck' has been documented in Nero's time. Buck Buck is a vigorous game usually played by boys which involves some children bending over, often next to a wall, and others jumping on their backs. Peter Breughel's famous painting, *Children's Games* (1546), shows over eighty games, including a form of Buck Buck, and most of these games are still played in Australia today.

The first Australian publication of Australian children's folklore seems to have been Robert Rooney's *Skipping Rhymes* (1956), a limited edition art book with five woodcuts and eleven rhymes. Similar to other versions chanted in the 1990s to accompany the skipping game, Rooney's 'Teddybear' is as follows:

> Teddybear, teddybear, touch the ground,
> Teddybear, teddybear, turn right around,
> Teddybear, teddybear, show your shoe,
> Teddybear, teddybear, that will do,
> Teddybear, teddybear, run up stairs,
> Teddybear, teddybear, say your prayers,
> Teddybear, teddybear, blow out the light,
> Teddybear, teddybear, say good-night.

The definitive Australian survey and description of children's playground games, albeit from only one city, is Lindsay and Palmer's *Playground Game Characteristics of Brisbane Primary School Children* (1981). During the years 1975 and 1976, twenty-one schools in Brisbane were randomly selected, one from each electoral ward in the city, and a total of 4824 children were observed playing games. Lindsay and Palmer's report gives descriptions of 255 different games and numerous photographs. The games are of ten types: ball games, pursuit games, skipping games, hopscotch, elastics, marbles, knucklebones/jacks, clapping games, counting-out rhymes and miscellaneous games. The verbal accompaniments to the games range from long life-cycle epics such as that used for the hand-clapping game 'When Suzy was a Baby' to the short, sharp verse used for the rhythmic jump game 'The Cat's got the Measles':

> The cat's got the measles,
> The measles, the measles;
> The cat's got the measles,
> The measles got the CAT!

The aims of Lindsay and Palmer's extensive study were twofold; first, to examine the structure and elements of children's spontaneous games in order to compare these traditional games with the formal school games syllabus, and second, to provide 'a descriptive record of children's games in Brisbane during 1975–1976'. Their analysis and conclusions are of great interest, particularly in so far as they found that in many important respects, traditional spontaneous games were of greater benefit to children both physically and socially than the formal school syllabus games.

Lindsay and Palmer were lecturers in physical education, and workers in other disciplines as well as folklorists *per se* have shown interest in Australian children's folklore (see, for example, musicologist Dr Hazel Hall's entry **Musical and poetic characteristics of children's folklore**).

One of the earliest records of children's games as played in Australia is from Lieutenant-Colonel Godfrey Charles Mundy's *Sydney Town 1846–1851*. Mundy affectionately referred to 'the minor games of the minor people' which 'come round in their seasons', and described hoops, peg-tops, marbles and cricket played in the streets as a hazard to horse-riders or passers-by.

Sir Joseph Verco's *Early Memories 1860–1870*, a manuscript held in the South Australian Library, includes some detailed recollections of 'School Games, 1860–67'. Verco writes about marbles:

> The boys were addicted to 'playing marbles' . . . In those days they were stocked in the shop windows and were a considerable article of commerce and could be heard to rattle in the trousers pockets of the lads. They were very varied. There were five (sic) different genera — 'agates', 'glass allies', 'stoneys' and 'commonies'.

Verco describes in detail the rules for playing a game called 'nucks' or 'nux' which involved firing a 'taw' into one of three holes in the ground, and other games called 'in the ring', 'in the hole', and 'on the line'. He also gives a detailed description of playing tops, where 'the top par excellence was known as the "gummy". It was turned out of a piece of the heart wood of a South Australian red gum tree and was very heavy'.

Verco, his younger brother and their friends played cricket and rounders in an empty paddock ('the acres') in the neighbourhood of their home. These games were not played at school, as there was not enough space. Some of the games played at school 'which the confined space would allow' were hopscotch or 'hoppy', 'cockfighting' (boys pick-a-back, attempting to push off an opponent's 'passenger') and High Cockalorum, a version of Buck Buck described earlier.

In the *Bulletin* of 26 February 1898 Victor Daley wrote about the songs and games of some girls heard from the 'open window of a cottage in a quiet suburban street'. He gives some of the words of ring games such as 'See the Robbers Passing By', 'In and Out the Window'. 'Poor Alice Weeping for Her Sweetheart' and 'Here come two Dukes a-riding':

> Here come two Dukes a-riding,
> A-ree, a-row, a-riding,
> Here come two Dukes a-riding,
> To greet your daughter Jane.

Circle or ring games of this kind are one of the few types of traditional games which have largely disappeared from children's spontaneous play during the twentieth century, although they were still played before the First World War. Those which are still played have largely been annexed by adults as party games such as 'drop the hankie' or school syllabus games.

Children's string games or 'Cats' cradles' are an ancient tradition still widely practised in Australia. It is not known whether nineteenth century white Australian children learned string figures from Aboriginal sources where contact existed, although string games and stories were common in Aboriginal communities and are well documented (as in Lyn Love's article in the *Anthropological Society of Queensland's Newsletter* 141, 1963). Some similar string figures, as well as the practice itself, appear in widely separate nations and cultures such as those of Australian Aborigines, Canadian Inuit (Eskimo), Maori, South American, Fijian and many other Pacific Rim peoples.

The most comprehensive overview of Aboriginal children's traditional play is included in June Factor's *Captain Cook Chased a Chook* (1988). She comments on the wide variety of descriptive material concerning Aboriginal life, including play, dating from the earliest years of European settlement, although advises caution in assessing its validity. She includes Lyn Love's 1983 categories of Aboriginal children's play, namely imitative play and make-believe; finger games; memory and guessing games; play with natural materials or animals; throwing, catching, chasing and finding games; games with toys; singing and dancing games; story-telling play; and verbal play. Love had noted that many of these games were seasonal, as in other cultures, and Factor comments that Love's categories are 'familiar to students of child life in many parts of the world'. Despite the similarities in types of play, the materials used by the Aboriginal children reflected what was locally available — sand, grass, bark, mud, gourds and other vegetation. Children in contact with Europeans frequently used manufactured materials for play. In a 1989 interview for the Children's Museum in Melbourne, Sandy Atkinson described trains and other wheeled toys made by children from food tins and wire, at the Cummeroogunja government reserve in the 1950s. There has been little study of the play of Aboriginal children living in urban environments, although the Australian Children's Folklore Collection at the University of Melbourne has received private funding to begin such a study.

The first attempts to study the traditional play of immigrant children and immigrant influences on Australian children's play were both carried out in Victoria. For her B.Ed. honours dissertation at Deakin University, Margaret Clark (1981) conducted a number of observations in school playgrounds in the city of Geelong, and found evidence of some non-English hand-clapping and skipping games and rhymes, some used by a group of girls of mixed nationalities. In 1982 at the Institute of Early Childhood Development in Melbourne, some postgraduate students made a video-tape of newly-arrived immigrant children at the Eastwood Language Centre. Games such as elastics and marbles as played by the Chinese, Vietnamese and Cambodian children were similar to those of Australian children, although there were some differences in both rules and techniques; hence the title of the videotape *Just a Little Bit Different*. Here too, some Australian children were incorporating some aspects of the immigrant children's games, in particular the spectacular 'high jump' in elastics, learnt from the Indo-Chinese girls. Elastics is a game where a long loop of household elastic is hooked around the legs of two children (usually girls) while a third executes jumps of increasing complexity in, out of and with the elastic. It is of particular interest in that it is one of the few children's games which cannot be identified in Australia earlier than the 1950s. Researchers in that period sometimes described the game as 'Chinese skippy', and it is often known in the United States as 'Chinese jump-rope' (e.g. in Knapp and Knapp *One Potato Two Potato*, 1976, p. 118). This terminology plus the familiarity with the game by Australian immigrant Indo-Asian children have led to speculation that it originated in Asia: certainly it has been observed in mainland China in recent years. The mystery is, however, compounded by the Knapps' note that 'Chinese jump-rope' is called 'Indian jumping' in parts of Philadelphia.

In 1984 Heather Russell carried out an intensive ethnographic study of one inner-city playground in Melbourne, a study published as *Play and Friendships in a Multi-Cultural Playground* (1986). This study took place over a six-month period, and involved observations, interviews, photographs and video- and audio-tape recordings. The school could be described as 'high-rise, high-migrant, high-turnover' and was called 'Hightown' in the study. Some 82 per cent of the children were from non-English-speaking backgrounds and 95 per cent lived in the high-rise Ministry of Housing flats adjacent to the school. The largest ethnic group was Turkish, closely followed by children from Vietnam (both Chinese and Vietnamese). Russell found, among other things, that girls preferred to play in monocultural groups; that verbal rhymes etc. accompanying games were very limited; that prejudice did exist in the playground and often took the form of excluding unwanted children/ethnic groups from a game; that game players (both boys and girls) were less likely to be involved in playground fights than other children; and that many children had adopted the 'Chinese flick' when playing marbles, and most used a counting-out ritual called 'Aw Sum' which had tenuous links with Turkish, Vietnamese and Chinese language and rituals. The intriguing ritual 'Aw Sum' is played thus: children stand in a circle and chant 'Aw Sum', putting one hand in the middle on the word 'Sum', either palm down or palm up. Those in the minority are 'out', and the ritual continues until one or two players are left. Two players 'play off' using another ancient and widespread tradition, 'Scissors, Paper, Stone'. 'Aw Sum' appears to be idiosyncratic to the Hightown School playground.

Folklore for children

Folklore for children includes nursery rhymes, fairy tales, jokes, cautionary tales and modern legends, family games, baby games, lullabies, rituals to do with Father Christmas and the Tooth Fairy, and traditional reprimands and evasions such as the time-honoured 'wigwam for a goose's bridle' (a common retort to children who ask inconvenient questions such as 'what are you making, Mum?'). All these forms of folklore for children are practised by adults for children's amusement or edification. Like children's own playground games, this type of folklore may be of great antiquity (see **Nursery rhymes**, **Folk tales** and **Women and folklore**). Two compilations of family sayings used in Australia, Davey, *Snug as a Bug!* (1990) and *Duck Under the Table!* (1991), include examples from non-English sources as well as a large selection of English-language sayings. There is a surprising similarity in the jocular and sometimes sharp parental retorts to children's persistent questions such as 'What's for dinner, Mum?'. A Greek mother might say 'My liver and kidneys!' and a Croatian mother 'Cakes with honey', whereas some English answers to 'What's for dinner?' are 'Duck under the table' or 'Bread and scrape'. There has been little scholarly attention to this interesting dimension of children's folklore, with the notable exception of Widdowson's 1977 study from Canada, *If You Don't Be Good: Verbal Social Control in Newfoundland*. Widdowson's work concerns mainly threats such as 'If the wind changes [while you're pulling that face, etc.] you'll stay that way'. This threat is also common in Australia, as also are warning rhymes such as:

> Don't care was made to care,
> Don't care was hung;
> Don't care was put in a pot
> And made to hold his tongue.

Traditional family sayings, however, have many other functions besides threats and warnings. In addition to sheer amusement and enjoyment of verbal display, other identifiable functions include evasion ('How old are you, Gran?' 'As old as my tongue and as young as my teeth'), maintaining adult mystique ('little pitchers have big spouts'), tension reduction ('What do you think this is — bush week?') and training in etiquette ('All joints on the table will be carved').

The great diversity of folklore for children in community languages as well as English became apparent through a two-year study in Melbourne in 1976 and 1977, ultimately known as the Multicultural Cassette Series (Davey 1979). The Australian Government Interim Children's Commission (now Office of Child Care, Department of Social Security) commissioned cassette tapes of songs, stories and rhymes for young children in what were at the time some of the most significant community languages spoken in Australia, namely Italian, Greek, Spanish, Turkish, Serbian/Croatian, Macedonian, Arabic and English. Essentially, the aim of the project was to provide some material which monolingual English-speaking teachers could use with young immigrant children, both to encourage maintenance of the family language and as a respite for the children from bombardment with English.

Although the Multicultural Cassette Series did not begin as a folkloric project, the shortage in Australia in 1976 of published material for young children in languages other than English led to an extensive programme of field recording among immigrant parents in Melbourne. Much of the material they supplied was traditional folklore, such as the following translation of a Turkish dialogue 'Komsu Komsu':

> Has your son come back? *Yes.*
> What has he brought? *Pearls and beads.*
> To whom? *To you and me.*
> And to who else? *The cat.*
> Where is the black cat? *He climbed up the tree.*
> Where is the tree? *The axe cut it.*
> Where is the axe? *It went to the mountain.*
> And where is the mountain? *It burnt down and finished.*

An interesting finding from this project was the high degree of similarity in the values transmitted through this folkloric material, in all the languages recorded. Apart from a greater emphasis on religion in the Italian and Spanish items, most dealt with family life, animals, wisdom and foolishness and humour. Many immigrant parents appeared to be selecting the traditional folklore which they passed on to their children, choosing themes which might assist their life in a new society, themes such as resourcefulness, humility, knowledge and realism.

Folklore about children

This is sometimes expressed in proverbs like 'Spare the rod and spoil the child' or 'Children should be seen and not heard'. Lest these axioms seem obsolete today, it is relevant to note the results of a Saulwick Age Poll published in the Melbourne *Age* on 13th July 1989: 66 per cent of those polled supported the right of schools to cane or strap children, and 70 per cent agreed that it did not hurt a child to be smacked from time to time. Only 7 per cent said that parents should never smack their children.

Some other folklore about children is commonly called 'old wives' tales' and expresses beliefs such as that 'If you tickle a baby's feet you'll make it stutter'. Beliefs about birth, pregnancy and infancy form a large part of such folklore about children, and exist in many different ethnic groups (see **Women and folklore**). The following were collected in Victoria in 1981 and 1982 and are now housed in the Australian Children's Folklore Collection at the University of Melbourne:

> Don't let a baby look at itself in the mirror or it won't grow properly.
>
> Cutting a baby's fingernails will make it stutter when it starts to talk.
>
> A fall into water while pregnant will make a woman have a cross-eyed baby. (Source: Anglo-Australian)
>
> Don't touch a baby on the soles of its feet or it won't be able to cross a bridge when it's old. (Source: Chinese)
>
> Old people shouldn't look at the baby before the Christening. (Source: Ukrainian)

Other folklore about infants involves a confusing array of beliefs and practices for predicting the sex of the unborn infant, including 'carrying high' or 'carrying low', and dangling a wedding ring or laundry peg on a string over the abdomen to see if it turns left (or right).

There is some other folklore about children expressed as common clichés rather than as traditional proverbs, phrases such as 'children are cruel', 'children don't feel the cold', and 'babies don't feel pain'. Like all forms of children's folklore, these folk beliefs are worthy of scholarly study, as many provide excuses for inertia or irresponsibility.

Children as apprentices in folklore

Some of this learning is through observation and imitation, but a great deal is through direct instruction. Children may be taught the family-approved methods of lighting a fire, washing dishes, hanging washing on the line, cooking, table manners, behaviour with adults and so on. Their apprenticeship may also include more serious aspects of adult folkways, and may include circumcision, first communion or confirmation, all rituals which may involve much preparation, ceremony and special head-dresses or costumes. Through such activities, children are inducted into the folk beliefs and values of their parents and their parents' significant group memberships.

In the publication *Blue Bags, Bloodholes and Boomanoomana: Folklore of the South-western Riverina* (1988), four Year 12 students from Finley High School (NSW) presented many examples of children's 'apprenticeship' into adult folkways. Sonia Ryan described some adult medical remedies known to her fellow students, such as ice, cut tomatoes and cold tea for burns and sunburn; garlic to 'keep the blood clear', and dandelion juice to get rid of warts. The Finley students also contributed some local weather beliefs, such as that ants actively running around or cows lying down are indicators that rain is coming; and conversely, that finding a yabbie in a waterwheel while irrigating is a sign of impending drought. This collection of regional folklore was supervised by Finley High School teacher, John Marshall (q.v.), who teaches folklore studies to Years 11 and 12. The 'Blue Bags' are a common treatment for bee stings, the 'Bloodholes' are used to catch the blood when a sheep is slaughtered in the traditional way, and 'Boomanoomana' is a place in the district associated with a local legend about the bushranger Ned Kelly (q.v.). These pieces of local folklore, incorporated in the title of their publication, were all known to and collected by the Finley students.

No discussion of children's folklore in Australia could be written without a tribute to the American Fulbright scholar Dr Dorothy Howard (q.v.). Her visit and field collecting in Australia in 1954–55 provided the impetus for a great deal of subsequent activity with regard to children's folklore. The work of a number of Australian collectors has also been of great importance, notably that of Alan Scott (q.v.), John Meredith (q.v.), Wendy Lowenstein (q.v.) and Ian Turner (q.v.). The first academic compilation of Australian children's folklore was Turner's *Cinderella Dressed in Yella*. Despite Turner's status in the History Department of Monash University, the publication of *Cinderella* in 1969 created somewhat of a furore when it was for a time banned from transmission through the Australian post on the grounds of obscenity. Vulgarities such as the ubiquitous 'Fat and Skinny' rhymes were apparently too much for 1969 sensibilities:

Fat and Skinny went to war,
Fat got shot by an apple-core,
Skinny went home to tell his mum,
All he got was a kick up the bum.

or

Fat was in the toilet,
Skinny was in the bath,
Fat let off an atom bomb
And made Skinny laugh.

Vulgar rhymes were only a part of the varied contents of *Cinderella*. Wendy Lowenstein later produced a specialist booklet called *Shocking, Shocking, Shocking* (1974):

Shocking, shocking, shocking,
A mouse ran up my stocking;
He got to me knee
And what did he see?
Shocking, shocking, shocking.

Ian Turner divided the contents of *Cinderella Dressed in Yella* into Rhymes for Games (Counting-Out, Skipping, Handclapping, Ball-Bouncing, Other Games) and Rhymes for the Playground (About School, War Cries, Taunts and Insults, Practical Jokes, Charms and Divinations, For Amusement Only). The book also contained a reprint from the *Bulletin* of 1898 of an essay on song-games in Sydney by Victor Daley, with comments from some other *Bulletin* correspondents, and a definitive essay by Ian Turner entitled 'The Play Rhymes of Australian Children'.

In 1978 a second edition of *Cinderella Dressed in Yella* was produced by Ian Turner, June Factor and Wendy Lowenstein. As well as the field collections made by the compilers, it included a great deal of material collected by June Factor's students at the Institute of Early Childhood Development in Melbourne. In 1979 June Factor and Gwenda Davey put together the field collections for *Cinderella* and the Multicultural Cassette Series to begin the Australian Children's Folklore Collection (q.v.).

See also: **Australian Children's Folklore Collection**; **HOWARD, DOROTHY**; **Musical and poetic characteristics of children's folklore**; **Nursery rhymes**.

Gwenda Beed Davey

Chinese folklore in Australia The Chinese have had a long presence in Australia, dating mainly, but not exclusively, from the 1850s when large numbers joined people of many other nations in search of gold on the eastern goldfields. While the pattern of single men working in Australia for a few years before returning, wealthier they hoped, to the family in China (mainly in Guang Zhou, previously Canton) was not conducive to permanent settlement, over time substantial Chinese communities developed across Australia. These were mostly in and around large urban areas, concentrated particularly in market gardening, laundries, restaurants and other service industries. The characteristic and colourful 'Chinatowns' of most large Australian cities are part of Australia's heritage from this period.

Since the 1970s substantial numbers of ethnic Chinese from south-east Asia have settled in Australia. Often many generations removed from direct contact with the Chinese mainland and its culture, these peoples nevertheless retain strong links with China, usually thinking of China as 'the motherland'. While there may be some tensions between the older-established Chinese Australians and the more recent arrivals, the sense of tradition is extremely strong amongst most Chinese Australians, whether descended from families resident in Australia for some generations or coming more recently from various parts of south-east Asia. Although overall numbers of Chinese in Australia have waxed and waned, and it is difficult to establish exact numbers with any certainty, it has been suggested that approximately 200 000 Australians are of direct or indirect Chinese descent. While numerous variations can be observed in the different national and geographic experiences of these Chinese Australians, there is a remarkable consistency of belief and observance.

The twelve-year lunar calendar is the basis for a good deal of the most significant Chinese folklore, determining the major celebrations and festivities of the Chinese year. These festivals are both of a public nature and of a more private, familial nature, involving various kinds of commemorations for the living and the dead, according to Buddhist or Confucian practice, which place great emphasis on the veneration of ancestors. Some of the better-known and most significant Chinese folk customs are *Chun Jie*, the Spring Festival (usually known as 'Chinese New Year'), and *Zhonqui Jie*, the mid-autumn Moon Festival. These customs generally involve public spectacles, such as the Dragon Dance, fireworks (though these are frequently banned these days), costume, music and dance. One of Australia's oldest dragons appears annually at the Bendigo Easter Festival. Since the early 1980s *Duanwu*, the Dragon Boat Festival, and its accompanying boat races have been revived in many capital cities and are attended by Chinese and non-Chinese Australians alike, as are the other public celebrations mentioned.

In common with folk customs of all national and ethnic groups, Chinese customs are complex combinations of belief, artistic expression (song, dance, music), costume, foodways, celebration and calendric observation. The Spring Festival, for instance, is spread over many days and nights, involving the public displays mentioned above, eating out with family and friends, often accompanied by traditional Chinese entertainments, such as noodle-making, conjuring, gambling, music and dance. Private family observances are also carried out during this period, including decoration of the house with lanterns and devices for bringing good luck in the coming year and Buddhist religious observances at the household altar, involving offerings of fruit and incense. Gifts of money are given to children. The Chinese calendar contains many other festivals, such as *Yuan Xiao Jie*, the Lantern Festival, and *Quingming Jie*, the Clear and Bright (or Pure Brightness) Festival. These may or may not be observed, depending upon the background of particular groups of Chinese.

Folk beliefs concerning good and bad luck are strong in Chinese tradition, and may determine whether or not a particular house is purchased, whether a job or journey is taken and so on. The colour red is symbolic of good luck, health and prosperity, and features heavily in Chinese costume and decoration. At Spring Festival children are traditionally given gifts of money wrapped in red paper, an example of the way in which Chinese folklore often manages to combine the symbolic and the practical. Methods for predicting the future are also an important aspect of Chinese folk belief, including astrology and the use of such esoteric techniques as those associated with the I Ching.

Chinese food has long been an established aspect of Australian culture. Chinese-Australian food is based mainly on Cantonese-style cooking, though it has been considerably Australianized over time, the average 'Chinese takeaway' bearing little resemblance to the superb cuisine of Guang Zhou.

An important element of Chinese life is traditional medicine and related techniques, such as acupuncture. Traditional Chinese medicine generally makes extensive use of natural products, such as herbs and other plants, many of which are subjected to various processes believed to increase their efficacy.

The extended family group, in combination with the lunar calendar celebrations and whichever dialect of Chinese is spoken, are the most fundamental determinants of Chinese folk tradition. Respect for elders, for tradition and for the concept of China, 'the motherland' herself, are strongly emphasized. One particularly powerful focus is the once-dominant figure of 'the Kitchen God', a protector of the home and family and divine reporter of the family's behaviour. Traditionally, each house would have a shrine to the Kitchen God, upon which various offerings would be made, particularly near the time of the Spring Festival. While such physical manifestations of the centrality of the family are no longer common in Australia, the extended family remains the fundamental reality of Chinese culture and folklore.

See also: **Ethnic folk heritage**; **Folk medicine**.

Feng Wei; Lin Zesheng; Graham Seal

Collectors and collections of folklore The collection of folklore is an essential element in the work of the folklorist. The language, customs and other practices and expressions that go to make up the folklife of a particular community or group must be obtained by various ethnographic techniques, archived and, if appropriate, made available for study, for analysis, for comparison with other folklore collections, and, perhaps, for dissemination. Folklore collection in Australia has generally been haphazard and spasmodic, depending on the work of usually unfunded individuals and voluntary organizations. A number of official institutions such as libraries and museums have, as an accidental effect of their central activities, accrued folkloric materials of various kinds and in varying volume. Because of the lack of co-ordinated collecting activity, we have only selective knowledge of Australia's folk heritage and of its contemporary folkloric expressions and activities. Nevertheless, some strong collections do exist.

In the area of rural folk song, music and dance the John Meredith (q.v.) collection at the National Library of Australia and also at the National Film and Sound Archive, is generally considered to be the most detailed and most important, particularly as it includes material collected in the 1950s, as well as material collected during the 1980s and early 1990s when Meredith returned to active fieldwork. Other significant collections of such material have been made by Chris Sullivan, Robert Michell, Stan Arthur, Norman O'Connor, Mr and Mrs Lumsden, Hugh Anderson (q.v.), Wendy Lowenstein (q.v.), Ron Edwards (q.v.), Warren Fahey (q.v.), Bill Scott (q.v.) and Alan Scott (q.v.), among numerous others.

Collections of mainly non-rural folklore have also been made, though to date have not attracted as much sustained interest as the collection of bush song, music, dance and recitation. Archives containing such materials include the Australian Children's Folklore Archive at the Australian Centre, University of Melbourne, and the West Australian Folklore Archive at the Centre for Australian Studies at Curtin University in Perth. Most collecting institutions of the states, territories and the Commonwealth hold folklore materials as part of their general collections, as do many regional and local history centres and museums, not to mention the holdings of numerous private individuals. As well, the Australian Broadcasting Corporation holds extensive folklore materials in its archives, largely as a fortuitous by-product of its documentation of Australian life in general.

There is a clear need for some form of national register of significant folklore collections and holdings to document folklore heritage for research, for education, for entertainment, and for economic activities such as cultural tourism and related leisure pursuits.

A number of individuals are known to have collected Australian folklore in one form or another, at various times and in various locations. Some did so as a by-product of other endeavours — for example, in the course of editing a newspaper column which included invited contributions or of undertaking anthropological or literary research. Those who have published or in some other way publicly presented their collected works (there are many others who have contributed their collections to local libraries and archives, or who may have collections in their possession) include: Alice Moyle (q.v.); Catherine Berndt; Ronald Berndt; Daisy Bates; Vance Palmer; Will Lawson; Roland Robinson; John Greenway (q.v.); Dave Johnson; Joseph Jacobs (q.v.); Rev. Percy Jones (q.v.); George H. Haswell; K. Langloh-Parker; A. B. Paterson (q.v.); A. V. Vennard ('Bill Bowyang') (q.v.); Brad Tate (q.v.); Colin McJannet; Maryjean Officer; Joy Durst (q.v.); John Manifold (q.v.); Bob Rummery; Mark Rummery; Graeme Smith; Thérèse Radic (q.v.); Peter Parkhill; Chris Woodland; Roger Montgomery; Brian Crawford; Dave De Hugard; Lloyd Robson; Russel Ward (q.v.); Barbara Gibbons; John Marshall (q.v.); Hazel Hall; Kenneth Goldstein (q.v.); Kel Watkins; Graham Seal (q.v.); Gwenda Davey (q.v.); June Factor (q.v.); Helen O'Shea; Jeff Corfield; Chris Buch; Ian Coggins; David S. Hults (q.v.); John Watson; Sue Watson; Peter Ellis; Bill Wannan (q.v.); Bill Beatty (q.v.); Sydney Baker (q.v.); Bill Hornadge; Geoffrey Ingleton; Edgar Waters (q.v.); Ian Turner (q.v.); A. W. Reed; Alex Hood; Gary Shearston; A. L. Lloyd (q.v.); Nancy Keesing (q.v.); G. A. Wilkes; W. H. Downing; Eric Partridge; Frank Clune; Frank Hardy (q.v.); John Lahey; Lionel Gilbert; Stephen Murray-Smith (q.v.); Don Laycock; R. Eagleson; J. S. Gunn; Rob Willis; Stathis Gauntlett; Anna Chatzinikolaou; Geoff Wills; Shirley Andrews (q.v.); Margaret Walker (q.v.); Phyl Lobl (q.v.); and Jill Stubington (q.v.).

Colonial Dancers, The, were formed in Melbourne in 1974 and grew out of a need for demonstration sets at public dances in the days before 'bush dancing' (see **Bush dance**) became as well known as it is today to both bands and the general public. Its general aims were to promote Australian dancing by displays and by collecting, recording and teaching dances and thereby fostering a knowledge and enjoyment of dancing as it was done in the nineteenth and the early twentieth centuries. The group promotes traditional social dancing from the very simple 'bush dances' to the more complex quadrilles and couples dances which were so popular in the 1800s.

The Colonial Dancers are one of the main groups trying to broaden the dance repertoire and improve the general standard of social dancing in Victoria. They have developed considerable expertise in organizing and running bush dances and colonial balls, teaching classes and workshops, and encouraging new dance composers, callers and dance musicians. The Colonial Dancers are also a well-known display team, giving performances of traditional Australian dancing in a theatrical form as a way of making its existence better known and appreciated. They have performed at all the major events/venues in Melbourne and are the only Australian dance group to have performed at the Edinburgh Military Tattoo.

Their current activities include conducting weekly classes, a monthly social and several colonial balls each year. They also organize workshops, dance teachers, callers and dance bands as well as dance displays.

Colonial or heritage dance These general terms have been adopted to cover the traditional social dances done in Australia from 1788 to about 1900. They include those English country dances and folk dances that were danced here in the earlier period; but the major part of our dance tradition was in the new dance styles of the nineteenth century. These were the quadrilles or 'sets', such as the First Set, Lancers, Alberts, etc., and the closed-couple dances, such as the waltz, polka, Varsoviana, etc. There were also dances which combined some aspects of folk dance with the new steps and figures. The dance we now know as the Waltz Country Dance, the Dashing White Sergeant, and the Tempest are examples of this type of nineteenth-century dance.

Many of the couples dances from the 1850s, i.e. Varsoviana, polka, mazurka and Schottische were what we now call sequence dances. In these a sequence of steps, usually to four, eight, sixteen or thirty-two bars of music, are repeated continuously. This style became popular again later that century and early this century with the barn dance, Veleta waltz, Boston two-step and many others. As so many of these, including the Pride of Erin from the 1920s, are sufficiently similar in style to the true colonial dances, they are often included on the programmes of colonial balls today, except for those functions that aim to re-create the colonial period exactly.

Many of the couples dances have survived longer than the sets of quadrilles in country districts, although their footwork has frequently been adapted to the modern dance style with feet parallel and longer steps with the heel leading. Some examples of Australian sequence dances from the colonial days are the Princess polka, still danced in several areas of Victoria and South Australia, and the Manchester Galop, Uncle Ev's barn dance and an unusual version of the Berlin polka, all of which are danced only at Nariel and Corryong in north-eastern Victoria. The Lancers, Alberts and Waltz Cotillon were still danced in some areas in recent times.

The quadrilles and the couples dances both developed many versions and variations in Australia. Now that many of these dances are being revived this variety is being repeated, with different groups reviving their favourite versions.

We no longer have a collective name for our traditional dances of the nineteenth century. In that century they were always known as 'Ball Room' or 'Ball-room' dances, as shown in the many instruction books published then. In the 1920s this name was taken over by the modern 'ballroom' dances, and earlier dances were referred to as 'old time'. Around the late 1930s, the name 'new vogue old time' appeared and now describes a more flamboyant style based on more modern steps.

See also: **Dance**; **Quadrilles**.

REFERENCES: S. Andrews, *Take Your Partner*, 1979; S. Andrews and P. Ellis, *Two Hundred Dancing Years*, 1988.

Shirley Andrews

Committee of Inquiry into Folklife in Australia, The, In 1985 the then Minister for the Arts, the Hon. Barry Cohen MP, invited a working party under the chairmanship of Hugh Anderson to provide definitions and draft terms of reference for an inquiry into folklife in Australia. Subsequently the Minister commissioned Hugh Anderson (q.v.) as chairman and Dr Keith McKenry (q.v.) and Gwenda Beed Davey (q.v.) as members of the Committee of Inquiry. The Committee was asked to report upon:

1 The nature, diversity and significance of Australian folklife.
2 Existing institutional and other arrangements for safeguarding Australian folklife, and the need for new arrangements, having regard in particular to: (a) collection, documentation, conservation and dissemination of folklife materials; and (b) support for the practice and development of folk arts.
3 Other measures appropriate to the proper safeguarding, dissemination and appreciation of Australian folklife.

The Inquiry was sponsored by the then Department of Arts, Heritage and Environment, the National Museum of Australia, the Australia Council and the former Australian Institute of Multicultural Affairs.

During 1986 and 1987 the members of the Committee travelled extensively within Australia and overseas, holding meetings and consultations with interested persons and with government and cultural institutions. Two hundred and forty-five written submissions were made to the Inquiry, something of a record for inquiries of this type. The Committee's report was presented to the Australian Government in August 1987 and was published as *Folklife: Our Living Heritage* (1987). The 300-page report included fifty-one recommendations, of which the principal ones concerned the establishment of:

- an Australian Folklife Centre, to provide a national focus for action to record, safeguard and promote awareness of Australia's heritage of folklife;

- an Australian folklife grants scheme, to be administered by the Australian Folklife Centre, to support urgently needed folklife collection, research and documentation, and the maintenance of traditional arts and craft skills within communities;
- a national collection of Australian folklife, under the control of the Australian Folklife Centre, to give identity and stimulus to the development and conservation of folklife materials within national collections;
- a Folk arts committee within the Australia Council, to provide needed expertise and advice on folk arts support; and
- a Folk arts grants programme, under the control of the Folk Arts Committee, to provide more adequately within the programmes of the Australia Council for the particular needs and circumstances of the folk arts.

Following the publication of the Report, the then Minister for the Arts, Mr Gary Punch MP, called for written responses to the Report's content and recommendations. The Australian government has made no official response. The Folklife Inquiry has nevertheless generated considerable interest and activity within Australia. Folk organizations and other interested persons have continued to press for implementation of the Report's recommendations, and in 1991 the Australian Folk Trust established the Australian Folklife Centre (q.v.) which is housed at the National Museum of Australia's Mitchell offices in Canberra. In 1990 the Victorian Folklife Association was formed by about thirty organizations with the aim of establishing an Australian Heritage and Folklife Centre in Victoria.

Concertina Magazine began with issue no. 1 in winter 1982, and is published quarterly under the editorship of Richard Evans and with the production skills of John Ramshaw. The magazine caters for players and researchers of the concertina in particular, but also of other free-reed instruments (see **Musical instruments**).

Contemporary legends: see **Modern legends**.

Country halls As the focal point of a district's social life, at which communal celebrations take place, the hall is a centre for folklife. Its role has diminished considerably since the 1960s brought, on the one hand, better access to towns and cities, and on the other, home entertainment such as television and video. Yet even in districts where the severely depleted population can scarcely support communal functions, people still regard the hall as essential to their sense of community identity.

What is the history of these halls? A close look at the building might reveal the official title of Mechanics' Institute or, in New South Wales, School of Arts, in deference to the movement imported from Britain during the nineteenth century which sought to foster a more highly skilled and educated workforce by providing technical education for the working man. In Australia, mechanics' institutes were established in the major population centres from 1827, and the movement flourished until the depression of the 1890s, by which time there were over four hundred mechanics' institutes in Victoria alone. Initially, the aim of their middle-class trustees was to promote the intellectual, social and moral improvement of the working man, by means of lectures and instructive books. Although this scheme met with little success, the middle class continued to run the institutes for their own recreation and education. From the late 1850s, the institutes' mission was to provide wholesome recreational facilities, particularly for the 'democratically inclined multitude' on the goldfields. During this period audiences for lecture programmes dwindled, but libraries and especially reading rooms were still in great demand.

By the end of the century, however, the mechanics' institutes had been appropriated for more popular pursuits such as billiards, dancing and card-playing. Reference libraries and reading rooms were superseded by circulating libraries issuing popular fiction. These changes in the role of the institutes reflect the changed social profile in rural areas following the Selection Acts of the 1860s. Where the large pastoral runs and the goldfields had supported a predominantly male population, the selectors were mainly families. As farming communities emerged in newly settled districts, the public buildings they erected were a reflection of family priorities: schools, churches and halls. The mechanics' institute became the district or public hall.

These halls were built with some assistance from government land grants and subsidies, particularly in the 1880s, but most of the funds, and the labour, were provided by local residents. Settlers in newly opened-up rural areas continued to build public halls well into the twentieth century, and in 1972 over three hundred were still in use in Victoria. Local historians have noted the central place of the hall in the life of a rural community, recording with pride and pleasure both the co-operative work that maintained it and the communal activities held there.

The buildings themselves reveal something of a district's affluence and aspirations. On the goldfields, imposing brick and stucco edifices with neo-classical façades reflected the pride of the newly prosperous. The selectors, on the other hand, were more likely to erect simple structures, most often a single-room weatherboard hall with a gabled roof of galvanized iron.

In the 1920s, as roads and transport improved and the popularity of dancing brought large crowds to dances, many halls were extended or rebuilt. After each of the world wars, more were built or renovated as memorial halls, assisted by a combination of patriotic fundraising and generous tax concessions. Where previously the billy had been boiled for supper on a fire outside, the more modern hall came with a supper room and kitchen built on. In most cases, the library became less important, and was finally abandoned. The building was lit by kerosene lamps, later by pressure lights like the 'Gloria' type, before electricity arrived, in some rural areas as late as the 1970s.

These improvements changed the social character of the hall considerably, particularly at large social gatherings such as dances. In the earlier days the simple structure encouraged movement between the hall and the world outside. Lanterns were brought in from sulkies to light the hall; supper was prepared outside; lavatories were at a distance from the building. As the size and facilities of halls increased, they were divided up into separate spaces with different functions: the supper room, the meeting room, the cloak rooms, the kitchen, the box office. The intimate atmosphere diminished further as the soft light of kerosene lanterns gave way to harsher electric light. At dances, people who were not dancing were separated according to their roles: women in the kitchen, supper room or cloak room, a man in the box office, the master of ceremonies and the band on the stage. While it was still acceptable for the men to wander in and out, respectable women were expected to stay inside all night.

If a settlement grew into a town, many of the organizations which used the hall would build their own meeting places. Country towns today still have these band halls, church halls, halls for the Country Women's Association, the Scouts, the Returned Services League, the bowling club, the racing club and the Masonic Lodge. In less populated rural areas, however, there was usually only one hall at the heart of a district. It was used for official functions: as a polling booth, a church, a temporary school; for meetings of civic organizations such as the progress association and fire brigade; and by sporting clubs and charities for fundraising activities which included dances, concerts and balls. At all these public gatherings, everyone in a district was literally on common ground, and almost always people were invited along by public advertisement, rather than by private invitation.

The country hall was also an important site in the community's folklife. Rites of passage, courting rituals and 'Christmas trees' all happened in the hall. This last custom appears to have developed out of the end-of-year school concert, which began in the later years of the nineteenth century. Organized by the teacher, the concert demonstrated the pupils' skills in recitation, exercise drills, dramatic sketches and choral singing. Afterwards, prizes were bestowed upon students whose conduct, attendance, and academic achievement had been exemplary. The school concert was held in the evening, and all the children in the district went, including those too young for school. Their parents and grandparents, older brothers and sisters, aunts and uncles and neighbours went, as well as the people from the store and post office. That amounted to most of the population in a district settled mainly by families.

December is a busy time of the farming year, and the concert and Christmas tree may provide the only opportunity for people of a district to get together for Christmas. For the children, the Christmas celebration is also the beginning of their six-week release from school.

Over the years, the school concert's celebration of the benefits of formal education was transformed progressively into a folk custom, but as the small rural schools have gradually disappeared so have the school concerts. The communal Christmas tree, on the other hand, remains as an annual celebration for country children.

Dancing was overwhelmingly the most popular form of social recreation in rural Australia (see **Dance**) and accompanied almost any social gathering, from card parties to picnic races, 'welcome homes' for returning soldiers, or farewells for people leaving a district. The local dance was a regular event in country halls between the 1920s and the 1960s. By offering opportunities for young people to mix socially with potential partners from their own or nearby districts, it played an important role in courtship. When individuals recall their youth in the rural Australia of this period, and when local historians record the history of their district's hall, it is these local dances which are most potent in their memories. They recall the anticipations of courtship, the pleasures of the feast, and the excitement of transgressing the everyday order.

Inside, the hall was decorated with flowers, streamers or gum branches, and the floor cleaned and waxed for dancing (a common method was to drag across the floor a bag of sawdust impregnated with kerosene, then sprinkle the clean floor with candle shavings).

The dance was supervised by adults, who would prepare the hall, sell the tickets, provide supper and clean up. Money taken at the door paid for the band and the hire of the hall; the rest would go to the organizing group, usually the hall committee, a sports club or a church organization. Ticket stubs were later drawn for a 'door prize', and there might also be a raffle. Prizes might be a basket of produce, or pieces of linen or china, which would end up in a young woman's glory box. Smaller prizes, such as a bar of chocolate, were awarded in novelty dances, and there were also prizes for waltzing competitions. Once the dance began, the master of ceremonies would ensure that dancers took the floor at the right time, that women were not left without partners, and that rowdy behaviour was quashed. The women kept an eye on the girls, noting any who were audacious enough to leave the hall during the evening.

On arriving at the dance, the young men took up their positions around the door, where they could get a good look at new arrivals, while keeping an eye on their potential dancing partners, the young women who sat along the sides of the hall. The men wore jackets and ties, the women frocks or skirts and blouses, nothing too elaborate. They wore lipstick and possibly rouge (or substitutes like squashed geranium petals), but anything more would be considered 'tarted up'. Married or engaged couples might sit together, and if children had been brought along, they would be put down to sleep later on, in the ladies' cloak room or outside in the car.

Both the dances and the music were a mixture of the old and the new. In many places the 'square' dances, like the lancers and the Alberts, still had a place in the programme. Sets were formed both by the older people who had danced them since childhood and by youngsters attracted by their sociable and potentially boisterous nature. Most of the dances, though, were foxtrots and modern waltzes. It was the same with the music. There was fierce competition among bands to be first with the latest song from the radio, but they also knew all the old favourites, and it was not uncommon for the dance hall to hum as the dancers sang along with the band (sometimes with parodies of the original words).

Outside the hall was the male world, where the dance hall's strict controls were customarily breached. Alcohol was banned from the hall, but drinkers would bring along their own supply. This would be hidden in a nosebag or tree trunk, or in the back of a car, and visited throughout the evening. Joe Lilley recalls a dance at Taroon, Victoria, at which the entire supply of grog was hidden in the back of a teetotaller's utility. One night the owner, unaware of his cargo of contraband, drove home early and left the drinkers dismayed.

Fights were common, and acceptable as long as they took place outside; but if they broke out inside the hall, they threatened not only that night's entertainment but the entire district's reputation. At Ayrford, Victoria, regular dances were abandoned in the early 1940s after one particularly bad fight in the hall.

Pranks were also a male prerogative. Usually they took place outside the hall — one trick was to reharness a horse on the other side of the fence from its buggy — but they might include a cowpat in the coffee urn or a chook through the hall window.

Supper was an important feature of the country dance, and the women who prepared it were aware that their district's reputation was at stake at the supper table. In preparing the food and working in the kitchen, local women were allocated a place in the pecking order and a particular speciality. There were those whose éclairs or sponges were renowned thoughout the district, and those whose contribution never rose above baked or, in more recent times, unbaked slices. There was also considerable rivalry among districts, each of which might have its own speciality. Although generally the supper for a weekly dance need be no more than milk coffee and, more recently, tea, with plenty of sandwiches, sponges and small cakes, it could rise to quite an elaborate spread. The promise of a fine supper was a drawcard, particularly for young men from other districts, who were understandably attracted to tables groaning with delectable iced sponges, cakes, scones, sandwiches, éclairs, cream puffs, napoleons, patty cakes, fruit cake, slices, trifles, and since the 1960s, pavlovas, cheesecakes and Black Forest cakes.

After the dance was over, there were strict rules governing the behaviour of couples on the way home. A girl was allowed to be escorted to the family car after a dance, and if she and her young man managed a quick kiss before her parents arrived, well and good; but to climb into the back seat for a cuddle was absolutely out of the question. If a boy saw a girl home from a dance, there was no commitment implied. Things only 'got serious' when he began to take her to the dance, and to pay for her entrance as well as his own.

At the dance, young people from about the age of sixteen learnt the social skills of dancing and mixing, and found partners for dancing, escorting home, and eventually marriage. The transgressive, but tolerated, activities that took place outside the hall were an equally important aspect of the dance, particularly for young males, who could prove themselves in an adult world through drinking, smoking, fighting, and some degree of sexual initiation. Young women, on the other hand, were more likely to be confronted with public disapproval upon leaving the sanctioned area of the hall; they went beyond the pale of respectability, and had to run the gauntlet of the men lounging around the doorway, and the watching women.

Although the young women's social activity and sexuality were strictly controlled, they did have their own special days. One was the debutante ball, a custom which derives from the English upper-class tradition of presenting marriageable young women at court (see **Customs**). In Australia this ceremony was appropriated by the middle class and usually took place at a civic ball. Since 1945, however, country girls have 'come out' at balls dedicated solely to the presentation of debutantes. Until the end of the 1960s these balls were often held in local halls, but since then they have been concentrated in country towns.

The debutante ball has in this relatively short period become well established in the social calendar of rural communities. They could be large affairs: several hundred people attended in some cases. The participants are no longer necessarily from middle-class homes, and although the outfits can be very expensive, the majority of girls in the Western District of Victoria, for example, make their debuts.

Strict conventions governed the elaborate but unsophisticated style of the debutante's long white dress, her modest demeanour, the master of ceremony's statement of her pedigree, and her low, straight-backed curtsey on being presented, and the couples' performance of 'old-time' and 'new vogue' dances. As with most social events held in the hall, the preparations — the clothes, the decorations, the food — for the debutante ball were made mainly by the women. The debutante ball included many of the features of a wedding: the long white dress; a page boy and flower girl who preceded the 'debs' at their presentation; a formal supper, with a wedding-type cake which was cut and distributed; speeches of congratulations and thanks, and toasts. A photographer and/or video operator recorded the proceedings.

Until the end of the 1960s, when the local dances were dying out, the debutante ball was organized by church parishes, hall committees and charity organizations, and was popular even in quite small rural districts. The ceremony retained something of its earlier meaning as a celebration of a young woman's sexual maturity and availability for courtship. At sixteen or seventeen, most country debutantes had left school, and their debut was the first of a number of dances to which they wore their white dress. In the 1980s, however, most deb balls were organized by schools for their students. There was no longer any relation between 'coming out' and being allowed out with boys, and the deb dress was rarely worn again. Despite this, the debutante ball remains the social event of the year in many rural communities.

During the 1960s another popular event in many country halls was the 'miniature deb ball', at which children were dressed as debutantes and their partners, and were trained to perform the same dances and presentation ceremony. In the 1980s another variation on the debutante ball appeared: the 'second-chance' deb ball for older women who had never made their debut. A further variation is the 'reverse deb ball', usually organized by a football club as a theme for a fundraising social. Here, the footballers form the deb set and teeter round in high heels and long dresses, outrageously made up and usually considerably 'under the weather'.

Another custom which focuses upon women, and which reaffirms the values of marriage and domestic responsibility, is the kitchen tea. When a country girl announces the date of her wedding, a kitchen tea (in some areas it was called a gift evening) is arranged by her bridesmaids, friends in the district, or by members of a club to which she belongs. A notice is put in the local paper and another at the store. In earlier times everyone who was related to her, or was on good terms with her or her family, would go along. This meant that virtually the whole district might turn out for the occasion. Since the 1960s the kitchen tea has changed its form, however, and is more likely to be held in a private house and attended by women only. Guests bring along a present of a domestic nature and sometimes a cake or plate of biscuits and enjoy a handsome morning or afternoon tea together, and possibly some party games or a dance. The bridesmaids-to-be will often unwrap and display the gifts, which are practical items designed to help the bride set up her household: mixing bowls, tablecloths, utensils, casseroles and so on. People came to kitchen teas to endorse the young woman's choice of partner, to give her some help in setting up her new home, to wish the couple well, and to show respect for her family. Some of the guests would not be invited to the wedding; for them the tea provided an opportunity to share in the celebrations surrounding the marriage.

A birthday or wedding anniversary might also be the occasion of a celebration held in the district hall, but unlike the dance, the debutante ball and the kitchen tea, these celebrations are not entirely communal. Invitations are selective, and the hall in this case becomes an extension of the family home. Although a local organization may be hired to help with the catering, it is the mother, wife or daughter who is the chief organizer, determining which dishes will be provided, and which guests invited. At this kind of celebration, as at kitchen teas, there will be dancing, speeches and a gift presentation (in the case of a twenty-first birthday party, a token 'key to the door'). Supper will probably be rather more elaborate than at a dance or kitchen tea, perhaps with hot dishes and desserts.

The country hall is also the venue for the various celebrations of sports clubs — cabarets, prize nights, victory dinners, pie nights, novelty fundraising nights — each of which has developed its own traditions.

Since the 1960s many halls have been closed down due to a web of factors, including a depleted rural population and improved access to mass culture and to larger population centres. Most of the entertainments that took place in country halls during the 1930s and 1940s have disappeared: concerts and 'nigger minstrel' shows, touring picture shows, roller skating, community singing and mouth-organ bands. Country church services have been rationalized. The dance circuit has been replaced in some areas by old-time dancing, in others by bush dances, but for the most part people look to the towns for their social life.

On the other hand, many rural districts still use their halls. Organizations which have survived into the 1990s — fire brigades, farmers' organizations, sports clubs and charities — use their hall for meetings and social activities. Fundraising activities are still held there, although fashions have changed from ugly-man competitions and popular-girl quests to casserole teas and demonstrations of craft techniques. People still gather in the hall for events such as Christmas parties for the district's children, kitchen teas for women about to marry, and farewells for people leaving the district, as well as family celebrations such as birthday parties and wedding anniversaries.

The people of a country district have access to a wider world now than ever before, and yet they retain a sense of local community. Essential to this are the many activities that take place in their communal hall, which is at the heart of the district and the heart of its folklife.

REFERENCES: Peter Ellis, 'Folklore Associated with Dance, and the Folk Instruments', *Proceedings of the 3rd National Folklore Conference*, 1988; Helen O'Shea, 'The Golspie Hall', *Meanjin* 4, 1988; Karen Twigg, 'The Role of the "Local Dance" in Country Courtship in the Nineteen Thirties', *But Nothing Ever Happened to Us . . . Memories of the Twenties and Thirties in Victoria*, 1986.

Helen O'Shea

Country music in Australia Until the 1950s country music was known as 'hillbilly music'. When this name developed 'hick' connotations, and the American Wild West was all the rage, it became 'country and western' music, but this evoked rather too much of the prairie and cowboy image, while ignoring mountain and other rural associations. Today 'country music' is the preferred term. For their part, 'crossover' artists who blend country music with other popular forms are still struggling to find a name that will somehow dispense with the 'hick' and embrace only the 'romantic' in country life.

Definitions of 'country' and 'folk' music often seem to coincide. In Australia and the United States the two forms have the same Anglo-Irish origins and historically have been the 'music of the people'. In the United States, however, the commercialization of country music as Nashville's

Tom Brittain and Bob Rado, timber workers who live among the magnificent Jarrah and Karri forests in the south-west of Western Australia, demonstrating their skills at the nineteenth National Folk Festival, April 1985. (Photograph, Ann Fitton)

Tony Jones, master stack-maker, and fellow workers making a haystack from sheaf hay in the Riverina district, NSW. This is an endangered skill in Australia since sheaf hay is now not commonly used. (Photograph, Mark Brennan, *Craft Australia*)

Welcome refreshments taken during a break in wheat harvesting in the Wimmera District, Vic. Mr Alphin Johns, his family and workers drink tea and lemonade brought from the homestead by Mrs Johns and her daughter. (National Library of Australia)

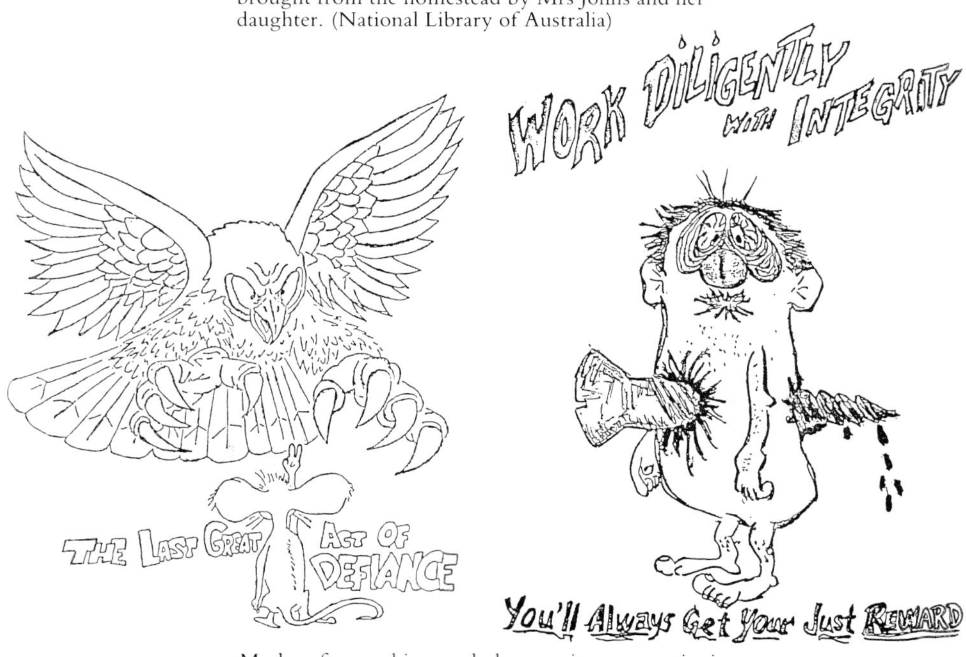

THE LAST GREAT ACT OF DEFIANCE
WORK DILIGENTLY WITH INTEGRITY
You'll Always Get Your Just REWARD

Modern fax machines and photocopiers are continuing a long tradition of anonymous 'reprographic folklore', which is circulated in offices and other workplaces.

'The Furphy's coming'. Soldiers in the First World War were said to have used this phrase about Furphy's water and sanitation carts — and to have created a phrase in folk speech, the 'furphy' referring to a rumour or false story. This fine specimen is in the old Fanny Bay gaol, Darwin, NT. (Photograph, Keith McKenry)

Trade union banners are a colourful and historic part of the folklore of work in Australia. Banner-maker Bergitte (Gita) Hansen marches in front of her banner for the Newcastle Metal Workers' Union, 1984 May Day Parade, Newcastle, NSW. (Photograph, Roger Broadbent)

Tom Larkins rests against one of his dry stone walls, in Victoria's Western District. Some dry stone walls, built 100 years ago without any cement or bonding material, are still faultless. (Photograph, *Age*, Melbourne)

dominance grew led to the increasing use of electronic instruments and a parting of the ways from the earlier, purely acoustic, fiddle sounds and wilder 'mountain' characteristics. This strain is preserved in bluegrass music which, as a result of a shared ancestry, somewhat resembles Australian bush music. In Australia country music is largely distinguished from folk music by its strong ties to its American origins, even though much of it is acoustic and has been grafted on to the Australian experience in the form of the bush ballad (which in country music is distinct in tempo, style and feel from other ballads about the bush played by some bush bands). Some folk musicians feel that the bush ballads of Slim Dusty, Shorty Ranger, Gordon Parsons and Stan Coster still bear too much resemblance to their hillbilly forebears to qualify as folk music.

The bush ballad was pioneered by Buddy Williams and Slim Dusty in the 1940s and was heavily influenced by literary balladists like Adam Lindsay Gordon, Henry Lawson and 'Banjo' Paterson. Until recently it has defined that musical form known as 'Australian country music'. However, since the mid 1980s, a distinct form of Australiana initiated by John Williamson has triggered a new direction among emerging artists.

Country music is a huge, probably growing, movement, but is generally ignored by the mainstream media or programmed at times when few people are listening. A country music festival is held somewhere in Australia nearly every weekend, there are hundreds of country music clubs and associations, and country music is performed regularly at most clubs and pubs (especially in the country). When Dolly Parton and Kenny Rogers come to town, they pull in larger audiences than international pop stars such as Madonna or Kylie Minogue. Slim Dusty, who has outsold these and every other recording artist ever released in Australia (including the Beatles), is rarely heard on mainstream radio.

Country music fans come from all walks of life, but mostly from the working class. This may account for some of the contempt in which country music is held by 'patrons of the arts', and for its continuous, almost underground, survival in Australia. (In the United States, because of the larger population, it supports a commercially viable industry.) Australian country music has also become an integral part of Aboriginal (or Koorie) culture, so much so in fact that in some outback communities jukeboxes offer a choice of only Slim Dusty, Buddy Williams, Rick and Thel or Brian Young — all Australian country music bush balladeers.

Every year for eleven days in late January, country music comes into its own at Tamworth in New South Wales, during the Tamworth Festival. Thirty thousand people bring millions of dollars worth of business to this rather reluctant city which, like Nashville, is at best apathetic about country music, despite its being Australia's 'country music capital'. Several country music recording studios and record labels operate out of Tamworth, and the city is home to a large number of country music artists; a national magazine, *Capital News*; a nightly radio programme, 'Hoedown' (heard from Queensland to New Zealand and Tasmania); a syndicated radio programme, 'Country Music Jamboree'; the Country Music Waxworks, and several other Festival memorabilia.

Tamworth was already beginning to attract country music artists and fans in the 1960s, owing to the long-time involvement with country music of local radio station 2TM, the establishment there of country music record companies such as Hadley, Opal, and later others, and the promotion of country music at venues such as Joe Maguire's Pub. In those days the forerunner of the Capital Country Music Association (CCMA) held an annual talent quest and jamboree over the Australia Day long weekend. It was ex-2TM manager Max Ellis who envisaged the long-term possibilities for country music and coined the slogan, 'Country Music Capital'. With a core of talented staff, he launched the Australasian Country Music Awards in 1973, and in subsequent years, with the establishment of BAL Marketing, fostered and wove together the many disparate elements that today constitute the Festival. Highlights of the Festival include the Starmaker talent quest, the Hands of Fame ceremony, the Bluegrass championships, free concerts for amateurs, and the Roll of Renown. All of these events are taken very seriously by artists, fans and the recording industry, and are seen as door-openers to careers.

Gradually the Tamworth Country Music Festival attracted buskers, the gigantic Saturday Country Music Cavalcade evolved, and other talent quests (in addition to Starmaker) sprouted in shopping plazas and hotels as every venue in the city began to take part. The duration of the Festival grew, straining the city's already stretched resources. Although BAL Marketing still co-ordinates the programme and most major events, the Festival is now largely its own animal. It offers music for all tastes, from bush ballads to Nashville songs, crossover, country rock, gospel, honkytonk, rockabilly, western swing, Texas dances, bush bands and even country punk. Gospel groups and Aboriginal people have run their own festivals-within-a-festival. There are busking, yodelling, bush poetry and recitation competitions, harmonica and bluegrass championships, a steel-guitar showcase, the National Battle of the Bushbands, trucking shows, 'family' shows, autograph sessions by the stars, a display competition among stores, and all-night 'shindigs' at the camping grounds. The climax is the presentation of the Australasian Country Music Awards, an event which is televised nationally.

The evolution of Australian country music

In the 1930s Australian folk and bush music, with its strong Irish Catholic heritage, remained largely undocumented, in written or recorded form. Attempts by 'Banjo' Paterson at the turn of the century and Tex Morton in the 1930s to collect old bush songs met with limited success; it was not until the 1950s that serious field documentation and recording began — and by then American music had taken a strong hold in popular imagination. With the increasing popularity, from the 1930s, of the wireless and gramophone, the plaintive recorded songs of North American hillbilly singers such as Goebbel Reeves, Jimmie Rodgers, Vernon Dalhart, and later, Canadians Wilf Carter and Hank Snow, touched a deep chord in the entertainment-deprived rural areas of Depression Australia. The lonesome-sounding tales of struggle on the land, of cattle, isolation, poverty, mother and religion evoked feelings to which rural Australians, with their own frontier history, could well relate.

A New Zealander by the name of 'Tex' Morton (real name Robert Lane 1916–83) is regarded as the founder of Australian country music in the 1930s. Although strongly influenced by the American hillbillies, particularly Jimmie Rodgers and followed by Australians who copied the American style, Tex Morton (and Buddy Williams) soon began to write lyrics adapted to the Australian experience. These songs included 'Wandering Stockman', 'On the Gundagai Line' and 'Murrumbidgee Jack' (Morton), or 'Happy Jackaroo', 'Where the Whitefaced Cattle Roam' and 'The Overlander Trail' (Williams). Occasionally cultural confusion reigned, as in Buddy Williams' 'The Australian Bushman's Yodel', where yodelling (Swiss), the coyote (American) and the dingo (Australian) somehow all came to be entangled in the same song. Other well-known pre-1945 performers were the Singing Stockmen, Gil Harris, Shirley Thoms, 'Smiling' Billy Blinkhorn and June Holm.

The post-Second World War period saw a resurgence of American cultural influences in country music. The growth of the United States as a superpower, the arrival of all-pervasive rock'n'roll, the rise of Nashville as a country music mecca, American ownership of all recording companies in Australia, and the well-documented Australian cultural inferiority complex, resulted in the domination of our radio waves by the American product. The latter phenomenon is still with us today. Through it all Slim Dusty stuck doggedly to his bushman's image and Australian accent.

By the end of the war, rural industry was in decline and young people were flocking to the cities in search of work. There was a resurgence of American record releases and an interest in the American Wild West. Gene Autry and Roy Rogers became Australian infatuations. As Smoky Dawson himself admitted, sponsorship and success in those days came from following the sponsor's line, and that line was American. In the 1950s a new generation of country and western singers emerged — in the cities — and for a while hillbilly music and 'singing cowboys' captured the public imagination. The Tim McNamara, Ted Quigg and Reg Lindsay shows launched the careers of dozens of stars; the teenage McKean sisters, Joy and Heather, had their own Sydney radio programme, 'The Melody Trail', and the unique Outback Country and Western Tours began, starring such legendary performers as Rick and Thel, Nev Nicholls, Chad Morgan, Kevin King, Reg Lindsay and, of course, Slim Dusty. Tex Morton and Buddy Williams combined singing, sharp-shooting, whip-cracking and rough-riding in their Wild West Rodeo shows.

Smoky Dawson perhaps epitomized the dilemma of the Australian country music artist. His dress was that of an American cowboy, he was sponsored by an American cereal company, and he appeared on Kelloggs Cornflakes packets using Western expressions, including 'Howdy, pardner!'. His radio programme, 'The Adventures of Smoky Dawson', made him the idol of a generation of young listeners. It replaced an American programme and dealt with the life of a stockman on an Australian cattle station called 'Jindawarrabel'. While Smoky Dawson was enjoying popularity in the cities, Slim Dusty and Buddy Williams continued to resist the American image, travelling the outback roads of Australia and quietly building up their followings. (Despite hit songs like 'The Pub with No Beer', Slim Dusty really only achieved superstar status in the 1970s, a period that finally saw the public emergence of Australian pride in Australian product.)

In the bush ballad, as with earlier bush songs, the lyrics are regarded as of primary importance, the tune often being repetitive. In the tradition of the writings of Lawson and Paterson and Australia's rather womanless early documented history the songs have traditionally dealt with working conditions on the land, mateship and bush characters, rather than with relationships between men and women. Unlike modern 'folk' music, they are usually written by working-class people. Their content goes some way to explaining the popularity of American love songs in Australia, especially with female country music artists for whom there has been a paucity of material. John Williamson, flag-carrier for the newer Australian country music, has been influenced by the women's movement and by a growing awareness of the difficulty experienced by Australian men in expressing emotion. As a result, he tries deliberately to write distinctively Australian love songs, a task he claims to have found second in difficulty only to learning to sing in his Australian accent. Songs such as 'I'm in the Mood' and 'Cootamundra Wattle' are alternatively saucy, moving, poetic and very 'Australian' love songs.

Traditionally Australian country music has been less concerned with tear-in-the-beer guilt and remorse than has American country music, and has paid more tribute to Australian history and the man on the land. Tribute has been paid to women of the outback by songwriters such as Joy McKean, Reg Poole, Norma Murphy, John McSweeney, Judy Small and Ted Egan, but, in general, the contribution of women to Australian history has until recently, been ignored. The status of female country music singers in Australia is poor compared to that of their American and New Zealand counterparts. Top Australian women artists, such as Jean Stafford, Anne Kirkpatrick, Evelyn Bury, Deniese Morrison, Judy Stone, Joy McKean and Norma Murphy, all find it a battle to be recognized equally with the top male artists. They cite long touring distances in Australia, media inaccessibility and prejudices against women artists as inhibiting factors.

The politics of country and folk music

With the popularity of radio and its domination by American and English music, folk bush music did not disappear altogether from the bush, but was relegated very much to the back stage. In the early 1960s the folk boom arrived, again from the United States, triggering a revival of interest in Australian folk and bush music which was now more accessible in published form. The movement of 'alternative life-stylers' to the land in the late 1960s and early 1970s did much to revive the practice of live colonial music in the bush. Bush bands and bush dances are popular now all around Australia and other bands, such as Bullamakanka, Redgum and the (former) Three Chord Wonders, have blended bush, rock, political and country music influences into creative new forms.

Generally speaking, however, since the 1970s a gulf has grown between folk and country music which, according to former Tamworth journalist and historian Dan Byrnes, reflects deep divisions in political and social values, embodied in both lyrics and mood. Overlap does occur, particularly at (tolerant) Tamworth, but broadly speaking there is little common ground between the two. Folk music is often seen as appealing to middle-class, intellectual or left-wing 'trendies', usually from the cities. Country music, on the other hand, is seen as 'redneck' or ultra-conservative music, espousing traditionally right-wing values. Ironically, most of the predominantly working-class and Aboriginal followers of country music are probably also traditional Labor supporters. However, environmental and land rights issues have crept into Australian country songs, particularly those of Eric Bogle and John Williamson, both of whom have taken a strong environmental stance, and of emerging Aboriginal artists like Kev Carmody.

Dan Byrnes argues that country and folk music both express themes from history, but differ in their attitudes to authority. Folk music has a strong Irish Catholic heritage going back to early colonial and convict days, and is based on anti-establishment, anti-English and anti-class traditions. The lyrics dwell on social change (the breakdown of the nuclear family, for instance), and the anguish of a country at war (Northern Ireland). When political in theme, folk music advocates an almost joyful defiance of authority and seeks to change society and its laws — hence the popularity of rebel heroes such as Ned Kelly, and the proliferation of liberation and protest songs in folk music. When rural in theme, folk music tends to be about sheep and shearing, Australia's earliest farming industry which attracted many of these self-same emancipated Irish convicts and gave birth to the union movement in Australia. The Bushwackers capture that rebellious spirit in Australian history in songs such as 'The Ballad of 1891', a modern composition (Jacobs/Palmer) which deals with the shearers' strike of that period.

Country music, according to Byrnes, has traditionally been submissive and deferential to authority, and when defiant of it, almost guilt-strickenly so. There is a certain admiration for the individual underdog, the rebel without a cause (as portrayed in many of Johnny Cash's songs in the United States), but he or she usually comes to a tragic end or is brought to heel by the laws of justice or revenge. Country music lyrics often seem to portray a feeling of personal helplessness in the face of problems or social change. Solutions are individualistic, escapist or fatalistic. There is sometimes an apocalyptic view that life is something over which ordinary people have no control. Bush Balladeer Stan Coster sings in 'The Muddy Old Barcoo':

> This mad old world is heading for disaster
> This I feel like a truck out on a highway
> With a madman at the wheel
> But I know that I can't change it
> And neither friend can you
> So I'll go catching yellowbelly
> In the muddy old Barcoo.

© Lyrics and tune, Stan Coster

The conservatism of country music also surfaces in its links with American gospel and evangelical (Protestant) fundamentalism, which has been growing in Australia, and with 'redneck' farmers whose values are held to be traditional, dogmatic and narrow. Political songs have existed in country music, but they usually protest at social or political change. Australian examples include Slim Dusty's 'Cattlemen of the High Plains' (about the prohibition of traditional grazing on the Victorian Alps) and Stan Coster's 'Your Country's Been Sold' (about foreign ownership). Given the current farming crisis, the development of the New Right and overseas rural trade blocs, these kinds of politico-rural themes are increasing.

Country music is also about cattle and horses, rodeos and cowboys, trains and that modern-day version of the restless cowboy maverick, the truckie — features of Australian life which, like country music itself, were introduced relatively recently in our history. Many Australian country music artists grew up on dairy farms and cattle stations, and identified readily with songs whose rhythms are said to resemble those of horses' hooves, trains and trucks. An ex-member of Redgum, John Schumann, feels that folk and bush music is 'dense' music, reflecting its origins, whereas Australian country music is more 'spacious', like cattle country itself.

Songs of poverty, 'drinkin', cheatin' and gamblin'' songs, a feature of traditional (particularly American) music, are often laced with themes of bitterness, despair and remorse, love gone wrong or infidelity. Sex roles often appear clearcut and unquestioned, in contrast to the concerns of more recent folk music. Women bemoan their lack of power and their ability to live without a man: 'All He Did Was Tell Me Lies' (Anne Kirkpatrick), 'One On the Way' (Loretta Lynn) or 'Joelene' (Dolly Parton). They seem confined to extreme role stereotypes: the passive, golden-haired angel, the evil temptress or the creature of mystery. Men are often portrayed as aimless drifters, driven by a search for manhood (or revenge), identity and freedom — traits a 'good' woman might never understand, but would always forgive. The contemporary feminist folk view would more likely see this hero as incapable of meeting adult commitments or of attempting to discuss his feelings. Country music is more concerned with individual experiences, family values, togetherness and old-fashioned solutions to current problems. Critics of country music say it lives in a past of long-gone glories — the space age, technology, the future and even war (with some exceptions) are usually ignored.

Fans say they like country music because it does not 'beat about the bush', but sings of everyday experiences to which ordinary people can relate, and avoids the 'boring and less rhythmic' tones of folk music.

Changes in country music

Originally country music grew out of the hardships, isolation and struggles of poor American whites on the land. The despair of country music reflects the despair people feel in the face of social movements over which they have no control. For this reason country music appeals more to people who have been victims of social change and who have been unable to cope with that change, be it rural disaster, migration to the cities in search of work, instability in the home or soaring crime rates. In Australia the people most affected by unemployment, the collapse of rural and manufacturing industries and a breakdown in traditions have been the long-term working class and the Aborigines, the groups which together constitute the largest following of country music. (It is not surprising that prisoners are avid country music fans.)

Despite its conservative nature, country music has not been immune from changes in the broader society, although it may have been slower to reflect them. In Australia, for example, the hilarious Geraldine Doyle album *Stand On Your Man*, written mostly by Australian Allan Caswell, was a 1980s rejoinder to the Tammy Wynette classic 'Stand By Your Man', and Norma Murphy's and John Williamson's songs increasingly deal with the inner strengths of women. Rising Aboriginal country music singers, such as Roger Knox, Kev Carmody and Archie Roach, and bands such as Warumpi and Coloured Stone, are developing some original sounds. They combine Western instruments, country music, gospel and rock with didgeridoos, clapsticks, chanting, traditional languages and political themes such as land rights.

The future of Australian country music

The post-war rise of the Nashville Sound in Tennessee heralded the arrival of a different form of country music, one which both launched artists' careers and changed their music as irrevocably as the new social environment in which they now worked. 'Nashville Sound' originally referred to a recording method whereby the sound arrangements were 'sweetened and polished' by the use of vocal harmonies and string sections, in contrast to the rawer, earthier sounds of old-time country music. In this way the music tapped a much broader commercial market and began the process of loosening its ties with traditional folk music. Certain forms resisted the Nashville 'whitewash': bluegrass, Western swing, honkytonk, rockabilly, the Cajun music of Louisiana, the Bakersfield movement, the initial 'outlaw music' of Willie Nelson and Waylon Jennings, and more recently a revival by 'new traditionalists' Ricky Skaggs, Emmylou Harris and Randy Travis of hard-edged acoustic guitar music. However, until the mid-1980s the Nashville Sound dominated country music around the world, eventually moving the music further into the 'crossover' mould, headed by Dolly Parton, Kenny Rogers and Willie Nelson. More and more Australian country music artists also see the incorporation of crossover — a blend of country music with other popular forms such as pop, rock and roll, middle-of-the-road, jazz and cabaret — into their repertoire as essential to their survival, in order to overcome the supposed 'hick' image of country music. Despite the lamentations of traditionalists, the trend is not as cynical as it seems. The extreme isolation and hardships of country people many years ago have diminished as improvements in living standards, mass communication, transport and entertainment have evolved. The lifestyles of country people now are not so different from those of their urban counterparts (who themselves increasingly hanker after a more natural life away from the rat race). The blending of country themes with love songs and city sophistication has enormous market appeal and may be the biggest-selling form of music in the Western world. To many young and city people, the bush ballad is outdated, associated with an older and disappearing Australia: Slim Dusty's image as a quiet, modest, solid bushman, surrounded by a loving and musical family, stands in stark contrast to the typical non-nuclear family household of today and the brash ockerism of media heroes such as Paul Hogan and Merv Hughes. Nevertheless the bush ballad form led by Slim Dusty, Stan Coster and talented new artists such as Brian Letton often still outsells the big American hits in Australian charts.

Holding the middle ground are the lower-keyed and more contemporary tones of John Williamson, bridging the gap between the city and country, patriotism and cringe, country music and folk, republicanism and the vestiges of colonialism. John Williamson is riding the crest of the burgeoning cultural groundswell he sensed some years ago. In six years he won eleven Golden Guitars at the Australasian Country Music Awards, and his album, *Warragul*, went triple platinum with sales in excess of 210 000 within a year of release. Country music traditionalists are still coming to terms with his unapologetic form of Australiana, while parallel movements are occurring in the folk scene. In the cities his name has made waves with hits such as 'True Blue' and 'Rip Rip Woodchip'. With his Mallee wheat-farmer background, republican and conservationist sentiments and original all-Australian flavour, Williamson has taken Australian indigenous music by storm. If anyone is able to eventually surpass Slim Dusty as 'King of Australian country music', that person will be John Williamson.

Another major player who has emerged since 1987 is young singer/songwriter, former stockman and rodeo champion James Blundell who, with Dusty and Williamson, is now among the top three Australian country music artists. He is undoubtedly the industry's biggest media success and heart throb, already dubbed the 'Mel Gibson of Australian country music'. Blundell has given country music a higher, more youthful profile and a broader, rockier — but still rural — content, as evidenced in albums like *Hand It Down*.

The other significant development in Australia is the emergence of acts from the inner cities. Foremost among these is the Happening Thang, an up-tempo honky-tonk-style band which has arrived very quickly at the top. Other city-based 'country punk' bands making an impact are the Danglin' Bros and Keith Glass and the Tumblers. The 'urbanization' of 'country' music is resisted by many diehard followers; others believe the musical form must broaden and evolve to ensure its survival.

In Australia a controversy rages, not so much between country and folk music supporters, but between those who look and sound American (and call their music 'international') and those who believe that Australia has enough of its own history and culture to sing about. Unfortunately the issue is less straightforward than just not singing about Texas in an American accent. Artists are dependent on public support for employment, and frequently feel unable to satisfy the conflicting demands of requests for American hits, media bias and the need to be original if they wish to escape mediocrity and eventual oblivion. The Tamworth Songwriters' Association (TSA) is an influential Australasian lobby group open to all songwriters. It asserts that artists would boost their own careers and prevent millions of royalty dollars flowing out of Australia and New Zealand if they would sing each other's songs, instead of American hits. Top artists like John Williamson, Eric Bogle, Ted Egan, Norma Murphy and Allan Caswell are successful *because* their material is original and Australian, yet few have made much money from it.

The Tamworth Country Music Festival and the Australasian Country Music Awards are the barometer of trends in Australian country music. The drift seems to be away from the American style of earlier award winners, such as Jean Stafford and Arthur Blanch, and towards artists forging new or more Australian sounds — artists such as James Blundell, the Happening Thang, Deniese Morrison, Reg Poole, singer–songwriters Allan Caswell, Norma Murphy, Colin Buchenan, John Williamson and, of course, the so far indomitable Slim Dusty. In fact the Australiana movement is attracting former 'folkies' increasingly into the country music camp.

It is unlikely that country music in Australia will ever compete with popular music to the extent of its American counterpart, even if it moves towards a crossover sound. Unlike their more urbanized and mobile equivalents in the United States and New Zealand, Australian artists still complete their country music apprenticeships on the lonely stretches of Australian roads. The vast distances and small population of the continent and the relative lack of interest in country music in the big cities have led to a parochialism in country music that is hard to overcome. It militates not only against the ability of more than a few artists to 'make it', but has also led to the consistent failure of any national country music organization, to attract a large membership and sponsorship and to promote the industry, as do the equivalent associations of the United States and New Zealand.

Nicky Erby, national country music radio programmer and a captain in the country music industry, has highlighted another, more insidious, problem in the industry. He contends that not only content but also production standards make most Australian country music uncompetitive with the American product. Australia's population cannot support a country music industry on the same scale as that in the United States. Nevertheless there is resentment among struggling artists at the apparent mercilessness of DJs and all-American-owned record companies in not promoting more Australian product and thereby creating public demand. Undoubtedly the international marketing efforts of the aggressive (American) Country Music Association (CMA) are also difficult for a disorganized Australian industry to contend with. Things are changing — the last fifteen years in Australia have seen the emergence of a real national pride; but this newfound pride has yet to manifest itself on a broad scale in the marketplace, in the media or in the quality of many Australian (not just country music) products.

The success of the Tamworth Country Music Festival and the Australasian Country Music Awards has done much to raise standards and foster an indigenous sound. The sorest need at present is for a national professional organization to weave together the artists, country music clubs, radio programmes, fan clubs and disparate musical forms into a commercially viable and competitive industry. The Tamworth Songwriters Association (TSA) is a growing Australasian organization that plays a vital role in promoting indigenous artists, and there are other songwriter organizations attempting to gain a national foothold. The fruits of these efforts remain to be seen.

See also: **Folk music in Australia: the debate**; **Folk song**.

Monika Allan

Craft As might be expected in any country less than two hundred years away from its pioneer days, Australia has strong traditions of making things, whether for use or beauty (see **Folk and popular art in Australia**). Women's handcrafts such as knitting, crochet, embroidery and dressmaking are some of the best-known examples, and Jennifer Isaacs's book *The Gentle Arts* (1987) is an excellent review of the field. The book includes a wide range of handcrafts ranging from lace-making and cake decorating to poker-work and trade union banners. *The Gentle Arts* also has strong emphasis not only on *what* women make but also on *why* they make the things they do, and Isaacs's chapter headings include 'Custom', 'Tradition and Necessity', 'Nourishment and Nurture', 'For a Higher Cause: Religious Arts' and 'Commemorating People and Events'. A work that also deals with women's crafts and their significance is *Generations* (1988) by D. Bell and P. Hawkes.

Men's traditional crafts are less widely recognized, although well documented in books such as Ron Edwards's (q.v.) numerous publications about bush crafts. For example Edwards's *Bushcraft 1: Australian Traditional Bush Crafts* (1975) discusses gates and fences, simple buildings, furniture, cooking, lighting, bush power, domestic crafts, leatherwork, plaited work, bridles, reins, saddlery, horse lore, transport, bush lore, traditional bush tools, knots and splices, fishing and fish traps and prospectors' equipment. Some specific examples with strong folkloric connotations include the Coolgardie safe, the wattle-and-daub hut, the squatter's chair, the camp oven and the post-and-rail fence. Ron Edwards's publications on bush crafts are notable not only for the amount of information provided (sometimes based on his own efforts at, for example, saddle making) but also for his evocative and detailed drawings. Male crafts are by no means confined to the bush, however, and there are few suburban backyards which have no examples of an owner's or tenant's efforts at making a letterbox, a dog kennel, a shed, a children's cubby-house or a fowl run (more likely known as a 'chook-house'). Such amateur efforts are frequently known by the generic description of 'bush carpentry' and are in a strong tradition of 'having a go'. Murray Walker's book *Making Do: Memories of Australia's Back Country People* (1982) details a tradition which is still powerful in the city as well as the 'back country', and in the city at least is more commonly described today as 'do it yourself' (DIY).

Although it is true that crafts in Australia are frequently gender-linked, there has always been considerable crossover between the sexes. Maritime crafts practised by sailors during long hours at sea not only included fancy ropework, scrimshaw and ship modelling (including the proverbial ship in a bottle) but also embroidery, knitting and macrame. Likewise, many female farmers are expert in bush crafts. Woodwork has always attracted some women as well as men, and colonial Australia knew some fine women furniture-makers. Possibly the outstanding private collection of

Australian furniture is the McAlpine Collection, which includes 'primitive furniture' of slab, stick and board construction, 'recycled/salvage furniture', 'plain furniture' made by the users or local workers with limited skills (such as wheelwrights), 'basic British-influenced country furniture' of an elementary nature but often made by trained cabinet-makers, and 'formal furniture' where the professional maker's skill is at its height. The recycled furniture collected by Lord McAlpine includes furniture made from wooden kerosene cases, a frugal style of living recommended to new settlers by organizations such as the New Settlers League of Australia who published an instruction manual entitled *Makeshifts* in 1924.

Many occupations — wood-working, blacksmithing, fishing and baking, for example — involve traditional craft skills.

Some are less developed auxiliary skills, as seen in the signwriting on butchers' shop windows, for example. Given that butchers (like Banjo Paterson's barbers) are often 'humorists of note' it is not surprising that some of these signs involve humour, such as those inviting customers to 'come here for tender meatings'.

The Report of the Committee of Inquiry into Folklife in Australia (1987) commented on the 'flourishing Aboriginal arts and crafts industry' observed in central Australia, and noted its economic value to the Northern Territory. It stated that 'some of the artefacts widely sold included carved animals, musical instruments, weapons, dishes, beads and paintings'. The Central Australian Aboriginal Media Association (CAAMA) also drew the Committee's attention to traditional iconography and sand drawing, body and ground paintings and traditional tools. The Committee noted that enterprises such as the silk-screen enterprise of the Tiwi people on Bathurst Island and the batik and silk-screen fabrics produced by Aboriginal women at Utopia 'provided an opportunity for a number of Aboriginal and Island communities to bridge the gap between traditional and innovative crafts practice'.

The enormous increase in migration to Australia since the end of the Second World War in 1945 has led to greater awareness of craft skills from persons of non-English-speaking backgrounds (see **Ethnic heritage materials**). Among the huge variety of crafts practised in ethnic communities in Australia, one is of particular interest; both the Hmong community from Laos and some Latin Americans use embroidery and appliqué to tell the story of their suffering and subsequent migration (the Chilean and Peruvian appliqué technique is known as *arpillera*). In 1989 a book titled *Australia's Hidden Heritage* was produced for the Office of Multicultural Affairs; it includes large colour photographs of many family treasures, both brought from overseas and made in Australia. These objects include costumes and other textiles, eating and cooking utensils, decorative embroidery, jewellery, pottery and metalwork.

Children's folklore (q.v.) also includes some traditional craft objects such as paper dolls, aeroplanes and the mysterious fortune-telling device made from folded paper. This device appears to have no name in the English language, although a graphic Calabrian appellation noted in Australia is *colo di gallino* (chicken's bum).

See also: **Australian Plaiters' and Whipmakers' Society, The**; **Building and folk architecture**; **Patchwork quilts**; **Wagga rugs**; **Women and folklore**.

Gwenda Beed Davey

CROOKED MICK, an occupational hero of shearers and outback workers who is bigger, stronger, faster and generally better at everything. He is usually located on the Speewah, said to be a station somewhere out beyond the black stump, where the pumpkins grow so large that people hollow them out and live in them and where the wool on the sheep grows so high you need a ladder to shear them.

See also: **Folk tales**; **Occupational folklore**.

Customs Folklorists generally consider customs to be the concrete expression of shared beliefs. Customs may be major public events, involving ceremony and spectacle of various kinds, or may be small, private commemorations or celebrations. Whatever their size and nature, these behavioural practices are usually complex combinations of belief, expression, action and socializing, symbolizing moments, places, or events felt to be important to those who observe them.

Customs usually fall into one of three categories: calendar customs, or 'periodic customs', celebrated at certain fixed or predetermined times of the year; individual or life-cycle customs involving the individual, such as christening, marriage and funeral customs; and occasional customs carried out whenever circumstances demand, such as drinking, gambling or celebratory customs of one kind or another.

Religious customs are generally determined by the official calendar of each church, with Christianity, Islam, Judaism, Hinduism and Buddhism being the major institutions in Australia. There are, of course, many other religious beliefs observed within the Australian community. While formal observances of religious significance such as Christmas (25 December), Passover (first full moon after the spring equinox), Spring Festival ('Chinese New Year' — usually in February, depending on the cycles of the Chinese calendar) and Ramadan (in the ninth month of the Islamic year) are not themselves folkloric, many folkloric practices, expressions and beliefs cluster around such times. These differ in each culture, but frequently include one or more of feasting, fasting, good-luck visiting, exchange of gifts and general well-wishing.

Secular calendar (or periodic) customs such as St Valentine's Day (14 February) and Halloween (31 October) may also be based upon religious calendars. More commonly they are based on historical events, as are Anzac Day (25 April), May Day (1 May) and Melbourne Cup Day (first Tuesday in November), or on folk belief, as are 'Black Friday' (any Friday that falls on the thirteenth day of the month) and April Fool's Day (1 April). The desire to commemorate and celebrate is often the force behind the establishment of such customs, but as they become established and increasingly folklorized, it is often the elements of festivity and conviviality that take precedence, as in student revels, occupational picnic days and New Year's Eve.

Life-cycle customs, sometimes termed 'rites of passage', are generally observed in private and by families and close friends, and focus on the progress of each individual from womb to tomb. Rites of passage include christening and other naming customs, birthdays, the passage from childhood to adulthood, betrothal, courtship, marriage, funeral and wake. Obviously the manner of observance varies from culture to culture and from group to group, but such occasions usually involve celebration, festivity and the convivial consumption of special foods, drink, the exchange of gifts or other tokens and sometimes the wearing of special clothes.

Occasional customs, as the term implies, do not fall at any particular time of the year, nor mark the stages of progress in an individual's life journey. Rather they are observed whenever occasion demands and include a variety of practices, including such customs as the 'topping out' of a newly-completed building, a custom involving the lifting of a tree or other plant to the top of the building, combined with an alcoholic celebration; 'shouting' a round of drinks in a public house; and determining the sex of an unborn baby by suspending a wedding ring or peg on a string over the expectant mother's belly and observing the manner in which the device swings. There is an enormous diversity of such customs practised in Australia, though little attempt has been made to collect and study these. In fact, very little research has been carried out into folk customs in Australia to date, but some indication of the range and variety of such practices can be glimpsed in the multicultural calendars and almanacs that used to be issued by Hodja Press. Information about the customs of the numerous cultural, ethnic and linguistic groups that make up the contemporary Australian community can be found in J. Jupp (ed.), *The Australian People* (1988).

See also: **Anzac Day**; **Ethnic folk heritage**; **Festivals**; **Folk belief**; **Greek-Australian folklore**.

D

DAD and DAVE are rural characters in the much-loved stories by 'Steele Rudd' (A. H. Davis). Numerous apocryphal tales circulate usually showing Dad and particularly Dave as country yokels. Not surprisingly, such yarns are told mainly by urban-dwellers.

Dance The Aboriginal people had been skilled and enthusiastic dancers for many thousands of years before 1788, and had made dance an important part of their everyday life. The new settlers, once established here, showed a similar enthusiasm for dancing. The few historians who have given any serious attention to the activities of the people's leisure time all agree that dancing was one of the favourite entertainments of all social classes last century. This is described very well by Margaret Kiddle in *Men of Yesterday* (1961).

> Balls of all kinds remained one of the favourite entertainments of all classes from the squatters to the town larrikins. Those given by the publicans were open to all, and held at their hotels if there was sufficient floor space. Men paid for their tickets but women went in free. A German band, all brass and bristling moustaches, was essential equipment, 'seated on chairs placed on a high table at the extreme end of the room and under the leadership of a large powerful German with a pair of unexceptional whiskers and a spotless white vest.'
>
> The publicans dispensed brandy from behind the bar and the room whirled with the 'white and light dresses of the ladies, and the gaudy flaunting colours (red and blue shirts) sported by the men.' At daybreak the dancers rode home, many of them to out-stations and farms miles distant . . .
>
> At all balls, whether under the auspices of the bachelors or the publicans, dancing was vigorous and lasted till daylight.

The main traditions of dance in Australia were not developed in the style we know as 'folk' dance, i.e. the style typical of those societies where life was based in villages, a style of dance handed down by tradition with no known choreographers. A somewhat romantic conception of folk dance sees it as a product of 'the folk', uncontaminated by commercial or outside influences. This was not strictly correct, even in earlier days when most people lived in villages. There was always some interchange between the dances of the upper classes and those danced at the village level, as well as an interchange of dances between countries. Joan Lawson, in *European Folk Dance* (1953), comments on the influence of armies of occupation, visits of royalty or nobility, etc. in introducing new styles into the local folk dance. In Great Britain, even while the majority of the population still lived in rural settings, folk dance had already been subjected to similar influences. Dancing teachers introduced new steps and styles of dancing. An English traveller, Arthur Young, was very impressed with the villagers' enthusiasm for dancing and recorded that

> Dancing is so universal among them that there are everywhere itinerant dancing-masters, to whom the cottars pay sixpence a quarter for teaching their families. Beside the Irish jig, which they can dance with a most *luxuriant* expression, minuets and country dances are taught; and I have even heard some talk of cotillions coming in. (Young, *A Tour of Ireland*, 1780)

Later authorities on Irish dancing (J. G. O'Keefe and Art O'Brien, *A Handbook of Irish Dances*, 1903) present evidence that the characteristic steps of Irish step-dancing were also 'the creation of the dancing-masters of the eighteenth and nineteenth centuries'. Similarly, itinerant dance teachers were common in the rural districts of Scotland in this period.

The interchange worked the other way as well. Both in Scotland and England many of the gentry had always danced the local dances of the common folk. There were also particular versions of folk dances that had been danced in the larger cities, such as those recorded in the Playford series of dance books (1651 to 1728). In the later part of the eighteenth century some of the country dances became favourites in upper-class circles in England, especially the longways sets for-as-many-as-will and the shorter ones such as Sir Roger de Coverley or the Haymakers' Jig. This overlapping of dances among social classes indicates that it is unwise to be too rigid in one's definition of folk dance, at least as it applies to the folk dances of the British Isles.

In any case the more traditional village dances did not have a lasting influence in Australia. A large proportion of our early migrants, both convicts and free settlers, were people displaced and uprooted by the enormous changes arising from the agricultural and industrial revolutions in late eighteenth- and early nineteenth-century Britain. Many English writers in the early nineteenth century gave a depressing picture of village life in decline. Traditional pastimes such as the community dancing around the maypole or celebrations at Christmas and haymaking time were rapidly disappearing in many areas, although they survived in the northern counties and in Devon and Cornwall, as well as in Scotland and Ireland. Those migrants who had much earlier been displaced from their villages to the slums of London and other large towns were hardly in a position to bring a wealth of folk traditions with them. Moreover, social dancing, being a group activity, does not transfer so easily as song or story to a new environment. Individual dancing (such as the various styles of step-dancing and clog-dancing) were brought out to Australia and flourished here, but there do not seem to be any reliable records of other folk dances from particular areas of the British Isles being brought out by migrants from that area and becoming well known here. Those dances known to have been done here in the early days after settlement were either those country dances popular in upper-class circles or else those dances widely known in different areas of the British Isles, such as the Haymakers' Jig, the Country Bumpkin and the Irish Trot (also known as Thady You Gander). The latter two are mentioned in the *Sydney Gazette* of 15 May 1803 as being danced at wedding celebrations held in a local hotel to the music of a fiddler.

Sydney residents in the earliest days were divided into two groups — the convicts, and the soldiers and other officials sent out to guard them. It is not likely that convicts had much opportunity for dancing, although the grim humour of the day referred to the punishment treadmill as the 'dancing academy'. Details are very sparse, with most accounts referring to what was considered to be the upper level of society, those associated with Government House and the military and naval officers. This section of society tried to transplant and maintain British traditions, including dance traditions. At first they were dancing exactly the same dances they had danced in the English society they left behind, such as English country dances, Scottish reels and the minuet. Elizabeth Macarthur recorded in 1791 that she had learnt to play Foot's minuet on that famous piano that had arrived with Dr Worgan in the First Fleet. Reports of balls appeared in early numbers of the *Sydney Gazette*, and on 16 August 1810 the programme at one of the race balls was said to include the 'Santeuse' (presumably the Sauteuse, which was popular in London at that time — it was a forerunner of the waltz, danced with springy turning steps but with partners held at arm's length). Because there were few suitable venues, dances were often held on ships in the harbour. The music was provided by military bands such as that of the New South Wales Corps.

The new settlers constantly looked back to their old society and so were very quick to adopt the new dances becoming fashionable there. These were of two types, the then somewhat shocking closed-couple dances with partners facing one another, and the quadrilles, danced by sets of four couples placed in a square, facing inwards. Dancing teachers had adapted European folk dances for dancing on smooth floors to create many of these couples dances, and music for them had been written in the style of the dance's original folk tune. This folk background was a major factor in the ready acceptance of these dances. Philip J. S. Richardson (*The Social Dances of the Nineteenth Century in England*, 1960) states that 'A careful study of the history of our social dance during the past two or three hundred years reveals the fact that a new dance, if it is to have a world-wide appeal, must come from a folk dance'.

The transition of the waltz from its folk form to a ballroom one was not easy. A fierce battle had raged against its adoption in Europe's fashionable ballrooms since 1750. It developed its early ballroom version in the suburban dance halls of Vienna and arrived in London about 1812. Here it was still strongly resisted and the *Times* on 16 July 1816 roundly condemned its first appearance at the English court. Apparently Australians were more tolerant: historian Marjorie Barnard (*Macquarie's World*, 1946) recorded that it arrived in Sydney in 1815 and became popular with some dancers. It was not welcomed at Government House — Governor Macquarie favoured Highland reels — but was well established by 1824, when the *Sydney Gazette* (1 July) recorded that the dances at a private ball consisted of 'country dances, quadrilles and Spanish waltzes'. Spanish waltzes were danced in the old country-dance formations but included the turning waltz step. The quadrilles now became popular as well and this mixture of the older country dances with the new dances continued for some thirty years, although detailed information as to *which* country dances were involved is very scarce. Personal references occasionally mention specific dances, as in the hand-written diary of Christina Cuninghame in the Mitchell Library. This rather snobbish lady, visiting Melbourne in 1841, was not impressed to see another lady dancing the Haymakers' Jig wearing a coronet.

Melbourne had a lively dance scene from its earliest days and dancing teachers set up there as early as 1840. The members of the Father Matthew Teetotal Society were typical of the enthusiastic dancers when they held their grand Easter Ball in 1846. Held in a large theatre with the pit boarded over to provide a large dance floor, the programme included jigs, country dances, quadrilles, galops, waltzes and polkas.

The galop and the mazurka, both fast dances based on Slav folk dances, were popular earlier, but the final takeover of the social dance scene started with the arrival of the polka in Australia in 1845. This ballroom version still retained much of the liveliness of original Bohemian folk dance and more importantly introduced a new and interesting rhythm to dance music with its characteristic accentuated rhythms. The *Illustrated London News* was widely read in Australia and the instructions, complete with music of Offenbach, appeared in it on 23 March 1844. This probably helped to start the 'polka craze' that soon followed. It was to become one of our most popular dances, and some Australian dances using polka steps, such as the Princess polka, have survived until today.

It certainly wasn't only the upper classes who danced the polka and the other new and fashionable dances. One upper-class visitor to Melbourne in the 1850s was very impressed by the scene at a 'shilling ball' where the working-class patrons (in their everyday clothes) 'in the most correct and orderly manner imaginable, were dancing quadrilles, polkas, waltzes, etc. generally with great precision and evident enjoyment'. Some printed programmes that have survived from the early 1860s show how completely these newer dances dominated the social dance scene. Dances listed included the quadrille (First Set), Parisian quadrille, Caledonians, Lancers, Prince Imperials quadrille and numerous couples dances such as the waltz, galop, polka, Polka Mazurka,

Redowa, Schottische, Highland Schottische and the Varsoviana. Those contemporary dances which combined country dance formations with new dance steps, such as Spanish waltzes, the Tempest and Circassian Circle, were also popular. One very old folk dance, Sir Roger de Coverley, remained as a traditional finish to an evening's dancing.

Many of the dances listed here were to remain popular in Australia right through into the present century, along with other favourites such as the Alberts, Waltz Cotillon, Fitzroys, Exions, Royal Irish, Barn Dance and many others introduced later. There were also many dances that enjoyed varying periods of popularity but did not become permanent favourites, such as the Kent and Lurline quadrilles, the original mazurka and the Coquette.

The practice of playing a bracket of tunes for a folk dance is comparatively modern. Previously folk dances were danced only to their own tune, and in fact, most were known by the name of that tune. When the longways country dances were so popular in fashionable society, it was the custom for the lady at the top of the set to 'call the tune' which meant that she chose which dance they were to do. In contrast to this, the quadrilles and couples dances could make use of much of the popular music of the day as well as the music written for them. It was customary for the popular quadrilles to be danced several times, using different music each time. An amazing variety of music was arranged for quadrilles, as can be seen in the long lists of dance music advertised in the newspapers last century.

The publication and distribution of sheet music was a very active business here, with numerous composers of dance music listed — from the big names like the Strausses, Offenbach, Lanner and Jullien to many local composers. Much of this popular dance music was picked up by ear and passed on to be picked up by collectors much later: David De Hugard related some of the original tunes for the Lancers (1817) to variants collected in New South Wales (*Proceedings of the 1st National Folklore Conference*, 1984), and the collections of John Meredith and others (*Folk Songs of Australia*, vols 1 and 2, 1967, 1987) and Peter Ellis (*Collector's Choice*, vols 1, 2 and 3, 1986, 1987, 1988) have numerous examples of such dance tunes.

In the second half of the nineteenth century many Australian versions of dances developed, such as our version of the Alberts with its two waltz figures that is still danced today. The Fitzroys and the Exions are not mentioned in any overseas publications and appear to be of Australian origin. The Metropolitan quadrille originated in Perth and the Triplet in Melbourne. Many couples dances, including Princess polka, Berlin polka, Highland Schottische and Manchester Galop, all have Australian versions.

Dancing played a very big part in everyday life. Any new business venture, from Dight's mill on the Yarra in 1843 or a new shopping arcade in Melbourne in 1854 to the new goods yard at Spencer Street station in 1872, was likely to be launched with a ball. The procession to celebrate the winning of the eight-hour day on 12 May 1856 finished with dancing at the Cremorne Gardens. In the prosperous times following the discovery of gold Melbourne gained a special reputation as 'the dancing capital of Australia', with overseas visitors lavish in their praise for the expertise of its dancers. Any special occasion was likely to be celebrated with a fancy-dress ball, and sporting events such as the country picnic races usually wound up with a ball. A tradition of elaborate suppers was early established, with quite gargantuan sit-down spreads being provided at some balls. This tradition has continued with the sit-down supper considered essential for special occasions. Similarly, dances in country districts have long been known for their impressive suppers.

One very special Australian tradition was the ball held at the end of the shearing. It was often held in the shearing shed with the shearers themselves providing the musicians and the MC. Shearers also had a high reputation as vigorous and expert dancers. Here, the conventions varied greatly in different localities. Nell Challingsworth records (*Dancing Down the Years*, 1978) that on some outback stations a chalk line was drawn across the floor to separate the workers from the squatter's family and friends. On the other hand folk-singer and ex-shearer Duke Tritton (*Time Means Tucker*, 1965) worked on stations in New South Wales where the two groups mixed and danced together.

One of the earliest traditions seems to have been that of dancing all night until at least sun-up. The *Hobart Town Gazette*, 21 January 1826, published a write-up of a ball in Launceston with eighty people present: 'Dancing commenced at 9 o'clock and was kept up with much spirit, and the greatest good humour, until 2 hours after daylight'. This was sometimes due to the dangers of rough and unlit roads. Balls were often held at full moon in the goldfields areas because of the extra hazards posed by open mine shafts.

Surprisingly, given the Irish influence on the modern folk revival, neither Irish dance nor music had much influence on social dance last century. Patrick O'Farrell (*The Irish in Australia*, 1986) attributes this to the antagonism which arose with the increase in numbers arriving after 1830 and which caused them to settle in groups and keep the practice of their national customs within those Irish communities. Reports of large public dances suggest that people of Irish descent followed the general social customs here. The account (reprinted in the *Age*, 24 March 1857) of a large ball held at Beechworth to celebrate St Patrick's Day states that the band played mainly music by 'Strauss, Musard, or Jullien' with only an occasional Irish tune on the programme. There were also many Irish musicians and dance teachers here who acted as MCs and taught quadrilles and couples dances to the general public. However, although step-dances and competition dances survived here to some extent, the Irish social dances have had to be revived here in recent years, just as they have had to be revived in Ireland itself.

There was a fashion for Scottish music and dance among English society in the first decade of the nineteenth century. This popularity continued into the Victorian era, and was mirrored in Australia. Scottish reels were popular on the earlier ball programmes and the occasional one could still be seen on programmes right through the century, especially in areas with many Scottish settlers. The tune, 'The Dashing White Sergeant', although written by the English composer Henry Bishop, was sufficiently Scottish in style to be used for a dance which combined the traditional set and reel with the round-the-room style of the nineteenth century. This dance, along with the Scotch reel, Highland reel, and the Eightsome reel — a latecomer, dating back only to the 1870s — was very popular. (This century Scottish folk dance has passed through its own revival period in Scotland, and this has passed on to Australia through the work of the Royal Scottish Country Dance Society and other enthusiasts for Scottish dance.) Nineteenth-century newspapers carried many advertisements of dance competitions at Caledonian and Hibernian gatherings. Prizes of one to three guineas are mentioned for dances such as the Reel of Tulloch, Highland Fling, sword dance, jigs and hornpipes, etc. Special competition versions of various dances (reels, hornpipes, flings and jigs) were developed for the very popular eisteddfods, but some were quite different from the original versions.

Clog-dancing was sufficiently popular for lists of sheet music to include titles such as the 'Melbourne Clog Dance' and the 'Geelong Clog Dance'. Some of the champion step-dancers last century issued challenges to one another and solicited paying customers to their contests. Other contests were more impromptu, often taking place at the local pub, where a door might be taken down to provide the required springy surface for their dancing. 'Banjo' Paterson has a good description of a typical contest in the last chapter of his novel, *An Outback Marriage* (1906). Many elderly dancers remember the step-dancing that was a popular item at balls, parties and concerts. David De Hugard heard many accounts in northern New South Wales of local people, bullock drivers, shearers and prospectors, who were also good step-dancers in the first two decades of this century. Some used a table as a suitable dance floor, and most danced to a single fiddle player.

The living dance tradition in the bush

Michael Cannon in his *Life in the Cities* (1975) gives a thoughtful analysis of the importance of dance in the new society as it developed in Australia last century. It was seen as still occupying this position early this century (Matthew Williams, *Australia in the 1920s*, 1984). Radio, films, and later television were to undermine this, especially in the cities. However, the importance of the dance as a major social function and a meeting place for the community was to continue longer in the bush. The music was often supplied by local musicians as a community duty or at a very low cost in small communities, and the tradition of excellent suppers was kept up in many places.

Last century the popular couples dances and quadrilles were known as ballroom dances, whereas this label is now applied to the foxtrot and other modern dances. The older dances have become known as 'old time' dances, and these remained popular much longer in the bush. Sometimes these were danced as complete programmes combined with the early sequence dances such as the Barn Dance, Veleta, Parma and Pride of Erin waltzes, Maxina and others. More commonly, they were combined with modern ballroom dances in '50–50' and '60–40' dances.

Over the years, many versions of the quadrilles developed. Not only were there differences between states but often places quite close to one another had different versions. It is usually quite impossible to trace how these differences came about, but when local MCs and dancers are interviewed it is very likely that they will insist that their version is the only correct and original version. Some most unusual versions have been collected by Rob Willis in New South Wales and by Maria and Mark Schuster in Queensland.

Styles of footwork, especially in the couples dances, were influenced by the modern ballroom styles; but many dances that were first introduced in the first half of the nineteenth century were still danced in the bush right through the first half of the twentieth, and are even danced today in some districts such as Nariel, in north-eastern Victoria, and the north-central areas near Bendigo. Their names appear on the many printed programmes and those accounts which have survived from the second half of last century and early this century, from all types of balls and dances ranging from those put on by the local squattocracy to simple 'hops' in small halls and private homes. The longest-lasting have been the First Set, Lancers, Alberts and Waltz Cotillon with the waltzes, polkas, Schottisches, Polka Mazurka and Varsoviana.

In the nineteenth century everyone learnt their dances as a normal part of growing up, knew all the steps and had acquired a reasonable expertise at least by the late teens. They may have learnt from parents, grandparents, family friends or teachers. This system still applied in those country areas which had maintained an interest in the older dances. In this way, dance traditions were passed directly down through the community. For example, the highly polished floors favoured by nineteenth-century dancers are still popular in country areas. Although some much younger dancers can find these difficult to negotiate safely, elderly dancers can be seen swinging very fast on them with complete confidence and neat precise footwork.

It is not only the dances themselves that survived in the bush but also many of the older traditions associated with conducting a social dance. Interesting details of these have been collected along with the dances and dance music. They are numerous enough to make up a separate section of folklore, but space doesn't permit them to be included here. Many are mentioned in Helen O'Shea's 'Dancing and Romancing in Country Dance Halls' (*Proceedings of the 3rd National Folklore Conference*, 1988). (See **Country halls**.)

Engraved horn, *c.* 1852, from NSW. (Australian National Gallery)

Motto, *c.* 1900. Ink stencil, pen and ink, water-colour on paper. South Australia. (Australian National Gallery)

Turning Away, *c.* 1982. Charles Aisen's tin sculptures depict the Jewish experience in Poland and Australia. (Collection and photograph, Australian National Gallery)

This shell house, *c.* 1890, is of South Australian origin. (Collection and photograph, Australian National Gallery)

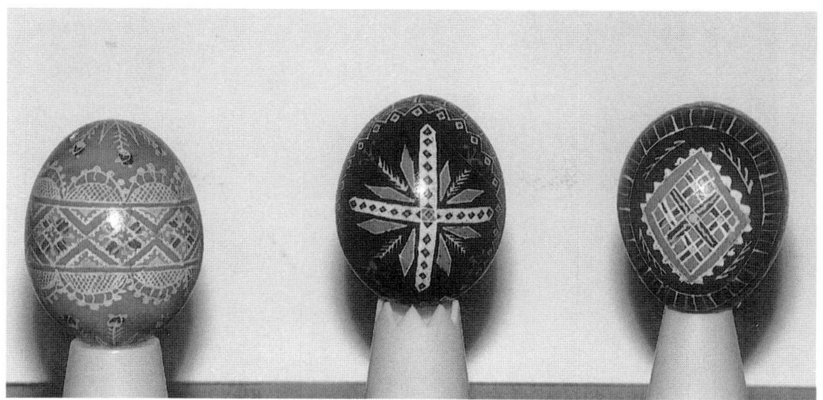

Traditional hand-painted Ukrainian eggs.

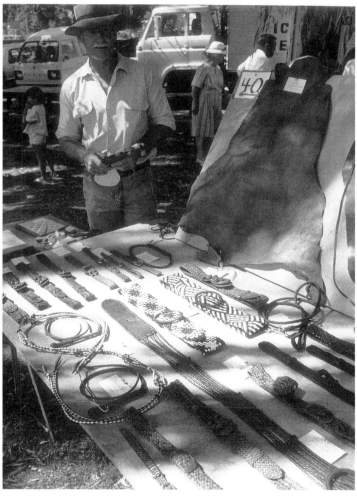

Examples of the traditional craft of leather-plaiting, Mudgee, NSW, 1985. (Photograph, Keith McKenry)

Mrs Leopoldina Mimovich, born in the Tyrol, migrated to Australia in 1949, bringing her traditional skills in wood-carving with her. This 1961 photograph shows her with examples of her work. (Photograph, Australian Foreign Affairs and Trade Department)

The dance revival

What appears to be the first dance collected as an Australian folk dance, the Jacaranda dance, would be more accurately described as a contemporary folk dance in an old style. It was choreographed by Mrs Leith Charleston in 1946 to her own music for the Jacaranda festival in Grafton, New South Wales, and danced there as part of the festival. It is in the English folk style and actually includes one step — a turn single — which was not used in English dance after 1760. The dance was collected by Margaret Walker (q.v.), leader of the Unity Dance Group; in 1951, the Unity Dance Group took part in the World Youth Festival, held in East Berlin, and the Jacaranda dance was the only Australian folk dance they had to present in their Australian concert there. At that time, folk dance was taught mainly through physical education departments and organizations such as National Fitness, and included quite a wide variety of British and European folk dances but nothing connected directly with our dance history. It was not surprising that people began to wonder about this heritage. The interest in folk song had been stronger, although some dancing was included in early productions of *Reedy River* (q.v.), and the Bush Music Club founded in Sydney in 1954 included some folk dancing at its functions, which were called 'singabouts'. The Club's members were co-organizers of the first Festival of Australian Folk Lore, held in Sydney in 1955, which included some dancing for audience participation with the music provided by traditional musicians. The dances were the Waltz Cotillon, Varsoviana and Circassian Circle. Shortly afterwards John Meredith collected the Armatree Brown Jug polka and BMC members learnt the Four Sisters barn dance from some elderly musicians. They remembered the dance from their youth but not its name. It has since been established as an Australian variation on the original Military Schottische from which the well-known Barn Dance also came. The Brown Jug polka is no longer danced as collected because the Armatree musicians played in a rhythm different from the usual polka rhythm (*Singabout* 1, 2). These newly collected dances were among those danced by the Unity Dance Group at the first public function of the Folk Lore Society of Victoria which had John Manifold (q.v.) as guest speaker. The Victorian Bush Music Club was set up in Melbourne in 1959 and initiated research into our dance history. From archival searches and from contact with country dancers, especially those of the Corryong and Nariel districts in north-eastern Victoria, the greater importance of the dances of the nineteenth century compared with that of the earlier folk dances became very obvious. However, it proved much simpler for the enthusiasts of the revival folk movement to promote the simpler folk dances for participation. This sort of dancing soon developed as a separate entity (see **Bush dance**). This revival-style bush dance features a mixture of folk dances, mainly from the British Isles. Some are those actually danced here in earlier days, and some are contemporary ones mirroring those styles. At the same time, others were more interested in the much wider coverage of our dance history that was being revived under the general names of 'colonial' or 'heritage' dance (see **Colonial or heritage dance**). Research into the nineteenth-century dances is still continuing, and has already enabled many of these dances which had not survived to be re-created from instructions published last century.

It may serve as a tribute to Victoria's dancers, much praised for their skill last century, that much of the pioneering research into their dances has been done in that state. Nell Challingsworth concentrated on research into dancing in upper-class society and re-created an elegant style for the Sovereign Hill Heritage Dancers in Ballarat. Shirley Andrews's research covered colonial society in general and, as had happened with bush dance, colonial dance was featured at national folk festivals, and through specialist dance groups in several states (e.g. the Colonial Dancers and the Heritage Dancers) these dances became much better known. In 1986 Nell Challingsworth took two groups of heritage dancers, from

Ballarat and from Sydney, to the Drummondville Folk Dance Festival, a major festival in Canada. Then in 1988, the Colonial Dancers from Melbourne, led by Lucy Stockdale, were part of the main programme at the Edinburgh Tattoo. Both groups were able to present a programme fully representative of our dance traditions, a big development since the Unity Dance Group's overseas visit in 1951. Although the standard of dancing in the general community today does not rate as highly as that of our ancestors, colonial dancing, in its revival form, does achieve the same happy atmosphere and sociability.

See also: ANDREWS, SHIRLEY; **Bush dance**; **Colonial or heritage dance**; **Country halls**; **Quadrilles**.

REFERENCES: Shirley Andrews, *Take Your Partners*, 1979; Shirley Andrews and Peter Ellis, *Two Hundred Dancing Years*, 1988.

Shirley Andrews

DAVEY, GWENDA BEED, graduated BA (Hons) in Psychology and Diploma of Education from Melbourne University and M.Ed. from Monash University for her thesis Folklore and the Enculturation of Young Immigrant Children in Melbourne (1983). She worked as a school counsellor for the Victorian Department of Education and as a lecturer in psychology and multicultural studies at the Institute of Early Childhood Development (now School of Early Childhood Studies, University of Melbourne) and Footscray Institute of Technology (now Victoria University of Technology). In 1979 Gwenda Beed Davey and June Factor jointly founded the Australian Children's Folklore Collection (q.v.) and the *Australian Children's Folklore Newsletter* (q.v.). In 1985 she was appointed to the Committee of Inquiry into Folklife in Australia (q.v.) and in 1990 to the Consultative Committee on Cultural Heritage in Multicultural Australia. She was a 1989 Harold White Fellow at the National Library of Australia, researching the Library's collections in Australian folklore, and was the foundation co-ordinator of the Australian Folklife Centre (q.v.) at the National Museum of Australia. In 1992 she became the first executive officer for the Victorian Folklife Association.

Gwenda Beed Davey has published two collections of traditional family sayings in *Snug as a Bug!* (1990) and *Duck Under the Table!* (1991), and *Jack and Jill* (1992). She is co-editor with Graham Seal of *The Oxford Companion to Australian Folklore* (1993).

Discurio was founded in Melbourne by Ruth and Peter Mann in the early 1960s and is still one of Australia's leading distributors of folk music recordings. In the mid-1950s the Manns launched the Score recording label and produced some of Australia's first recordings of Australian folk music as well as jazz, classical and Aboriginal music. Score also produced some pioneering recordings of Australian spoken word and contemporary culture. Some of the earliest Score recordings include *Bill Harney and Alan Marshall Go Talkabout* and Joan and Miles Maxwell's *Australian Folk Songs* as well as Barry Humphries's *Wild Life in Suburbia*. In the late 1950s Ruth and Peter Mann began importing records from Folkways and Bluenote in the United States, which were initially wholesaled from their home until Discurio was opened in York House, Little Collins Street. Some important recordings produced by Score in the 1960s were Declan Affley's (q.v.) *Rake and a Rambling Man* and *Bullockies, Bushwhackers and Booze*. Ruth and Peter Mann were both born in Germany and came to Australia as refugees in the late 1930s. The establishment of the Score recording label, Discurio and its subsidiary, Fine Music, are examples of the major contribution to Australian cultural life made by Jewish and other immigrants. Discurio is currently housed in McKillop Street, Melbourne and is run by Tim Mann. Although Score no longer exists, Discurio continues its tradition of supporting folk music in general, particularly through concert promotions and record sales of groups such as Sweet Honey in the Rock and Balkana.

Dog on the tuckerbox, The, is a well-known landmark on the Hume Highway between Melbourne and Sydney, 'five miles from Gundagai'. The bronze statue was cast by Gundagai's resident sculptor Rosconi and unveiled by the then Prime Minister, Joseph Lyons, in 1932. This popular icon has generated considerable commercial activity and is often thought to date from Jack Moses's poem, 'Nine Miles from Gundagai'. The story is, however, much older, with a number of earlier ballads dating from the nineteenth century. It is frequently a subject of debate as to whether the dog was a loyal mate guarding his master's property or a wretched cur who spoilt the 'tucker' by sitting (or worse) in the box. Various ballads depict both possibilities, although the Gundagai statue enshrines the dog's loyalty.

DURST, JOY, was born and grew up in New Zealand and came to Australia as a young married woman. She was living in Sydney in the early 1950s and became involved with the Bush Music Club and the original Bushwhackers' Band. This band had become so popular that it couldn't cope with all engagements offered. She was a member of a small group of musicians, including Alan Scott, set up to help with this work.

In the mid-1950s Joy Durst was holidaying with friends, the Bracegirdle family, who lived in Nambour, Queensland. They took her to Buderim to meet John O'Neill who had learnt a non-standard version of 'Waltzing Matilda' (q.v.) from his father. O'Neill sang this tune to Joy Durst, who transcribed it to musical notation. (The same family also introduced John Manifold [q.v.] to O'Neill around the same time.) She was very excited with this discovery and took this song, now known as 'the Queensland version', back to Sydney and to Melbourne.

In 1955 her husband's firm transferred him to Melbourne and here Joy Durst soon joined the Billabong Band. This band had been started by Frank Nickels in a format similar to that of the Bushwhackers, and it was featured in the second Melbourne production of *Reedy River* (q.v.). *Singabout* 3, 4 has a photograph of the band, taken in 1960, on its front cover.

Joy Durst, along with other members of the band, became very involved in organizing 'singabout' nights to popularize Australian folk music. Held at first in private homes, these nights led to the foundation of the Victorian Bush Music Club (now the Victorian Folk Music Club — VFMC) in June 1959. In 1960 Joy became the musical director of the Club. She had a dedicated interest and love for her subject and built up a wide range of contacts, introducing many of the professional singers of folk songs to Australian songs.

She was also very interested in collecting and did some work with Ron Edwards and later with members of the Folk Lore Society of Victoria. Her collecting some songs from foundry worker 'Ocka' Hill was probably a 'first' for songs dealing with an urban workplace.

Joy Durst took a major part in organizing Melbourne's first festival of Australian folklore in May, 1963, a joint project of the VFMC, the Folk Lore Society of Victoria and the Council of Adult Education, which was most successful. In 1964 the Club was asked to provide musicians again for a folklore concert at Moomba, but members were already committed to go to the Corryong–Nariel district for one of the earliest of the folk festivals there. In a very short time, Joy organized some of the younger members into a band which proved to be the show-stopper at the concert which included most of Melbourne's leading folk singers. Several years later, this VFMC Bush Band was to be a leading performer and organizer in the first two National Folk Festivals.

It was a great loss to the folk scene when Joy Durst died suddenly on 27 April 1964. She had been working on a collection of Australian folk songs for publication. The VFMC, with Frank Pitt's assistance, published the *Joy Durst Memorial Australian Song Collection* in four parts between 1968 and 1970; this popular collection has remained in print ever since.

E

EDWARDS, RON (1930–), was born in Geelong, Victoria. After graduating from Swinburne Institute of Technology in Melbourne with a Diploma of Art, he has devoted his energies to the field of Australian folklore and to the perpetuation of the Australian tradition. In the early 1950s Ron Edwards established the Rams Skull Press at Ferntree Gully in Victoria, and between 1951 and 1955 he collaborated with John Manifold to publish and illustrate a series of single-sheet broadsides entitled 'Bandicoot Ballads'. In 1956 he produced the first *Overlander Songbook*, which has been reprinted numerous times. Of particular significance is his *Index of Australian Folk Song* (1971 and 1972), reprinted in 1988 as *200 Years of Australian Folksong: Index 1788–1988* and continuously updated by the author. He has produced numerous other books concerning Australian folklore, including *The Big Book of Australian Folksong* (1976).

Ron Edwards is himself a noted craftsman and has devoted considerable attention to the ingenuity and innovative skills of the early bushmen and settlers as they built homes and fences, shearing sheds and furniture, farm implements and kitchen utensils. His *Australian Traditional Bush Crafts* was first published in 1975 by Lansdowne Press, and has run to ten editions, the later reprints being done by Rigby and Rams Skull Press. Rigby and Rams Skull Press also published Edwards's *Skills of the Australian Bushman* (1979, 1983). These books are notable, as are all Ron Edwards's publications, for his fine line drawings, full of humour and accurate detail. In the late 1970s and early 1980s Rigby published two books dealing with traditional folk narrative, *The Australian Yarn* (many reprints) and *Yarns and Ballads.* Ron Edwards's most recent book is *Fred's Crab and Other Bush Yarns* (1990).

Stockmen's Plaited Belts was the first of a series of small booklets dealing with various aspects of leather-craft. The first eleven were later put into one single volume called *Bush Leatherwork* (1984). These booklets dealt with such subjects as Bushmen's Belt Pouches, Bridles Plaited and Plain, Braided Belts, Homemade Leather Tools, Whipmaking (1 & 2), Saddle Repairs, Making a Stock Saddle, and Stockmen's Plaited Belts. All are illustrated with Ron Edwards's clear and detailed 'how-to-do-it' drawings.

All Ron Edwards's publications are based on extensive field research, and in the case of crafts, on the practice of the crafts described. He has travelled extensively, particularly in Queensland where he has lived for some years. Between 1966 and 1970 Ron Edwards edited the *National Folk* journal (formerly *Northern Folk*, q.v.), and founded the Australian Folklore Society (q.v.) in 1984. In 1985 he travelled to Britain on a fellowship from the Australian Folk Trust, searching for broadsides from the 1800s. This resulted in the discovery of a number of hitherto unknown ballads which he published in two books, *The Convict Maid* (1986) and *The Transport's Lament* (1986).

In recent years Ron Edwards has made a number of trips to China, sketching and learning something of Chinese folk medicine and the craft of mud-brick making and building. Several books have resulted from these experiences.

Ron Edwards represented the Queensland Folk Federation on the Australian Folk Trust for a number of years, and is currently a member of the Program Advisory Committee for the Australian Folklife Centre at the National Museum of Australia. In 1992 he was awarded the OAM for services to folklore and publishing through his Rams Skull Press.

EGAN, TED: see **Folk performers**.

ELLIS, PETER (1946–), was born in Bendigo, Victoria. He trained in chemistry and works at La Trobe University College of Northern Victoria (Bendigo). He has a strong interest in Australian native plants and is a life member of the Bendigo Field Naturalists' Club. He was a contributor of both photographs and texts to the Club's bicentennial publication *Wildflowers of Bendigo*.

Peter Ellis is a gold medallist in 'new vogue' dancing (see **Colonial or heritage dance**), but his preferred interest has been in collecting and performing the traditional old-time dances and folk dance and music which were the true heritage of the pioneers of rural Australia. He is a self-taught musician and plays a variety of instruments, principally button accordion, Anglo concertina, mouth organ and tin whistle. He has been MC, caller and musician with the Wedderburn Old Timers for over twelve years and was instrumental in forming what is now the Bush Dance and Music Club of Bendigo and District, and the Emu Creek Bush Band. The three volumes of *Collector's Choice* (1986–88) were based on collections of music derived from both traditional ear players and from printed sources which were used for the dance club's programme. This programme was based on a blend of traditional old-time, British folk and Australian social dances, and groups such as the Nariel Creek musicians, the Gay Charmers from Kerang, Harry McQueen's Old Time Band and the Wedderburn Old Timers were used as a model. The three volumes of *Collector's Choice* were published by the Victorian Folk Music Club, and Peter Ellis was awarded life membership of the VFMC for his efforts. He has conducted numerous dance and music workshops at national folk festivals, Traditional Social Dance Association of Victoria functions, Heritage Dance Group sessions and continuing education classes. Peter Ellis co-authored with Shirley Andrews (q.v.) *Two Hundred Dancing Years* (1988).

Ethnic folk heritage The term 'ethnic' is derived from the Greek noun *ethnos* (meaning nation, people) and from its adjectival derivative *ethnikos* (meaning pertaining to nation or people). Strictly speaking all Australians, whether they happen to be of Aboriginal descent or whether they claim their ancestry in the British Isles, continental Europe or Asia, can be described as 'ethnic': they all belong to an ancestral nation or a people. However, colloquial usage has applied the term to minority groups and cultures of other than Anglo-Celtic or Aboriginal origin. This practice has assumed a special significance since the Second World War, during which period the population of Australia increased from seven to seventeen million. About five million people have settled in Australia since 1945 and net migration has accounted for more than 40 per cent of the increase in Australia's population during this period.

When the Commonwealth government launched its programme of large-scale immigration in 1945–46, the Labor Minister for Immigration, A. A. Calwell, announced the government's commitment to the principle 'that our population shall remain predominantly British', adding that 'for every foreign migrant there will be ten people from the United Kingdom'. This goal, however, could not be achieved and Calwell and his successors were forced to raise the 'foreign' element well above the level of safety which political rhetoric advocated.

Over the last four decades or so, immigration to Australia has been characterized by a growing diversity in source countries. The introduction of a global non-discrimination policy in 1973 and the extension of assisted immigration schemes to refugees from eastern Europe, the Middle East, south-east Asia and Latin America fostered even greater diversity. Australians today are drawn from some one hundred countries around the world. Their ethnic roots are no longer predominantly Anglo-Celtic as they were when the Second World War ended. The population census of 1986 revealed that just less than half of the Australian population is of pure Anglo-Celtic ancestry and a quarter of the population has no such ancestry. Another quarter represents the rapidly growing segment of the population of mixed Anglo-Celtic and non-Anglo-Celtic descent — the product of the ever-increasing rate of intermarriage. The extent of this transformation and the speed with which it has been achieved is virtually unprecedented in the history of international migrations to developed countries.

Given the numerical prominence of the non-Anglo-Celtic and mixed-origin population in Australia, the question must be asked as to what kind of folk culture — customs, conventions, values, skills, arts, modes of living and, above all, social institutions — do the ethnic groups introduce to Australia, and how is this heritage transformed in the inevitable and ongoing interaction with the Anglo-Celtic and Aboriginal cultures and the Australian environment? In short, how have A. A. Calwell's assumptions of an essential and continuing Anglo-Celtic homogeneity been affected by the experience of the ever-increasing cultural diversity, and what has been the outcome of this process on the transplanted and the transformed heritage of the ethnic minority groups?

The first of these questions — the impact on Australian society — has been the subject of a substantial body of sociological research (comprehensively presented and evaluated in the encyclopaedia edited by Jupp, *The Australian People*, 1988) and in the ongoing debate about the ideology and policy of multiculturalism. The other issue, however, has largely escaped the attention of sociologists, social anthropologists and people interested in the systematic study of folklore. Also, with one exception, none of the social history museums in Australia saw fit to include in its collection and research activities a systematic study of the material culture of Australia's ethnic groups. This entry is, consequently, no more than an attempt to map out the main contours of what is largely an uncharted territory.

The first step in the process of sketching out the area of enquiry called 'ethnic folk heritage' is a conceptual one: the definition of the term 'heritage' and its application to 'folk cultures'. For the purpose of this entry, heritage includes more than 'objects of cultural importance that have been handed down from the past', to quote a standard definition given in the UNESCO handbook by Kenneth Hudson, *Museums for the 1980s* (1977); and it is certainly more than the colourful costumes, jewellery, pottery or devotional objects included in Judith Winternitz's *Australia's Hidden Heritage* (1990). In this entry the term 'heritage' has a much wider meaning. It stands for those culturally defined values, ideas and practices — customs, conventions, religious and secular rituals, handicrafts, foodways — together with their reflections in the person's material culture which determine who we are and in what sort of community we find affinity, security and attachment. Defined in this way, the heritage of immigrant groups represents what the anthropologist and the sociologist sees as culture or a set of socially inherited elements in the life of a people, both material and spiritual. Included in this definition are items that are tangible, such as dress, utensils, styles of domestic architecture, ornaments, to mention a few obvious examples — together with the intangibles of custom, speech, gesture, and other inherited patterns of behaviour and social practices that are regularly and continuously repeated, are sanctioned and maintained by social norms and are at the foundation of the community's social structure.

The task of folkloric research must be, consequently, to study, record and preserve the transplanted heritage of all Australians whose ancestral roots are to be found outside the country. This includes in the first place the Anglo-Celtic majority or the descendants of those who migrated to Australia in the nineteenth and twentieth centuries: the Yorkshire farmers of West Riding dispossessed from their rural holdings in the Pennine Hills by the Enclosure Acts of the parliament at Westminster, the Irish peasants from Kilkenny County fleeing from the potato famine of 1846, the Cornish tin miners who sought better living conditions in South Australia and many, many more. These have been written about at some length; but the folk culture of those of other origins — the Viggianese street musicians in North Carlton in Victoria, the landless peasants of Lettopalene in Newcastle, the Castellorizian Greeks transplanted to Kingsford in New South Wales, the German-speaking Templers forced to abandon their settlements in Palestine and shipped to the Tatura detention camp in Victoria during the Second World War, the Hmong tribespeople from northern Laos settled in Tasmania, and so on and so forth, has not been the subject of folkloric research.

The folk culture of the transplanted heritage initially, at least, distinguishes one community from another before the inexorable forces of mass culture begin to destroy its uniqueness. But even then, as van Gennep reminded us in *The Rites of Passage* (1909, 1960) the practices so designated are not just quaint reminders of the past but aspects of living culture. In a literate, industrial society like Australia, what is here designated as folklore or the process of learning by word of mouth and imitation becomes simply one fragment of culture which includes the total body of learning.

This anthropological view of the process of transplantation and survival of folklife or a living heritage of cultural traditions is at the centre of the report emanating from Australia's national Inquiry into Folklife (*Folklife. Our Living Heritage*, 1987). The significance of folklife as the 'inherited consciousness of all communities that make up or contribute to the makeup of the Australian psyche' has been acknowledged in the Inquiry as something which is a living element of Australian society today and as one 'we have inherited not only from earlier generations of Australians but *also from our forebears in other parts of the world*' (emphasis added). Given this acknowledgement it is also important to realize that folklife in all its manifestations enhances the group's separate identity and its uniqueness. As Srebrenka Kunek put it to the Inquiry, 'People come here with their pasts in the form of embroidery, crafts, myths and as memories, feelings, beliefs. In Australia they often experience a time warp — their pasts become more real than their present'.

While there is a good deal of evidence (discussed later) of ethnic folklife becoming fossilized in a 'time warp', there is also some indication that this transplanted ethnic folk heritage can be transformed and adapted to the Australian environment and the social and material norms of the receiving society. In the case of material culture there is, as Winternitz shows, a whole spectrum of possibilities to be found, ranging from attempts literally to duplicate or replicate items from culture of origin or descent through to free-flowing adaptions and hybrid forms showing the distinct influence of the Australian environment. Transformed heritage also includes those institutional practices, customs and forms of group behaviour that perpetuate in Australia the ancestral values and norms, but which change in the context of inter-group relations. There is an evident preoccupation in the existing Australian folk museums and parks with material culture — those objects and items that are tangible. It is important in this context to use material culture to extend historical understanding. An example of this general approach is Peter Stanley's essay on the interpretation of artefacts of British soldiers in colonial Australia (*Australian Folklore* 3, 1990). Museums dealing with ethnic folk heritage should likewise aim to present specific instances of migration, settlement and social, political and economic adjustment, with objects and material culture being used to illustrate, explore and explain these issues. As Donald Horne put it in *The Great Museum* (1984), 'What can be most intellectually debilitating in a museum is the senseless reverence given to objects merely because of their authenticity . . . we are confronted by objects without social processes'. This is very important for the members of ethnic minority groups in Australia who would quickly see that an institution which collects objects *quo* objects does not address issues of real concern such as cultural maintenance, marginality, discrimination, social advancement and so on. In the past, community and government concern for heritage protection in Australia has been primarily focused on the material heritage — 'the things you keep' — or physical objects shaped according to 'culturally dictated' norms, such as eggs painted at Easter in the Ukrainian community, Greek women's embroidery and tapestry, Maltese lace, Italian leatherwork, Latvian weaving, German wood carvings, Polish kilim rugs. But there are other segments of the physical environment of equal significance to the folklorist and the collecting institutions in the portrayal of minority ethnic groups: personal belongings, household utensils, musical instruments, work tools, religious objects — all examples of people's transplanted or imported heritage. Then there are corresponding items fashioned in response to Australia's environment and to the Australian culturally dictated patterns that comprise the immigrant's transformed heritage. All of these objects have a legitimate place in social history museums. To quote from T. S. Eliot's *Notes Towards the Definition of Culture*: 'Even the humblest material artefact which is the product and symbol of a particular civilization is an emissary of the culture out of which it comes'.

But are these items of tangible heritage — both imported and transformed — capable of portraying, documenting and interpreting the lifestyle defined by a culture? The answer to this question cannot be an unqualified 'Yes'. There are serious methodological difficulties in using artefacts as cultural explanation: the difficulties of inadequate, haphazard item survival; the problems of accessibility and verification of material objects; and above all, the danger of an exaggeration of human efficacy as portrayed in a beautiful work of craft, an example of a household utensil or a set of tools. In all of these instances, museums and other collecting institutions might be accused of promoting a view of social history as a story of success and achievement and neglecting the 'downside' of human history particularly relevant to the immigrant experience — the struggle for economic security in an alien environment, the experience of isolation, monotony, racism. An American scholar, Malcolm Arth, put this issue in a keynote address titled 'Interpreting Cultural Diversity' to the 1985 Conference of the Museum Education Association of New Zealand and Australia:

> A question we must face is, 'To what extent are material artefacts capable of communicating the essence of any cultural tradition?' How well can an exhibition, no matter how well done, convey an appreciation of the 'soft' elements of culture through music, the warmth of affection for kin or the humor that made us smile? A pot, a skirt, a cape, or a crucifix speak volumes to those for whom they have contextual meaning, but how do we get that meaning across to others?

There are even broader questions that can be asked when it comes to documenting that dimension of 'lifestyles defined by a culture' which manifest themselves in the social institutions of extended kinship, ethnic schools, churches, national societies and so on. In all of these examples the museum's exclusive reliance on the material heritage — 'the things you keep' — might present at best an incomplete picture of the social, cultural and political reality that surrounds any social institution.

We must, then, look beyond the ethnic heritage collections of the National Museum of Australia, the exhibitions of the Migration Musem in Adelaide or in the more specialized collections held by the Chinese Museum in Melbourne, the Australian–Hellenic Historical Society in Sydney, the Jewish museums in Sydney and Melbourne, the Lithuanian and Latvian museums in Adelaide, the Ukrainian Folkloric Museum in Melbourne — to mention some major institutions active in the 1990s. Clearly an effort has to be mounted to extend the systematic search for ethnic heritage from the tangibles to the intangibles of culture that one prominent scholar, Gwenda Davey, referred to as 'one of the great gaping holes in research in Australia'. A number of possible categories of folklife might be mentioned; but perhaps the two main areas of Australia's ethnic folklife that might offer relatively easy and intellectually rewarding points of entry into this uncharted territory are the family and religion. The reason is that the existing social science research (primarily sociological — and literary) in Australia already contains a record, albeit incomplete, of such practices. Several studies of Australia's postwar migration, for example, have focused on the social institution of the extended family and the practices, customs and rituals that surround this institution. The extended family is conventionally defined as 'a social unit comprising parents and children and other more distant relatives, perhaps including grandparents or uncles and aunts, living under one roof'. Housing in Australia is rarely designed to provide permanent shelter for so many people, but the fact remains that a modified form of an extended family transplanted to Australia survives in the form of a looser set of relationships in which a three-generation family becomes a residential unit, while its members receive practical assistance in a variety of circumstances not only from close relatives but also from other people. The remarkable persistence of the institution of spiritual kinship (*koumbaros*) in Greek communities in Australia has been recorded by Bottomley (*After the Odyssey*, 1979) in Sydney and Mackie (in C. Price [ed.], *Greeks in Australia*, 1975) in Melbourne: the *koumbaroi* (ritual co-parents) are recognized as belonging to the family in virtue of having been wedding sponsors, and the relationship continues when the *koumbaroi* act as godparents of the children of the marriage. Much the same pattern has been observed in Yugoslav families in Australia (Tisay, in D. Storer [ed.], *Ethnic Family Values in Australia*, 1985) where the *kumstvo* relationship serves to augment if not wholly replace the fragmented kinship network through migration: the family of *kum* and *kuma* become adoptive members of the Yugoslav kinship network in Australia. They give gifts to the child at baptism and pledge future responsibility regarding the life of the child. This custom is also to be found in other Slav communities such as the Ukrainians (Zubrzycki, *Settlers of the Latrobe Valley*, 1964). A somewhat similar pattern of adoptive kinship is the *kiorelik* relationship among Muslim families from Turkey and Lebanon. This is a form of ritual co-parenthood established through the Islamic rite of circumcision. It involves the sponsor of the circumcision ritual in shared responsibility for the boy's circumcision expenses, education and marriage.

The shared expressions of life in the family and particularly the rites of passage of its members are inextricably linked with religious rituals introduced into Australia with the arrival of each group of migrants and each new religious denomination. As Lidio Bertelli and Robert Pascoe put it in their chapter in Ata (ed.), *Religion and Ethnic Identity: An Australian Study* (1988), 'Different cultural groups not only have varying modes of ethnic organization, they also draw on different metaphors for their collective expression as a group'.

In the case of the Polish community in Melbourne, the emphasis is placed on sharing and exchange. Families meet for the traditional Christmas Eve meal, the *wigilia*, and go through the ceremony of breaking small wafers baked from the unleavened bread (*oplatek*), exchanging wishes and greetings. Again, at Easter, boiled eggs blessed by the priest (*swiecone*) are shared in the process of exchanging wishes and greetings. During the period of *zapusty* — the last three days before Lent — girls present their lovers with decorated eggs wrapped in a fine linen handkerchief hand-embroidered with the man's initials. Greek families in Sydney exchange at Easter circular sweet cakes known as *christopsomon* — the bread of Christ — marked with a contracted form of the Easter greeting: *Christos anesti*, Christ is risen.

For the Italians, major feasts of the church calendar such as Christmas, Epiphany, Ash Wednesday, Good Friday, and Easter, together with the feast days of patron saints of specific villages and towns, are occasions for festivals. Invariably the *festa* involves the parish priest and his congregation, which for special occasions like the Festival of the Three Saints in Silkwood (north Queensland) swells by thousands of participants coming from afar. The pilgrimage in honour of the Blessed Virgin of the Rosary at Donvale and the annual Viggiano festival of the Madonna in Carlton (both in Victoria) are other examples of feasts in honour of the patron saint in their home town attracting *compaesani* from various parts of Australia. In Sydney patron saint days are celebrated by local Italian communities in Drummoyne, Leichhardt and Fairfield. Another colourful example of celebration in the Italian community is the blessing of the fishing fleet, an annual event which combines a religious ceremony with the atmosphere of a fun fair. The blessing of the fleet in Fremantle, Western Australia, where the Portugese refugees from East Timor join forces with the older-established Italians, is described by Bertelli and Pascoe:

> The fishermen carry a replica of the original statue from Molfetta, having adorned Our Lady of Martyrs with a crown of gold, a flowing cape, diamond earrings, and a blue sash on which is pinned gold ornaments and votive offerings. The procession comprises various secular elements, such as motorcycle police escorts, marching bands, and a gaggle of notables. Other Italian and Portuguese groups have joined the procession over the years, adding to its length. It snakes around Fremantle's streets, making its way from the church to the pier. All the fishing boats are at the moorings, decked in flowers and adorned with the fluttering pennants of all nations. The statue is taken aboard the pilot ship and the flotilla of craft circle around the harbour, their crews singing Marian hymns. Then the procession rewinds its way back, taking the statue in the reverse direction back to the church, after which the participants make merry the rest of the afternoon at a fun-fair set up on the Esplanade.

This particular celebration imitates an annual event in the Italian fishing town of Molfetta on the Adriatic coast of Italy. The same applies to other ceremonies of the blessing of fishing boats: the Sicilian people operating the Woolloomooloo fleet in Sydney, the Lipari islanders in Ulladulla and the fishermen of Calabria in Wollongong, New South Wales.

Many Maltese festivals take on a secular appearance as they are linked with horse racing, folk singing and various Maltese foods. Frendo, in Ata (ed.) *Religion and Ethnic Identity* (1988), describes how the feast of St Peter and St Paul (L-Innaria) held in the Royal Melbourne Showground attracts thousands of participants, as does the celebration of St Grigor held annually at Portarlington on Port Phillip Bay in Victoria, which also includes the greasy pole competition (known as il-gostra) and other games and races. In Adelaide the *festa* of San Katarina, patroness of the town of Zejtun, comprises folk dancing and poetry recitals (Frendo: 195). Strictly religious ceremonies surround Lenten observances such as the penitential Good Friday procession (*il purcissjoni tal-gimgha l-kibra*), while the recitation of the rosary on the visits to seven churches (*il-vizti tas-seba knejjes*) precedes Easter Sunday. A similar blend of the religious and the secular is seen in the Dutch celebration of the Feast of St Nicholas (December 6) which includes a traditional street procession in which the 'saint', dressed in bishop's robes, is accompanied by two Black Peters (*zwarte Piet*).

While public celebration of a religious feast is the norm in the Italian, Maltese and Dutch communities, comparable celebrations in other communities may be restricted. This is particularly important in the Islamic communities where the law of the Koran restricts participation to co-religionists, especially in the ceremonies that surround the Ramadan — the period of fasting during the holy ninth month of the Muslim year. The celebrations ending Ramadan (*Id-al-adha* and *Id-al-fitr*) involve ceremonial breakfast and the giving of sweets and money to young children.

An example of the celebration which combines the religious and the secular as well as the public and the private spheres is the *Tet* holiday (Vietnamese lunar New Year). This major cultural event unites all people of Vietnamese origin in Australia, overriding differences of religion, class and generation. Tet actually starts on New Year's Eve with the ritual of *Tet nien*, in which the dead are invited to come for a few days and share with the living members of the family the joys of Tet. At midnight the ritual of *Giao Thua*, which marks the transition from one year to another, is celebrated with firecrackers in the home and drumbeats in temples. On the first day people flock to temples to say prayers; the day is reserved for the worship of ancestors. The second day is reserved for close relatives and the third for the dead who are honoured before they return to another world. Throughout the Tet holiday a bamboo pole two to three metres high is erected in front of the house adorned with a few gilded paper bears and a paper carp, which, according to a popular legend, will turn into a dragon to take 'the Genie of the Hearth to Heaven where he reports to the Emperor of Jade on the family's activities and behaviour during the year'.

The examples of ethnic folk culture in Australia described above give only a few glimpses of a myriad of unique folklores that reflect the identity of local, regional and national groups. Each of these groups has its own distinctive sense of self that expresses itself in the language or dialect, religion, a specific pattern of family obligations and rituals as well as clothing, leisure activities including dance, crafts and food.

There is a great need for systematic folkloric research among Australia's ethnic minority groups. Two major goals of research should be highlighted. In the first place, we need to know much more about the content and form of ethnic customs and rituals and more descriptive and analytical research into the process of folk learning or the transmission of knowledge by word of mouth and imitation. This includes learning about such things as speech, dress, music, arts and modes of living that sustain mutual obligations in a social group and provide the means of sharing deep rooted attachment to a particular culture.

In the second place, we need comparative research into the process of transformation of both the tangible and intangible ethnic folk heritage. We need to know how items of material and spiritual culture introduced into Australia can be transformed through the interaction with the receiving Anglo-Celtic majority and the Australian environment. Literature is capable of providing useful insights into this process, as has been shown by Gunew (*Migrant Women Writers*, 1985, 'Varieties of Migrant Dreaming', 1986). A recent collection of stories about Latvians in Australia by Aina Vavere (*The Blue Mountain in Mujani*, 1991) focuses on three generations of immigrants who reminisce nostalgically about their homeland but also attempt to come to terms with a culture that seems to threaten Latvian traditional values. In 'Sol', for instance, the young Latvian girl is seduced by the attraction of student life to abandon the traditions of her grandparents:

> You should be singing folksongs about the beauty of your homeland, you should be dancing in the dress of your mother's country with a wreath of marguerites on your head, you should be going to confirmation classes, take your first communion all in white.

It does not take specialized knowledge to realize that this expectation of the grandparent is unrealistic in the face of inexorable assimilation. It is important to realize that ethnic crafts ('tangibles') and social institutions ('intangibles') rarely persist in an unchanged, fossilized form — the 'time warp' is an exception, not the norm. The norm is a hybrid form of folk culture. As Gardini and Skardoon put it in a paper to the Federation of Ethnic Communities' Councils of Australia in 1985, 'however much a transplanted minority culture strives to keep its links with the official culture of its original homeland, it soon creates something that is unique . . . it creates folklore that belongs to this country'. The study of the contribution of transplanted ethnic folk heritage to Australian folklore is a most worthwhile topic for research.

See also: **Chinese folklore in Australia**; **Ethnic Music Centre of Western Australia, The**; **Festivals**; **German folklore in South Australia**; **Greek-Australian folklore**; **Greek-Australian folk song**.

REFERENCES: Margaret Anderson, 'Collecting and Interpreting the Material Culture of Migration', in Patricia Summerfield (ed.), *Proceedings of the Council of Australian Museum Association Conference*, 1986; Kimberley Webber, 'Material Culture and Australian Studies: Limitations, Contradictions and Paradoxes', in Committee to Review Australian Studies in Tertiary Education, *History and Cultural Resources Project*, Part 2, 1986; J. Zubrzycki, *Ethnic Heritage in the National Museum of Australia: An Essay in Museology*, National Museum of Australia, 1992.

J. Zubrzycki

Ethnic Music Centre of Western Australia, The, was established in 1983 and is a non-profit organization assisted by the government of Western Australia through the Department for the Arts, the Australia Council, the Perth City Council, subscription and sponsorship. It is housed in part of a historic building, the former North Perth Town Hall. It aims to develop and highlight Australian multiculturalism through the promotion of ethnic music and related ethnic arts activities. Each year the Centre organizes a wide array of concerts, workshops, educational seminars and festivals for the WA public and for practising artists and members. Children's activities are arranged in school holidays, and each Friday night talented ethnic musicians (local and imported) perform at the Cafe Folklorico. One of the highlights of the EMC calendar is the annual Ethnic Arts Festival of *Etna Arta Festivalo* which celebrates Australia's cultural diversity.

The Centre aims to assist ethnic musicians in a number of ways. It maintains a multicultural arts directory which lists both professional and amateur ethnic musicians, and provides names and addresses of available performers on request. All performers nominated by the EMC must be paid union rates or higher. The Centre has an extensive library of resource material which is available to teachers, musicians, members of the public and EMC members, and also produces a monthly newsletter.

F

FACTOR, JUNE, is a writer, critic and folklorist. For a number of years she was Senior Lecturer in English at the Institute of Early Childhood Development in Melbourne, and in 1989 became Senior Research Fellow at the Australian Centre at the University of Melbourne, where she is now an Associate. She is the co-editor of a number of books for children, including the *Big Dipper* series (1980, 1982, 1984) and compiler of the popular poetry and children's folklore collections, *A First Australian Poetry Book* (1983), *Far Out, Brussel Sprout!* (1983), *All Right Vegemite!* (1985), *Unreal, Banana Peel!* (1986), *Ladles and Jellyspoons* (1989), and *Real Keen, Baked Bean!* (1989). Her picture books include *Micky the Mighty Magpie* (1986) and *Summer* (1987). As well as writing for children, June Factor has published a number of short stories and poems and a variety of literary and cultural reviews and essays.

Dr Factor has been involved in research into children's folklore since she co-edited *Cinderella Dressed in Yella* with Ian Turner and Wendy Lowenstein (qq.v.) in 1978. She is one of the very few people in Australia who has looked closely at the culture of childhood in this country, and is the author of a number of papers and books on this subject. Her study of children's play traditions, *Captain Cook Chased a Chook: Children's Folklore in Australia* (1988), was awarded the Opie Prize in the US in 1989 for the most outstanding book on the subject of children's folklore to be published in the preceding year. *Australian Childhood: An Anthology*, co-edited with Gwyn Dow, was released in 1991.

Dr Factor is the Director of the Australian Children's Folklore Collection (q.v.), the only archive of children's folklore in Australia, housed in the University of Melbourne archives. She is also co-editor of the bi-annual *Australian Children's Folklore Newsletter* (q.v.)

A significant part of June Factor's time is devoted to community organizations. She is past president of the Victorian Council for Civil Liberties and the Australian Council for Civil Liberties. She represents civil liberties interests on the national Privacy Advisory Committee. She has been the Australian Society of Authors' representative on the National Book Council, a member of the Victorian Premier's Literary Awards Committee, and a member of the Victorian International Year of Literacy Committee.

FAHEY, WARREN (1946–), was born in Sydney and trained in advertising at the NSW Institute of Technology. In 1970 he formed a 'one-man-band organization', the Australian Folklore Unit, to collect, research, write about and popularize Australian folklore. In 1973 he commenced a thirteen-month field programme which took him to most of the eastern coast of Australia 'knocking on doors' to collect folklore. The project was highly successful and yielded a major collection which is housed in the National Library of Australia. Warren Fahey has 'never stopped' collecting, researching and promoting this material.

Warren Fahey formed Folkways Music (q.v.) in 1974, Larrikin Records (q.v.) in 1975 and the Larrikin Booking Agency in 1988. He has produced numerous Larrikin Music Festivals since 1977. Since 1974 he has produced a variety of publications, including *Eureka. The Songs That Made Australia* (1984, rev. edn 1989); *Joe Watson: Traditional Singer*; *The Balls of Bob Menzies* (1989); and *How Mabel Laid the Table* (1992).

In addition to the above books and numerous articles contributed to Australian and international magazines on the subject of folklore, Warren Fahey has produced over 150 recordings of Australian music, including many featuring Australian folk music. All have been published on the Larrikin label. Notable recordings include *Bush Traditions*, *Man of the Earth*, *A Garland for Sally (Sloane)*, *Navvy on the Line*, *Tales of My Uncle Harry* (Keith Garvie), plus the recordings of Eric Bogle, Bernard Bolan and other 'observers' of Australian social change.

A number of successful folklore series on radio have been written and produced by Warren Fahey in association with the Australian Broadcasting Corporation; they include 'The Australian Legend', 'Navvy on the Line', 'While the Billy Boils' and 'The Song Carriers: Traditional Singers in Australia'. Warren Fahey also acted as co-ordinator for the national television programme 'That's Australia!'. He is currently working on two books for the State Library of New South Wales, to be titled *The Folklore of New South Wales* and *The Australian Songbook*, plus ongoing research work and collection in the field of political folklore and Australian humour.

As a singer Warren Fahey has a career that spans over twenty years. He formed the Larrikins in 1971 to present authentic Australian folksong, dance music and folklore, and the group has toured for the Australian government to the South Pacific region, Indonesia, New Zealand and Malaysia. In 1987 they represented Australia at the Commonwealth Arts Festival in Edinburgh, and in 1989 the group participated in the Vancouver Folk Festival. The Larrikins have also toured extensively in Australia for Musica Viva and the Arts Council network. Warren Fahey has always seen himself as a 'shower singer' but has managed to sing on television, in films and radio. His recordings include *While the Billy Boils*, *Pint Pot and Billy*, *Limejuice and Vinegar*, *Live in Concert*, *Navvy on the Line*, *Man of the Earth*, *Game as Ned Kelly*, *On the Steps of the Dole Office Door* and *Billy of Tea*.

Warren Fahey was a foundation member of the Australian Folk Trust and the New South Wales Folk Federation, and was producer of the Port Jackson Folk Festival in 1974. He was a member of the Woollahra Municipal Bicentennial Celebrations Committee from 1987 to 1989, and formed the Paddington Improvements Working Committee in 1988 as a bicentennial contribution. He has been a Justice of the Peace since 1968 and president of the Paddington Chamber of Commerce since 1979. Warren Fahey was awarded the Advance Australia Award in 1988 for his services to Australian music.

Festivals In February 1974 a researcher discovered the skeleton of a long-dead Aboriginal man at Lake Mungo in south-west New South Wales. Wind and rain had exposed the skeleton which had remained buried there for over 40 000 years — buried, we now know, with some ceremony, for the body had been ritually sprinkled with powdered ochre as it lay in its shallow grave.

Ceremonial performance and celebratory ritual have been a feature of Australian social life for thousands of years. The religious traditions of the six hundred or more linguistically and politically separate landowning groups that occupied all the Australian lands prior to 1788 trace the origins of these ceremonies and their ritual to the great creative ancestors. The spiritual effectiveness of ceremonial performance depended on the quality of each performer's contribution. The great creative ancestors, in creating each ceremony, set the expected performance standards.

Contrary to popular belief, the life of the original Australian people was not dominated by the daily hunt for food. In many parts of Australia people lived in comparative affluence, and only in the harshest climatic conditions did they need to spend more than two to three hours in catching and harvesting the day's food. Consequently, people had time for artistic activity, for planning ceremonies and for skilled crafts. No calendar of religious and secular celebrations has been recorded. We do know from the senior initiated men and women of the present and the recent past that there exist both public and private ceremonies which serve a range of religious and social purposes. Private ceremonies are usually restricted to either male or female initiates and have particular religious significance. Public ceremonies may also be religious and demand solemnity and high-quality performance. Others, while drawing on religious tradition, may be more light-hearted, spontaneous and just for enjoyment.

From time to time, not so frequently now as in the past, the seasonal availability of surplus supplies of food made it possible for large groups of people to meet. These large gatherings would bring together, at an agreed neutral meeting point, all the clans of one group and all those of their neighbouring groups. They provided an opportunity for trade, for arranging marriages, for settling disputes, for performing religious ceremonies and for general festivity. The bunya-nut season in south-east Queensland, the Bogong moth season in the Snowy Mountains, the run of fish in the bays and river estuaries, and an abundance of eels and yabbies in inland streams and ponds were occasions for these large gatherings.

The *Canberra Times* for 30 September 1989 cited evidence that as late as 1862, many hundreds of Aborigines from as far away as the south coast of New South Wales, from the lower Lachlan and Murrumbidgee Rivers, had met on the present showground site in Queanbeyan. There was ample food in the nearby rivers and marshy ponds. The site was close to the meeting point of Ngunawal and Ngarigo lands, and was a convenient place for the Walbanga from the coast and Wiradjuri from further inland to gather for ceremonies and for celebrations with the Ngunawal and Ngarigo people.

Regular festivals of this kind have become less common. In those parts of Australia where forms of traditional culture are maintained, however, people still gather to perform ritual ceremonies and to keep alive the celebratory traditions of the past. Thanks to the ready availability of motorized transport, quite large groups can assemble for these occasions.

A variation on this ancient celebratory tradition in recent years has been the gathering together of people for cultural festivals. One such is the spectacular Barunga Festival, hosted by the Jawoyn people near Katherine in the Northern Territory. Hundreds of people from all over the Territory meet there to display, through artistic performance and presentation, the continuity of each group's traditional arts and skills.

Contrasting with the ancient ceremonial and celebratory traditions of Australia's original folk, are those of the new folk. In the newly developing colonies of Great Britain, established in Australia after the naval invasion of 1788, there was at first little time for frivolity. Celebrations were confined to royal occasions, the major Christian religious festivals, market fairs, agricultural shows, annual balls and sporting events. The *Sydney Gazette* of 5 June 1803 reports on the previous day's celebration of the King's birthday. The ceremonies began at 9.00 a.m., with the raising, for the first time in the colony, of the royal standard at Dawe's Point. The guard from the New South Wales Corps fired three rounds and the Dawe's Point battery thundered a royal salute over the harbour. To this, Royal naval ships, and East India Company and sundry other merchant vessels then in harbour, replied. The salute thundered again at noon. From 1.30 p.m., the Governor received the compliments of Army and Navy officers and other dignitaries of the colony and their wives. At 3.00 p.m., they all dined at Government House. In the evening there were fireworks, followed by a ball and supper to end the official engagements for the day. The rest of the colony's population was not forgotten. Free pardons were extended to some thirty-three military and naval prisoners; sixty-seven other convicts gained conditional emancipation; all gaol gangs were to be liberated; all NCOs and privates had half a pint of spirits issued to them.

The *Gazette* made reference to two other celebrations that year, the birthdays of the Queen and the Prince of Wales. In 1805 it added a description of the behaviour of children on Guy Fawkes Day and of Christmas celebrations. By 1820 Macquarie had given official recognition to St Patrick's Day celebrations for Irish emigrants and convicts.

Over time official celebrations have tended to enter on to, or disappear from, the annual calendar as politics and social change dictated. Visitors to Australia from overseas comment on the number of public holidays Australians celebrated. A diary for 1989 listed nine public holidays celebrated nationally, with at least one extra day celebrated in each state or territory and others, like Melbourne Cup, being celebrated locally. Christian religious festivals are listed on the national calendar, but increasingly, non-Christian religious groups and people with non-English cultural traditions are seeing their festivals — Ramadan and Chinese New Year, for example — officially recognized.

One of the more persistent local celebrations is the Royal Agricultural Society Show. The first Royal Agricultural Society came into existence in Sydney in 1822. In New South Wales alone there are now over 200 agricultural societies and they run major annual agricultural society shows in some twenty or more country centres — apart from the Easter show in Sydney. The story is the same for all the other states.

Sports events also began to appear early on the colonial social calendar. Horse racing contests were reported as early as 1800 at the Hawkesbury and soon after at Parramatta. Officers of the 73rd Regiment ran a three-day racing carnival at Hyde Park in 1810, and by 1825 the first Sydney Turf Club was formed. An early version of Australian Rules football was being played in Melbourne in 1858. A team of English cricketers played in Melbourne in 1862, a year after the running of the first Melbourne Cup. There is no doubt that Show Day in country towns and important annual sporting events attract the support of large numbers of people and are accompanied with great festivity. More formal public celebrations like Australia Day, Labour Day, Anzac Day and the Queen's Birthday do not seem to attract much folk support, mainly because they are days structured by the performance of official rituals rather than by events aimed at encouraging public festivity. It is not in such official celebrations that one looks for the beginnings of a festival tradition in Australia, however, but rather to local and regional celebrations which seek to reflect and to promote the real life experiences of people in that place. Such, for instance, could have been the Bendigo Easter Fair around the 1870s which, according to Roper in the *Journal of Australian Studies* (17, 1985),

> consisted of a public procession, Ladies Charity Bazaar, Richardson's Show and numerous other curiosities and amusements . . . the mayor . . . led the procession in a carriage, trailed by members of parliament and councillors of the Sandhurst, Eaglehawk and surrounding Shire Councils.

At the end of the procession the mayor mounted a rotunda in the park and gave a speech followed by the singing of the Anthem. Less solemn was the frolicsome participation of the people at its tail end, with the fair committee inviting novelty groups like 'performing aboriginals' to join on. The rest of the day, too, was given over to public and private entertainments, with the Ladies Charity Bazaar a popular feature of the fair and people's participation encouraged. The increasing importance of the local Chinese community's participation in the procession is to be noted. 'By the 1880s, the [Chinese] display had become the undoubted main attraction of the fair'. Clearly the Bendigo Easter Fair had in it the beginnings of a genuine folk festival tradition. That the tradition didn't blossom is attributed by Roper to the fact that 'Bendigo's fair was organised from above. It did not evolve from a "shared oral culture". Rather, it was created by a committee which employed the fair tradition to foster social unity in an increasingly diverse, industrial area'.

No burgeoning folk festival of the nineteenth century survived. Possibly the longest-running present day festival originated in Grafton in 1934. It was an early summer floral festival, celebrating the flowering of the jacaranda trees that line Grafton's wide streets. Its origins could have had something to do with the opening of the rail link across the Clarence River and the conjoining of the town with its satellite South Grafton on the other side of the river. Whatever the reason for the festival, it has persisted now for over fifty years, and has become a standard feature in the town's annual calendar.

A survey conducted in 1986, and again in 1987, established some 500 events throughout Australia which could be classified as festivals, using as a broad definition 'regular local events that are labelled festivals or that consist of a number of inter-related activities held as a celebration of something significant to the local community'. These festivals fall into eight major categories: seasonal, harvest, food, sport, historical, cultural, religious, ethnic/folk.

Grafton's Jacaranda Festival is one of the many which celebrate the harvest or the flowering of a natural local product — mineral, agricultural, pastoral, floral or arboreal. Particular events at harvest festivals will vary, but the focus of the festival is always the bloom or the product that provides the harvest theme. In Grafton, according to a festival brochure,

> Floral displays decorate shops and offices throughout the week ... Jacaranda garden competition prize-winning gardens are open for public inspection ... Some 10 000 trees of many varieties form avenues for the most part of Grafton's 150 streets ... Grafton is transformed into a mauve coloured city for several weeks during October and November.

There are street fairs, a procession, dinners, fireworks, band performances, art and craft displays, a festival queen, entertainment for children and young adults and a variety of sporting events. There is also the statutory public event that goes with most festivals — the official opening.

Although there are other long-running festivals like the Barossa Vintage Festival and Toowoomba's Carnival of Flowers, many of the current major festivals in the separate states had their beginnings in the 1950s. First of the state capitals to have its own festival was Perth. It grew out of the annual summer school of the University of Western Australia, which was extended in 1953 to include evening concerts and plays. By the 1970s the festival was attracting hundreds of thousands of people to venues all over Perth, where there were art exhibitions, musical performances, plays, films and ballet. Capital city festivals followed in succession in Melbourne (Moomba, 1955), Adelaide (Festival of the Arts, 1960) Brisbane (Warana, 1961), Sydney (Festival of Sydney, 1977 and Carnivale, 1981) and Canberra (Canberra Festival, 1980).

Writing in the *Canberra Times* in 1965, Jean Battersby commented, 'Festivals are booming all over Australia and claiming unique attractions with a fervour equalled only in France'. She went on, however, to state that the claim on uniqueness, with the exception of Adelaide, was unfounded. 'All over the land, be it Melbourne's Moomba madness or Gunn's Gully Gala Day, Australian festivals are a uniform mad whirl of crepe paper, bikinied amateurs, marching girls and volunteer brass bands.' The evidence certainly suggests that there is a sameness in the types of entertainment put on for the folk who attend. Nancy Cato, referring to a frenzy of festivals (*Canberra Times*, 1966), thought that the over-supply could be resolved by picking the best features out of each one, 'like Adelaide's art shows and open-air poetry readings, and Melbourne's Book Fair', and moving them to the shores of Lake Burley Griffin. 'I have a wonderful idea for co-ordinating Australia's frenzy of festivals', she says, 'transfer them all to Canberra'.

Nobody took up Cato's suggestion, but others continued to express concern through the 1970s at the number of festivals and at the repetitiveness of their content. One writer in the *Australian* spoke of a 'plethora of festivals' leading to an eventual state, for one who attended them all, of cultural exhaustion. A solution to this was proposed by Mary Gage in the *Sydney Morning Herald*, 1980. She thought that if all the major festivals could agree to arrange their dates in sequence, with no overlapping, it would be possible to assemble a programme of festival entertainment, under the title of the

Festival of Australia, and to take the festival in turn to all the major festival centres across the country. Pointing out 'that festivals are more than fun: looked at together, they make up a multi-million dollar industry', Gage saw an agreement between the various festivals on cost-sharing, leading to increased profits for all of them, and to an improvement in the standard of artistic performances. While her proposal also was not taken up, some sharing of costs does occur, at least between Adelaide and Perth, and international performers often extend their season beyond the festival period so as to be able to travel to and to perform in other cities.

In all these major new festivals there is a tension between what may be called the commercial and the folk elements. It is true that they are very expensive exercises. Festival organizing committees generally try to bring the best entertainment they can afford to their city or country town. To pay for this entertainment, they must rely on sponsorship from major businesses and on admission charges to the entertainment. Festivals like Adelaide's biennial Festival of the Arts have a budget of millions to contract international and Australian performers to appear at the festival. Ticket prices are often high, but fortunately there are a number of events on the festival programme that are free and which offer an opportunity to the ordinary citizens of Adelaide to share some of the major attractions of the festival — a night of opera in the park, a range of art exhibitions, writer's week, a park concert, a food fair and street performances. But these 'free' events also have to be paid for, by sponsorship preferably, and the cost of entertaining the folk has to be covered in the festival budget.

It is a reality that festivals that consistently run at a loss don't survive. The pressure is therefore on all festival organizers to run a festival that draws large numbers of people to enjoy and to pay for the attractions. In this sense the conclusion that 'Australian local and regional festivals are spectator events' and that they are festivals for the folk rather than folk festivals has some validity. Len Evans drew much the same conclusion in 1972, when he wrote in the *Bulletin*, 'sometimes I feel there is too much looking on and not enough participation . . . When we learn to be more spontaneous, festivals might be more festive'.

What then is a folk festival? Australia's several hundred festivals are certainly meant to be entertainments for the people and they do attract people to attend and to enjoy them. What makes a festival a folk festival rather than a festival for the folk? Roper's assessment that Bendigo's Easter Fair did not have the same quality as the traditional fairs of rural England was based on the fact that it lacked spontaneity, in that it was 'organised from above', 'created by a committee', designed 'to foster social unity' and 'did not re-affirm existing social bonds'. From a study 'of long-lasting, highly popular, spectacular and much documented festivals on the world calendar of festivals' Alex Barlow concludes that

> the most important principle about folk festivals is that they are directed to the people of the community, they are planned for them and with them and they seek their support and involvement . . . They are celebrations of the folk of the community, for the folk of the community, and by the folk of the community.

It is the participation of the community in a celebration *of* the community that seems to separate folk festivals from spectator festivals.

In a detailed study of Melbourne's Moomba Festival (*Journal of Australian Studies* 17, 1985), Alomes proposes that the differences to be seen in Australian festivals may be attributable to 'the role of festivals, symbolism and leisure in a new and transplanted society'. He sees harvest festivals as being of

> little relevance to the itinerant workers of the Australian bush . . . The imperatives which once produced Carnivale in Europe and still do in South America were generally lacking . . . the social inversions of Carnivale . . . the inversion whereby the poor took on the roles of the rich for a day was culturally, if not always eco-

> nomically, inappropriate in a pioneering and modern society which was often sceptical of elaborate ritual in manners and ostentation in dress.

In short, Alomes asks, 'Have the new festivals been much more than local and leisure events, being instead an aspect of changing contemporary capitalism and the internationalisation of culture in Australia?'. Although his analysis of Moomba leaves this question basically unanswered, it does raise some interesting ideas. It may well be that a European, Asian or South American perception of a folk festival/carnival is not relevant in contemporary Australia, in which the role of business, tourism and the media promoting, supporting and organizing festivals has come to be accepted as appropriate and necessary. On the whole folk participation, least of all unplanned, spontaneous folk participation, is not encouraged. People's role in the festival is as spectators, there to enjoy the orderly parade of entertainment.

> The tension between organization and spontaneity (or accident) [Alomes points out], is part of the story of every public ritual. There is evidence to suggest that Moomba is a very modern event with everything increasingly planned: the permanent organization; the large budget; the choice of the period in March, on meteorological advice; the successful designer who assists with many floats and the role of a professional promoter as the first director of Moomba; the early involvement of the *Sun* newspaper; the role of business and government in providing floats from the very first; and the 100 marshals now necessary to control the parade and the spectators.

In time Australia may develop its own concept of a folk festival that is appropriate to the Australian context. There is some evidence that this is, in fact, starting to happen. Aboriginal Week, celebrated nationally in July, provides Aboriginal and Torres Strait Island people with an opportunity to celebrate their communities and their history. Even if no distinctive form of celebration has yet emerged, the festival events are organized and planned with and by local communities. There is a national Aboriginal Week committee, but its role seems to be to advertise the festival through the publication of a poster and to encourage Aboriginal and Torres Strait Island communities to organize their own celebratory events.

A number of other culturally distinct Australian people have also begun to establish their own community festivals (see, for example, **Chinese folklore in Australia**; **Ethnic folk heritage**; **German folklore in South Australia**; **Greek-Australian folklore**). The festivals are not intended to be exclusively for their own recently established migrant communities; drawing on the cultural traditions of their own motherlands, they seek to display and to share those traditions with all Australians. There is much more of the folk festival tradition in them than in other Australian festivals, and spontaneity is encouraged, as is the full participation of the community in planning and presenting the events. The Glendi Carnival in Adelaide, held in 1990 as part of the Adelaide Festival of Art, offered 'a weekend of feasting, music, dancing and family delight Greek style . . . Glendi celebrates the Greek way of life'. A number of Oktober-fests celebrated around the country from Darwin to Canberra recall the famous Munich festival and, no doubt, rely more on spontaneous folk frivolity than on planned events to provide the festival entertainment — along with the consumption of large quantities of beer.

Apart from these 'ethnic' festivals, there are a number of other festivals which label themselves 'folk festivals' or feature folk events. For the most part they provide opportunity for local people to put on show folk arts and crafts, to perform folk music and dance, to demonstrate folk skills and to take part in traditional (or recently invented) folk sports like brick-throwing, cane-toad racing and tossing hay bales. Some, like the Ben Hall Festival in Carcoar (NSW), the Thunderbolt Festival in Uralla (NSW), the Shinzu Matsuri Festival in Broome (WA) and the Fisher's Ghost Festival in Campbelltown (NSW) are based on events in local history and encourage re-enactments and representations of those and other events in the community's past.

Every year until 1992 the Australian Folk Trust took its National Folk Festival to one or other of each of the states and territories. In 1992 the Trust decided to fix 'the National' in Canberra. In 1990 the venue was the picturesque rainforest township of Kuranda in North Queensland. Following custom, the people of Kuranda planned the festival as a celebration of their local folk traditions as well as of national and international folk cultures. A major contribution to the programme came from local Aboriginal and Torres Strait Island folk artists. The programme itself was exhaustive and exhausting, with activities scheduled from as early as 8.30 a.m. and as late as 11.00 p.m. Events ran concurrently through the day. At one time, for instance, one could choose from a session on dance heritage, a concert, classes in playing the penny whistle, or a solo performance of song, yarns and verse on 'Our Sporting Heroes'. Children had their own festival programme, running daily from 8.30 a.m. to 1.00 p.m. Concerts, cabarets, workshops, dances, history sessions, folklore, poetry, spontaneous music performances, storytelling, displays, craft and a number of special events were all on offer. Such festivals, whilst essentially for folklorists and practitioners of folk arts, do provide an opportunity to display Australia's folk traditions, with an emphasis on those of the local region. As general celebrations of Australian folklife combined with schools in the history and practice of folk arts, they belong in the category of folk festivals.

Most of the festivals coming under this heading of ethnic/folk have in them the potential for the development of a different type of Australian festival — one that is much more a festival 'of' rather than just 'for' the folk.

There will always be festivals put on as entertainments for people, of course. The 'bread and circuses' approach to popular public festivities has existed from the earliest historical times and will continue here in Australia as it does in other parts of the world. In time these festivals may develop unique characteristics which could distinguish them from other such festivals on both the national and international calendar. Canberra's newly established *Floriade*, with its concept of laying out to an annual theme newly designed flower beds, and combining the floral display with open-air performances, theatre and art exhibitions, has the potential for turning the 'bush capital' into a giant springtime flower garden. It could also provide the occasion for displaying all Canberra's artistic talent — musical, dance, drama, literary, painting, sculpture, crafts, the arts in all forms. Such a festival could effectively combine the best features of both the festival for, as well as the festival of, the folk. The more formal, organized, 'official' floral display could be supplemented by community groups in the many Canberra suburbs collaborating with the ACT Parks and Wildlife Service in preparing floral displays in their own local parks and in their home gardens. Similarly, the official artistic displays and performances could be supplemented by more or less spontaneous, voluntary performances and displays mounted, with official encouragement, in suitable venues made available by public authorities and private companies.

This is one way a folk festival tradition may develop in Australia. Another is through festivals featuring local 'folk arts' and using an event in local history as a focus for celebrating the life and history of the community. Whether a *national* folk festival, like the Festival of American Folklife held annually in Washington, is possible or desirable is an interesting question. The American festival is not restricted to music or dance, but embraces crafts, recreations and traditional skills. Each American state is featured in turn in the annual festival and indigenous skills from that state are displayed. The proposed National Museum of Australia is intended to provide a permanent record and exhibition of the folklife of Australia, including regular demonstrations of indigenous folk skills. Maybe an occasional national look at Australian folklife, at times like the recent bicentenary, is all that is really necessary. For the rest, it could be left to the folk in Australia's urban and rural communities to celebrate together their own folk arts and skills.

There is an assumption here that there is an on-going, distinct, Australian folklife that has its roots in local communities and that this is recognized and valued by the people of those communities. 'On-going' in this context does not mean 'unchanging'. It means rather that people of a community have a perception of themselves as a 'folk' — people who share a common tradition, who speak an 'in' language with overtones of meaning that acknowledge that tradition and who share many aspects of daily experience. Many Aboriginal and other non-Anglo-Australian communities in urban settings constitute a folk in this sense, but so do many other Australian people, especially in country districts and towns.

Under the influence of press, radio and television, the folk life lived by the parents and grandparents of these present-day folk has changed, along with folk skills. Because of the current revolution in communication through highly developed electronic media, even in a country as large and as far spread as is Australia, it is not easy to sustain folk individuality. It is almost inevitable then that local folk festivals will become increasingly similar in content. The only real differences to mark one from the other may be the theme — harvest, historical, religious, folk culture — or the differences in the way of life and economy and the consequent valued skills of the community. None the less, these elements are still sufficient to make possible genuine folk festivals, provided that that is what the festivals are intended for. To do that, it is important that each community develop a content and a style for its festival that specifically recognizes and identifies the folk elements of community life. The folk must be able to see it as their festival by seeing themselves in the festival. This presupposes a deliberate intention on behalf of the community to create a festival about themselves, one that reflects the life, the history, the concerns, the events of their everyday life. It should be seen with hilarity. Festivals should be about laughter, about being part of it, about sharing, about singing, about having fun, about celebrating the folk.

It may take time for any Australian festival to take on its folk-festival form. *Festnacht* in Basel, one of the great folk festivals of Europe, had its origins in early mediaeval traditions and has undergone many changes in developing its present-day format. Alomes claims that 'in the contemporary world, tradition is something which grows readily over surprisingly short periods of time'; perhaps not over the twenty-five-year period, though, to which he allots the development of some Moomba 'traditions'. If some of the many festivals which are now celebrated around Australia survive until Australia's tricentenary of European immigration, it may be that they will by then represent a real Australian form of folk festival, with their own well-established and unique sets of folk traditions.

Alex Barlow

FISHER'S GHOST, probably Australia's best-known ghost, is said to be the restless spirit of Fred Fisher, murdered by his business partner at Fisher's Creek, Campbelltown, NSW, in the 1820s. A full survey of Australian ghost-lore remains to be carried out, but for a nation often said to be decidedly non-spiritual, there is a considerable folklore of ghosts and hauntings of natural features and locations (the Nullarbor Nymph; the Black Horse of Sutton in the Monaro district of NSW; the Min-Min Light of Queensland; the Ghosts of the Glen, Kiama, NSW); shipwrecks (the *Alkimos* off the West Australian coast, the Mahogany Ship off Warrnambool, Vic.); monasteries (The Blue Nun of New Norcia, WA); innumerable past and present public houses throughout the country; public buildings including theatres, courthouses and historic ruins of every kind. There are ghosts of bushrangers both famous and obscure, of convicts, of Aborigines and of sundry other unfortunate souls who came to a violent or otherwise unnatural end. As well there is a 'Phantom Mail' at Hay, NSW; 'phantom hitchhikers' on many roads, said to be the victims of road accidents; ghosts or mysterious happenings near cemeteries, bridges, railway/freeway flyovers; headless or otherwise phantom motor-cyclists, not to mention the extremely common 'haunted house'. This is hardly an exhaustive list, but gives some indication of the breadth and diversity of the Australian ghost and other unexplained phenomena, such as UFO sightings and related events.

Folk and popular art Australians are to a large extent unaware of and ill-informed about the existence of popular art traditions in their own country. Scholarly interest, exhibitions and publications have focused almost exclusively on 'high' art, and there are few collections, public or private, devoted entirely to popular art. Australia's 'cultural cringe' has been responsible for the development of a cultural outlook which, in an attempt to establish an extensive history of cultural achievement, has tried to ensure that popular art is accepted into the canon of 'high' art under the tag 'colonial', 'naive', 'primitive' and 'folk'. The descriptions 'folk artist' and 'folk art' seem inappropriate in Australia. It is better to use the label 'popular' artist/art which can encompass the amateur and the professional of varying degrees of skill, as well as the artisan who works in a trade that we might easily describe as an art.

Those who have produced popular art in Australia have come from a wide variety of backgrounds: convict and free settler, professional and amateur. Although their culture was most often Anglo-Celtic, the Chinese and German settlers of the nineteenth century brought and developed other traditions. In the twentieth century a much wider variety of cultural and religious backgrounds has been introduced into a culture still predominantly Western European. Popular art, unlike fine and 'high' art, belongs to no specific economic stratum of society. The leisured gentry of the nineteenth century produced water-colours to fill scrapbooks and pursued an endless variety of hobbies, including embroidery, shellwork, fernwork, fretwork, etc. (see **Patchwork quilts**, for example). Although examples of the hobbies and arts of the poor and less leisured have seldom survived, scrimshaw, primitive furniture and some convict crafts give evidence of the existence of a lively interest in popular arts pursued as either relaxation from arduous occupations or as a means to supplement a meagre income. Economic distinctions, however, often determine the nature of popular art: a wealthy nineteenth-century gardener was likely to build (or direct the building of) and/or decorate a grotto or rustic arbour, the poorer gardener would have settled on a more modest scheme, often decorating the outside of the house or filling the garden with topiary, gnomes and other sculptural creations, sometimes of his own making. These distinctions continue to exist in the twentieth century.

Many of the first European artists working in Australia were members of the naval and military forces, exploring or stationed in the colony. Some had training as topographers and cartographers and as such were familiar with some artists' materials and techniques. Some, although skilled, were obviously amateur artists; their interest was to record for themselves, their family and friends the infant colony, its landscape, flora and fauna, its native inhabitants and their customs. Of these, the works of the sailor George Raper (*c.* 1768–1797), Captain John Hunter (1777–1821), and perhaps Joseph Banks (1743–1820) are representative. Many of the early popular artists remain unidentified; one such is the painter of *Captain Bligh Hauled from Under His Bed*, *c.* 1808. Because of their anonymity these works have been little noticed in any account of Australian art.

A larger and better documented group of early popular artists were convicts. Of these R. Browne (1776–1824), W. B. Gould (1803–53) and C. H. T. Costantini (*c.* 1801–*c.* 1860) are among the better known and most notable. Only a few had received some sort of training in the medium in which they worked in Australia. Engravers and china painters and others whose trades and skills associated them with painting were pushed into service to fulfil the needs of the infant settlements in Sydney and Hobart. Of these W. B. Gould is an excellent example. A china painter by trade, he executed a remarkable series of botanical watercolours while assigned to Hobart's Colonial Surgeon, Dr James Scott, but is best known for his still lifes of game and flowers which tend to reveal his limited ability as an artist. An exhibition of his work, accompanied by a catalogue, *William Buelow Gould*, was mounted by the Burnie Art Gallery, Tasmania, in 1980. The first book to pay particular attention to such work was Hackforth-Jones, *The Convict Artists*, 1977.

The population explosion that occurred as the result of the discovery of gold in 1851 greatly altered the nature of Australian society. A much more sophisticated population meant a greater demand for more sophisticated art and it became possible for professional artists to make a living from their work. The existence of a larger artistic community with a much greater production has meant that accounts of art in Australia in the last half of the nineteenth century have almost totally ignored the great proliferation of amateur and popular artists. Of the popular artists S. T. Gill (1818–80) must be the best known and documented. The subject of Dutton, *Paintings of S. T. Gill*, 1962 and Appleyard, Forgher and Radford, *S. T. Gill: The South Australian Years 1839–1852*, 1986, an exhibition catalogue, Gill's work best exemplifies the concept of a popular artist. He had some formal training, and is best known for his scenes of daily life, especially those on and around the goldfields. His work is characterized by accurate observations, a delight in commonplace detail, wit, and a sometimes overwhelming sense of the ridiculous.

Among numerous other artists who made a living from popular imagery, especially house and homestead 'portraits', Henry Gritten (*c.* 1818–73) and William Tibbits (1837–1906) are notable for the quantity and quality of their work. Others such as Rene Raimonde (active 1880s) and C. G. S. Hirst (active 1870s and 1880s) are known by only a few works.

At the same time other artists, most often amateur and less competent, were working, but few are remembered by anything more than a single work in a provincial art museum. Paintings by Horatio Bruford (1846–1920) *The Scene of the Wreck of the Loch Ard* *c.* 1878–79; Cuthbert Clarke (active *c.* 1850–65) *View of the Disastrous Fire in Moslyn St Castlemaine on Sunday 15th April 1860 by Which Five Lives Were Sacrificed*, 1860; Daniel Clarke (*c.* 1837–1928), *Tower Hill*, 1867; Charles Hill (1824–1916), *Family Group*, late 1860s; Theodore K. King (active *c.* 1855), *The First Parliamentary Election in Bendigo*, 1855; and T. G. Moyle (active *c.* 1865–85), *A View of Sergeants Hicks', Williams', Smith's, Wilson's and the Washington Quartz Gold Mining Company's Claims, Redan, Shipton Street, Ballarat, January 1881*, 1881 are all memorable images. Their work was presented to a larger audience for the first time in a 1988 exhibition and its accompanying catalogue, Hansen, *The Face of Australia: The Land and the People: The Past and the Present*.

The existence of another group, nineteenth-century Aboriginal artists working in a Western tradition, has more recently been considered. William Barak (*c.* 1824–1903) and Tommy McRae (*c.* 1836–1901), both working in Victoria in the late nineteenth century, produced paintings and drawings such as Barak, *Corroboree*, *c.* 1880s, and McRae, *Sketchbooks* *c.* 1890, that are masterpieces of naive art. Further research will undoubtedly reveal other Aboriginal artists of this period and another aspect of Australian art will come to light.

Following the arrival of the Europeans, Aboriginal life and culture was disrupted and altered in innumerable ways. The Aborigines were and remain, however, receptive to new ideas and ways of expressing themselves in their art, as well as ways of making money. As contact with the Europeans increased, works of art were made by Aborigines especially to suit European tastes. The Tasmanian Aborigines and their descendants made and sold shell necklaces, which enjoyed a special vogue in the 1920s and 1930s. The Aborigines of the Lake Eyre district of South Australia produced dog sculptures and the enigmatic 'toas' (ostensibly 'way-markers') most probably to meet the demand of collectors, a fact which is well documented in the South Australian Museum catalogue *Art and Land: Aboriginal Sculptures of the Lake Eyre Region*, 1986. Feather flowers; feather, nut and seed jewellery; emu-egg and boab-nut carvings; painted and poker-worked wooden objects; shell work and numerous other art objects have been derived from traditional arts to produce something for sale in the market for souvenirs. Some became and a few remain forms of popular art of great importance, but they are as yet poorly documented.

From the earliest times of European settlement — particularly in the nineteenth century and less so from the early twentieth — women were the most numerous of artists working in a variety of popular art forms. As well as executing paintings, sketches, miniatures and silhouettes, sometimes showing great skill and sensitivity, they also kept scrapbooks and albums. Although little has been published on such work, Dr Joan Kerr of the Power Institute of The Arts, University of Sydney, has done noteworthy research.

An interest in naive Australian artists and their work developed in the 1960s, perhaps as collectors who had seen the work of the English naive, Alfred Wallis, or heard of the efforts of the much publicized American Grandma Moses, sought an Australian equivalent. In Sam Byrne (1883–1978) and Lorna Chick (born 1922) collectors found remarkable examples. The immense popularity of the work of Pro Hart (born 1928), and artists such as the Brushmen of the Bush is a phenomenon of another kind, more closely linked with contemporary publicity and the mythologizing of an Australian bush tradition.

A wide variety of Australian popular art objects have been better served by history than those designated 'fine' art. Although seldom collected and exhibited by art museums, much has survived in private collections and in history and folk museums. More recently, the approach of the bicentenary of European settlement and associated plans for special bicentennial exhibitions in 1988 fostered a greater public awareness of our more humble origins and some of our 'less impressive' traditions. As a result the rich variety of Australian popular art has been shown in several exhibitions and has featured in publications specifically concerned with the subject. In some cases examples of popular art have been included in exhibitions devoted primarily to other media, notably 'Victorian Vision: 1834 Onwards' at the National Gallery of Victoria in 1985; and in 'Shades of Light: Photography and Australia 1839–1988', and 'Drawing in Australia: Drawings, Watercolours and Pastels from the 1770s to the 1980s', both held in 1988 at the Australian National Gallery. In such instances the inclusion of select examples of popular art has added another dimension to the exhibition.

One of the earliest exhibitions devoted specifically to Australiana, and including a wide variety of popular arts, was 'Australian Antiques: First Fleet to Federation'. Organized by the Women's Committee of the National Trust of Australia (NSW), the exhibition opened at Darling Point, Sydney, in October 1976. The book of the exhibition, published the following year, has proved an invaluable and popular guide for the interested public. Another exhibition, 'Colonial Crafts of Victoria: Early Settlement to 1921', which was researched, selected and catalogued by artist, historian and collector Murray Walker, and held at the National Gallery of Victoria from November 1978 to January 1979, proved enormously popular with the public. The exhibition and catalogue were divided into ten sections: Aborigines, By Road and Track, On the Land, Hearth and Home, Recreation, Gold, Children, Town and Metropolis, Ceremony, and Pictorial Exhibits, and explored and exhibited a great diversity of skills, crafts and popular arts and demonstrated the richness of the Australian experience. The catalogue of the exhibition was a precursor to Walker's *Pioneer Crafts of Early Australia*, 1978. In 1986, the exhibition and catalogue *Colonial Crafts of South Australia*, prepared by

Judith Thompson, Curator of Australian Decorative Arts at the Art Gallery of South Australia, Adelaide, surveyed popular art in that state. Most notable were examples of furniture, model-making and fancy work, much of which showed the strong traditions of German settlers.

More recently, a section of the exhibition 'Australian Decorative Arts 1788–1988', held from October 1988 to February 1989 as part of the Australian National Gallery's bicentennial exhibition programme, was devoted to popular art of many different kinds and in many different media, from all parts of Australia and from all periods of our history since European settlement. To coincide with that exhibition a souvenir book, McPhee, *Australian Folk and Popular Arts in the Australian National Gallery*, 1988, was published. The Australian National Gallery's collection of popular arts, while concentrating on work of outstanding artistic merit, is the best collection of its kind in a public art museum. Its quilt collection includes the Rajah quilt (1841) made by convict women en route to Van Diemen's Land and the Westbury quilt (1900–03), and it holds outstanding examples of primitive, rustic and decorated furniture, as well as scrimshaw, models and toys.

Various magazines and books devoted to the collecting of Australiana in general have given some attention to popular art. However, there have been few serious assessments of any particular medium or aspect of the tradition. Cuffley and Carney's *A Catalogue and History of Cottage Chairs in Australia*, 1974, devotes some attention to primitive furniture. Hooper's *Australian Country Furniture*, 1988, explores the subject further. Cornall's *Memories, a Survey of Early Australian Furniture in the Collection of the Lord McAlpine of West Green*, 1990, documents an extraordinary private collection. Accounts of individual commercial potteries that flourished in the nineteenth and early twentieth centuries, and Ioannou's *Ceramics in South Australia 1836–1986: From Folk to Studio Pottery*, 1986, take into consideration the popular art aspects of pottery production. However, the most thorough publication devoted to popular art in any medium is Rolfe's *Patchwork Quilts in Australia*, 1987. Isaacs's *The Gentle Arts: 200 Years of Australian Women's Domestic and Decorative Arts*, 1987 is similarly thorough in its account of many aspects of women's popular arts.

No account has yet been written of, or an exhibition devoted to, the popular art traditions of a specific ethnic group within Australia. Various German settlements (especially in South Australia), Chinese communities, and more recently, southern European, Middle Eastern and Asian immigrants have all brought with them distinct cultural traditions, some of which have been perpetuated in Australia. Winternitz's *Australia's Hidden Heritage*, 1990, briefly examines one aspect of these traditions.

While some forms of popular arts such as scrimshaw, fernwork, seaweed and shell pictures and illustrated addresses and mottoes have virtually disappeared, many others continue and are frequently not so different from their original form. The traditions of union banners, Sharman's boxing tent signs, and the Skipping Girl neon live on in the Big Banana or the Big Merino; and the street decorations and parades of the nineteenth century, very often designed by artists such as Augustus Earle (1793–1838), are perpetuated in the Melbourne Moomba and Sydney Mardi Gras parades and floats — the latter frequently 'choreographed' and designed by Peter Tully (born 1947) — and the Macquarie Street, Sydney, decorations for the bicentennial celebrations in 1988. Nineteenth-century fireworks displays have been equalled by displays to mark the Queen's coronation visit in 1954 and more recently with the torching of the Sydney Harbour Bridge as part of the fireworks on 26 January 1988.

While harvest festival decorations in churches are now not as elaborate as they were, butcher's shop windows are no longer painted weekly and fruiterers seldom bother to amuse their customers with artistic (sometimes erotic) arrangements, department stores now vie with each other in increasingly elaborate and spectacular window dressing, and the cake decorating and the Grand Parade at Sydney's Royal

Easter Show still ensure the continuation and development of the popular arts in the twentieth century.

See also: **Australian National Gallery**.

John McPhee

Folk belief is often referred to as 'superstition'. As this term has pejorative connotations, one person's belief or conviction being another person's 'superstition', few folklorists today use it. The term, however, is still widely used in everyday speech. Belief is often an integral element of many kinds of folkloric activity and may take a variety of forms. The performer of an artistic folklore item, such as a song or poem, may believe the contents to be a factual record of events, people and places, even though authoritative sources might provide evidence to the contrary. Alternatively, even when there is no factual information conveyed in an item, the performer(s) may feel that the item is emotionally 'true' and appropriate to their shared values and attitudes.

The observance of certain rituals and ceremonies also implies a high level of belief amongst those who participate. Some of these events have a formal religious component, such as the various Blessing of the Fleet ceremonies. Others, such as the initiation ceremonies inflicted on newcomers in some trades and occupations, have little or no formal religious element. Such activities (see **Customs**) are usually strongly traditional, and failure to observe them may be considered a serious oversight by members of the group or groups involved.

Good and bad luck are obvious areas where folk belief is involved. While some may laugh or be sceptical about a person's reluctance to walk beneath a ladder or to open an umbrella inside the house, many people genuinely believe that such acts will bring 'bad luck' upon them. There are innumerable methods for warding off or counteracting 'bad luck' or 'evil': for example, throwing a pinch of salt over one's left shoulder, wearing garlic cloves or other natural or manufactured amulets, charms and tokens, crossing fingers, 'touching wood' and a host of other appropriate gestures and practices, varying from culture to culture (see, for example, **Chinese folklore in Australia**).

Prediction and prophecy of various kinds also involve folk belief. Typical practices in these areas include predicting the weather, the outcome of a race or match or lottery, likely marriage partners, the course of one's life, and fluctuations in the stock exchange to name only some. Numerous, often ancient, methods may be employed for such prophecies, ranging from reading the Tarot to reading the pattern of leaves at the bottom of a teacup. Likewise, some traditional aspects of agriculture involving the right time to plant, harvest or otherwise process crops or animals, imply strong levels of belief in the beneficial and malign powers of the moon, sun, stars or climatic patterns.

Certain kinds of folk medicine (q.v.) also involve belief, particularly those preventatives and cures derived from natural sources. Instructions about the most efficacious or otherwise appropriate time to pick, treat or utilize herbs and natural products may involve the cycles of the moon or other heavenly bodies, appropriate locations, such as crossroads or churchyards, or spells of various types and the presence of a 'healer' or other special person who is believed to possess efficacious and unorthodox powers.

Belief is an important element of some types of narrative folklore, such as myths, tall tales and legends, particularly those involving hauntings or otherwise unexplained phenomena (see **Modern legends**; **Folk tales**). Children's folklore (q.v.) frequently involves high levels of belief, and belief also motivates the negative and derogatory stereotypes that underlie the various kinds of racist folklore that circulate widely in Australian society. The beliefs about 'others' that impel such folklore are deeply embedded in cultural tradition and are related to the occasional large-scale outbreaks of mass belief — or disbelief — such as those associated with the Azaria Chamberlain case.

Some works that deal with aspects of folk belief are Nancy Keesing, *Just Look*

Out the Window (1985) and Graham Seal, *The Hidden Culture: Folklore in Australian Society* (1989).

Folk commerce is a term which might be applied to many types of informal buying, selling and exchange of goods and services. The ancient practice of *barter* is widespread, although often conducted on an indirect *quid pro quo* basis where a neighbour's gift of surplus lemons or vegetables will be returned in kind some time later. Direct exchange or swapping of goods is less common unless between hobbyists such as stamp or coin collectors or among children swapping marbles or swap cards. Services such as baby-sitting or household maintenance may be exchanged between friends, colleagues or neighbours, sometimes using a token or points system to account for the services provided.

For many years an Australian wanting to raise money by the disposal of goods had little choice but to patronize the pawnshop, second-hand dealer or newspaper classified advertisements, depending on the seller's degree of prosperity. Today the individual has more choice; the *garage sale* has become a common part of Australian folklife. Garage sales were introduced to Australia from America after the 1950s, and provide an opportunity for an individual or family to raise a few dollars and get rid of unwanted possessions, usually at rock-bottom prices. Sometimes the holder of a garage sale will advertise in a newspaper, but the most common form of advertising is the hand-made sign propped outside the seller's residence or tacked on nearby telegraph poles. 'Car boot sales' are common in Perth. *Trash and treasure* markets were also introduced from the United States of America. These markets are often large and organized by a charity in an open-air location such as a car park where 'stall-holders' may simply spread their wares on the ground. They may sell plants, handcrafts and commercial products, but principally feature second-hand goods and 'junk' offered for sale in the belief that one person's poison may be another person's meat.

Local markets are often held. They mainly sell local produce, food and new handcrafts, and stall-holders are a mixture of amateur and professional sellers. These small markets are often held monthly, and visiting them is a popular weekend outing for families and individuals alike. Some stall-holders 'do the circuit' and go from market to market as well as (in many cases) selling from home.

Street stalls and *bazaars*, unlike all other types of folk commerce previously described, are not run for the benefit of individuals but usually for local schools, churches or charities. They are often held annually. They are run and stocked by volunteers and traditionally sell handmade and knitted goods, homemade cakes, jams and biscuits, 'stick-jaw' toffees and chocolate crackles, and usually have a 'white elephant' stall for unwanted household goods. In recent years second-hand books have become increasingly common at stalls and bazaars, and larger bazaars or fêtes may have new clothing provided by manufacturers. Some bazaars or fêtes include lucky dips and amusement stalls where (for example) you can test your throwing skills or unload your frustrations by smashing crockery.

Some of the practices characteristic of folk commerce have been evident in more commercial areas. An old-established custom is the hand-written advertisement stickytaped to the window of the corner shop or 'deli'. This type of community noticeboard has been extended in some large supermarkets. Folk commerce has become the basis of a successful commercial enterprise in newspapers such as the *Melbourne Trading Post* and similar large-city publications which offer a kind of media marketplace for advertisers to buy and sell all kinds of goods. The *Trading Post* runs largely on an honour system where the advertiser pays a commission only after a sale is made. Similarly, the house party organized among a group of friends or work colleagues has become a forum for selling many commercially produced goods, of which the best known are the plastic kitchen products known as Tupperware. The host of the party (usually female) receives a commission on sales made. The phrase 'Tupperware party' has passed into Australian English with a variety of meanings other than the literal one. They are not always complimentary, and may refer to the somewhat boring or fatuous party games played, rather than to the quality of the goods.

Folk heroes and heroines are found in all the folklores of the world, with Australian folklore having its fair share — of heroes, at least. A folk hero or heroine has attained a popularity that becomes enshrined in some form of communal, informal tradition (i.e. in folklore). This excludes many individuals such as Caroline Chisholm, the colonial social reformer, and William Farrer, developer of a rust-resistant wheat strain, who are justly celebrated in our history but do not figure in our folklore. Nor do explorers such as Leichhardt, Burke, Wills, Wentworth, Flinders, Mawson and so on.

Ambivalence is often a characteristic of folk heroes — while they are heroes to some, they are frequently villains to others. In Australian tradition we have only to think of the numerous bushranger (q.v.) heroes to bear out this observation. Politicians are another example of the frequently ambivalent folk hero. Jack Lang, Labor Premier of New South Wales during the Depression, was both celebrated and reviled in depression folklore for his political and economic views and actions. Some other politicians who fall into the same category are William Morris Hughes, (the 'Little Digger'), Sir Robert Menzies, (variously known as 'Ming the Merciless' and 'Pig-Iron Bob'), and most recently, Robert Hawke (variously 'the Silver Bodgie', the 'Mild Colonial Boy', 'Hawkie' or simply 'Bob'); Sir Joh Bjelke Petersen ('Joh', 'Bjelke-Joh' and latterly, 'Sir Joh'), and the Prime Minister, Paul Keating ('Mr Cheating'), to name only some (see **Political songs**).

Bushrangers and politicians are not the only heroes in Australian folklore. The tough, independent bushman pioneer is a recurring image in much of our folklore and is celebrated in a very large body of songs and poems. There are also occupational heroes, such as the mythical Crooked Mick of the Speewah (see **Wannan, Bill**) and the Aboriginal stockman or rouseabout Jacki Bindi-eye, a trickster figure about whom many tales are told in the northern parts of the country. Historical occupational heroes include Jackie Howe and Jack Dunne, the champion shearers. Closely related to these is the figure of the digger, the sardonic volunteer footsoldier (see **Anzac Day**; **Wartime folklore**). Sporting heroes include cricketer Sir Donald Bradman ('The Don'), boxer Les Darcy, a number of nineteenth-century jockeys, and the race horse Phar Lap (see also **Sporting folklore**).

Few women are celebrated in Australian folklore. While there is no shortage of women with the appropriate qualifications, such as Grace Bussell of Western Australia, the 'Grace Darling of the West' who saved survivors from a shipwreck, Dawn Fraser the swimmer and MP, and Daisy Bates the anthropologist and writer, to name only a few, none of these women are known to figure in folk songs, verse or narrative. The scarcity of heroines in Australian folk tradition is no doubt related to the general masculinity of Australian culture, and of the cultures from which many of its traditions are derived or influenced. As well, the emphasis on the collection of 'bush' traditions that has often characterized Australian folklore research has been inevitably biased towards the masculine concerns of rural life, pioneering, and occupations traditionally associated with males, such as shearing, overlanding and itinerant rural labour in general. The concept of a 'swag-woman' is, in Australian folklore, a nonsense. When women feature at all in folk song, verse and narrative it is either as incidental or background characters, such as barmaids, wives, landladies, for example, such as the 'Eulo Queen', or as the objects of the considerable body of male bawdry (q.v.). The usually anonymous women so depicted are never 'heroic'. One of the few named women to appear in Australian folklore is Ned Kelly's (q.v.) younger sister, Kate. Her role is to aid her and their friends and seems to derive mainly from the conventions of Victorian melodrama.

It has often been pointed out that many Australian folk heroes are losers, a fact that is usually noted in connection with the celebration of such notable defeats as the Eureka Stockade and Gallipoli. Reasons given for this rarely rise above the pedestrian, even the inane, and generally involve some variation of the 'colonial cringe', such as modern Australia having had insufficient time to develop a 'proper' national history, and the alleged superficiality of Australian culture. Such views also ignore the fact that the folk heroes of many cultures are losers, whether they be tragic heroes doomed from the outset, courageous individuals of one kind or another, tricksters, or outlaws who are almost invariably hunted down by the dominant forces they oppose and threaten. What is essential for the attainment of folk hero status is not success but struggle. It is the spectacle of men or women battling against what they know to be the overwhelming power of fate or authority that generates the folklore about most heroes and heroines. In losing to such irresistible forces, the individual ultimately wins by displaying the courage to resist them.

See also: **Bushrangers**; **Folk song**; **Folk tales**; **Women and folklore**.

Folklife Inquiry: see **Committee of Inquiry into Australian Folklife**.

Folk Lore Council of Australia, The, was founded in August 1964 by Francis Ashburner, Margaret Ashburner, Maryjean Officer, Rose Sayers, William Sayers and Barry Woods (all of Melbourne) as a non-profit making company under the *Companies Act* (Victoria) 1961. The Council was established to collect, study and disseminate folklore and oral tradition, particularly matters pertaining to the Australian heritage but also the folklore and tradition of other countries; and to promote literary, musical and cultural research in this area.

Early in the FLCA's life, the main activity was providing speakers and music programmes, with the special aim of getting people to recall the songs, ballads, stories, working skills, etc., of earlier days. Archives were built up to assist members, researchers and school children, local and overseas. An important recent link has been with the Storytelling Guild of Australia (Victoria). The FLCA has financially assisted many allied groups and projects, including funding issue 44 of *National Folk* (q.v.), 1971, and restoration of Gulf Station, Yarra Glen, Victoria.

FLCA publications include *A Collection of Australian Folk Songs and Traditional Ballads*, compiled by Rose Sayers and Maryjean Officer (1966); *Australian Folksongs of the Land and its People* (1974; revised paperback edition compiled by Rose L. Sayers 1979); the *Folk Lore Gazette* (1980–84); and newsletters.

Folklore occasional papers, is a series of irregular publications of unknown origin and uncertain ownership which deals with various aspects of Australian folklore. Many of Australia's most prominent folklorists have published in this series which, at the time of writing, appears to have come to rest at the Rams Skull Press, the latest paper, number 22, having been published by that organization in 1992.

Folk medicine is of two major varieties. The first employs 'natural' or 'rational' cures and preventatives that are derived from the natural environment. The medicines employed are taken from flowers, plants, herbs, animals and other elements of the natural world. The widespread use of lemon juice in hot water, sweetened with sugar (and possibly some alcoholic spirit) for colds is one example of this type of folk medicine, as is the use of eucalyptus oil. Other medicines may not be derived from nature, but may be adapted from commercially available chemical products. An example of this is the use of 'Blue', a form of washing whitener, to alleviate the pain of insect and other stings, particularly those of 'bluebottles' and other marine hazards that bedevil Australian beaches in summer. Many folk medicines of this type are, or have been, marketed in commercial form. These include 'goanna salve' ('the Australian bush ointment') and the various Rawleigh products, and those of other companies, once sold door-to-door. When folk medicines are commercialized, the range of ailments that they are claimed to be able to cure often expands significantly. One such product advertises its ability to relieve sores, eczema, piles, lumbago, sciatica and dandruff.

In the second major category of folk medicine are various 'non-rational', magical or occult techniques and beliefs in preventing illness or curing injury, disease or possibly unexplained sickness. A simple example is the wearing of garlic on one's person to deflect evil or bad luck. However, many 'non-rational' folk medicinal practices involve elaborate preparations, execution and timing according to any number of belief systems and calendars. Thus, various items of the human body (hair, tissue, faeces, etc.), of clothing or personal possessions may be required, as may the ritualistic sacrifice of an animal at a particular place and at a particular time, usually in accordance with some phase of the lunar or solar calendars. Charms, spells, prayers and other incantations may also be associated with these activities. While this may conjure up images of the Hollywood versions of 'Satanism' and 'black' magic, most non-rational practices are as simple as 'selling' a wart to someone kind or silly enough to pay for it, or washing one's face with early-morning dew to remove unwanted freckles or other so-called 'facial blemishes'.

Clearly there is considerable overlap between the 'rational' and 'non-rational' forms of folk medicine. What binds them together into a related system is belief (q.v.), the willingness of both the patient and, where applicable, the folk medicine practitioner to accept that the treatment will be effective. A considerable body of anecdotal and even documented evidence suggests that folk medicine of various kinds may in some circumstances have the desired effect. This applies both to the natural approaches, which form the basis of herbalism and naturopathy, and to the non-rational varieties involving such practices as 'faith healing'. While medical science investigates or ignores such beliefs and practices, folk cures for common ailments and inconveniences as colds, warts, hiccups, nose-bleeds and so on will continue, as will the more elaborate forms of folk medicine associated with serious, sometimes terminal, illnesses and afflictions. (See **Aboriginal folklife in central Australia**; **Folk belief**).

Folk music in Australia: the debate In the twentieth century Australia, like many other countries, has been grappling with the after-effects of a severe attack of nineteenth-century colonialism. The most crucial of these is a totally changed population profile, going from several hundred thousand black people to several million predominantly white people. The political and environmental changes in the country during the last two hundred years have been among the most radical, disruptive and stressful to have taken place anywhere in the world during the period. The social, intellectual and artistic changes have been just as important, though less visible. Folk music provides a useful example of the process of 'Australianization' which has been occurring during the last two hundred years in many fields of intellectual and artistic activity. On the one side, Australians are uncovering and making sense of what has gone on in the past: the attempted destruction of an indigenous population with rich and varied cultures, and the imposition of others from the opposite side of the globe with cultural beliefs and practices which often proved unsuitable in the new environment. On the other side, Australian culture is continually stretching to accommodate new arrivals from still different parts of the world. Australian solutions will probably be of a similar kind to those arrived at in other countries with comparable experiences, but because no two sets of circumstances are exactly parallel, they will have a peculiarly Australian slant.

The terms 'folk music', 'folk song', 'folklore', 'folklife' came into use in European languages during the late nineteenth and early twentieth century. Their etymology is important, but need not be rehearsed here. In 'Making Folk Music' (*Meanjin* 44, 4, 1985) Graeme Smith has given a brief account of Australian usage of the concept. A romantic linking with agrarian life, connections with nationalist sentiments, and later (twentieth-century) links with left-wing political life can all be found in the Australian folk music ethos, but do not by any means account for it entirely.

Also influential but not deterministic were the large collecting programmes being carried on in Britain and the United States in the last two hundred years. The stories, legends, myths and songs uncovered by these programmes now form a large body of material, whose provenance is in some cases highly questionable, and whose representativeness is severely impaired. Dave Harker documents this in *Fakesong* (1985). Similarly, at a recent festival workshop in Port Fairy, talking about those late nineteenth-century romantic ideals of folk music, Danny Spooner quoted Hubert Parry's characterization of folk songs as containing nothing unclean or filthy and having no false glitter or vulgarity. Danny Spooner pondered aloud about which songs Sir Hubert had heard, and proceeded to sing a collection of songs which did not conform in any respect to his description. It seems clear that the only songs published by the collectors were those which fitted their expectations; any which did not were altered so that they did, or left out entirely. Early Australian collectors, working in the 1950s and 1960s, very soon abandoned the censorship which such adherence to rigid definitions of folk music would demand. Spooner, a British-born singer who lives and works extensively in Australia — and who does on other occasions sing Australian songs — still draws on and is engaged with British material and thought. The paradoxes inherent in this situation are those which Australians live with and manipulate constantly, and are indeed vital to the present richness of Australian culture.

The European and American definitions of folk music rest heavily on the work of Cecil Sharp. His characterizations of folk music provided the basis for the International Folk Music Council's 1955 definition in which oral transmission and the 're-fashioning' and 're-creation' of music by members of a community are crucial elements. Some of the difficulties raised by this definition are discussed by Charles Seeger and Klaus Wachsmann in their respective entries on 'Folk Music' in the *New Grove* (Sadie, *The New Grove Dictionary of Music and Musicians*, 1980). The concepts might, as Charles Seeger suggests, have been scrapped, but for the redefinition and expansion of their use during the first half of the twentieth century. Albert Lloyd (q.v.), in *Folk Song in England* (1967), defines 'folk music' as the music of the working class, and extends the application of the term to 'industrial songs'. An extension of this usage provides the identification with oppressed minorities and the development of the notion of 'protest' songs.

In Australia, the use and definition of the term 'folk music' is still being debated. Many Australian publications contain discussion of this topic, and suggestions almost always provoke heated reactions. Much of the debate centres on what folk music *ought* to be, and therefore involves the values and attitudes of the contributors as much as the results of empirical investigation. Many of the strictest definitions of 'folk music' are vigorously defended by people who nevertheless publish material under that title which does not conform to their definitions. The extent to which any song or dance tune has been passed on by means of oral transmission, whether it reflects a community's experiences, whether it exists in several versions, and whether the performer knows who the composer is are all matters which have been suggested in Australia as providing criteria to assess whether or not a piece of music is 'folk music'. These are matters which may or may not turn up in the course of research, and may or may not help to explain something about what music means or how it works in particular circumstances. They do not provide a readily identifiable corpus of material with which to begin research.

Instead of searching for a definition of 'folk music', we might do better to take a holistic view of the musical life of the folk clubs and of the substantial section of the Australian population which has an interest in them. ('Music' here refers to vocal and instrumental performances. Isolating these from closely related forms — the dances which they often accompany, and other performance genres such as recitations and yarns, for example — involves a considerable distortion. For simplicity's sake, however, the conventional distinctions are preserved here.) Of course folk music does not only take place in folk music clubs; but these clubs provide a good starting point for research.

This social definition of the genre excludes several kinds of music which are sometimes called 'folk music'. Foremost among these are certain kinds of Australian Aboriginal music. Aboriginal music in its original setting uses musical forms, instruments and languages whose origins predate European settlement in Australia. It is diverse, rich and little understood in contemporary Australia outside the groups which practise it. In those groups music is a much more central part of life than it is in European-derived cultures. It is crucial to the expression of religious beliefs, social identity and physical place. The term 'folk music' cannot be usefully applied to any of this music. However, in the continuing struggle to establish an Aboriginal Australian identity in the late twentieth century, some Aboriginal groups are arranging performances of their music for display to non-participating audiences. Some performances of this kind take place at folk festivals. Australian Aboriginal music which uses European-derived musical forms and instruments has also been of interest to folk music clubs. These present quite different uses of Aboriginal music and fall within the sphere of folk club music.

Also not included are the extensive and well-supported clubs which owe allegiance to particular countries of origin and are concerned to preserve and teach songs, dances and customs of another country. These clubs have particular social functions often related to special problems of identity and are different in focus from the 'folk clubs' to be described here. They are of crucial importance in contemporary Australian society: their artistic contribution is often recognized, but their social and political importance is often underplayed, probably because of the inherent conflicts which admission of them would reveal.

This account, therefore, is of Australian folk club music. The term is not intended to be pejorative. Like 'church music' and 'chamber music', folk club music is defined by the place where it is usually heard.

The folk club culture in Australia is extensive. Politically motivated interest in 'folk music' was evident in Australia during the 1950s, and several people, independently, began to 'collect' music. Some of the results of this 'collecting' were published in songbooks. In *Australian Tradition* (q.v.) of October 1966 Ron Edwards gives a list of songbooks, published between 1950 and 1966, as an indication of the 'pattern of revival'. Nearly forty years later the 'revival' has produced an extensive network of clubs and associations throughout the country. Individual clubs form and fold, but at any one time there are dozens of folk clubs in the country. They band together to form state folk federations, which in turn elect the trustees of the national body, called the Australian Folk Trust (q.v.). Activities undertaken by clubs include mounting regular workshops, sponsoring performances, providing classes in music and dance, organizing dances and balls, providing a booking service for performers, sponsoring collecting activities, publishing songs and dance tunes and other publications of folkloric interest, publishing newsletters and journals, providing resources for radio programmes, theatrical performances and films, maintaining libraries of material and archives of folk music activities, providing a performance circuit for performers, organizing festivals, and sponsoring and undertaking research.

Margarita Prosilis shows the string figure variously called *The Bridge*, *Jacob's Ladder* and many other names. String games are popular among Australian Aborigines and Europeans, and among many other nationalities. (Photograph, Kel Watkins)

Varieties of the game of marbles are almost universal. An Indo-Chinese boy displays his marbles collection, Debney Meadows Primary School, Melbourne, 1984. (Photograph, Australian Children's Folklore Collection, University of Melbourne)

Wind toy, *c.* 1930, Wyong district, NSW. (Australian National Gallery)

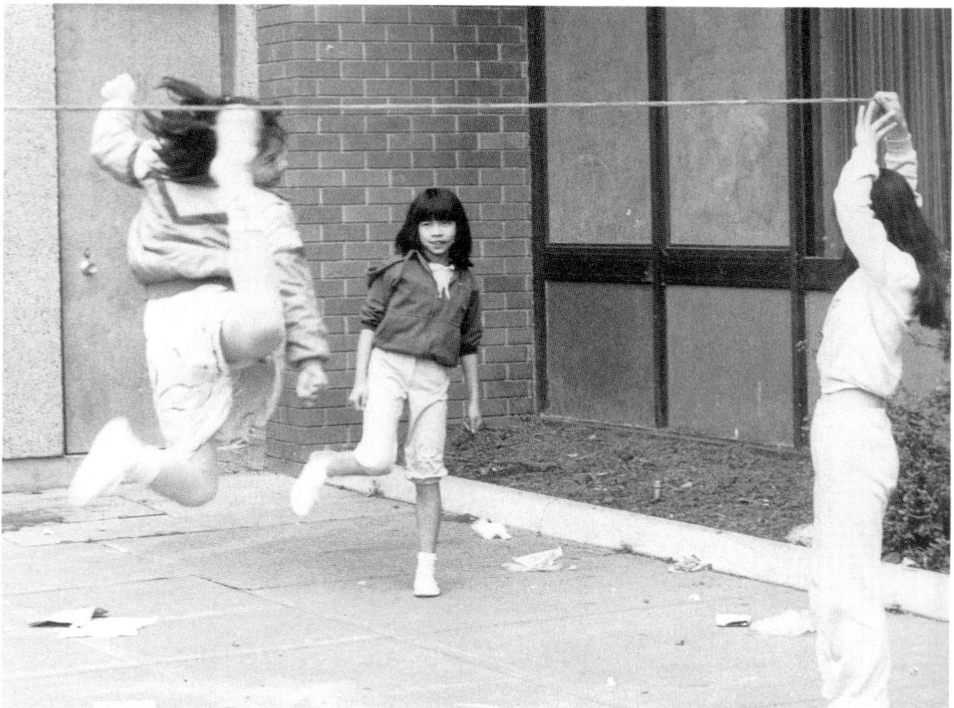

Vietnamese girls playing elastics at Debney Meadows Primary School, Melbourne, 1984. The spectacular 'high jump' is done with the elastic held at a height of 'one plum'. (Photograph, Australian Children's Folklore Collection, University of Melbourne)

Children playing handclapping games, Brentwood Primary School, Melbourne, 1983. (Photograph, John Scott Simmons FRPS)

Home-made toy koala from Tasmania, *c.* 1950. (Collection and photograph, Australian National Gallery)

The organizers and participants have very strong and forcefully articulated ideas about what 'folk music' is, and the extension into the community through both research and performance is an important part of the philosophy of the clubs (not that the folk clubs have a unanimous point of view in regard to any matter: on the contrary, debate and dialectic are central to their intellectual life). These views about the meaning and purpose of music take precedence over musical aesthetics and criteria of excellence. This is in contrast to most other loci of musical activity in Australian society, where aesthetics are discussed to the exclusion of meaning and purpose. This reversal in the articulated priorities is a source of both strength and weakness for folk club music.

Although it will not be described here, folk club music has an identifiable musical style. Indicators of the style include the use of particular musical instruments and particular combinations of instruments, a preferred vocal timbre and singing style characteristics, particular vocal and instrumental forms and performance conventions. Just as important as musical style is the ideology that goes with it. Folk music club ideology draws a sharp distinction between the 'tradition' and the 'revival'. The distinction often refers to performers rather than to performances or material. The 'traditional' musician is defined as one who has learnt musical skills and material from older musicians orally and by virtue of a particular kind of social connection with them. A family relationship is probably the most commonly accepted credential for a 'traditional' singer: however, regional or occupational connections may also be established. The 'revival' singer, on the other hand, is one who has learnt skills and materials from a variety of sources outside his or her own immediate family and social setting. The distinction assumes that the 'traditional' musician is part of an unbroken line of musicians operating within a restricted social and community setting, learning skills and materials immediately available. The 'revival' musician, on the other hand, has had access to a much wider diversity of sources.

The extent to which so-called 'traditional' musicians collected material from outside their local setting has not yet been examined; indeed, early reluctance among collectors in Australia to accept material of this nature from 'traditional' singers mitigates against this concept being tested. Vic Gammon's study of British singers (in 'Not Appreciated in Worthing?', *Popular Music* 4, 1984) suggests that while the 'folk singer's' audience may have been restricted, his repertoire was wider than that of the other singers examined. Fortunately, Australian collectors very quickly abandoned attempts to censor the repertoires of musicians they were working with, and it may be that there is among Australian collections sufficient material for detailed studies of the musical life of early twentieth-century Australian communities, at least. Such studies have not yet been undertaken.

The term 'traditional' is also used to refer to material. Songs, particularly, are said to pass into the tradition. In this case, what is referred to is a 'canon', a list of recognized 'authentic' songs. Of course, such a list exists only notionally, and every member of the folk club community would have a different list. One of the criteria for inclusion in the list, however, would be the extent to which a song had been sung by different members of the community and in different places, and altered in the process. The 're-fashioning' and 're-creation' of music described by Cecil Sharp is essential in this concept of the 'folk process'. The concept of the 'folk process' is in conflict with the ideal of authenticity, however. Although one of the tests of whether a song is traditional is whether it exists in various versions, an expected result of the 'folk process', there is nevertheless a prejudice in favour of 'original' performance practices, rather than modern ones. A. L. Lloyd (q.v.) was criticized for changing the words of Australian songs which he collected. If he had simply sung the changed words instead of publishing them, it would have been accepted as part of the 'folk process'. A similar contradiction occurs when material becomes so well accepted that it passes into the realm of popular music. In that case, its performance in different versions does not help its absorption into 'the canon'. Nevertheless, it is accepted that there is a body of 'traditional' material. Many contemporary songwriters in the folk club community express an ambition that their authorship of particular songs might one day be forgotten, and that the songs might 'pass into the tradition'. The primary source of material, however, for folk club music, and for the existence of the folk club community, is the songs and social dances, learnt often by oral transmission and used in Australia in the past, and 'collected', either by means of sound recordings or film or by being taught to younger performers. The collectors of the 1950s visited elderly musicians and with tape recorders made sound recordings of what the musicians could remember of their repertoire.

'Collecting' is still an important part of the activities of some of the folk music clubs. Others concentrate on performance. The contrast between the two activities is emphasized in the ideology. The fact that performers must of necessity be 'collectors' in order to assemble a repertoire is a paradox not widely acknowledged. The main financial supporters of collectors are the Australian Folk Trust, the umbrella body of the folk club movement, and the National Library of Australia and the National Film and Sound Archive (qq.v.). Many of the Trust's devolved grants are made to collectors; it also lends them high-quality sound recording equipment. The Australian Folk Trust has been closely involved in attempts to convince the Australian government of the need for adequate support and institutional backing for research into folklore.

The results of this collecting activity take different forms. It was because the collectors wanted to promote the material they recorded that the folk clubs and the early performance groups were started. Enthusiastic about the material being recorded, they wished to share this material with other musicians. Publications often resulted; the three-volume set of Peter Ellis's *Collector's Choice* by the Victorian Folk Music Club (1986, 1987 and 1988) is a recent example. Collectors sometimes facilitate meetings and sometimes regular contact between older and younger musicians. Dave De Hugard's organization of a visit by two accordion players to the National Folk Festival at Sutherland in 1987 is an example. The activities of the Horton River Bush Band, which played with and made recordings of Charlie Batchelor, a fiddle player, is another. The rejuvenation of the failing Nariel Creek dances because of the interest of the Victorian Folk Music Club in the 1950s is a further example.

The experience of collectors has led to a relaxing of the attitude to oral transmission and the use of published material. Many definitions of folk music rely heavily on oral transmission of material, but Australian collectors have shown that the written sources were also important. There was an extensive sheet-music publishing industry in Australia in the last century, and collections of this material, in the National Library and other depositing institutions, are rich sources, particularly for dance tunes. Together with the recorded collections, this material provides the opportunity for a study of the close interweaving of oral and written transmission of the material, a potentially exciting area for research in Australia.

Performance of the material collected was, and to some extent still is, an important goal of collecting activities. Collectors performed themselves and facilitated other performances by publishing notations of songs and dance tunes. What kinds of performance are appropriate is the subject of argument. One view maintains that contemporary performances of 'traditional' material should always be as close as possible in spirit and in sound to that of the 'original' performances. Another argument says that because modern resources were not available in the past, it doesn't mean either that performers of that time would not have used them if they had been available or that contemporary performers should shun them simply because they are recent.

Musical instruments used in the last century included fiddles, mouth organs, button accordions, concertinas, tin whistles and banjos (see **Musical instruments**, and Scott and Bolton's unpublished work, Traditional Australian Bush Instruments, 1984). The widespread use of the guitar to accompany singing and the addition of a stronger bass line to performances of dance tunes are two of the recent changes which are mostly unchallenged; but the use of electronic instruments is often decried on the grounds of their inauthenticity.

There are several criteria which are used frequently in talking about and evaluating 'folk' songs. The texts are considered to be the most crucial; in choosing which songs to perform, the musical form of the song may be just as important, or more so, but this is not emphasized in discussions. It is argued that the texts should refer to Australian experiences. Songs automatically having the 'right' subject matter are those about rural life in the nineteenth century: songs which refer to shearing, droving, bushranging, pioneering, being a rural itinerant worker, being a convict, and particularly, being on the other side of the world from places familiar to you, and not seeing family and friends again. Twentieth-century 'folk' songs often have texts which refer not so much to disembodied emotions — there are not many love songs, for example — but directly to events and issues which concern Australians. The great number of political songs, cynical, biting, but ephemeral, come into this category, as do songs which refer to contemporary issues, for example, the environment, the depression, the world wars.

This concern for the relevance of the text is related to two other issues. One is the identification with 'working-class' values and left-wing politics, seen more strongly in the British and American folksong revivals, but nevertheless of relevance here. Its importance is perhaps emphasized by the challenges which are from time to time thrown out by contemporary songs which have a right-wing rather than a left-wing political stance. Comment pointing out the traditional alignment of folksong with protest movements sometimes provokes the counter-comment that folksong simply reflects social concerns and these may be right-wing as well as left-wing. The second issue is the concern that the folksong repertoire should reflect Australian social history accurately. The collected repertoire shows a distinct lack of songs which refer to women's experiences, for example, and contemporary songwriters and performers have sometimes deliberately set about filling some of those gaps.

The concern with relevance is seen in the performance of folk songs. In workshops, and to some extent in concerts, folk club performers often introduce their material by explaining the historical setting of the songs, giving as a background to their performance the social setting and the social issues which the song addresses. This is not a token introduction either; detailed and extensive research often goes into the preparation of this material.

The social conscience of the folk club community, evident in its concern for social issues in Australian history, is also operative in its definition of what is 'Australian' and its attitude to the music of more recently arrived communities in Australia. Folksong ideology has always been associated with nationalistic ideals in Australia as much as anywhere else. Taking into account that white settlement in the country is a bare two hundred years old, and that from the beginning settlers came from many different cultures, the delineation and formation of something particularly and individually 'Australian' becomes a problem.

In practice the folk festivals and folk clubs may well serve as performance venues for music of non-English-speaking cultures, even when that music meets no criteria for folk music: the performance of a Chinese classical orchestra at the National Folk Festival in Melbourne is a case in point. Graeme Smith has suggested that marginalization in Australian society is the operative criterion in this case. Some recently arrived communities, of course, prefer not to take part in Australian folk club activities; indeed, an important part of their ideology may be the preservation of an 'other' identity, in which case absorption into Australian social settings may be anathema to them. Such antagonisms often rest on fundamentally mistaken views: on the one hand, 'traditional Australian' material originated in many different parts of the world — it has been strongly 'multicultural' from its beginnings; on the other hand, songs and dances of other countries have usually originated in completely different social settings and with completely different functions from those now associated with them. These arguments can be politically motivated, as when particular definitions of music determine its eligibility for funding. The prejudice in national funding bodies against artistic forms not included in traditional definitions of 'Western high art' or 'fine art' is only gradually being removed. While this prejudice remains, antagonism between people involved with different artistic forms is encouraged, and a healthy attitude to music and dance in the whole community is obstructed.

Interwoven with the concern for music which accurately reflects social issues is the view that music is a form of expression which should be available, not just to a highly trained group of specialist musicians, but to the whole community. Folk music clubs, therefore, are particularly interested in encouraging participation. However, there is a dilemma involved, albeit one that is rarely articulated. Many 'traditional' and 'revival' performers were and are very talented and highly trained musicians, even though their training in many cases was not by means of formal music lessons. The conflict between participation, on the one hand, and a high musical standard, on the other, lies uncomfortably close to the surface of many discussions; but it is very unusual for criteria for musical excellence to be discussed openly, and there are several reasons for this. One is the strongly held view that participation is important, and the concomitant wish not to discourage beginners. A second reason is to be found in the complicated set of views and reactions surrounding the process of learning and training in an oral tradition. 'Traditional' musicians often received no formal music lessons in the way that classically trained musicians normally do; there is also no institutional avenue for gaining recognized qualifications in folk music. (This state of affairs enables funding bodies which are fundamentally prejudiced against kinds of music other than Western 'art' music to maintain that excellence is not to be found within folk music traditions.) In emphasizing the informality of the learning situation, the folk music club ideology encourages a tendency to deny that learning took place at all. This is fundamentally mistaken. The teaching and learning of musical techniques, skills and material in Australia in the last century and in the first part of this century may have taken place through a process of sharing rather than through formal lessons, but it certainly took place. The existence of identifiable singing and playing styles, and a common repertoire guarantee that. The ready identification of master musicians among 'traditional' singers and players is another sign.

The emphasis on participation and music as an experience open to the whole community raises other difficulties. The reluctance to discourage beginners entails a reluctance to train musicians in presentation and professional attitudes towards performance. Indeed in some places there is a deprecating attitude to professionalism. People who make a living from folk songs may be looked at askance, and similarly, folk songs which become successful popular songs may also come under suspicion. The conflict is not an easy one to resolve. It exists because of the clash between the essentially humanitarian view that personal experience, including the experience of making music, is important, and the materialistic view which has increasingly valued achievement and forced increased specialization in all areas of human activity.

The extreme demands for specialization and professionalism in our society have led to another doubled-edged sword: the commodification of music. Sound recordings, made in the field by collectors, are central to the very existence of folk club music. Most of the repertoire has come into circulation and appropriate singing and instrumental styles have been learnt by means of recordings. Sound recordings are products which can encourage transmission by the same oral means which the folklorists have used as a defining feature of folk music. At the same time, sound recordings are the means by which music has become a commodity; they have allowed the development of a music 'industry', depersonalized and exploitative, the antithesis of folk music club values.

One of the most central and very distinctive parts of folk club music ideology is that music is important. In Western society music has become trivialized. No longer valued as central to a society's health, it has become an 'optional extra'. Even professional musicians believe this. While they often maintain that music is important for them personally, they nevertheless agree that public funding, that crucial indicator of a society's values, should be first spent on 'more important' things like health and education. Like many non-Western musical ideologies, folk club music ideology maintains, on the contrary, that music is essential to health, both social and personal. The high value placed on participation in folk music club ideology is predicated on this view of the significance, meaning and function of music. When the balance between participation and excellence is tipped in the other direction, towards excellence, it leads to an emphasis on highly developed skills and training; the possibility of most members of the society using music for their own expressive purposes is then severely diminished. In Australia folk club music ideology is one of the very few music ideologies arguing strongly for general participation. This puts it at a disadvantage when competing for funds with other musical interests whose ideologies emphasize excellence and therefore specialization. If and when it becomes recognized that the trivialization of music has been impoverishing for Western society and that the situation needs to be redressed, a balance between these two essentially polarizing views may be found.

See also: **Country music**; **Folk song**; **Folkways Music**; **Musical instruments**.

Jill Stubington

Folk narrative: see **Folk tales**.

'Folk on Wednesday': see **Radio and television programmes**.

Folk performers Notable performers of Australian folklore — song, music, poetry, dance, tale, and so forth — may be divided into two groups. The first, or 'traditional', group includes those performers who are, or were, primary bearers of folk traditions learned or acquired by them in the course of pursuing their everyday lives — in the family, at work, within their communities. Such performers include Harry Cotter; Ina Popplewell; Gladys Scrivener; Jack Parveez; Charlie Batchelor; 'Hoopiron' Jack Lee; Joe Watson; Harry McQueen; Cyril Duncan; Joe Cashmere; Bill Harney; Ted Simpson; Simon McDonald (q.v.); Sally Sloane (q.v.); H. P. ('Duke') Tritton (q.v.); the Klippel family of Nariel Creek; and the Wedderburn Old Timers. These performers are part of the social group that has kept the folklore alive. The second group, by contrast, comprises performers who were mostly brought to their particular art(s) as a consequence of their interest or involvement in what is generally known as 'the folk revival'. These performers acquired their artistic repertoires as a consequence of a conscious effort to study and learn about traditional material. They provide entertainment for folk revival audiences with interests similar to their own at venues such as folk clubs, concerts and festivals and in various media forms such as radio, recording and television. The second group also contains performers who may or may not perform 'traditional' folk material, but who in any case use folkloric themes, motifs or styles in contemporary compositions. While the same material is being utilized by both traditional and revival performers, the rather different circumstances of its acquisition and performance must be taken into account in any assessment of meaning.

Prominent revival performers include the original Bushwhackers band; the Bushwhackers and Bullockies Band; Warren Fahey (q.v.) and the Larrikins; Phyl Lobl (q.v.); Alex Hood; Danny Spooner; Martin Wyndham-Read; Gary Shearston; Lionel Long; Denis Gibbons; Brian Mooney; Leonard Teale; Dave De Hugard; John Dengate; Ted Egan; Eric Bogle; Margaret Roadknight; Declan Affley (q.v.); Colin Dryden; Shirley Andrews (q.v.); Stan Hastings; Judy Small; Cathy O'Sullivan; Mike and Michelle Jackson; Redgum; Sirocco; and Rakes. It should be recognized that these are — or were, some being now deceased or defunct — only the most notable of a large body of highly skilled instrumentalists, singers, dancers, reciters, yarn-spinners and other artists who regularly perform, record and broadcast aspects of Australian folklore.

In Britain, America and elsewhere such distinctions have come to have less value as the growing interest in traditional arts and crafts has led to many authentic traditional performers becoming professional entertainers/educators, thus maintaining their folkloric roots while performing or instructing others about them. This development has only recently taken place in Australia, though can be expected to gain pace as the blossoming folklife movement seeks to more fully and meaningfully integrate traditional arts and crafts and their practitioners with the older and fading concept of the 'revival folk festival'.

The folk revival movement in Australia has, since its origins in the 1950s, developed an extensive array of organizations, publications, festivals, recorded and published works and radio and television programmes and series that have contributed a great deal to the cultural life of this country and to the lives of many individuals in it. Each state and territory of the Commonwealth has at least one folk federation or other body that counts most of the folk dance organizations, performers and others among its affiliates. Each state body is affiliated with the national peak organization, the Australian Folk Trust (q.v.). Between them these amateur groups organize dozens of local, state and national folk festivals each year, run folk clubs and concerts, present radio shows, and, through performance, add to the overall cultural life of their particular part of the country.

Folk poetry is generally characterized by simplicity of rhyme and rhythm, the utilization of traditional symbols and images, and a strong tendency to narrative rather than lyrical content. What is valued in folk poetry are generally those literary devices and attributes scorned by 'literary' poets and critics. These include the use of proverbs, traditional similes and comparisons and cliché. While innovation and originality are usually highly regarded in literary poetry, folk poetry is often composed of stock phrases and deals with relatively mundane matters. Folk poetry tends to concentrate on the external and the public expression of emotion, rather than on internal, private emotions, a characteristic that springs from its function as a form of group or community expression. The commemorative mode, often nostalgic in mood, is commonly employed to mark such things as disasters, the disappearance of old ways and the common experience of momentous events, such as war. Other common themes and topics include social and political protest, occupational dissatisfactions, local characters, events and environment, as well as all manner of current issues.

Folk poetry is usually oral though it is also found in non-oral folk forms, such as reprographic (q.v.) or photocopy lore and graffiti (q.v.). Common forms of folk verse are traditional epitaphs, limericks (usually crude), autograph rhymes ('Roses are red, violets are blue, Sugar is sweet and so are you'), nursery rhymes, toasts ('Damn the teamsters, damn the track, Damn Coolgardie, there and back, Damn the goldfields, damn the weather, Damn the bloody country altogether'), playground rhymes (see **Children's folklore**) and folk poems and ballads of all kinds (see **Folk poetry and recitations**). While most of the above-listed forms may appear of a trivial nature, each category contains an extensive and continually developing repertoire, often formed by parodying originals (or parodying parodies . . .).

Folk verse topics are as varied as any other form of folkloric expression, though they are frequently humorous, scatological, vulgar and satirical. Some reasonably well-known examples are 'The Bastard from the Bush', 'The Spider from the Gwydir', and 'The Great Budget Debate'. There is very little distinction made between verse and song in folk tradition, ballads often being rendered as songs and vice versa.

Folk poetry and recitation In Australia the term 'folk poetry' is most often associated with popular rhyming verse. In keeping with age-old tradition, many Australians find rhyming verse an accessible and satisfying medium of self-expression. We are inconsistent in our use of language, however. We describe as 'folk craft' a known craftsperson's hand-embroidered tablecloth or hand-made table, where these accord with traditional forms and result from the application of skills and knowledge acquired informally from within the maker's own family or community. Songs and poems are treated differently. In conventional usage the label 'folk song', or 'folk poem', is reserved for items which not only satisfy the criteria for acceptance as folk craft, but also have passed into oral tradition, i.e. they have passed informally between members of the community, and preferably also between generations, showing some variation in form. Convention therefore denies the writers of popular rhyming verse status as 'folk poets', even though they practise an age-old tradition.

Many people also find self-expression in writing free and/or blank verse, or verse which otherwise falls outside the metrical and rhyming confines of popular verse. While verse in this form is widely published, it is not often committed to memory and is rarely encountered in oral tradition. For this reason such verse is not normally regarded as 'folk' poetry.

Beginnings

The Aboriginal peoples of Australia maintain oral traditions that go back tens of thousands of years. These traditions include myth and legend, drama, song and ceremony, and have an immensely rich inherent poetry.

One of the most distinguished interpreters of Aboriginal myth and legend, Roland Robinson, commented in *The Feathered Serpent* (1956):

> The themes of aboriginal mythology all stem from, and serve to illustrate, the eternal source and meaning of the Spirit and of Life. It is the source of a pattern of social behavior in harmony with the Earth-Mother . . .
>
> The Aborigine has never possessed a written language. His mythology has been handed down through the centuries by means of the great song cycles and rituals and by oral explanations of ritual objects which are features of the landscape created by ancestral beings.

The first widely circulated renderings into English of Aboriginal myths were those by Mrs K. Langloh Parker, whose book *Australian Legendary Tales* was published in 1896. Like many which followed, her attempts at retelling in English the myths conveyed to her by Aboriginal people were deeply flawed, and conveyed little of the poetry inherent in Aboriginal narrative. As Ronald and Catherine Berndt have noted,

> From almost the time of first European settlement in this country, various people have written down fragments of myths told to them, in various circumstances, by Aborigines. Many of these were indifferently recorded, their content considerably anglicised in the process, and often dramatically censored . . . It is only in recent years that Aboriginal oral literature has been taken seriously by people other than Aborigines . . . (Berndt and Berndt, *The Speaking Land*, 1989)

It is still the case that published accounts of Aboriginal mythology, with its rich store of natural poetry, rely at least in part upon translation into English by non-Aboriginal people. There are, however, a number of publications, such as those by Robinson and the Berndts, in which every effort is taken to communicate the atmosphere and sense, as well as the rhythm and structure, of the original spoken versions.

The British and Irish heritage

Early English-language poetry in Australia cleaves neatly into two groups. First, there was that which aspired to be high poetry, emulating the English masters. Witness, for example, the following, written in 1805 by John Grant, a convict:

> Nature! there dwells in these Australian lands
> Thy faithful Copyist whose Art expands
> Thy novel Beauties o'er our ancient Globe
> Who to far distant climes thy Charms derobe.

Then there were the broadsides, the street literature, hawked around the towns and cities of Britain and Ireland, and recited or sung to popular tunes of the day (see **Broadside ballads**). Along with hangings, murders, last confessions and the like, transportation to the colonies was a popular topic. Many broadside ballads travelled from the British Isles to Australia, and a few originated here. Some, like 'The Convict Maid', 'Van Diemen's Land' and 'Bold Jack Donohoe', entered oral tradition, thereby becoming the earliest Australian folk ballads. Some were still in oral tradition in Australia in recent times. An example is 'Caledonia', which John Meredith collected from a retired shearer, Joe Cashmere, in the mid-1950s (*Singabout* 6, 2)

> My name is Jimmy Randall, in Glasgow I was born,
> All through a sad misfortune, I was forced to leave in scorn.
> From my home and occupation, I was forced to gang awa',
> And leave those bonnie hills of dear old Caledonia.

At least one of the Donohoe ballads was the work of the Irish convict Francis MacNamara, better known as Frank the Poet (q.v.), a prolific and fearless rhymester, renowned for his capacity to extemporize. Frank was a true folk poet, a number of his works entering the oral tradition. He is best known for his magnificent Swift-like satire 'The Convict's Tour of Hell', which he wrote in 1839 while working as an assigned servant shepherding stock, and for the ballad 'The Convict's Arrival'. This latter piece circulated widely and was eventually transformed into one of the finest and best-known Australian folk ballads, 'Moreton Bay'.

> I am a native of the land of Erin,
> I was early banished from my native shore,
> On the ship Columbus went circular sailing,
> And I left behind me the girl I adore.
>
> Oh, the bounding billows which were loudly raging,
> Like a bold sea mariner my course did steer,
> We were bound for Sydney, our destination,
> And every day in iron wore.
> (in Hugh Anderson, *Time Out of Mind*, 1974)

The gold rushes of the 1850s attracted professional entertainers from northern climes, veterans mainly of the music-hall circuits. These entertainers brought to Australia popular melodies, songs and recitations from overseas, which having landed here often took hold, some serving as rootstock for Australian adaptations, others providing models for entirely new works. Some, like the 'inimitable' Charles Thatcher, commented in song and verse on colonial society, and their best compositions circulated rapidly through word of mouth, printed sheets, periodicals and songbooks.

> The morning was fine,
> The sun brightly did shine,
> The diggers were working away;
> When the inspector of traps,
> Said now my fine chaps,
> We'll go licence hunting today.
> (in Hugh Anderson, *Goldrush Songster*, 1958)

Not all the newcomers wrote in English. In 1860, for example, Ludwig Becker published in Melbourne his *Ein Australisch' Lied* (An Australian Song), which told of the tribulations of a German migrant in Australia.

> *Zwolf Chyser er gesunken hat —*
> *auf einmal hat er's Diggen satt;*
> *Wird Bullockdriver jetzt genannt,*
> *mit Ochsen zieht er durch das Land.*
>
> [Twelve chasers he has sunk
> And suddenly he's fed up with digging.

He is called now a bullock driver
And travels with oxen about the country.]

If the language was German rather than English, the form was rhyming verse in classic ballad metre. Becker, an artist and naturalist, was later to perish on the ill-fated Burke and Wills expedition.

By the early 1860s there was firmly established in Australia a body both of (mostly execrable) colonial verse modelled on the style of the 'major' foreign poets, and of popular verse and song evolving primarily from the broadside and folk traditions of Britain and Ireland and reinforced by the repertoire of the music hall. This was at a time when adult literacy among white Australians was attaining high levels, and the boundaries of settlement were spreading into the country's inhospitable interior along a rapidly expanding pastoral frontier. The scene, therefore, was set for the establishment in Australia of a truly naturalized school of Australian verse, i.e. for the emergence of the bush ballad.

The bush ballad

The ballad is an ancient form, and in the mid-1800s even the most eminent English poets were seen to use it. Small wonder then that colonial poets, even in respectable circles, tried their hand as balladists. They soon discovered the racy, narrative style fitted colonial life, and especially bush life, splendidly. Further inspiration came from across the Pacific in the form of contemporary poetry and stories by Bret Harte and others about the Californian goldfields and the American West. The bush ballad quickly became established as an acknowledged literary form, the epitome of Australia's emerging national consciousness.

While the bush ballad had no single parent, the way was led by the expatriate Englishman, Adam Lindsay Gordon. Although most of his work harked back to English themes, Gordon wrote a small number of ballads set in Australia and one in particular, 'The Sick Stockrider', showed how the ballad could be used to good effect in the bush context. In the years that followed Gordon's untimely death in 1870 his reputation as a poet grew immensely. His work was recited both around bush campfires and in the drawing rooms of high society, and was published in serious poetry anthologies and popular reciters alike.

The trail blazed by Gordon and others rapidly became a highway as Australians everywhere, burning to give expression to their own experiences and literary aspirations, took up the muse. As Peter Hay states in the introduction to his excellent collection of folk verse from Victoria's Western District, *Meeting of Sighs* (1981):

> Few of these bards had literary pretensions. For most, turning a rhyme was a means of rendering an everyday experience for the delectation of a small circle of people locally familiar with that experience. Some was not even committed to paper . . .
>
> The most significant publishing outlet for many poets was the local press. Local audience was the limit of their literary aspirations, and many became local celebrities of a sort, their creations eagerly awaited by the local readership, and often inspiring considerable controversy. Some fought verse duels with each other . . . Most Western Victorian newspapers carried verse as a matter of course — the *Port Fairy and Belfast Gazette*, for instance, at one time seems to have aspired to the status of a quality literary review . . .

Western Victoria was not alone; by the late 1870s many rural towns had their own newspapers eager for contributions, however rustic, and for the more ambitious rhymesters there were the larger city publications like *Bell's Life* and the *Australian Town and Country Journal*.

The establishment in 1880 of the Sydney *Bulletin* gave new impetus to this trend. Rapidly becoming known as the 'Bushman's Bible', the new weekly actively encouraged a new generation of bush balladists, and provided them with an audience that stretched from Maoriland (New Zealand), in the east to Western Australia, and to every bush homestead, cattle camp and township in between. Many of the balladists it nurtured far surpassed Gordon both as poets and as apostles of bush life.

The subject matter of the bush ballads (horses, drought, mateship, love, toil, etc.) was familiar to all social classes, and the most popular pieces entered the repertoire of reciters at all levels of society. As a result the bush ballads achieved a grasp on the national consciousness that could never be matched by their more rustic counterpart, the bush *song*.

One of the finest bush balladists was Will Ogilvie, a Scot who came to Australia as a twenty-year-old in 1889. He spent twelve years on outback stations droving, mustering and horse-breaking, his prolific pen providing *Bulletin* readers with a constant flow of finely crafted, rhythmic ballads romanticizing bush life. His work is still widely recited in the bush.

The prominent balladists came from all walks of life: Barcroft Boake was a drover and boundary rider, Harry Morant a horse-breaker, George Herbert Gibson an inspector with the NSW Lands Department, Patrick Hartigan a Catholic priest, Thomas Spencer a building contractor and W. T. Goodge a rural journalist. Spencer and Goodge were masters of bush humour, and delighted *Bulletin* readers with well-crafted comic ballads, a number of which remain popular as recitations. Goodge's 'The Great Australian Adjective' entered oral tradition, and a number of 'anonymous' versions have been recorded (some as far afield as the United States). Hartigan (who wrote as 'John O'Brien') is best known for his classic ballad of bush optimism, 'Said Hanrahan'. To this day, '"We'll all be rooned", said Hanrahan' is a popular saying in the bush.

Although all the best-known balladists were men, many women also wrote bush verse. The best-known ballad by a woman was probably 'Bannerman of the Dandenong' by Alice Werner. This told a heady tale of friendship and self-sacrifice during a desperate ride to escape a bushfire, and was a certain winner in school reciters.

The practice of publishing newly written bush ballads with attribution in widely circulated newspapers and periodicals such as the *Bulletin* has resulted in only a few orally circulating bush ballads being without known authors. Where authorship *is* unknown, often the ballad in question has become well known only after being republished without attribution, say, in a popular reciter. The publication in the 1930s of 'The Boss' Wife' and 'Flying Kate' in Bill Bowyang reciters is a case in point.

Today, one of the best-known 'anonymous' ballads is 'Nine [or Five] Miles to Gundagai'. This bullocky's diatribe probably dates from around 1859 and a number of versions circulated widely in oral tradition before a text was published in a Bill Bowyang reciter in the 1930s. In its published versions the poem is almost always sanitized:

> Twas gettin' dark, the team got bogged,
> The axle snapped in two,
> I lost me matches and me pipe,
> So what was I ter do?
> The rain came on, 'twas bitter cold,
> And hungry too, was I,
> And the dorg — he sat in the tucker box,
> Nine miles from Gundagai.

That the dog shat in, rather than sat on, the tucker box was widely but not universally understood. The poem became so well known that in 1932 a statue of the dog sitting on the tucker box was erected on the Hume Highway the requisite distance outside Gundagai, and unveiled by the then Prime Minister, Joseph Lyons. In his *Dictionary of Australian Folklore* (Fearn-Wannan 1970) Bill Wannan has described it as Australia's most popular national monument, and today it supports its own ecosystem, a community of souvenir shops and roadhouses.

Paterson and Lawson

While the *Bulletin* nurtured dozens of bush poets, two have an eminence which sets them apart: A. B. ('Banjo') Paterson (q.v.) and Henry Lawson (q.v.).

There was no finer exponent of the bush ballad than Paterson. Even the London *Times* was forced to concede that his work compared 'not unfavourably' with that of Rudyard Kipling, the then literary darling of the British Empire. And while Kipling was indeed an influence on Paterson, as he was on other bush balladists of the time, in no way did Paterson belong in his shadow. Paterson's ballads brought the bush to life, giving Australians characters to make them laugh and stories to excite their imagination. He took his readers shearing and droving, and to bush race meetings, and shared with them his vision of the joys and disappointments of bush life. Such was his importance that when Douglas Stewart and Nancy Keesing compiled their major anthology *Australian Bush Ballads*, published in 1955, Stewart lamented he could not include Paterson's *Collected Verse* virtually in its entirety. 'The Man From Snowy River', 'Clancy of the Overflow', 'The Man From Ironbark', 'Saltbush Bill', 'Mulga Bill's Bicycle', 'The Bush Christening' and many more of Paterson's ballads have entered oral tradition. 'Waltzing Matilda' (q.v.) and 'A Bushman's Song' not only entered oral tradition but also acquired tunes.

Henry Lawson was born in 1867, three years after Paterson, in a tent on the goldfields at Grenfell in New South Wales. His parents were poor, his bush childhood a far cry from that fondly remembered by Paterson. Encouraged by the *Bulletin*'s editor J. F. Archibald, he carried his swag outback. He became a short-story writer of genius, and a poet whose unromantic vision of the bush provided a stark contrast to the lyrical images presented by Paterson. He wrote with feeling about his own circumstances, and about the loneliness and hardship of bush life.

Significantly, it is Lawson who has retained the greatest affection of the Australian people, and who properly should be acknowledged as Australia's national poet. Lawson, destitute and alcoholic, struck a chord with Australians who loved and recited Paterson's ballads but knew little of the man. Lawson wore his heart on his sleeve for all to see; Paterson kept his to himself. When Lawson died insolvent the NSW government was shamed into according him a state funeral, and tens of thousands attended; when Paterson died, in respected old age, he merely received glowing tributes.

Many of Lawson's verses have entered oral tradition as recitations. These include 'Ballad of the Drover', 'Outback', 'The Fire at Ross' Farm', 'The Teams', 'Faces in the Street', 'The Roaring Days', 'When Your Pants Begin to Go' and 'The Lights of Cobb and Co'. One in particular, 'The Captain of the Push', has gained notoriety through a bawdy version, 'The Bastard from the Bush', which some consider Lawson wrote himself. Moreover, it was Lawson's verse, not Paterson's, that cried out to be sung. True, Paterson wrote the words of 'Waltzing Matilda', which became Australia's unofficial national song, and several other of his poems either acquired tunes or were set to music, but songs crafted by Australians from Lawson's verse number in the hundreds, and the list is still growing. The best-known songs from Lawson include 'Andy's Gone with Cattle', 'Freedom on the Wallaby', 'Reedy River', 'The Tent Poles Are Rotting (On the Wallaby)' and 'The Shearer's Dream'.

C. J. Dennis

While by the late 1800s most Australians were city-based, the bulk of our popular verse continued to focus on the bush and outback, the locales of the Australian Legend. The key exception to this was *The Sentimental Bloke*, created by C. J. Dennis, whose debut in 1915 took Australia by storm. In a sequence of delightful verse sketches, Dennis told of the meeting, courtship and marriage of the Bloke, a Melbourne larrikin, and Doreen, a fair maid who worked in a pickle factory. In 'The Play' the bashful Bloke takes Doreen to see a performance of Shakespeare's *Romeo and Juliet*:

'Wot's in a name?' she sez . . . An' then she sighs,
An' clasps 'er little 'ands, an' rolls 'er eyes.
'A rose,' she sez, 'be any other name
Would smell the same.
Oh, w'erefore art you Romeo, young sir?
Chuck yer ole pot, an' change yer moniker!'

This is a masterful work, and one of the classic Australian recitations. But even here the bush ethos wins out at the end as the Bloke and Doreen leave the city to realize their dream and own a farm.

Such was the success of *The Sentimental Bloke* that it has twice been made into a feature film. The first, a silent film made in 1919, is itself an Australian classic.

Dennis's sequel to *The Sentimental Bloke*, *The Moods of Ginger Mick*, was also a great success. These and a number of other books by Dennis, as well as books by Paterson and Ogilvie, were even published in special pocket editions for the trenches.

Australians at war

The onset of the Great War in 1914 coincided with the high point of popular interest in Australian verse. Small wonder then that scholarly discourses on Australian poets featured in soldier's magazines like *Aussie*, and that when Australian diggers sought to give expression to their feelings, very often they did so in verse, usually adopting the familiar and beloved metres of the bush ballad. The pages of C. E. W. Bean's *The ANZAC Book* (1916), compiled in the field at Gallipoli, are filled with soldiers' verse.

There are numerous examples of soldiers writing verse in the slow hours before battle. Some, like gunner Frank Westbrook, awaiting dawn before the landing at Gallipoli, were contemplative, yet patriotic:

And now I make ready for death or his master,
This thought as the moments in flight hurry by,
If I live 'tis my privilege all for my country,
For Australia to live, for Australia to die.
(Westbrook, *Echoes of Anzac*, 1916)

Others, like signaller Tom Skeyhill, writing the night before he was blinded in the attack on Cape Helles, were bitter and direct:

Well, I've picked up me old Lee-Enfield,
And buckled me webb about;
I'm only a bloomin' private,
And I've got to see it out . . .
But if I do get shrapnelled,
Though I die without a groan,
Well, the one who's really killed me
Is me Brother Wot Stayed at 'Ome.
(Skeyhill, *Soldier Songs From Anzac*, 1915)

Westbrook and Skeyhill and their many counterparts may not in some eyes qualify as folk poets, but they certainly were exponents of a long-established tradition (see also **Wartime folklore**).

Passing of an era

In the decade following the war Australian society changed rapidly. However, Australians everywhere still read for enjoyment and the love of rhyming verse, and especially of bush ballads, remained, although increasingly the new crops of ballads and balladists harked back to the past, referring nostalgically to times many had never known.

There was no shortage of poets. The *Bulletin* continued to provide an outlet for publication of popular verse, as did regional periodicals such as the *North Queensland Register*. A number of ballads from these publications gained wide distribution when they were included (without attribution) in *Bill Bowyang's Bush Recitations*, a series of six very popular booklets published in north Queensland during the 1930s.

The new hillbilly or country music (q.v.) soon began to incorporate local themes, with lyrics often indistinguishable in metre and content from bush verse. Indeed in some respects this music often was simply bush verse rendered into song. Many of the finest country-music artists like Slim Dusty openly acknowledged their love of the old bush ballads and set these to music, a practice which continues to this day.

In the Second World War the tradition of soldier's verse continued. In 1942, for example, sapper Bert Beros penned on the Kokoda Trail his tribute to the 'Fuzzy Wuzzy Angels':

. . . they haven't any halos,
Only holes slashed in the ears,
And with faces worked by tattoos,
With scratch pins in their hair,
Bringing back the wounded,
Just as steady as a hearse,
Using leaves to keep the rain off
And as steady as a nurse . . .
(Beros, *The Fuzzy Wuzzy Angels and Other Verses*, 1943)

This poem circulated widely, both orally and in print, and in Australia helped create an image of the 'fuzzy wuzzy angel' that has lasted for generations.

Australia also had its amateur poets at home, including some, like Oswald

Volkmann, whose land of birth was then unfashionable:

> We have been Hitler's enemies
> For years before the war,
> We knew his plan of bombing and
> Invading England's shore,
> We warned you of his treachery
> When you believed in peace,
> And now we are His Majesty's
> Most loyal internees . . .
> (quoted in Emery Barcs, *Backyard of Mars*, 1980)

Came the peace, and Australia settled into a period of unprecedented prosperity and growth. However, as Peter Hay states:

> The climate which nurtured a tradition of folk verse suddenly dissipated in the 1950s, a period which saw major advances in media technology — particularly the advent of television. No longer were local communities so self-reliant; the nature of regional newspapers changed, poetry disappeared from their pages, and the people lost their natural mode of creative folk expression. One poet was moved to comment that he gave up reciting his verse at local functions in the 1950s, when he suddenly realised that he and his kind had come to be looked on as 'bits of fools'.
> (Hay, *Meeting of Sighs*, 1981)

Despite these changed times, bush verse is still popular. Every year, for example, many hundreds of Australians enter the Bronze Swagman competition for bush verse. These entries come from all over the country (and even from overseas), from people in all walks of life, and generally echo in sentiment the bush verse of yesteryear.

Further evidence of the continuing propensity of ordinary Australians to seek self-expression through the writing of rhyming verse is easy to find. In 1985, for example, Roger Montgomery collected within a three-month period over 300 original items, predominantly rhyming verse, from people living in the iron-ore mining communities of Western Australia's remote Pilbara region.

Unlike bush song, bush verse has never needed a conscious revival and there is every reason to expect it to continue to prosper. Its natural habitat has never been threatened, nor has it faced serious competition from exotic forms of verse. Further, the popularity of Australian country music has if anything reinforced the bush verse tradition. Some comparatively recent compositions have already received the ultimate accolade of entering oral tradition, knowledge of their authorship sometimes becoming lost in the process. Examples here include Edward Harrington's 'The Swagless Swaggie', John Manifold's 'Incognito', Claude Morris's 'A Grave Situation', Rob Bath's and Andrew Bleby's 'McArthur's Fart', and Rob Charlton's 'Bloody Sheilas'. The concerns of modern bush poets are little changed from those of times past: they focus on rural themes, are tentative about love and human relationships, and are openly nostalgic and passionate about horses, the pioneer spirit, sport and mateship. While there is also some bawdy and racist verse, some of which is very unpleasant, this remains a minor element of the whole. The *form* of the verse — its rhythm and metre — hasn't changed either: simple four-line stanzas are perhaps most common, although the numbers of lines per verse or syllables per line is not really an issue, provided the pattern is regular. It often happens that people compose verse for a social occasion such as a speech at a birthday, or possibly for a wedding telegram. In these instances simple rhyming forms are almost always preferred. Very often, too, the verse will take the form of a parody of a popular poem. Parodies of 'The Man From Snowy River', for example, are legion.

Despite its undeniable public appeal, popular rhyming verse is openly denigrated by the literary establishment, largely (one suspects) because it *is* popular and therefore, by a perverse logic, deemed to be sub-literary. As a result, poets writing in this form have little access to government arts funding or to publication in literary journals. Faced with this situation, some poets publish their own work, while others seek out a commercial publisher. Good rhyming verse has a ready audience, and in recent times a number of publications of rhyming verse by 'non-established' poets have been financially successful. *Ripples on the Wannon*, for example, by Stan Dunn of Warrnambool, has sold over 5000 copies, a sales figure which would be the envy of many of Australia's most prestigious poets (and publishers).

Not all contemporary rhyming verse is written with performance or publication in mind. Some finds its way onto greeting cards or autograph books; some adorns toilet walls; some is used as inscription on gifts; and much is retained undisclosed by its authors.

Recitation in Australian popular culture

Recitation is the art of performing verse from memory. A highly developed and widely practised folk art in Australia, it is not to be confused with mere verse *reading*. Performance styles vary. Most reciters use a plain, unaffected delivery style, but some adopt an expansive approach that may best be characterized as mime with words, complete with movement, gestures and facial expressions. In any event, reciters differ from yarn-spinners and storytellers in that they desire their audience to remain silent while they perform a set piece.

The practice of reciting verse for enjoyment was transported to Australia from the drawing rooms, schools, taverns and music halls of the British Isles, the goldfields of California, and other places besides. In the days when entertainment in Australia was mostly home-made, recitation formed as natural a part of an evening's social programme around the campfire or in society parlours or the local Mechanics' Institute hall as did song, dance or instrumental music. Indeed, it survived being taught in the schools to become one of the most characteristic aspects of this country's popular culture.

The repertoire of Australian reciters was drawn initially from foreign sources, although by the 1880s Australian ballads were coming into their own. At the height of their popularity, in the period between the 1880s and the early 1920s, bush ballads were usually published in reciters alongside 'humourous, pathetic and dialect' verses and monologues from overseas. Indeed, the emphasis on British and American authors was such that when *Dymock's Australian Reciter* was published in Sydney in the 1890s it contained no Australian items at all!

It is easy, too, to associate reciting exclusively with the popularly promoted 'bush' image and ignore the fact that recitation cut across class barriers and was practised in a wide variety of social circumstances. In this regard the cover of the immensely popular *Bulletin Reciter* (1901) is instructive. It depicts a kangaroo in formal evening wear reciting on stage before an audience of native fauna similarly attired. This is not to suggest that such audiences consisted solely of the upper or monied classes — they didn't; in those days it was customary for *everyone* to dress up for a social evening out.

The early 1920s saw the spread across Australia of a whole range of marvellous new entertainment technologies — the wireless, moving pictures, phonograph recordings — and these combined with normal social change to undermine the old ways, and in particular the ways in which Australians relaxed. Poetry recitals and monologues declined in popularity. Like our counterparts overseas, we also became more a nation that expected to be entertained by professionals, and less one where family and community members entertained each other. Fewer and fewer people gave attention to the maintenance of their own performing skills in order to be able to contribute, say, a song, piano piece or poetry recitation to an evening's social programme.

Although the numbers of reciters diminished, recitation retained its status as a form of popular folk art, indulged in by enthusiasts for their own and other's private enjoyment. In this capacity it has remained very much part of the Australian scene.

While most reciters continue to learn their material from published texts, it is instructive to note what has happened when no published text has been available. In 1974 two Adelaide students, Rob Bath and Andrew Bleby, wrote 'McArthur's Fart', a satire of the bush ballad, for a university revue. Their ballad told how the townsfolk of Bungadell sent for the mighty bushman McArthur to save the town from drought.

At last McArthur came
And the people gathered 'round
To see the man whose fart
Would send the waters down.

He came on two big horses
With half his bum on each,
A bum so wide a man could drive
A tram between the cheeks.
(in R. Marshall [ed.], *Down the Track*, 1985)

So successful was their effort that within eight years, without ever being published or commercially recorded, the 'Fart' passed into the repertoire of reciters throughout Australia. In some instances the text was written down verbatim during or after the performance, or it was recorded from the audience on portable hand-held cassette recorders and transcribed from the recordings thus obtained. In one instance cassette copies of a festival workshop containing the poem were made available for purchase and, of course, photocopies of texts (however obtained) have been legion. Most carriers of the poem have been unaware of its authorship.

Unfortunately, opportunities for reciters to perform in public are comparatively few. There is no periodical dedicated to the dissemination of popular rhyming verse, and no society of reciters. For these reasons, reciters operate largely in isolation from each other, each in his or her own community. The closest approach to a *community* of reciters is found in the music-oriented folk revival and country music movements. Here reciters find an appreciative audience, both when they perform alongside singers and musicians in concerts and when they give workshops and recitals in their own right.

See also: **Broadside ballads**; **Folksongs**.

REFERENCE: Keith McKenry, 'The Recitation in Australian Oral Tradition: Past and Present', in J. Ramshaw (ed.), *Folkore in Australia: Proceedings of the First National Folklore Conference*, 1985.

Keith McKenry

Folk revival, the: see **Appendix: A Bibliographic guide to Australian Folklore**; **Australian Folk Trust**; **Collectors and collections of folklore**; **Commercial folklore recordings**; **Country music**; **Dance**; **Festivals**; **Folklore and popular culture**; **Folk poetry and recitations**; **Folk song**; **Magazines and publications**; **Musical instruments**; **Performers**; **Radio and television programmes**; ***Reedy River***; **Wattle Recordings**; see also the names of individual performers, publications and organizations.

Folk song Many attempts have been made to define folk song. Readers who wish for an introduction to the literature on this subject might well begin with the first chapter of A. L. Lloyd's *Folk Song in England* (1967) or George Herzog's article, 'Song: Folk Song and the Music of Folk Song', in *Funk & Wagnall's Standard Dictionary of Folklore, Mythology and Legend* (1972). Suffice to say here that almost the only point on which all definitions agree is that some kind of oral transmission should be involved. It might seem then that the folksong collector might be well advised to record any song which shows signs of having been transmitted orally; to publish all (or most) of the material collected; and leave each folksong scholar to select for study those songs which meet his or her definition of folk song. This field recording strategy seems, however, to have been adopted only occasionally.

Systematic recording of folk songs in Australia began only in the early 1950s. One of the first into the field, and the most persistent and successful collector of all, was John Meredith (q.v.). For four or five years he collected songs (and other material) from singers in Sydney and in a limited region of central-western New South Wales. Meredith and Hugh Anderson put the results of that fieldwork into a book called *Folk Songs of Australia, and the Men and Women Who Sang Them*, first published in 1967. Meredith has an interesting comment on his field recording strategy in his book on the singer and songwriter 'Duke' Tritton, whom he met early in his collecting career:

> He told me that he had a few songs he had written himself, but being a bit of a folksong purist at the time, I told him that I wasn't particularly interested in them, that it was the traditional ones I wanted. So it was left to Alan Scott to collect 'Shearing in a Bar' and 'The Gooseneck Spurs'.

He does not explicitly say so, but it would seem that after this early period of 'folksong purism', he took to collecting everything, or almost everything, that singers offered him. He and Anderson selected for publication a representative sample of almost all the kinds of songs recorded; there seems to have been some suppression of material with sexual reference that some readers might have found offensive.

Meredith resumed fieldwork in the 1980s. Songs collected in the 1980s were printed in *Folk Songs of Australia*, volume 2.

Meredith and Anderson said, in the preface to the first of these volumes, that

> most collections . . . limit their material to those items that have an Australian content or flavour . . . *Folk Songs of Australia* marks an important departure by including a number of songs of British and North American origin.

Others have not followed their example, and the two volumes of *Folk Songs of Australia* remain the only books from which a good general overview of the Australian folksong tradition can be readily obtained. The folksong enthusiast cannot get a better overview without delving into the pages of journals difficult to find even in major libraries, and beyond that, into the unpublished field recordings held by the National Library and the National Film and Sound Archive (q.v.). Meredith and Anderson claimed, in the preface to *Folk Songs of Australia*, that 'the material presented is representative of the whole continent'. The songs published so far only in journals, or still waiting on transcription from the field recordings, show that this claim needs some qualification. This does not affect the main point made here: the Meredith books are the best place to begin an acquaintance with the Australian folksong tradition.

Folk songs from the British Isles in Australia

It might seem logical to begin an account of Australia's folksong tradition by looking at the songs that settlers from the British Isles carried with them. It might be better to begin with a comment or two on what they did *not* carry with them.

The folklore of many parts of Europe is still rich in pagan ritual, more or less imperfectly integrated into Christian religious practice; or, from the point of view of the religious reformers of the Reformation, more or less hideously corrupting Christian practices. In most parts of the British Isles these rituals, and the songs that went with them, came under strenuous attack in the rationalizing, revolutionary years of the seventeenth century. They hung on here and there to be recorded by nineteenth- and twentieth-century folklorists. Nothing of this kind reached Australia, or if it reached Australia, survived, with the seemingly solitary exception of the song of 'The Derby Ram'. Many versions of this song have been recorded in Australia, most of them keeping at least some reference to the magical sexual potency of the ram. But nothing remains of the ritual that it once accompanied, though it was preserved in attenuated form into the twentieth century in parts of northern England.

Ritual songs apart, many Anglo-Scottish songs, more or less probably of mediaeval origin, have continued in oral tradition until today, not only in England and Scotland, but also in Ireland and in North America. Very few songs of this age seem to have reached Australia or, perhaps, to have survived transplantation. There are versions of 'Our Goodman', a 'favourite joking song' wherever songs are sung in English. There are versions of 'Barbara Allan', a favourite sentimental ballad wherever songs are sung in English. There is a solitary version of a riddling song folded into a ballad, which is convenient to call 'Captain Wedderburn's Courtship'; Harry

Lewis Lampton playing the role of Bungul Jarma in *Chosen Ones*, Aboriginal Islander Dance Theatre Student Ensemble, 1988. (Australian Association for Dance Education, ACT)

Doing the 'Chook' dance, Sydney, 1990.

Canberra Morris Men, 1989. (Photograph, Ross Gould)

Young students of Irish dance perform in Martin Place, Sydney, 1976. (Photograph, Australian Foreign Affairs and Trade Department)

Aboriginal dance musician, Jim French, with three-row accordion, Moree, NSW. (Photograph, National Library of Australia)

Ball at Cootamundra Town Hall, 1920s. (National Library of Australia)

Greek families dancing at a beach picnic, Torquay, Vic., 1987. (Photograph, Elizabeth Gilliam)

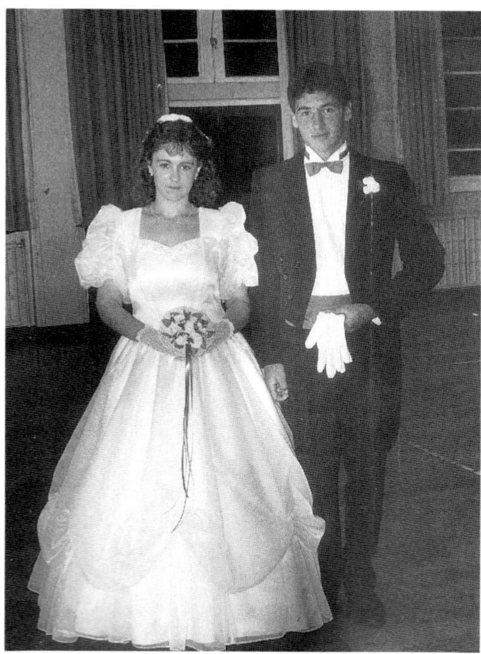

'Coming out' at the Timboon Debutante Ball, Vic., 1989. (Photograph, Helen O'Shea)

Dicks, who sang it for Meredith, called it 'The Chicken and the Bone'. There is a single record of 'The Trees They Do Grow High' ('My bonny boy is young, but he's growing'), a fragment at that. The list could be extended, but many songs on the list would be like the last mentioned: a single record of a fragmentary version. Even then, it is a short list compared with one that could be made for North America.

The four songs mentioned above have all, at one time or another, been referred to as *Scottish* songs. In fact, like a great many other folk songs commonly thought of as Scottish, they were widely sung in England; conversely, many folk songs commonly thought of as English are widely sung in Scotland. Furthermore, many Anglo-Scottish songs have been in the Irish folk-song tradition since a time before white settlement in Australia began.

An Irish folklorist, Brian O'Rourke, said in his book *Blas Meala: A Sup from the Honeypot* (1985), subtitled 'A Selection of Gaelic folksongs with prose translations and verse equivalents':

> For a long time now, songs in English, both native and imported, have been popular in the Gaeltacht [that is, those parts of Ireland where a considerable part of the population still uses Gaelic for everyday speech] . . . It is certainly a fact that the exponents of songs like 'Barbara Allan', 'The Banks of the Nile', 'The Bonny boy', 'John Mitchel', 'Skibbereen', and 'A stòr mo chròi', have included some of the pillars of the *sean-nós* singing style [the 'old style' Gaelic tradition].

The first three of the songs mentioned by O'Rourke are Anglo-Scottish. The first, 'Barbara Allan', seems to have been fairly widely sung in Australia. The second, 'The Banks of the Nile', seems not have been recorded in Australia; but a derivative ballad, 'The Banks of the Condamine' ('The Banks of Riverine' in some versions), has been recorded from Victoria to the Northern Territory. 'The Bonny Boy' is presumably the song cited above as 'The Trees They Do Grow High'.

Anglo-Scottish songs probably began to spread widely in Ireland only after the settlement of considerable numbers of English and Scottish in the seventeenth century. It may well be that the adoption of Anglo-Scottish songs into the Irish folksong tradition proceeded most quickly in the southern parts of the country. In Ulster the Anglo-Scottish settlers, especially the descendants of Presbyterian Scots, have maintained a separate identity. In other parts of Ireland, English settlers — even soldiers who had served in Cromwell's campaigns against the Irish — showed a strong tendency to go native, reverting to the Roman Catholic faith and becoming absorbed into the Irish population. Presumably this process helped in the spread of English songs as well as the English language.

By the seventeenth century English presses were beginning to pour out ballad texts printed on broadsides (see **Broadside ballads**). Most of these ballad texts disappeared almost as quickly as most of today's pop songs. But ballad singers hawked the broadsides around country towns and at places for social entertainment, such as fairs and public houses, so a few of the street ballad texts passed into oral circulation. At the same time a few songs that had previously existed only in oral tradition were learnt by the ballad hawkers. Texts were carried to London, printed on broadsides and then carried back to the countryside in print. Some songs — 'Barbara Allan' is one — owed their widespread and long continued life in oral tradition to the constant interaction between printed and oral traditions. More important, the street ballad came to form a model for country songmakers creating new texts. As the use of English spread in Ireland, the English street ballad style became the model for Irish songmaking in English. Broadside ballad printers in Dublin and other Irish towns helped to spread both Anglo-Scottish ballads, and the street ballad style, in Ireland. And when the tide of migration began to change direction, to run from Ireland to Britain, Irish ballads were spread into England and Scotland.

In the eighteenth and nineteenth centuries, itinerant farm workers who moved across the Irish Sea for the harvest in England and Scotland must have been important in this spread of Irish ballads. But as the new industrial cities developed, Irish peasants began moving into them in large numbers, to become industrial workers in the appalling slums of cities like Manchester and Glasgow (this was happening on a considerable scale long before the great famine of the 1840s). Irish ballads became as common in English broadside print as English ballads in Irish broadside print.

In short, when settlers began to arrive in Australia from the British Isles at the tail end of the eighteenth century, there was already a considerable stock of folk songs common to all the English-speaking parts of the British Isles. The stock that was held in common undoubtedly became larger during the nineteenth century as internal migration mixed up the various peoples of the British Isles even more thoroughly than they had been mixed already. The mixing went on in countries of British and Irish settlement overseas.

When Henry Lawson made a trip up-country, and wrote a jaundiced account of bush life afterwards, A. B. Paterson chided him in verse for presenting such a gloomy picture. He asked Lawson:

Did they rise up Willie Reilly by the campfire's cheerful blaze.
Did they rise him as we rose him in the good old droving days?

Lawson replied:

Yes, I've heard the shearers singing 'Willie Reilly' out of tune;
Seen them fighting round the shanty on a Sunday afternoon.

'Willie Reilly' is an Irish ballad; Lawson and Paterson seem to agree that it was one of the songs generally known among bush workers. This is the sort of evidence which in the past has been used to argue that the main influence on the Australian folksong tradition has been Irish. But the text of 'Willie Reilly' had been put into print in the nineteenth century by a number of English broadside printers, including Catnach of London whose sheets sold far and wide and in great numbers. It turns up, though infrequently, in oral tradition from the south-west of England to the north-east of Scotland. In short, it is unlikely that 'Willie Reilly' was carried to Australia solely by Irish singers and passed on by Irish-Australian bush workers to bush workers of English or Scottish or Welsh descent. It is reasonable to assume that it was known to some singers from Britain as well as Ireland before they migrated to Australia. Perhaps even more important, it belonged in a ballad tradition that was common throughout the British Isles. 'Willie Reilly' was just one of dozens of similar ballads carried to Australia. We will never know just how many, because it is obvious that songs of this kind were already disappearing from a moribund folksong tradition when Meredith and others set out to record in the middle of the twentieth century.

All this may be illustrated by considering the songs of Sally Sloane (q.v.). Sally Sloane was the finest singer of all those recorded by John Meredith. Only a few of the singers recorded by other collectors sang anything like as well as Sally Sloane, and no other singer had a repertoire of songs as big as hers. Sally Sloane sang a number of songs passed on, through her mother, from her grandmother. That grandmother, Sarah Alexander, arrived in Australia from County Kerry, in the south-west of Ireland, in about 1838. Many of the songs passed on to Sally Sloane from her grandmother are, of course, Irish in origin, but a number are certainly English or Scottish in origin: 'The Sprig of Thyme', 'The Lowlands of Holland', 'The Trees They Do Grow High', 'Lovely Nancy'. There are others which might be either Irish or Anglo-Scottish in origin: 'The Cherry Tree', 'The Wee One', 'The Banks of Claudy', 'Molly Baun Lavery' (obviously Irish with a name like that?; maybe — it appeared in English broadside print as 'Molly Whan', and was much sung in parts of England as 'Polly Vaughan' and 'The Shooting of His Dear').

It seems that Anglo-Scottish songs were being sung in the Gaeltacht long before Brian O'Rourke made the comment cited above.

The continued mixing of traditions in Australia is illustrated by the fact that Mrs Sloane learnt a couple of other songs, 'The Green Bushes' and 'The Rambling Sailor', from a migrant English railway worker. Both are almost certainly of English origin, though 'The Green Bushes' has also been found in Irish oral tradition.

The most interesting singer that Meredith recorded in the 1980s was Harry Dicks, whose grandfather had been transported from England in 1825. Some of the songs he sang were passed on in the family. 'It's a good song if you can sing it', he said of one song: 'I had it from my father, and he had it from his father'. The song was 'Skibbereen', one of the native Irish songs in English which O'Rourke refers to in a passage cited earlier. It ends with a cry for Catholic Irish vengeance on Protestant landlords. Harry Dicks learnt his version of 'Captain Wedderburn's Courtship' ['The Chicken and the Bone'] from Hilton Hogan of Oberon about 1940. Meredith refers to Hogan as 'an Irishman [Irish-Australian?], living in the middle of an old time Irish [Irish-Australian?] enclave'. Dicks learnt the song from Hogan while shearing: 'He used to pick up shearin' and pick up wool and that around the sheds, and I used to hear him singin' it'.

Three points may be made about this account of Harry Dicks's songs. First, an Anglo-Scottish song survived in an 'Irish enclave' to be passed on in the middle of the twentieth century from a singer of Irish descent to one of English descent. Second, an English convict learnt an anti-British Irish song and passed it on to his Australian son, and he in turn to his son. Convicts were not segregated, or at least not systematically segregated, by religion or nationality. This kind of transfer must have gone on quite widely among the transported convicts; and not only while they were serving their term of punishment. Ex-convicts of all kinds mingled in Sydney and Hobart and up-country. Third, enclaves such as the one that Meredith mentioned grew up mainly in regions where the great sheep and cattle runs of early colonial times were broken up for closer settlement. In such enclaves, obviously the mingling of traditions must have been slowed down. But men were constantly forced out of these enclaves, where the farms created by free selection did not prove profitable, to take other jobs, often as pastoral workers:

> There's brand new chums and cockies' sons,
> They reckon that they are great guns;
> They fancy they can shear the wool
> But the buggers can only tear and pull . . .

The mingling of traditions went on.

In the first volume of *Folk Songs of Australia* the proportion of songs carried from the British Isles, as against songs made (or significantly altered) in Australia, is about one to two. If the claim made by Anderson and Meredith in 1963, that the selection of songs printed in *Folk Songs of Australia* was 'representative of the whole continent', is true, then presumably the proportions would have been similar if Meredith's collecting had been done in Western Australia or the Northern Territory, rather than New South Wales.

Anderson and Meredith also noted in 1963 that 'none of Sally Sloane's children have inherited her love of folk song'. Meredith's later work showed that it was not just a family tradition that was dying. In 1987 Meredith commented on the results of his fieldwork in the 1980s:

> Instead of singing and making music country folk watch TV or a video; home-made entertainment is no longer an integral part of daily life . . . During the 1950s I recorded dozens of singers. More recently, over a period of four years, I have encountered only two performers who were good enough to warrant a return visit.

In the 1950s and early 1960s, systematic field recording of folk songs was carried out in Victoria by N. and P. O'Connor, Maryjean Officer, R. Michel and other members of the Folk Lore Society of Victoria (q.v.). Some of the most interesting of these songs were published in the Society's journal, *Australian Tradition* (q.v.) (which was edited by Wendy Lowenstein (q.v.), who was also a collector), but many of them have still not been published. The field recordings are now held in the sound archive of the National Library. A little later, over a period of several years, systematic field recording was carried out in south-eastern Queensland by R. and M. Michel, S. Arthur and others. Many of the songs recorded in this region also have not been published, but copies of a considerable part of the field recordings are now held by the National Film and Sound Archive. Systematic recording has been done in the far north of Queensland, most of it by Ron Edwards; he has published many of the songs collected. His field recordings have apparently either not been preserved or are at present inaccessible. Very little collecting has ever been done in other parts of Queensland, or in the Northern Territory, Western Australia, South Australia or Tasmania. We will, in fact, never know whether the claim that the material in *Folk Songs of Australia* was 'representative of the whole continent' is true. All that can be said is that the material was representative of the folksong tradition of the south-eastern mainland of Australia, from Victoria to southern Queensland, as it existed in the third quarter of the twentieth century. A considerable proportion of the songs in the tradition at that time had been carried to Australia from the British Isles. It is probable that the proportion had been higher in the century between 1850 and 1950. It is probable also that the songs carried from the British Isles had undergone a selection process in Australia. We may suppose with some confidence that that process removed from the Australian folksong tradition fairly quickly songs in Gaelic (and Welsh); songs in dialects of English unfamiliar to Australians speaking the remarkably uniform kind of English that evolved so quickly; and those macaronic Irish songs which had texts partly Gaelic and partly English.

Folksong making in Australia

We know very little about folk songs of the early colonial period in Australia. We have a solitary recording of a song about the 'convict times' in Tasmania: 'The Seizure of the Cyprus Brig'. There is a poem, known from a manuscript, of this title, said to have been written by the Irish convict known as Frank the Poet (q.v.). The song was recorded from an old whaler named Davies in the 1960s. It makes use of bits of the text from the manuscript poem, adds an introductory verse from a broadside ballad called 'Van Diemen's Land' and sets it to one of the tunes used for that song. 'Van Diemen's Land' is about poachers transported as convicts to Tasmania. Differing texts locate the poachers in Ireland, in England, and in Scotland. The tunes often have a vaguely Irish feel about them, but of course, as Samuel P. Bayard, the American authority on the song tunes of the British Isles, put it, 'We can often reasonably infer that a given version of some widespread air is Irish or Scottish, for example, but we cannot therefore claim that the air itself was of Scots or Irish origin'. Mr Davies's version of the tune, and his singing style, both suggest English rather than Irish influence. Frank the Poet's manuscript looks Irish; Mr Davies's song sounds — to use a term that gives offence to some — decidedly Anglo-Celtic.

A squatter from the country around Goulburn in New South Wales printed, in a book of reminiscences about life in the first half of the nineteenth century, the text of a song about an English convict transported for poaching. The tune used for this text, 'Jim Jones at Botany Bay', was 'Irish Molly-o', he said; seemingly another Anglo-Celtic song.

There is a song about the hardships of convicts at the penal settlement on Moreton Bay and the spearing to death of its commandant, Captain Patrick Logan, in 1830. The bushranger Ned Kelly (q.v.) seems to have been quoting it in part of his rambling 'Jerilderie letter', written in 1879. It purports to be the testimony of an Irish convict, and it has been attributed to Frank the Poet: text and tune both seem to be decidedly Irish in character. The only complete version recorded from oral tradition came from a fine singer, Simon McDonald (q.v.), recorded by members of the Folk Lore Society of Victoria.

There is a cluster of ballads about the young Irish convict turned bushranger, Jack Donohoe (see **Bushrangers**), who was shot dead by mounted police troopers (the 'horse police') in 1830. Most of them are in a street ballad style that has nothing particularly to distinguish them as Irish; but some have texts that voice Irish defiance of British tyranny: 'He'd scorn to live in slavery or be humbled to the Crown'. There is plenty of literary evidence to show that ballads about Donohoe were much sung by bush workers, Irish-Australian or otherwise. From Donohoe, shot in 1830, to Ned Kelly, hanged in a Melbourne gaol in 1880, many of the notable bushrangers were Irish or Irish-Australian. (Those who like to think that Australians are a rebellious lot, and like to attribute this to Irish influence, sometimes dwell on this circumstance. It happens that a lot of the police who fought the bushrangers were also Irish-Australian. One of the policemen who finally captured Ned Kelly was a Senior Constable Kelly.) The bushranger ballads, not unexpectedly then, often express Irish-Australian hostility to the British crown and hence Australian colonial authority. It is worth noting explicitly that they do not express hostility to fellow-Australians descended from other ethnic groups. Joseph Cashmere, a bush worker who had spent most of his working life in the south-western corner of New South Wales, told collectors in the 1950s that bushranger ballads had there been called 'treason songs'; a term once used in Ireland for anti-British songs. A policeman had once been offended when one of Cashmere's friends sang 'The Wild Colonial Boy', and locked up the singer for the night. But Cashmere did not suggest that a liking for bushranger ballads was in any way confined to Irish-Australians. There is evidence aplenty to show that attitudes to the bushrangers — and the singing of bushranger ballads — were determined by social class rather than ethnic origin. Bob Michel collected a version of 'Bold Jack Donohoe' from a singer in Queensland, who had learnt it from his brother. His brother had enrolled as a special policeman during the shearers' strike of 1891, and had gradually been won over to the shearers' side. He had learnt 'Bold Jack Donohoe' from some of the shearers. They sang it as an anthem of defiance, but no longer of Irish defiance of the British crown and its colonial representatives. Now the song was used to voice the feelings of a militant Australian working-class group and its defiance of home-grown squatters and the elected government of Queensland.

The gold rushes that began in the 1850s brought an enormous sudden increase in the white population of Australia; but the settler population was already beginning to take off, and most of the population growth in the second half of the nineteenth century came from natural increase, not from immigration. Cultural trends, like demographic trends, grew from foundations laid before the gold rushes and were not affected greatly — or, at least, were not *directly* affected greatly — by the influx of gold seekers.

The gold seekers, a lot of people crammed densely together, many of them with money to spare for luxuries, attracted a following of professional entertainers: opera singers, ballet dancers, music-hall singers, 'black minstrels' straight from the gold diggings of California. The professional entertainers wrote the songs of the Australian goldfields. Like most of the songs produced by professional stage entertainers, they were ephemeral. A lot of them have been printed in collections of Australian 'folk' songs. There is very little evidence that more than two or three ever passed into oral tradition. It is much more important that from now on a constant flow of music hall, vaudeville and 'black minstrel' entertainers came to the Australian cities from London and Dublin and San Francisco; the Australian 'poor relations' —

professionally speaking — picked up their songs and jokes and ways of dressing and spread them, through travelling shows, to country towns and even shearing sheds.

A lot of the song texts that can be dated to the second half of the nineteenth century, and the early years of the twentieth, belong to the old street ballad style: lugubrious songs about the untimely ends of jockeys and boxers, about railway disasters and shipwrecks. *Folk Songs of Australia* has its fair share of them. 'The Wreck of the Dandenong', transcribed from the singing of Mary Byrne, looks dull, bathetic, in print. Listening to Mary Byrne singing it, with the dry voice, the uncertain intonation of a woman of seventy-three, on John Meredith's tape, a sympathetic listener may find more, and understand a little better what these songs once meant. Some of these songs were printed on broadsides, others in newspapers of one kind and another.

Literacy rates in the Australian colonies seem to have been relatively high even in the first half of the nineteenth century. Colonial governments began to introduce free and compulsory primary schooling in the last quarter of the nineteenth century. Some of the goldfields entertainers had published 'songsters': small, thin books containing song texts. In the second half of the century, some of the songsters published were aimed at an audience of bush workers. Some texts from these songsters passed into oral circulation; in other cases, a version printed in a songster seems to have influenced texts already in oral circulation. Many singers seem to have taken to writing down song texts newly learned, relying on memory only for the tunes. Singers who had forgotten a song text took to writing to newspapers, asking if another reader could supply it; such requests were often rewarded, and the practice is not quite dead yet. Some singers kept scrapbooks, containing texts that they had written out and other printed texts cut from newspapers. Thus, written and printed sources began to influence the form of songs which had at first circulated only orally.

Conversely, print helped many songs originating with urban popular entertainers to an uncertain life in oral tradition. So, for example, an Australian music-hall performer, Lance Lenton, wrote a song about Woolloomooloo larrikins, to the tune of an Irish stage song, 'Killaloe'. Collectors found the song (or, more usually, fragments of it) remembered by old singers in Tasmania and Victoria, as well as New South Wales, in the 1950s and 1960s. Some English, Irish and American texts originating with professional stage entertainers passed into Australian oral transmission in the same way. An adaptation of 'The Wild Colonial Boy' by a Dublin entertainer, Percy French, even passed into Australian as well as Irish oral tradition.

The biggest group of native Australian folksong texts was produced by bush workers. They took their models, their tunes, scraps of text from all kinds of sources. Two or three examples will have to do. The dialogue song known usually as 'The Banks of the Condamine', but sometimes as 'The Banks of Riverine' (i.e. the Riverina), has been collected from Victoria to the Northern Territory. The dialogue is between a woman and her lover, who is about to set off to a shearing shed or to a horsebreaking camp. It is made over from an Anglo-Scottish broadside ballad of the Napoleonic wars, 'The Banks of the Nile'. The setting has been transferred to Australia with astonishingly little alteration to the original. This is one of the songs mentioned by O'Rourke, cited earlier, as sung in the Gaeltacht, but it seems to have been a special favourite in the north-east of Scotland, and many of the tunes used for 'The Banks of the Condamine' have a Scottish sound about them.

One of the most widely sung of many shearers' songs, 'Click Go the Shears', borrows its tune and a little of its text from an American popular song of Civil War reference.

'The Drover's Dream', recorded in many versions from Victoria to northern Queensland, is set to the tune used for the Irish stage song 'Killaloe'; the same tune that Lance Lenton used for his song about

Woolloomooloo. The text of 'The Drover's Dream' also has something of a music-hall or minstrel-show style about it.

The speaking of verse seems to have been a kind of entertainment as popular with bush workers as the singing of songs. A few of the texts of the better-known bush balladists, notably Paterson and Lawson, were set to tunes by bush singers, and have been recorded from oral tradition. Many bush workers wrote verse, as well as reciting it, and often had it published. 'Duke' Tritton, a singer and versifier recorded by Meredith in the 1950s, wrote a lot of verse, and a song called 'Shearing in the Bar'. This has been popular with urban singers of the 'folk song revival'. Tritton explained that he wrote the words, with a bit of help from his mates, and only later thought of setting a tune to the words. Thus, the style of the bush balladists came to influence bush songmaking. But the popularity of verse recitation probably helped to enfeeble the folksong tradition. The wider spread of the products of the music industry — aptly so-called — made possible by the electronic media have now killed it almost completely.

For an introductory overview of the Australian-made folksong texts, and their tunes, the two volumes of *Folk Songs of Australia* should be supplemented by *Snatches and Lays* (2nd edn, 1973) and the publications of Ron Edwards, preferably *The Big Book of Australian Folk Song* (1976). The first edition of *Snatches and Lays* was published under a spurious imprint and by pseudonymous editors Sebastien Hogbotel and Simon Ffuckes. The introduction to the first edition said that

> The songs and verses in this collection have, in some cases, been sung and declaimed for centuries. It is proof of the vitality of this form that the most recent of them is less than three years old; by the time this book has appeared the store of unprintable, 'dirty' songs will have been further added to.

To this they added in a foreword to the second edition of 1973:

> Just in case there is any misunderstanding about the matter, this is a serious collection of songs as sung by sailors, soldiers, airmen, students and the like in and around Australia between the years, say, 1940 and 1960.

Ron Edwards brought together in *The Overlander Song Book* (1971) material published in a number of earlier, smaller publications. *The Big Book of Australian Folk Song* is virtually a revised edition of *The Overlander Song Book*. It includes a considerable number of songs which probably never entered oral circulation. It contains very few of the songs carried from the British Isles; on the other hand, it shows that a rich folksong tradition existed in north Queensland, with much local material. It suggests that such local material may have existed in other regions which did not attract field collectors. It also contains a bibliography which is still useful, though out of date.

Aboriginal songs in English

Very little was recorded from English-speaking Aboriginal singers in the 1950s and 1960s. An exception was the recording by an anthropologist, Jeremy Beckett, of songs written and sung by an Aboriginal itinerant worker, Dougie Young. A few of Dougie Young's songs were published on a gramophone record by Wattle Recordings in 1963. One of Dougie Young's songs, 'The Land Where the Crow Flies Backward', was taken up by white singers of the folksong revival movement, but the Wattle recording soon went out of production.

Collecting in the 1980s, especially work done in parts of New South Wales and in Brisbane by Chris Sullivan, shows a strong tradition of songs in English. Some parts of this tradition are rather unexpected. Aboriginal singers have preserved some songs carried from the British Isles that have not been recorded from white singers; for example, a version of an English broadside ballad known as 'The Indian Lass' (the 'Indian' in this case being Hawaiian). Aboriginal singers have also preserved Australian-made white song texts, such as 'The Old Bark Hut'. Most important of all are the songs made by Aborigines and dealing specifically with Aboriginal life. Sullivan and others found some of Dougie Young's songs had been passed on to other Aboriginal singers. Sullivan also recorded songs about Aboriginal life written by other Aboriginal songmakers quite recently, but already passed into oral tradition. The style of these songs derives largely from acquaintance with the sound recordings of popular professional entertainers; the very thing that was helping to end folksong making by white Australians.

The outsider wishing to learn more of this impressive tradition will have to seek access to the field recordings held by the National Library, or the Australian Institute of Aboriginal and Torres Strait Islander Studies, or wait for publication of the songs in one form or other. The Library and the Institute are preparing a joint publication of some of the material.

Edgar Waters

Folk Song and Dance Society of Victoria, The, incorporated in 1981, pursues its various activities under the registered name 'Folk Victoria'. The aims of the Society are the collection, preservation, presentation and promotion of all aspects of the folk arts, and the Society represents the interests of all Victorian folk organizations on the Australian Folk Trust. FSDSV activities include the weekly Melbourne Folk Club, concerts, dances, workshops, festivals and master classes, and a regular newsletter is published for members.

Folk speech is a collective term describing a wide variety of speech forms, most of which stand in an unofficial, informal, exclusive or non-respectable relationship to standard or respectable language. Folk speech includes slang; swearing; jargon (see **Occupational folklore**); argots (speech of social groups, such as criminals, leisure groups, etc.); dialect words and regionalism (what is a 'deli' in Western Australia is a 'corner shop' in New South Wales); folk names (nicknames and contractions such as 'Paddo' for Paddington or 'hostie' for air hostess; 'car' for automobile); proverbs ('A bird in the hand is worth two in the bush'); similes and comparisons; ('Silly as a two-bob watch'); secret languages ('Pig Latin'); curses and spells; prayers; stump speeches and parodies of other verbal forms such as sermons; linguistic adaptations from English into another language ('billicano' is an Italian-Australian term in the New England district of New South Wales) and vice versa ('imshee', a First World War 'digger' term derived from Arabic 'falafel', is a more recent multi-cultural arrival); and humorous sayings. Folk speech also encompasses the linguistic and social dynamics of everyday conversations and a range of idiomatic expressions continually generated by people's ingenuity.

See also: **Folk speech: community languages**; **Folk speech: English language**; **Koorie languages and folk speech**.

Folk speech: community languages It is estimated that there are approximately 100 non-Aboriginal languages other than English in daily use in Australia. Such languages have been part of the Australian scene for most of the period of European settlement, and they were prominent in the mid-nineteenth century, especially in Victoria, South Australia and Queensland. They have enjoyed a broken history in Australia due to restrictions on their public use for much of the twentieth century and the dominance of the assimilation policy. Since the mid-1970s, the legitimacy of the languages has been implicit in their designation as 'community languages' or 'community languages other than English' (CLOTEs)'. They are promoted, especially in education and the media, through national and state language policies.

According to the 1986 census, there were in that year twelve community languages used at home by over 50 000 speakers, a further three with over 30 000 home users, and another three spoken at home by over 20 000 people. Italian and Greek are the most widely employed community languages throughout Australia, followed (in rank order) by the language(s) variously designated as Serbo-Croatian/Croatian/Serbian/'Yugoslav', then Chinese, Arabic, German, Spanish, Polish, Vietnamese, Dutch, Maltese, French and Macedonian. The numbers for the older established languages (especially German, Italian, French and Polish) would be much higher if the use of languages in domains other than one's own home were taken into account. Arabic and Chinese are becoming increasingly common among school-age children in some parts of Australia, while Macedonian and Maltese have benefited from 'ethnic revivals'. The ethnolinguistic composition of different states and cities varies, with Victoria and South Australia building on their earlier post-war multilingualism, and a number of new and expanding speech communities (Arabic, Spanish, Chinese, Vietnamese, Filipino, Indonesian) centring in New South Wales.

Meanwhile, the shift from community languages to the use of English only is continuing to rise in some linguistic groups, especially among speakers of Dutch, German, Italian and Hungarian. Language groups with relatively low shift rates are Vietnamese, Macedonian, Greek, Turkish and Arabic speakers. Language shift increases markedly in the second generation, particularly where one parent is a monolingual English speaker, a typical state of affairs in most ethnic groups today.

Many community languages are becoming mainly languages for communication by and to the elderly. This is due to the language shift mentioned above and also to a tendency for many older bilinguals to revert to their first language and culture as they age. The language of peer-group communication among the young in most ethnic groups is English, and when they form their own nuclear families, the main domain of community language use is the home of their parents.

Other important domains for community languages are private and group-specific: they include religion, especially where it is closely linked with language and culture (e.g. Eastern Orthodox and Eastern-rite Catholics), secular ethnic community organizations and the community language press. Through the areas of education, radio, television and libraries, on the other hand, governments have, in recent years, supported the public use and maintenance of community languages. There are now over thirty languages other than English examined at final secondary-school examinations, including Khmer, Ukrainian, Turkish and Latvian. The government education systems of Victoria and South Australia have a total of seventeen languages taught in their primary schools, in some cases in bilingual programmes. However, most community language groups run their own part-time classes to teach the language to the younger generation. There is a government television channel (SBS — Special Broadcasting Service) transmitting films in languages other than English, with English sub-titles. Ethnic and multilingual radio stations broadcast in over fifty languages, some of them being access stations enabling groups and individuals to make their own programmes. Public libraries in urban centres stock large holdings of books, periodicals and cassettes in community languages. Many services in community languages are now available in the public domain, and there is a telephone interpreter service. Community languages are also employed frequently in shops which sell 'ethnic' foods or books and records, cafés and restaurants, and continental guest houses in holiday resorts. These become venues where people come to interact in their first language or other languages they spoke before they came to Australia.

Due to the exigencies of living in Australia, the influence of English, and isolation from the countries of origin, community languages have undergone some changes in Australia. There is a great deal of variation in the ways in which people adapt their languages to Australian use: homogeneous changes require closed networks of the kind that existed in some German-settled rural areas of South Australia and western Victoria in the nineteenth century. However, some general tendencies in the grammar and vocabulary of community languages can be detected. Renewal in vocabulary to express the new reality in Australia includes 1) neologisms, new words based on the traditional word-formation devices of the language, e.g. *milkbaraki* (little milk bar, Greek), *fensje* (little fence, Dutch), *farmista* (farmer, Italian), *farmerieren* (to farm, literally 'to farmer', Barossa Valley German); 2) semantic expansion of archaic words, e.g. *Eisschrank* (ice chest) used by German-speaking pre-war refugees for 'refrigerator' (*Kühlschrank*), or *Luftschiff* (air ship) employed for 'aeroplane' (*Flugzeug*) by descendants of nineteenth century settlers in old German settlements; 3) words directly transferred from English, e.g. beach, paddock, tea-tree; toolmaker, overtime, payday; chemist, newsagent; spelling, class-mark, play-lunch; mudguard, oil-change.

Words transferred from English are often substituted for items that require complex constructions or because there are a number of equivalents to choose from in the community language. For instance, the transference of 'remember' into German avoids the use of a reflexive pronoun, the preposition *an*, and the accusative case after it. The transference of 'put' in the form *putten* or *geputtet* saves a choice between a number of verbs based on the often necessary criterion of the position into which something is put (standing, sitting, lying, hanging).

Unlike English, most community languages mark their gender (masculine/feminine/neuter; or neuter/non-neuter; or masculine/feminine) through articles (equivalent to *the*) or endings. Thus, words transferred from English into such languages need to be assigned to one or the other gender, usually because of the meaning or the sound of the word. So 'ticket' is *il tichetto* in Italian and 'beach' is *la biccia*, 'car' is *to karo* in Greek and 'stock' is *o stokos*. Similarly we hear *le job* (French), *die beach* (German) and *de shop* (Dutch) in Australia. In some languages most transferred nouns are allocated to a particular gender. So most English nouns are integrated into the non-neuter (*de*) gender in Dutch and to the neuter (*to*) in Greek. As will be seen from the 'put' example, community languages develop rules for integrating English verbs into their grammatical systems to form past tense and the forms for different persons (German *er shrinkt* 'he is shrinking', *sie shrinken* 'they are shrinking').

Sometimes the meaning of an English word is transferred to a similar-sounding word in the community language, or one that has some overlap in meaning. So Italian *bosco* meaning 'bush, shrub' is extended to cover 'bush' in the sense of 'Australian forest', and German *Gummibaum* 'rubber-tree' takes on also the meaning of 'gum-tree'. There can be a general shift in meaning with, for instance, *farma* replacing the standard Italian word for 'farm', *fattoria*, which, in Australian urban centres, has widely assumed the meaning of 'factory'. Idiomatic expressions are often translated word for word. For instance, 'to sit for exams' is translated into German as *für Examen sitzen* when the normal German way of expressing this is *Examen machen* 'to make exams'.

Gradually, too, the grammatical systems of community languages are changed, especially those of languages which are not maintained very well. Italian, French, Dutch and German all use 'have' and 'be' as auxiliaries for different verbs while English has only 'have' as an auxiliary. There is a tendency for second-generation speakers of all of these languages in Australia to generalize the 'have' auxiliary (e.g. Italian *ho andato* 'have gone' rather than *sono andato*). In standard Dutch, verbs take different endings in different persons in the singular and plural (*ik houd* 'I hold', *we houden* 'we hold'). Some second-generation Dutch speakers in Australia delete the *-en* in the plural, while children of Italian migrants often do not make agreements between adjectives and nouns (*la camera vicina* 'the neighbouring room' becomes *la camera vicino*). The *-s* plural ending is applied to Dutch nouns even when another ending (such as *-en*) would be appropriate. The position of the verb is fixed after the subject as in English by many Dutch speakers and some German speakers in Australia, but this change takes about a generation longer in German than in Dutch.

Many bilinguals switch from one language to another depending on who they are talking to, what they are speaking about, where they are, and how formal their speech is. Some will switch in the middle of a sentence to quote what someone has said in another language or because they have lost their linguistic bearings. This occurs around a trigger word, a word that is used by the speaker in both languages. Names, words that sound much the same in the two languages, and words transferred from English into the community language function in this way. For example:

Die matron hat alles decorated prachtvoll artificial decorations, she puts lovely big vases of gladdies 'n all that in too, but still an' I bet she makes it a lovely Christmas dinner. (The switch here appears to be triggered off by 'artificial decorations' which, incidentally, is preceded by a hesitation pause which increases the time lapse since the last unambiguously German item.)

Ze zijn gedeeltelijk waitress 'n and the others are staff. (Not only is 'waitress' transferred, but *en*, the unaccented form of English 'and', is the normal word for 'and' in Dutch. This overlap is sufficient to trigger a switch into English.)

Trigger words are sometimes anticipated, as in *Hij staat* on the bridge (where the speaker switches from Dutch into English at the beginning of the phrase of which the trigger word 'bridge' is part).

English trigger words cause switches from one community language to another, for instance from Dutch to German, from Russian to Yiddish, or from Hungarian to German.

Some of the phenomena described above have found their way into novels, short stories and poems in community languages as well as into cabaret set in an Australian environment. This is intended to add a local flavour, to portray characters who are Australians of the appropriate ethnic background, to appeal to readers from such a background, and/or simply to exploit the combined potential of the two languages.

In the 1920s this technique was employed by two Barossa Valley German writers who called themselves 'August von der Flatt' and 'Fritz vum Schkrupp'; their folkloristic prose was intended to capture the language and culture of the readers' grandparents. In the period between the wars, an Albury newspaper published a regular column in which the writer created a language variety which appeared to be half-German and half-English, which readers were expected to identify with elderly people in some of the nearby German enclaves such as Walla Walla and Pleasant Hills. More recently, the Italian–Australian writer Pino Bosi has incorporated English transfers into his Italian novel *Australia Cane* (1957), the German pre-war refugee Walter Adamson used them to describe his earlier experiences in Australia in *Das australische Einmaleins* (1973), and they play a role in much recent Greek–Australian and Ukrainian–Australian literature, to name but a few examples. Songs and ballads of non-English-speaking ethnic groups contain instances of words transferred from English, often strung together in an exaggerated way, and some actually switch entirely from one language to another. The first phenomenon was common in oral traditions of German speakers in the Barossa Valley, as shown in this example:

Meine Mutter, die tut grumbeln
daß ich marryen will mein Johnny
und die sagt, ich tu mir humbeln
und er poked mit mir bloß fun!
Glaub nich', daß er's tun wird.
No, no, mein Johnny ist kein Flirt.

The humour of bilinguals in Australia is rich in puns. Many jokes and anecdotes feature the language problems of interference experienced by recently arrived migrants, drawing from autobiographical recollections or from family stories. There is, for instance, the story of the gasman who appeared on the doorstep of a new arrival from Germany asking 'Where is the meter?'. The lady of the house, identifying the question as '*Wer ist der Mieter?*' (Who is the tenant?) promptly replied 'I am ze meter'! Another anecdote concerns an elderly Austrian refugee who was confronted by a policeman in the early years of the Second

World War rounding up firearms which enemy aliens were not allowed to have in their possession. 'Any guns, madam?', he asked politely. '*Eine Gans?*' (A goose?), she responded, '*Komm mit mir!*', dragging him into the kitchen where a most delicious fowl was roasting in the oven. Such stories are passed on from person to person. Bilingual puns are still productive, as in the deliberate word-for-word translation of an idiom, *jemandem eine Ring geben* 'to give someone a ring' (as a present), used instead of 'to ring up'.

Bilingual humour has contributed substantially to plays, revues and cabaret which present a light-hearted approach to problems of settlement and adaptation and sometimes a critique of the old and new homelands. This is a tradition developed by German-speaking refugee repertory groups, such as the *Kleines Wiener* Theater with its 'in-group' mixture of Viennese German, Australian English and Yiddish elements. Today, similar problems are the focus of theatre in many community languages. But the themes recur in English-medium productions such as *Wogs Out of Work*, where creative use is made of transference from other languages. This type of play appeals to the now vast group of second-generation Australians of different non-English-speaking backgrounds whose common language is English.

The future of Australia's very enlightened national policy on languages may not be very secure. However, the underlying multiculturalism is irreversible. While most community languages in Australia do not survive beyond the third generation under normal dynamic urban social conditions, we are beginning to see the positive effects of more supportive policies and attitudes, both in more established languages and (most especially) in the newer ones. Australia will continue to be one of the linguistically most diverse and interesting countries in the world, even if there is a progressing attrition of some languages. Australia's present awareness of its language needs and language resources is quite unprecedented.

REFERENCES: Camilla Bettoni, *Italian in North Queensland*, 1981; Michael Clyne (ed.), *Australia Meeting Place of Languages*, 1985, and *Community Languages — The Australian Experience* 1991; James Jupp (ed.), *The Australian People*, 1988; Joseph lo Bianco, *National Policy on Languages*, 1987; A. Pauwels (ed.), *The Future of Ethnic Languages of Australia*, 1988; S. Romaine (ed.), *Language in Australia*, 1991; Anastasios Tamis, The State of Modern Greek as Spoken in Victoria, Ph.D. thesis, 1986.

Michael Clyne

Folk speech: English language Like other national varieties of English, Canadian English or New Zealand English for example, Australian English is distinguished by features of accent, idiom, and vocabulary which are in part a reflection of the origins of the people and in part a record of their experience in a new country. So there are words like 'back o'beyond', 'bowyang', 'chook', 'dinkum', 'jumper', 'larrikin', 'mud-fat', 'skillion', 'staggering bob', and 'taw' which can be traced back to British regional dialect vocabularies, and words like 'backblocks', 'boomerang', 'bush', 'dim sim', 'jackeroo', 'koala', 'outback', 'prospector', 'squatter', and 'top-ender' which are non-British in provenance, being either new uses of old words, coinages or borrowings from the indigenous languages, other immigrant languages and other immigrant varieties of English (like American).

Words which came into English in Australia, which have a greater currency here than elsewhere, or which have a special historical significance in Australia are called Australianisms, and it is words like 'billabong', 'bonzer', and 'bush' that come to mind when we think of Australian English. But they are a small proportion of the whole, by far the greater number of the words we use being shared property, held in common with other varieties of English. Again, most Australianisms are nouns — naming words called into existence by new things in a new land — and the largest group of these is of names attached to specimens of Australia's extraordinarily rich flora and fauna. Many of these are simply descriptive compounds made up of a distinguishing epithet like 'native', 'red', 'white' or 'wild' and the name of an 'old world' species like apple, carrot, cedar, cucumber, pine, poplar, etc. Many are Aboriginal words — like *boodie*, *boongarry*, *dalgite*, *dunnart*, *euro*, *kowari*, and *mapi* — which are tied to the species they name. Words like these are in a sense inert — they do not have the vitality and flexibility of colloquialisms like 'bonzer', 'brumby', 'cobber', 'dingo', 'flaming', 'jumbuck', 'mate', 'perish', 'shickered', 'shivoo', 'tucker', and 'walkabout'. While they add to the numerical strength of Australianisms, they have not entered significantly into the popular consciousness and do not have the special quality that a word like 'billabong' has of evoking Australianness. It has to be realized, therefore, that the number of Australian words which do have this quality or power is small, and makes up only a small part of Australian English.

Folk speech is not the language required by a new society in response to formal needs like those of naming unnamed things, species of flora and fauna, features of the environment, etc., but the language chosen by such a society to express its informal needs, its response to the environment and to the circumstances of the individual and collective encounter, its realization of the social bonds forged in the encounter, and its engagement in the associated occupations and pastimes. Folk speech is traditional in that it is handed down from generation to generation, almost as a confirmation of the society's view of itself, but it has always a contemporary manifestation in that it may modify or build on the traditional in response to social change or simply to the passage of time. Folk speech may belong to so small a group as a family but is at its most significant when it represents the self-perceptions of a people. It is possible to find elements of folk speech used by any discrete part of a community — any occupational, racial, religious or regional grouping — but it is most revealing to examine that folk speech which has the endorsement of the community as a whole.

Australian folk speech in English is a reflection of popular emphases on aspects of Australian social history. It was at its most creative in the nineteenth century, when the quest for national identity was strongest, and therefore in its main components reflects an Anglo-Celtic and exclusivist view of society. The remainder of this article will look at a number of key words or key groups of words, with a view to establishing these components (quotations cited are drawn from the relevant *Australian National Dictionary* entries). There are, obviously, many pockets of folk speech in Australian English related to subjects as varied as children's games, shearing, social drinking and sport, but this article is deliberately selective.

Currency lads and lasses

The name of the continent itself, preferred by Flinders and endorsed by Macquarie, was essentially a choice of the people. The first Australians were the Aborigines, but the name was early annexed by the colonists, clearly because those who had been sent to the ends of the earth ('antipodeans' and later 'the men from down under') sought to establish an identity with their new home, the unexplored and uninterpreted 'great south land' beneath the new but soon to be potent symbol of the 'Southern Cross'. The colonists made other distinctions in their claim to identity, dispossessing the Aborigines of words like 'native' and 'native-born', taking pride in being 'colonial', and asserting their status as 'free' rather than 'bond', 'settlers' not 'emigrants' (or 'convicts'). An ironic distinction was made between 'sterling', originally used of British money circulating in Australia, and 'currency', originally any local medium of exchange, and from the 1820s 'currency lads and lasses' became the somewhat self-conscious but favoured darlings of popular Australian song and verse.

Colonial experience

More serious was the growing insistence on colonial experience and the jocular denigration of those who did not have this, the 'new chums'. A 'new chum' was in English and early Australian use a newly admitted prisoner or newly arrived convict;

its use from the late 1820s, outside the convict context and to describe an immigrant who was newly arrived and therefore lacking in colonial experience, is perhaps indicative of a breakdown in the otherwise fiercely maintained distinction between bond and free and the beginnings of the recognition of a new social order. The camaraderie of the colony replaces that of the gaol and entry into the society is virtually trial by endeavour.

Until Federation in 1901 Australia was a loosely-knit string of British colonies scattered around the periphery of a vast and inhospitable continent. Each had ties with the 'mother country', the 'old country', (later 'cold country') or 'Old Dart', with 'home', each had its own colonial administration and, increasingly, its colonial industries, activities and character. Throughout the nineteenth century, and lingeringly even now, as a result of the tension between these two poles, there was a crucial ambivalence in the meaning of the word 'colonial'. In 'colonial' ship or vessel, 'colonial' tweed or wool, 'colonial' secretary or treasurer, its connotations are neutral. In 'colonial ale', 'colonial aristocracy', 'colonial language', 'colonial society', 'colonial tobacco', 'colonial twang', and 'colonial wine' there is an implication of inferiority or provinciality. In 'colonial experience' there is a strong assertion of antipodean values:

> the knowledge and practical experience of agriculturists from the other side of our globe is of little advantage in too many respects when brought to bear on Australian affairs in general, rural in particular. There is such a topsy-turvy reverse in almost everything, that no-one need wonder why 'colonial experience' is an article or an attribute in constant demand; and a pretty mess even colonial experience frequently makes of it (*Maitland Mercury* 20, No. 1847, 2/5).

The bush

To a very large extent colonial experience, that which a newcomer (or from the 1870s a 'jackeroo') needed to acquire, equates with 'bushmanship', 'the ability to travel through, or live in, inhospitable country, especially that which is unfamiliar or unsettled, without getting into difficulty'. And in a very real sense the qualities of the bushman are those we have come to regard as distinctively Australian. Here the sense development of 'bush' itself is instructive. 'Bush' is first used in Australian English as it had been in American and South African to refer to natural vegetation, or a tract of land covered in vegetation, having particularly the connotations that such land remains in its natural state and has resisted settlement. From this it comes to mean the country as opposed to the town, and particularly country remote from settled districts or, in a transferred sense, rural as opposed to urban life. So the phrase 'to take to the bush' is used first of a convict escaping into the wilds, later of any person who leaves the town for the country or an Aborigine who returns to tribal life in the inland. The strongest of the several attributive or adjectival senses of 'bush' is that which designates the strength (or weakness, according to whether the perception is rural or urban) of artefacts and constructions which are ingeniously (or crudely) improvised by people in the outback, people who may lack an urban sophistication but who possess those values which are correspondingly prized in country people. So there are numerous simple combinations like 'bush bed', 'bush biscuit', 'bush blanket', 'bush carpenter', 'bush cook', 'bush experience', 'bush fare', 'bush fashion', 'bush hospitality', 'bush hut', 'bush life', 'bush lore', 'bush nurse', 'bush parson', 'bush picnic', 'bush pub', 'bush rider', 'bush road', 'bush school', 'bush style', 'bush tea', 'bush traveller', 'bush tucker' and 'bush worker'. The list is rich, diverse, and revealing. The bushman is the person who can 'bush it', the sense development following that of 'bush'. So the bushman is first one who is skilled and experienced at travelling through the bush (one of the early senses also of 'bushranger'), then, broadly, one who lives in the country as opposed to the town and displays the manners and practical skills of a country-dweller.

Other special senses of 'bushman' develop: it becomes a synonym for 'swagman', 'timber-getter', and 'rouseabout' and, later, is used of one of the contingent of men who volunteered for service in the Boer War. In each of these new senses what is being stressed is a person's possession of self-reliance and ingenuity, of the skills requisite for survival in a range of unpropitious circumstances.

The outback

Preoccupation with the bush, that 'nurse and tutor of eccentric minds', meant preoccupation with inland as opposed to coastal (and, in time, settled) Australia and this is interestingly displayed in the number of words which express distance from the comparative security of settled districts. To 'go back' from the settled fringe was, as early as 1800, to be in frontier country, in fear of danger either from Aborigines or bushrangers. An important second sense emphasized the necessity of access to water in areas of low rainfall, especially in the distinction between 'frontage' (in full, 'river' or 'water frontage') and country which lay 'back of' or distant from such access. A third sense developed as more of the country was occupied by squatters (people who leased large areas of Crown land to graze livestock), with a general reference to the sparsely populated interior, country remote from towns and settled districts. Originally American, 'back country' followed essentially the same sense development. How close the 'back country' was initially is revealed by Matthew Flinders's writing in 1802 that 'no hill, nor any thing behind the shore could be perceived, but it does not certainly follow that there are no hills in the back country, for the haze was too thick to admit of the sight extending beyond four or five leagues'. Water soon asserted its importance, the explorer T. L. Mitchell writing in 1839 that 'grass is only to be found on the banks of the river. . . . None may appear in the back country'. As settlers moved inland in search of pasture the 'back country' became that land immediately beyond a settled district or adjacent to a station, which might be appropriated as supplementary pasture. And this general sense broadened by the 1880s so that the reference was to any part of the sparsely populated interior. From the earliest days of settlement land was allocated in 'blocks', so that 'back block' was a logical development, referring in the singular to a specific parcel of land, in the plural to 'the back blocks', to remote land generally, to the 'back grounds' or 'back land' as it was also known for a time.

John McDouall Stuart recorded in 1860 that 'from [his] observations of the sun . . . [he was] camped in the centre of Australia' and from the 1890s 'Centre' was used, with somewhat less precision, to refer to 'the great alluring, mysterious Centre, the heart of a continent'. 'Centralia', originally proposed as a name for the colony of South Australia, remains a synonym. Numerous phrases like 'back of' (a town or place), 'back of beyond', 'back of Bourke' and 'back o' sunset' testify to the challenge and the growing sense of romance attaching to the slow reaching inland from the settled eastern coast towards this centre.

Those who stayed on the settled fringe remained, somewhat perversely, 'inside', or in 'fence country'. Those who ventured 'outside the fences' or into the 'Inland' (which is not used until the 1920s) went 'beyond the boundaries', 'beyond the limits', or 'beyond the limits of location', ultimately 'beyond the black stump'; they went 'out', 'outback', 'out bush', 'outside', or 'out west'. Again, the proliferation of collocations, such as 'out-district', 'out farm', 'out gang', 'out-settlement', 'out-settler' and 'out squatter' testifies to the significance attaching to such a venture. One who lived in the 'outback' was an 'outbacker' and briefly in the late nineteenth century an 'outsider'. Fanciful variations lived comparatively long lives: the use of 'far' in collocations like 'far bush' and 'far interior' led naturally to the formation of 'far-back' and 'far-out', but 'further back', 'further out', and 'furthest out' seem mainly literary in use. And this would be true also of both 'way back' and self-conscious collocations like 'red Centre' and 'great (Australian) outback'. More than the immensity of the land is conveyed by this range of terms: they become sentimentalized in reference from the 1880s or 1890s and part of a popular stereotype of life on the land, but prior to that seem to be conveying a familiarity, almost an identity with an environment which is perceived as elemental. There is no sense of triumph over natural majesty, rather of a simple heroism in the face of an environment that gives no accommodation.

The battler

The men who came to terms with the 'back country' have become the stuff of Australian legend: the 'bullocky', the 'bushman', the 'cockatoo' (or 'cocky'), the 'drover' or 'overlander', the 'rouseabout', 'selector', 'squatter', 'stockman' and 'swagman'. In all of these, there is an element of the 'battler', the person who has few natural advantages but earns respect by working doggedly on, with an uncertain prospect of reward, the person who accepts the chanciness of the struggle for a livelihood, who displays courage in so doing. 'Battle', the verb, and 'battler' are very much words of the 1890s, reflecting perhaps an interest in interpreting or assessing the quality of life, and it is significant that there are specific applications of the word to unemployed itinerants and, in the urban context, people living by their wits, small-time punters and prostitutes. These are people who are seen as being characterized by their resourcefulness and equanimity in the face of adversity, and by a rough and egalitarian code of honour which keeps them true to themselves and their fellows but allows the advantaged as fair game. No other code could have glorified the 'swagman'.

Water

The principal hazard for the traveller in the outback (and 'traveller' has as long a history as 'swagman'), the principal hindrance to profitable occupancy of the land, was the frequent uncertainty of the supply of water. And there is a considerable cluster of words the use and frequency of which testify to the importance placed on reserves of water. Rivers all too often dried up to leave only the occasional 'anabranch', 'billabong', 'chain of ponds', 'hole', or 'waterhole'. Natural depressions of various sorts held water for a time after rain and were variously named, re-employing British dialect words like 'soak' and 'waterhole', using simple descriptive coinages like 'crabhole', 'clayhole' and 'claypan', borrowing Aboriginal words like *gilgai* and *gnamma*, or recognizing the ability of Aborigines to locate reserves of water, as in 'native soak' or 'native well'. 'Dams' or 'tanks' were sunk ('dam' being a British dialect word for the body of water confined as distinct from the holding barrier, and 'tank' of Indian origin). *Mickeries* (the word is Aboriginal) were constructed for the watering of stock. 'Bores' (in full, 'artesian bores') were put down by individuals and 'bore water' carried in 'bore drains' or 'bore streams' from the 'bore head', but the role of government is acknowledged also in combinations like 'government dam', 'government tank' and 'government well'. On the Western Australian goldfields salt water was made potable with a condenser until such time as 'scheme water' was provided by the Goldfields Water Supply Scheme.

If there was no water the consequences were obvious. 'Wet', with reference to rainfall, is used in Australia only of 'the wet' or 'wet season', the rainy season in northern Australia, whereas 'dry' is used more extensively. 'The dry' is the corresponding dry season of the north, the summer, but also a tract of waterless country over which one has to travel, as in 'the "seventy-five mile dry"' of *We of the Never Never*. As an adjective 'dry' occurs in a cluster of collocations: 'dry camp', a camp (for travellers or travelling stock) at which there is no water, 'dry country', 'dry farmer', one who farms in arid country, 'dry feed', 'dry heart', another name for the 'Centre' (which is also known as the 'dead heart'), 'dry season' or 'dry spell', a period of drought, 'dry stage', a part of a journey which is through waterless country, and 'dry track', a track through such country.

The traveller who found no water 'perished': the word is used repeatedly to describe dying of thirst and so constantly present was the threat of this that the noun 'perish' became used, especially in the phrase 'to do a perish' and almost it would seem in flirtation with death, to signify a period of extreme privation, especially as caused by lack of water.

The bludger and the larrikin

It seems significant that there is no comparable list of urban 'heroes', no sense in which the urban experience contributes positively to the Australian's perception of self. There are colourful words for a range of petty criminals who engage in sharp practice, which might be seen to provide a negative contribution, but they take on no wider connotations. Two words, however, 'bludger' and 'larrikin', do acquire a special status. 'Bludger' is used in 1856 by Mayhew and is apparently originally a British slang shortening of 'bludgeoner', a pimp, a person armed with a bludgeon who lives off the earnings of a prostitute. It becomes a generalized term of abuse by 1900, especially as applied to a person who appears to live off the efforts of others. And it acquires specific applications — to a white-collar worker, as distinct from one engaged in manual labour, to any idler or loafer, to a person who does not make a fair contribution to a cost or enterprise. The word is now ubiquitous and may be said to connote values the reverse of those displayed by 'battler'. 'Larrikin' is rather different. A British dialect word, 'larrikin' is applied in both Australia and New Zealand from the late 1860s to a particular sort of dissident, initially a young, urban rough, a member of a street gang who flouted the accepted codes of behaviour as members of the 'cabbage-tree mob' did in the 1840s. By the 1890s it had broadened its area of reference and could be applied to any person who acted with apparently careless disregard for social or political conventions. As 'bastard' can be used equivocally so can 'larrikin', indeed there is a sense in which the larrikin as rebel attracts the reluctant admiration of the more circumspect. But the pejorative connotations of both words are strong and their importance in Australian English lies in their provision of a partial obverse to the values discussed earlier: no Australian would want a bludger for a mate and most would keep the larrikin at arm's length.

Mateship

Much of colonial experience centres on the notion of man alone, to borrow the title of John Mulgan's New Zealand novel of 1939. But always present was a need for companionship, for a 'mate'. As Henry Lawson, the apostle of 'mateship' wrote, 'The everlasting stars . . . keep the mateless traveller from going mad as he lies in his lonely camp on the plains'. The OED's first sense of 'mate' is defined and commented on as 'a habitual companion, an associate, fellow, comrade; a fellow-worker or partner. Now only in working-class use'. The latest quotation (1885) is from an Australian source. The new edition makes no change to the definition, but adds supplementary nineteenth- and twentieth-century quotations, most frequently of Australian provenance. The secondary use of 'mate' as a form of address, especially as by 'sailors, labourers, etc.', is attested from the fifteenth century but appears to be in recent usage mostly Australian. Given that it is the most frequently used form of address in Australia, and given that 'mateship' has taken on special values, its prominence needs to be examined. Early quotations make it plain that a 'mate' was a partner: Maconochie wrote in 1838 that 'men when they contract to do heavy work, as clearing, fencing, etc. almost always do it in parties of two, or more . . . requiring always the assistance of "neighbours", or "mates", or "partners", as they are severally called'. Griffith in 1844 amplified this: 'two generally travel together, who are called mates; they are partners, and divide all their earnings'. Corfield, reminiscing about the latter half of the nineteenth century, says: 'I have alluded several times to "partners", or "mates", which was the more popular term. These partnerships were quite common amongst carriers and diggers in bygone days. It was simply chums, owning and sharing everything in common, and without any agreement, written or otherwise'. The sense of partnership exists strongly in the use of the phrase 'to go mates', as when Idriess writes: 'none of us liked going mates with a man unless we could pay our own way'.

There was contemporaneously a weakened use, applying to 'an acquaintance, someone engaged in the same activity', which undercut the first: 'mates in the bush' (1849); 'all are mates' (1853); 'his mates in one of the little cemeteries on the Western front' (1919); 'not me mate; me mate's mate' (1963); 'and now, what about my little mate?'. Associated with this was the use of the word as a mode of address, implying equality and goodwill but used with increasing frequency and irony to a casual acquaintance: 'When they call you mate in the NSW Labor Party it is like getting a kiss from the Mafia'. The sense of a 'mate' as a partner develops towards the end of the nineteenth century into the sense of 'mate' as a sworn friend, but is seldom free from sentiment: 'Where his mate was his sworn friend through good and evil report, in sickness and health, in poverty and plenty, where his horse was his comrade, and his dog his companion, the bushman lived the life he loved' (1891). So 'mateship' is never really the 'bond between equal partners', as implied by the primary sense of 'mate', but an ideal, whether in a union exhortation — 'look here, lads, cultivate the spirit of mateship; love one another' (1898) — or in a Lawson reverie of 1913 — 'River banks were grassy — grassy in the bends, Running through the land where mateship never ends'.

War

The common man as hero found one theatre in the outback, and another in the trenches of the First World War. 'Billjim' was an apparent equivalent to the British 'Tommy Atkins': 'The word Billjim has become attached to the Australian Light Horseman as an expression for his type' (1919). But the word had already connotations which made it mean more than 'the man in the street', as this first presaging shows: 'The harrowing tale of the lost Bill or Jim in the Australian desert whose eyes are picked out by the crow almost before his death-struggle ceases' (1896). 'Billjim' was the battler turned soldier; he was to become the 'Anzac'. Australians were not alone in enlisting for service in the First World War, but they uniquely saw their service as proving their special — colonial or Australian — character to the world: 'The deathless name of "Anzac" That thrills from pole to pole' (1917); 'The marvellous spirit and genius of the Australian and the immortal name of Anzac' (1918). 'Anzac', a telegraphic code name for the Australian and New Zealand Army Corps found instant and heavy use — as an abbreviation of 'Anzac Cove', where the first landing was made at Gallipoli, as a name for the Corps or a member of the Corps, and as a name for the campaign. Though it properly referred to both Australians and New Zealanders it early acquired an emblematic significance, symbolizing the virtues displayed by those who served in the Gallipoli campaign, and was to a considerable extent annexed by Australians who sought to identify these as national characteristics: 'This is . . . the surest guide to an understanding of things Australian — that the Anzac, the 'Digger' in his best and worst qualities alike, is a fair type of his fellow countrymen'. Quotations which reinforce this identity are frequent: 'My thoughts go back to that chill morn When Anzac's soul first saw the dawn by lonely Samothrace'; 'Hundreds and thousands of heroes of the Empire . . . gladly come out to these smiling "plains of promise" . . . and multiply and prosper, and breed up a race of future Anzacs'. There is no New Zealand equivalent of the latter-day stereotype of the 'big bronzed Anzac'. The First World War saw also the coining of 'Aussie', a jingoistic identification of the Australian with his country: 'We consider the term Aussie or Ossie as evolved is a properly picturesque and delightfully descriptive designation of the boys who have gone forth from Australia' (1918).

Australians were quick to remark the qualities of those who 'went over there' and of those who 'stayed behind'. 'Six bob a day tourist' was a collocation applied, initially in a derogatory fashion, to those who volunteered. But the 'six bob a day tourists' earned respect as did the 'dinkums', 'the men who came over on principle to fight for Australia — the real, fair dinkum Australians', wrote Bean in 1917. In-service humour labelled the 'rainbows', reinforcements who 'came after the storm', and those colleagues of the 'bludger' who did not enlist were known as 'coldfooters', 'shirkers', and 'slackers'. But the Australian word most would associate with the First World War is 'digger', a word which joins the earlier 'cobber' and 'mate' and is used to refer to, at times to celebrate, an essentially male comradeship in a time of adventure and adversity. 'Cobber' is a word of the 1890s, probably of British dialect origin. At best it means something more than 'friend', at its most ordinary, and most frequent, it is a term of address like 'mate', implying equality of circumstance and goodwill. It is thought of as denoting something Australian: 'Who would not prefer that poetry-in-one-word — "cobber" — to the vapid equivalents "old bean" or "old thing" or "old top" of England?', asked the *Bulletin* in 1922. 'Cobbership and a cool approach to all problems — if there is a more Australian aspect of war it would be difficult to find', wrote Brodgen in 1944.

'Digger' was used on the goldfields from 1849 but apparently not as a mode of address. Its application to a soldier, first recorded in Melbourne in 1916, is therefore puzzling. It cannot reflect the experience of the trenches, though it may be a product of a training for that experience. Initially, like 'Anzac', it applied to Australians and New Zealanders alike; increasingly it was annexed by the Australian and, through its use as a mode of address, acquired the connotations of 'cobber' and 'mate': 'Digger, a friend, pal, or comrade, synonymous with cobbers; a white man who runs straight' (1916); 'G'day, dig. Me ol' mate, me cobber' (1954). The connotations are of courage: 'Another grave . . . which bears on the headstone: "Here lie two Huns who met a Digger". And near it, in terse and forcible language is inscribed on a second headstone: "Here lies the Digger"' (1919). But they are also egalitarian: 'My father's a Captain . . . and you're only a common Digger!' (1965). And they are Australian: '"Diggerism", that sane satisfaction of a social brotherhood, without political rancour, expresses most happily the idea of the social order most capable of realisation in Australia' (1945) (see **Wartime folklore**).

Australians are now a predominantly urban people and the nearest many come to 'doing a perish' is being short of a 'cold one'. The 'bush' has been tamed and the 'outback' made accessible in four-wheel drives. The 'digger' comes out on a variously observed 'Anzac Day' but is otherwise a figure of the past; 'cobber' is obsolescent, 'mateship' the subject of scepticism. 'Colonial' now most frequently denotes a style of domestic architecture which bears only a superficial resemblance to that of the nineteenth century; 'native' is used most frequently of indigenous plants cultivated in suburban gardens. But the searchings of the past have not lost their potency and these all remain, in the popular consciousness, important components of the Australian identity.

REFERENCES: W. S. Ramson (ed.), *The Australian National Dictionary: A Dictionary of Australianisms On Historical Principles* (1988).

W. S. Ramson

Folk tales Aboriginal secret–sacred myths and legends do not fall into the category of folk tales. Such narratives occupy a place in traditional Aboriginal cultures roughly comparable to that of the gospels in Christian cultures, or the holy books of other religions. To describe them as 'folklore', that is as essentially informal and unofficial expressions and practices, is therefore both inaccurate and, given the connotations of triviality and untruth that the term 'folklore' sometimes (inaccurately) has, potentially demeaning. There are, however, some aspects of Aboriginal narrative tradition that can be described as folklore, usually those elements where there has been some interaction with the traditions of recently arrived groups. The stories of the water-dwelling monster known to English-language folklore as the 'bunyip' (q.v.) is one example of this process. The process also operates in the opposite direction, with Aborigines adopting and adapting elements of non-Aboriginal culture to produce various new amalgams, particularly in music and art. In narrative one example is the portrayal in recent Aboriginal traditions of the West Australian warrior, Yagan, in terms similar to those of the outlaw hero of other cultures. In Australia, as in the Americas, it is possible to argue for a distinction between Old World and New World folklore origins and patterns. Thus, the folk traditions of North and South America — as is the case in the Australian continent — are a strange mix of the imported lore of the invading settlers, the Aboriginal lore of the tribes in their traditional lands, the lore evolving in the colonizing period, and that of many refugee groups, particularly since the mid-1930s. In all these cases the narrator is male, although women may well be included in the story as figures to be supported or ones whose favours are to be won.

Non-Aboriginal folk narratives deal with a wide variety of subjects, including ghosts and hauntings ('Fisher's Ghost', the 'Pearlers' Light' of Broome and the 'Min Min Lights' of northern Queensland being some well-known examples) — J. S. Ryan surveys the field in *The Possum Stirs* (1987); strange experiences and phenomena, including such things as showers of frogs, spiders or other creatures (for which there may well be a factual basis), and UFO and 'close encounter' accounts. Narratives of these types are closely allied with another important area of folklore scholarship — the study of belief (q.v.).

There are also wonder tales, as in Alexandra Hasluck, *Of Ladies Dead: Stories Not in the Modern Manner* (1970), and stories about historical figures, such as bushrangers (q.v.), sporting celebrities (see **Sporting folklore**) such as the pugilist Les Darcy, military heroes (q.v.) such as 'Simpson' and his donkey of Gallipoli fame, as well as other heroes and, much less frequently, heroines (see **Folk heroes and heroines**). Stories about animals are common, whether real, such as Phar Lap the racehorse, or imagined, such as the 'Nannup Tiger' of Western Australia, as are stories about fabled characters performing unbelievable feats in the never-never lands, like Crooked Mick of the Speewah, stories about lost or hidden treasures (Lassetter's Reef; bushranger's hidden treasure), and 'tall stories', such as 'The Split Dog' whose owner stuck the two halves together with two legs up and two legs down and saw the dog catch a 'roo and bite both ends at once — 'Fair dinkum!'. There are also 'numskull' stories, including those about the Drongo; stories about 'the World's Greatest Whinger'; stories about the asssimilation of newly-arrived immigrants (dubbed 'German Charlie' stories by folklorist Bill Scott); humorous tales, often obscene, as in Lennie Lower's *Here's Luck* (1930) or Bill Wannan's *A Treasury of Australian Humour* (1960); and tales about memorable characters, about places, natural features, events, and indeed about anything at all, told wherever there are people to 'spin a yarn'.

Folk narratives also include 'contemporary legends' (see **Modern legends**) 'jokes' of various kinds, 'shaggy dog' stories, fishermen's tales of the ones that got away, occupational stories (see **Occupational folklore**) of 'the hardest job I ever had', 'the toughest boss I ever worked for', and an almost endless variety of personal-experience stories that fall into a traditional narrative framework. Such narratives are not only told in English, but are related in the scores of languages spoken in Australia. Each of those languages bears its own narrative traditions specific to that particular language and its culture or cultures. Many narrative traditions, of course, are common to many cultures and language groups, including 'fairy tales' and other narratives for children (see **Children's folklore**), hero traditions, myth traditions and many others, as examination of any collection of international folk tales will show.

While we can classify Australian folk tales in various seemingly obvious narrative ways, it is not unhelpful to look to the 'European' system devised in 1910 by Antti Aarne and much refined by Stith Thompson as the vast *Motif-Index of Folk-Literature* (1955), since it can help us with identification of motifs, those details out of which full-fledged narratives are composed. Stories may contain elements of (Aboriginal) myth; be adaptations of European or bush ballads; enshrine elements of humour, exaggeration or joke; be *exempla* or illustrative/cautionary moral tales; or have as their core some local legend. This 'European' structure, however incompletely developed, has the merit of showing the universal mythic base of Aboriginal tales, and of illustrating the development or drastic modification of the European tradition in other continents. The groupings progress from the mythological and supernatural towards the realistic and, often, the humorous. Australian examples have been incorporated here.

A creation, the nature of the universe, and especially, the earth; the beginnings of life; the establishment of the animal and vegetable world;

B remarkable animals, such as the Rainbow Serpent; truth-telling animals like the crow; animals with human traits such as the too-thirsty frog; helpful or grateful creatures; those which marry humans, etc.; early totemic motifs;

C motifs concerned with tabu, and so the earliest philosophies of kinship, as well as of hunting prohibition, will occur here;

D transformation, (dis-)enchantment, magical objects and their employment, and other manifestations of supernatural power;

E concerns about the dead; resuscitation, ghosts and reincarnation;

F journeys to other worlds; demonic spirits; wondrous places and marvellous events;

G dreadful ogres, witches, malevolent giants (these three sections tend to be linked);

H tests of recognition, marriage, cleverness, prowess, vigilance, etc.; quests and endurance;

J tales probing the individual's (acquisition of) wisdom, foolishness, talkativeness, etc.;

K matters of deception — theft, trickery, escape, seduction, adultery, disguise and illusion;

L reversals of fortune, particularly for the outsider, or modest, weak or humble person; humbling of arrogance;

M human ordaining of the future by irrevocable prophecies, decisions, parts, promises or curses;

N luck, chance or (evil) fate, especially gambling; lucky accidents; accidental encounters;

P customs as between the social classes, or (army) ranks; trade/professional/governmental relationships;

Q rewards and punishments;

R captives and captivity; rescue, escapes, fugitives, pursuits and recapture (this field is largely important in slave or penal societies and is often linked with the next);

S unnatural cruelty from relatives; revolting murders, cruel persecution, especially of children;

T love, marriage, celibacy, illicit sexual relations, care of children, family life;

U (fables) told to illustrate the nature of life;

V religious customs, places, charity;

W stories designed to illustrate traits of character;

X telling of incidents whose purpose is entirely humorous, especially about discomfiture, social classes, other nations, sexual matters, exaggeration, etc.;

Z various minor motifs.

As will be obvious, Aboriginal culture should be well represented in A, B, C, E,

F, etc.; the convict period is related to H, K, L, R and S; life in the bush and/or in colonial Australia is close to H, J, L, T, U and W; the life of the urban poor will generate motifs in J, K, M, T, U and X; and post-1945 migration and later relative affluence, and squalor, will generate stories covering most of the more realistic categories.

Folk tales are, by definition, normally transmitted by word of mouth and those now written down are largely still the words of the relatively unlettered to those 'not much accustomed to reading and writing'. Those who read and would codify such tales have purposes very different from those who listen to or participate in them and so may well not understand the ongoing need for atmosphere and internal consistency of tone for a tale to be authentic or true to its cultural antecedents. It is helpful to look at the work of famed tellers of tales both in person and in print, since they, like Henry Lawson or 'Steele Rudd' earlier, have both kept the tale alive and, in large measure, determined the ongoing types and the themes likely to be treated in following generations. Indeed, they act as shapers of (modern) tale-telling, much as was postulated by Vladimir Propp in his *Morphology of the Russian Wondertale* (1927) and *Morphology of the Folktale* (1928). Thus W. E. ('Bill') Harney (1895–1962), long a drover, bushman and patrol officer in the Northern Territory, who had only two years of formal education at a local school, acquired the rest of his considerable knowledge of bush life 'from campfire discussions with explorers, engineers, drovers, road builders and beachcombers'. He liked 'yapping' and became one of the most successful broadcasters both in Australia and for the BBC. With A. P. Elkin he wrote *Tales from the Aborigines*, and his late programme, 'Harney's War', was published as text in 1983 as *Bill Harney's War*, with an introduction by Manning Clark. Various authorities have testified that his yarns were always in finalized form, being 'set-pieces that had been polished and composed so that each was a small masterpiece in itself and it was always told in the same way'. Yet other younger yarn-spinners continue the tradition: Dal Stivens (1911–) from 1936 on; Richard Magoffin (1937–) with his *We Bushies* (1968) or *Chops and Gravy: More Bushies* (1972). He and such contemporaries as Frank Hardy (q.v.) and Keith Garvey, as with his *More Tales of My Uncle Harry* (1987), have attracted considerable followings through their live narration for, and subsequent publication by the ABC.

Many of the still-familiar Aboriginal tales are those told by Irish-born Roland Robinson (1912–), as in his collection, *The Man Who Lost His Dreaming* (1965), and are 'tales in which mythical beliefs, tribal customs and magico-religious practices still played an important part'. They came to him from 'some remarkable aboriginal raconteurs, Percy Mumbulla, Alexander Vesper, Ethel Gordon, and Fred Biggs'. Unlike the material told in his *Legend and Dreaming* (1952) or *The Feathered Serpent* (1956), these are not myths or legends, but rather tales of entertainment for listeners, and described by Sidney J. Baker as 'aboriginal equivalents of magazine stories designed to amuse or harrow the western listener'.

In the second half of the twentieth century there have been three redoubtable collector-codifiers of vernacular lore and tales, namely Bill Wannan (William Fielding Fearn-Wannan, 1915–) (q.v.) for his collections *The Australian* (1954), *Hay, Hell and Booligal* (1961), *A Treasury of Australian Frontier Tales* (1961), *Australian Folklore* (1970) and *Folklore of the Australian Pub* (1972); Bill Scott (1923–) (q.v.), both the author of Aboriginal mythic stories like *Boori* (1978) and *Darkness Under the Hills* (1980) and compiler of the *Complete Book of Australian Folklore* (1976); and Ron Edwards (1930–) (q.v.), author of *The Australian Yarn* (1977) and compiler of *Yarns and Ballads* (1980) and *Fred's Crab and Other Bush Yarns* (1990). Of these the third is, perhaps, the best known theorist of the folk tale, being so identified in the 1987 Report, *Folklife: Our Living Heritage*. The Report refers to various types of personal-experience narrative, identifying those about fishing, farm animals, 'shearers, publicans, ethnics and sport' as well as 'joke yarns', and then continuing:

> A particular form of the folk tale is the 'tall tale' or 'tall story', a yarn intended to 'put one over' the listener, whether mate or stranger. Some exaggerate from the outset, others only gradually, but for credence such a tale must develop its own laws of logic, so that its hearer can at least savour the artistry of the whole. Early laconic and 'dead pan' narrations are often said to have been free from convict taint and to have come to the eastern states with American 'forty-niners' or Australian miners back from California. Yet many of these, like Bill Wannan's 'A Muster of Tall Jests' (in his *Crooked Mick of the Speewah and Other Tall Tales*, 1965) have much earlier local analogues and versions. The same collector pointed out that 'the earliest tall stories told in Australia were probably those dealing with Irish convict attempts to escape to China' on foot. The next phase was concerned with bush crafts — the need to train dogs for impossible tasks; the amazing size of mosquitoes; the remarkable feats of bullock-drovers — the last on leading to heroic achievements by 'ringers', amazing repression by station bosses, poverty of cockatoo farmers. Many of the last tales would illustrate a mordant sense of humour not often sentimentalised or moralised over as Henry Lawson would often do.

In 1977 Ron Edwards published his volume *The Australian Yarn*, a collection rare in its authenticity as it is a personal and vital record from the outback, 'touching on every aspect of the tellers' lives', and all presented to 'an immediate group of listeners'. After recounting his various tales, Edwards concludes his volume with a long essay (pp. 229–55) entitled 'The Australian Yarn — An Analysis', in which he puts the spectrum of spoken entertainment by recital as moving from anecdote to yarn, to tall story, and so he comes to a series of interesting definitions as to the core of the modern Australian bush yarn. The length of a yarn, he found, ranged between 300 and 400 words, this being 'the time period that one person can expect to hold another's attention during ordinary conversation'. There would usually be a biographical (and often geographical) connection between speaker and the events or location of the yarn. The purpose of a yarn may not just be to entertain or instruct, but also to warn of transgression of local custom, or — a very old theme — to establish a person's position in the society by giving sufficient detail of (social) identity, necessary credentials or (relevant) work experience. The transition to the tall story may be via the 'embroidered yarn', the recurring yarn (of international provenance), or the yarn written down and embellished. The exaggerated spoken yarn was often told as a form of testing of a stranger, while an older narrator would often be concerned to relive the past, e.g. by citing the achievements of his own more joyous youth.

Edwards suggests that the need for dramatic emphasis in a yarn is a form of reassurance for lives lacking the various social and occupational supports of urban security. Thus, gold miners or the pastoral workers need assurances as to their ability to 'strike it lucky' or triumph over flood, fire, and drought. (The urban worker is more likely to find his 'security' in lotteries, or in identifying with sporting heroes, rather than in creating a personal mythology of prowess.) He argues that the folk tale has withered to 'yarns with a quasi-historical setting . . . outside of hotel bars and places of employment it would seem that the yarn has little use in urban society where constant entertainment may be found'.

While this is not the place to pursue the argument that there are many (seemingly unrecognized) forms of urban folk tale (see **Modern legends**), it may be suggested that these too are concerned with such general behavioural forces as personal power, social restraint, the psychology of rumour and, above all, with the watcher's perception of the ironies and follies of industrial and technological society, not to mention the cultural contradictions and injustices of capitalism. Facing the fact that the past has 'died' for them, many Australians will invent 'new' traditions and illustrate them in their own anecdotes. Interestingly, the main themes are in some senses a continuation of earlier lore, as in the concerns with family unity and its breakdown; urban rituals; welfare/poverty experiences and the politicians; inner-city black lore; (new) hero-worship of achieving sportsmen; army folklore (that of Vietnam is now history); tales of Asian migrants, etc. Other less serious subject matter may include police dogs; 'bash-lore'; watchmen; diet and fast food; parodies of inter-office memos; xerox lore; office graffiti, etc. Many of these are included in newspaper columns of anecdote, human interest reports, etc., particularly if they enshrine sexual misadventure.

As many have pointed out, literary forms of yarns and tall stories came to print early, with various tales by Marcus Clarke, and later those by Barcroft Boake, 'Steele Rudd' (A. H. Davis), and the many contributors of exaggerated yarns to the Sydney *Bulletin* from 1880 to *c.* 1960. Wannan's fine collection *Crooked Mick of the Speewah* (1965) opens with the (colonial) recollection of a yarn:

> the tall trees out in them parts — a cooee and a half high, they were. You had to lay on your head to see the lower branches. The locals had the top half of every tree hinged, so they could be let down to allow the sun to get past.

Of course tall stories are still being written and read today, retaining many of the traditions of the last century but welded with the technology of today. Some excellent examples, like Lilith Norman's 'My Simple Little Brother and the Great Aversion Therapy Experiment', may be found in Anne Bower Ingram (comp.), *Too True: Australian Tall Tales* (1974). Many of Australia's lies, or apocryphal tales, have got into print in the last quarter of the twentieth century and perhaps half of them are part of our modern folklore and as much central to it as jokes or the gossip to which Australians are particularly addicted. All this matter is presented as true, coming now from a friend, now from a newspaper, or a magazine. There is never any suggestion that this particular anecdote is other than factual and easily able to be verified. It is quite beside the point that it may be an old tale in a new guise, or one long known and recorded in other countries. Now it is given characters, motive and setting and made relevant to the hearer and to his or her experience of our life.

Yet another aspect of the folk tale is the retelling here of 'distant' tales by the nation's more than 100 distinctive ethnic groups. They are in part represented in the 1979 volume, *Folktales from Australia's Children of the World*, covering some thirty-three groups, the stories also being told in the appropriate community language. And as the then Minister for Immigration pointed out, these children and their backgrounds, from African and American, through Croatian and Estonian, Lebanese and Polish to Thai and Ukrainian, are important since they will be 'future leaders and builders' of the richly varied society that Australia is becoming.

See also: **Legends**; **Narrative literature and folklore**.

J. S. Ryan and Graham Seal

Folkways Music, established in Sydney by Warren Fahey (q.v.) in 1973, is a retail outlet offering a wide range of recorded folk music and publications dealing with the folk arts. It issues a regular mail order catalogue and is considered a resource centre by many government departments, especially for the supply of reliable traditional recordings that can be used in multicultural programmes. It is also an information centre for musicians.

Foodways is a term widely used in international folklore studies to describe not only *what* is eaten by a particular group of people but also the variety of customs, beliefs and practices surrounding the production, preparation and presentation of food within that group. Michael Symons' *One Continuous Picnic* (1982) discusses many of these issues in his exhaustive account of the 'history of eating' in European Australia.

The period following the arrival of the

First Fleet in Australia was one of near starvation, and for some years afterwards meat, flour, sugar and tea were staples particularly in outlying areas. The song 'The Old Bark Hut' dates from the nineteenth century, but might well describe the monotonous and inadequate diet of early colonial times, although the meat could have been mutton:

Ten pounds of flour, ten pounds of beef, some
  sugar and some tay,
That's all they give a hungry man until the
  seventh day;
If you don't be mighty sparing, you go with a
  hungry gut,
That's one of the great misfortunes of the old
  bark hut.

It is perhaps not surprising in view of this history of privation that one of Australia's favourite icons is the Dog on the Tuckerbox (q.v.), immortalized in balladry, song and in sculpture at Gundagai in New South Wales. In the Australian folk tradition of canonizing losers, outlaws and failures the dog is remembered for its villainy in sitting (often regarded as a euphemism for shitting) in its owner's food container, rendering the contents inedible.

Neither colonial nor contemporary European Australians have made much use of indigenous foods in everyday cooking, with the exception of fish and sometimes birds or kangaroo. Nevertheless, many kinds of local fauna were sometimes eaten in earlier times, including the ubiquitous possum, and the Pioneer Women's Hut in Tumbarumba, New South Wales, is collecting old recipes such as 'Stuffed Belly of Blue Crane'. Contemporary interest has grown through the highly popular ABC television programme 'Bush Tucker Man' about the work of the former Australian Army Major Les Hiddens in studying indigenous foods as part of the Army's bush survival course. There have been a number of recent publications on this topic, by Isaacs, and by Low, for example, and it will be interesting to see whether more indigenous plants can be used or cultivated for nutritional and medical purposes.

Today most Australians live in large coastal cities, where an enormous variety of foods is available in shops, markets and supermarkets, produced by a gigantic food industry and heavily influenced by large-scale immigration. Australians today eat Chinese, Vietnamese, Indian, Greek and Lebanese food, for example, in both restaurant and 'take-away' form, as well as traditional English fish and chips and American hamburgers. Chicken has become a moderately-priced staple rather than a Christmas treat as it was one or two generations ago, and multinational 'fast food' outlets such as McDonald's and Kentucky Fried Chicken are widespread throughout Australia. A persistent favourite afternoon snack is the 'Devonshire tea', an Australian adaptation of the British 'cream tea' which in Australia usually consists of hot scones, strawberry jam and cream, served particularly at holiday and picnic resorts such as the Blue Mountains of New South Wales, the Adelaide Hills or Melbourne's Dandenong Ranges.

One of the most interesting features of *One Continuous Picnic* is Symons's discussion of the power of commercial enterprise to create a ubiquitous product (and subsequent mythology) such as the food spread Vegemite. Australians have taken this product to their hearts (and palates) to such an extent that it has become a national and international symbol, and one used in 1990 on posters produced by the Australian Office of Multicultural Affairs to promote the government policy of multiculturalism — 'it's as Australian as Vegemite'. Vegemite, lamingtons, pavlovas, damper and meat pies are sometimes regarded as Australia's major contributions to international cuisine. It is interesting to note that in Victoria at least, one major manufacturer of meat pies has drawn on traditional folklore for its brand name, Four 'n Twenty (from the nursery rhyme 'Sing a Song of Sixpence'), as also has the pie's constant companion, tomato sauce, which may be Tom Piper, among other brands. The meat pie acquires a particular Adelaide significance when combined with mashed peas to make a 'floater'. Apart from the Adelaide floater it is hard to name particular regional foods in Australia, although there are considerable regional differences in names given to some products. A particular type of luncheon sausage is known as Devon, Strasbourg or Polony in different parts of the country, and a rockmelon is a canteloupe in Melbourne at least. Ethnic foods may be found not only in city restaurants but also in rural communities settled by a particular nationality, such as the Barossa Valley in South Australia (see **German folklore in South Australia**).

Despite commercial influences, much domestic eating is truly folkloric in that cooking skills and recipes are principally learned in face-to-face situations, usually within a family. These directly transmitted skills may be supplemented by cooking classes or cookery books such as those produced by the Country Women's Association and the *Australian Women's Weekly* magazine. Enormous numbers of cookery books are available in Australia, an indication of the centrality of food in Australian folklife. Australian families will make special efforts for special occasions such as Christmas, where a traditional hot roast dinner in the middle of the day is widely served no matter how warm the day, featuring roast turkey, chicken, pork or ham and a hot plum pudding. In colonial days settlers needed to be inventive. Cusack's *Australian Christmas* (1966) includes an account written by Mary Thomas of her first Australian Christmas dinner near Adelaide in 1836, where the meal included plum pudding and 'ham and a parrot pie'. Children's parties also require special food such as bread and butter with 'hundreds and thousands' (sometimes called 'fairy bread'), small 'cocktail frankfurts' with tomato sauce and chocolate crackles (a chocolate sweet made with a little help from commerce, namely Kellogg's Rice Bubbles). Outdoor eating is popular with Australians of all ethnic backgrounds. Barbecues are popular both within families and as a means of entertaining friends, and the barbecue may be a permanent construction in brick or stone made by the householder or a portable cooker of commercial origin. The 'back yard' which most Australian households possess may include a vegetable garden as well as a barbecue. (The 'front garden' — not 'yard' — will be given to flowers and a lawn, and an immigrant family may be easily identified if vegetables, vines etc. are grown in the front, contrary to established Australian custom.)

Picnics are a popular pastime, and the food will range from simple sandwiches and cakes to elaborate cold meats, salads and wines, kept cool in an insulated box generically known as an 'Esky', after one popular brand name. The picnic may also involve a barbecue of chops, sausages or steak, and if a fire is lit the picnickers may make the traditional billy tea, where the tea is thrown into a billy can of boiling water and swung round overarm several times to settle the tea. It is interesting to note the association between billy tea and Australia's national song, 'Waltzing Matilda' (q.v.).

Good cooking is a source of pride to many adults, and increasing numbers of Australian men are engaging in domestic cooking, in both the everyday meals, and the traditional preparation of jams and preserves and new and fashionable items such as paté, quiches and terrines. The domestic freezer has to some extent removed the need for bottled preserves, although there is still considerable skill and preparation involved in freezing food. The Australian 'big spread' which caused Mary Mahony so much humiliation in England in *The Fortunes of Richard Mahony* continues to be popular in both country and city for special occasions such as weddings, engagements, kitchen teas and dances. Weddings, twenty-first birthdays, silver and golden wedding anniversaries and other ceremonial occasions will usually have an elaborate fruit cake, perhaps in tiers, with intricate iced decoration. Many Australian women and some men are expert at the art of cake decorating, as seen at 'The Show', the Royal Agricultural Society's annual show in Melbourne or at shows in Perth, Sydney and elsewhere. 'The Show' also has competitions for other kinds of domestic cooking, such as cakes, scones and buns. This type of traditional fare — such things as small iced patty cakes, lamingtons, shortbread, chocolate rum balls, afternoon tea loaves, oatmeal biscuits (sometimes called 'Anzacs'), iced butter cakes and cream sponges — can also be seen (and tasted) at local bazaars and fêtes.

No discussion of Australian foodways would be complete without a reference to drinking customs, and there has been considerable change in recent years, including Australians' increasing preference for wines rather than beer or spirits. Many Australian 'pubs' have become considerably more refined than they were in either colonial times or the days of six o'clock closing, although these rough and tough days have left a few linguistic reminders in well-known phrases such as the 'six o'clock swill', 'shouting' (paying for others' drinks, usually in turn) and the 'chook raffle' (now used generically for a wide variety of somewhat suspect or ramshackle activities, perhaps with a predictable outcome). For many years 'hard drinking', particularly of beer, was a part of Australia's mythology about mateship and masculinity, and many songs immortalized the practice: two such are 'Wild Rover No More' and 'Click Go the Shears', the final verse of which is

Shearing is all over and we've all got our cheques,
Roll up your swag, boys, we're off on the tracks;
The first pub we come to it's there we'll have a spree,
And everyone who comes along, it's 'come and drink with me!'

There are many persistent customs which survive in Australian foodways. On some relatively informal occasions guests might be asked to 'bring a plate'. More than one bewildered immigrant has (or is alleged to have) arrived with an empty plate, rather than with the contribution of food required. Some food occupations have powerful folkloric traditions, both in the work practices themselves and in surrounding features. Many stallholders in Australia's food markets still cry their wares, in a practice which is centuries old, although the content of their cries may have changed. Fyshwick Market in Canberra is a lively source of traders' cries, and one butcher at Melbourne's Queen Victoria Market imitates a sheep to advertise his best-quality 'La-a-a-a-a-m-b chops! La-a-a-a-m-b chops!'

The preparation and consumption of food and its surrounding customs are among the most central aspects of folk culture in Australia, and many ethnic food customs are still widely practised. Food also provides the clearest example of Australian multiculturalism in practice; one may question whether Australians could survive without Chinese food, for example!

See also: **Aboriginal folklife in central Australia**.

Gwenda Beed Davey

FRANK THE POET Of the 50 000 or more Irish men and women who were transported as convicts to Australia in the years from 1787 to 1867, Francis MacNamara was the most notable poet and maker of ballads. 'Frank the Poet', as he was known by his fellow convicts in New South Wales and Van Diemen's Land in the 1830s and 1840s, achieved widespread fame in his own time for his powerful evocation of the convict system. Nevertheless, the oral tradition faded and only two of MacNamara's poems were published in the nineteenth century: 'A Dialogue Between Two Hibernians in Botany Bay', in the Sydney *Gazette* of 8 February 1830 and another, 'A Convict's Tour to Hell', in 1885.

Geoffrey Ingleton republished 'A Convict's Tour' and 'Seizure of the *Cyprus* Brig' in *True Patriots All* (1952) and Douglas Stewart and Nancy Keesing included 'A Dialogue', 'Seizure ...' and 'Labouring with the Hoe' in *Old Bush Songs* (1957). However, it was not until 1979 that John Meredith and Rex Whalan's *Frank the Poet* collected all the works attributed to him and provided the first documented account of his time in New South Wales and Van Diemen's Land. A remaining and perhaps insoluble problem is the authentication of MacNamara's authorship of ballads recorded by other hands. While it is likely that much of his work has been lost, it is also likely that he has been credited with some of the work by other convict poets within the oral tradition whose names have faded. 'The Ballad of Martin Cash', for example, seems likely to have been written by James Lester Burke. That MacNamara's facility was by no means exceptional is clear from the newly-arrived Bishop W. B. Ullathorne's anecdote about an Irish convict whom he encountered in Sydney in 1833 after conducting a meeting to reunite the warring Catholic factions in the colony:

> Passing from the meeting to my residence, I was met at the door by a poor ragged Irishman, the only man in tatters I had yet seen. He asked me if I would please listen to what he had to say. 'Well', I said, 'What is it?' In reply he poured out a stream of hexameter verses, in perfect metre and harmony, describing the meeting and all its incidents, winding up with a touching thanksgiving for the peace restored to the Catholic body. I asked my Irish troubador, with some astonishment, what reduced a man of his ability and elevation of mind to such a condition. He replied, 'I am a child of nature, your Reverence; and I cannot refuse the drink which my countrymen give me in their generosity'. Some years later, when in the interior country, I called upon a wealthy Catholic magistrate, who pressed me to stay for dinner, promising me something interesting afterwards if I would do so. So I consented, and after dinner in rolled my troubador from the farm in a fat fine condition, smiling all over his face. Standing by the door, he resumed the history of my transactions from the time of the meeting, rolling out a stream of sweet and harmonious verses without halt or fault for an hour. He was a self-taught man, and spoke in tones the tender sweetness of which I completely recall at this hour. I never saw him again.

Although the convict records in Sydney and Hobart enabled Meredith and Whalan to piece together a surprising amount of information at the Australian end, they could not find out anything about Francis MacNamara's Irish background. All they knew was that he had been convicted at the Kilkenny assizes in early 1832 for the theft of a piece of cloth, sentenced to seven years' transportation to Australia and shipped off on the *Eliza* transport from the Cove of Cork in May of that year. There was no definite information about his place of birth, which he gave variously as Wicklow, Limerick and Tipperary; his religion, which he gave variously as Catholic and Protestant; or his occupation, which he gave variously as miner, mariner, labourer and shepherd. A County Clare connection is certainly suggested by the fact that it was a tradition of the Clare MacNamaras to name the eldest son Francis in commemoration of Lochlainn MacNamara's *c.* 1402 endowment of a monastery at Quin which was licensed by Pope Eugene IV in 1433 to receive Franciscan friars. On one occasion at the prison settlement of Port Arthur in Van Diemen's Land on Christmas Day 1842 he entertained his fellow prisoners, including Martin Cash, to some extempore recitations, after announcing himself in bardic style:

> My Name is Frank MacNamara
> A native of Cashel, County Tipperary,
> Sworn to be a tyrant's foe,
> And while I live I'll crow.

The epigram is unreliable about his place of origin, Tipperary almost certainly being chosen for the rhyme, but it does provide us with a useful statement of how he saw himself. High-spirited and not at all disposed to endure without protest the bullying and cruelty of the convict system, MacNamara thereby brought down upon his own back its worst extremes.

On his arrival at Sydney in September 1832, the convict authorities carefully recorded him as: Protestant, single, literate and a miner from Co. Wicklow. His height was given as five foot four and three quarter inches, his complexion 'ruddy', hair 'light brown' and eyes 'grey'. His features were 'broad and full' and there was a scar on the outer corner of his right eye.

Irish convict records for the period from 1791 to 1836 are extremely patchy and there is nothing in the Dublin archives about Francis MacNamara. However, newspapers often published details of the more interesting assizes cases and in the issue of 18 January 1832 the *Kilkenny Journal* printed the following description:

> As we before observed, there was nothing of importance or out of the usual course of petty crime, incidental to a large and populous city. Of the five persons sentenced to transportation not one was a native of Kilkenny; they were all strangers and quite unconnected with the neighbourhood.
>
> Mary Bagnall, for stealing muslin out of the shop of Mr Finn of High-street . . . seven years transportation . . .
>
> Same sentence on Francis MacNamara, a real Corkonian, for breaking the shop windows of Mr John McDonnell, and stealing therefrom a piece of worsted plaid.
>
> The cross examination of two witnesses by this prisoner afforded much amusement to the Court; his peculiar accent, cutting remarks, and mode of delivery, were both quaint and forcible.
>
> 'Please *your* Word-ship, as to Mr Prince the constable, his oath should not be thought much of against me. He may know the weight of that book in pennyweights and scruples, but of its awful meaning and substance he knows nothing, often as he kissed it. He should have the eye of a hawk, and the vigilance of a cat, to see me do what he swears. By the virtue of *your* oath, young man (to the shop man), did *you* get directions from any persons as who *you* were to swear against me'.
>
> Answer — 'By virtue of my oath I did not'.
>
> To the constable he put the same question and received a like answer.
>
> 'Now your wordship, I must prove them both perjurer, did not that decent looking gentleman*[sic] sitting under your worship, in a loud and distinct manner, that nobody could mistake, direct them to swear the truth, the whole truth, and nothing but the truth'.
>
> On the verdict of guilty being returned, sentence was immediately passed, and he was ordered from the dock.
>
> Prior to his leaving it he flourished his hand, and with a cheerful and animated countenance, and in a loud voice said,
>
> I dread not the dangers by land or by sea
> That I'll meet on my voyage to Botany Bay;
> My labours are over, my vocation is past,
> And 'tis there I'll rest easy, and happy at last.
>
> *It will be perceived he meant Mr John Watters [clerk of the court?]

At the age of twenty-one MacNamara was already demonstrating his dramatic flair and his skill in extempore composition, although his prognosis of life in Australia was absurdly optimistic as we shall see.

It is worth saying something about what was going on in Kilkenny at the time, with the thought that while MacNamara was convicted of 'grand larceny', a petty theft in fact, he may have been involved in the movements which were then convulsing the county. It may be that Frank MacNamara was suspected of being involved in agrarian protest activity with the 'Whiteboys' or of helping to organize the tithe protests. It was not uncommon for people to be convicted and transported for relatively petty crimes on suspicion of involvement in something far more important. There is no direct evidence that we can point to, but it does seem significant that in the convict transport *Eliza* in which he was shipped with 197 other convicts to New South Wales in May 1832, there were twenty-six men sentenced as Whiteboys, ten convicted of 'unlawful oaths' (probably association with Whiteboy activity or Ribbonism), four of carrying firearms and one of riotous assembly.

Further evidence has recently come to light which suggests that MacNamara's conduct during the trial may have resulted in a more severe sentence than he would otherwise have received. James Gordon, an officer on the *Eliza*, while evidently impressed by the poet's talents, also believed that he had used them to his own detriment. On 18 June 1832 when the *Eliza* was about four weeks out of Cove, Gordon recorded in his log:

> Today gave MacNamara (one of the convicts) 2 dozen [lashes] for bad conduct. This fellow is a sad scamp and yet far above the common herd in some

> respects. He has considerable abilities, has written some very palpable lines on his trial and sentence since he came on board and has a very extensive knowledge of the Scriptures. He it appears was tried for a very slight offence but his conduct on his trial was so bad that he was transported for 7 years. He recited a mock heroic poem of his own composing in which he ridiculed judge jury and other officers of the Court that had tried him. This of course enhanced his offence and added to his punishment.

The shipboard journal of Thomas Bell, surgeon and superintendent of convicts, also reveals that Frank MacNamara was one of the Protestant convicts to whom religious books were distributed.

Thanks to the careful keeping of convict records by the authorities in New South Wales and Van Diemen's Land, we know a good deal more about MacNamara's experience as a convict than about his early life. It is a grim story in many ways, reflecting the response of a spirited and courageous man to the bullying and cruelty of the convict system. To a great extent it is the record of his punishment, which can be summarized as follows: flogged fourteen times, receiving a total of 650 lashes; served three and a half years in iron gangs doing hard labour on roads, dock works and in quarries while wearing heavy leg irons and chains; served thirteen days of solitary confinement and three months on the treadmill. His final punishment was seven years at the penal settlement of Port Arthur in Van Diemen's Land. Although his experience could not be called typical — only 10 per cent of all convicts were sent to places of secondary punishment — the point is that the convict system made it entirely possible.

On his arrival at Sydney in September 1832 Frank was assigned as a servant to a Mr John Jones but within three months he had been sentenced, for an unknown offence, to six months' jail which he served in an 'ironed' gang on Goat Island in Sydney Harbour. Within a month of being returned to the Hyde Park Barracks he was given fifty lashes for absconding. Insubordinate conduct then earned him a month on the treadmill at Carter's Barracks on the site of what is now Sydney's Central Railway Station. After completing this he absconded again, but unlike Bold Jack Donohoe, he was not cut out to be a bushranger and was soon captured and sentenced to twelve months in an ironed gang. During this time he was accommodated with 200 other convicts on board the hulk *Phoenix* in Darling Harbour, probably spending the day ashore on harbour works under a military guard. On 1 February 1834 he was given twenty-five lashes for having a stolen shirt and on 3 March he was given seventy-five with the cat-of-nine-tails for 'insubordination'. Returned to Hyde Park Barracks in 1834, he was soon in trouble again: three days solitary confinement on 26 January for being absent from duty; twenty-five lashes for disobedience of orders on 18 February; one hundred lashes for obscene language on 9 March. And for assaulting a constable on 16 April he was given another twelve months in an ironed gang. That his spirit was unbroken we can see from the continuing catalogue of punishments: thirty-six lashes on 16 May for 'refusing to work and insolence'; seventy-five lashes on 8 August for destroying a government cart and fifty lashes on 14 December for refusing to work. The next two years he spent on the *Phoenix* again. Within nine days of going on board he was given twenty-five lashes for neglect of work and on 15 August 1835 was sentenced to ten days in the cells for being found drunk. To Captain David Murray, superintendent of the *Phoenix*, he addressed this epigram petition:

> Captain Murray, if you please
> Make it hours instead of days,
> You know it becomes an Irishman
> To drown the shamrock when he can.

He tried to abscond from the *Phoenix* on five occasions but was caught and punished each time, once by two months on the treadmill. Refusing to mount the wheel after nineteen days of this punishment, he was given fifty lashes for 'disobedience of orders'. He was still on board the *Phoenix* when the convict muster was taken on 31 December 1837 but thereafter disappears until we find him again in early 1838 shepherding sheep at Goonoo Goonoo, the Australian Agricultural Company's pastoral station on the Peel River. From Goonoo Goonoo he was transferred south to another station at Stroud and in early November 1839 to the Company's coalmines at Newcastle where there was a serious shortage of labour. An idea of the conditions can be gathered from the description given by the convict novelist James Tucker whose hero Ralph Rashleigh served there at some time in the 1820s or 1830s.

It was Frank's refusal to work underground that earned him another twelve months in irons in November 1839 which he served at Woollomooloo Stockade in Sydney and at Parramatta Gaol. His time at Newcastle is commemorated in three poems, the first of which was in the form of 'A Petition from the Australian Agricultural Company's Flocks at Peel's River on Behalf of the Irish Bard'. It included these verses:

> Why should the poet be sent down
> To toil in a coal pit
> Such service best suits a clown
> But not a man of wit.
>
> We yet shall hear his merry songs
> On fair Killala's plain
> Kind Heaven shall avenge the wrongs
> Of our much injured swain.

This was followed by another poem based on a traditional Irish Gaelic epigram form. Called 'For the Company Underground', it was prefaced with the message 'Francis MacNamara of Newcastle to J. Crosdale Esq greeting', William Croasdill being the supervisor of the Company's colliery at that time:

> When Christ from Heaven comes down
> straightway,
> All his Father's laws to expound,
> MacNamara shall work that day
> For the Company underground.
>
> When the man in the moon to Moreton Bay
> Is sent in shackles bound,
> MacNamara shall work that day
> For the Company underground . . .
>
> When cows in lieu of milk yield tea,
> And all lost treasures are found,
> MacNamara shall work that day
> For the Company underground . . .
>
> When Christmas falls on the 1st of May
> And O'Connell's King of England crown'd,
> MacNamara shall work that day
> For the Company underground . .
>
> When the quick and the dead shall stand in array
> Cited at the trumpet's sound,
> Even then, damn me if I'd work a day
> For the Company underground.
>
> Nor over ground.

The third poem is also a petition: '. . . From the Chain Gang at Newcastle to Captain Furlong The Superintendent Praying Him to Dismiss a Scourger Named Duffy From The Cookhouse And Appoint A Man In His Room'.

What happened to the poet at the expiry of his sentence at Parramatta is not known but it is likely that he was sent to Towrang stockade at Goulburn and from there put to work building the road from Braidwood to Nowra on the coast. We know from an anecdote related by a fellow convict, Conn Flynn, that he worked with a road gang who greatly appreciated his extempore versifying. This rare vignette also suggests that Frank the Poet was a man of some religious faith:

> Oh! Those were the days when the
> grasses wide
> Wet a horseman's feet on his morning
> ride
> When the mighty bullocks, their work-
> days past,
> As rations to convicts were served at last.
> And the tale is told how it chanced one
> day
> That a smoking round came the convicts'
> way.
> Then Frank the Poet, as large as life,
> Stood and tapped the beef with his
> carving knife
> And intoned this verse, so the tale relates

To the crowded board of his grinning
mates:
'Oh! Redman, Redman, how came you
here?
You served the Gov'ment many a year
With blows and kicks and with much
abuse
And now you are here for convicts' use'

Harry the Forger spoke up:

'Faith! Frank, I don't know for sure what you will be remembered most for, your verses or for that natural cross in the rocks, on top of the Devil's Pinch, where you always doff your cap when you pass and which they name Frank the Poet's Cross.'

'Must a man be remembered at all?' asked Frank.

'That goes without saying,' replied Harry. 'Tis a poor soul indeed who leaves no signature on history when he departs. Significant and permanent, if it's something great that he has done, or temporary and fading if he has merely passed that way and left no more than a track in the clay, like a straying beast. 'Tis not an epitaph, but his 'signature on history', a symbol to keep his name in the minds of men, as long as the caprice of Fate decrees'.

In May 1842 Francis MacNamara was reported to have absconded with some other prisoners while being escorted by constables from Berrima to Picton to be tried in Sydney and was recaptured carrying arms at the foot of Razorback mountain. An interesting detail is that at the time he was under warrant by the Braidwood Bench of Magistrates for forgery, an incident which probably gave rise to the folk tale that he had been originally transported for forgery. Absconding and carrying arms were regarded as very serious offences and he and his four comrades were sentenced to transportation to Van Diemen's Land for life. For some reason the sentence was commuted to ten years. There is only one report of his being punished at Port Arthur — seven days' solitary confinement for disobedience in September 1843 — but there is reason to believe that other punishments went unrecorded. According to another convict, Hugh Leonard, who was at Port Arthur when Frank and his comrades arrived there on 29 October 1842, the poet was one of the leaders of a successful campaign of resistance against the work system of carrying shingles in which, as Meredith explains, 'a lightly-laden prisoner ran at the head of the gang and the rest, carrying full loads, were supposed to keep up with the pacemaker'.

Everything went on well, until one morning the brig *Isabella* came into the bay, and I was told that it was loaded with Sydney men from Cockatoo Island. Next day they came ashore, and were put into the carrying gang. The runner, as usual, started off, but the Sydney men all kept in their ranks, and walked on steadily.

The overseer shouted for them to close up, but they took no notice of him, so that other men were coming back for loads before they had reached the first resting place. We each had bundles of shingles to carry, and the Sydney men said, 'Now do you think these bundles are overweight?' Some of them replied, 'Yes.' They then emptied some of the shingles out, and tied them up again, so that by the time they had done that, the other men had been into the settlement with their loads. When they got into the settlement they were marched in front of the office, and the commandant came. The loads were weighed, and found to be underweight.

The commandant then ordered the flagellators to be sent for, and the triangles; and the clerk took down all their names. Amongst the rest there was the notorious Jackey Jackey, Frank the Poet, and Jones, and Cavanagh.

There were upwards of thirty pairs of cats and four flagellators, and the surgeon, a young man named Dr Benson, who kept laughing and joking, and playing with his stick as unconcerned as though he was in a ballroom. When their names were taken, every other man was called out, and received thirty-six lashes. A fresh flagellator giving every twenty lashes, and they try to see who can give it the worst.

Next morning they went on just the same. They were then ranked up, and all were flogged and sent to work again; the

> overseer still snapping at them and if they could have got him in the bush they would have killed him. Next day they were all brought up again, and received seven days' solitary confinement on bread and water. When they came out they did just as before and were brought up before the commandant again, and he listened to what they had to say this time, and ordered that there was to be no more running in the gang.

Frank seems to have been well-behaved after this. His sentence was first reduced to seven years; he was given a ticket-of-leave on 19 January 1847, and received his certificate of freedom on 12 July 1849. Ironically, this is the last reliable record we have of his life in Australia, although tradition has it that on his release he took ship for Melbourne, delivering from the paddle-box of the Launceston steamer his 'Farewell' epigram:

> Land of lags and kangaroo,
> Of possums and the scarce emu,
> Squatters home and prisoner's hell,
> Land of Sodom, fare thee well!

It seems likely that after spending some time in Melbourne, Frank decided to try his luck at the Devil's Hole goldfield at Mudgee in New South Wales. In the Mitchell Library, Sydney, there is a copy of a family history written in fine calligraphy on 1 March 1861 for an innkeeper named John Calf whose son later recorded that it was by 'a convict well known in the early days as "Frank the Poet" — I believe his name was MacNamara'. There is no record of his death in New South Wales, however, and he probably returned to Victoria. Perhaps the last glimpse we have of Frank MacNamara is in Marcus Clarke's description of low life in the pubs of Melbourne in 1868:

> One of the men around the table, a little Irishman, with a face that would make his fortune on the stage, is a well known character in the low public houses. He is termed the 'Poet', and gains his living by singing his own compositions in the bar rooms. At our request, he favoured us with an improvised ditty — which, despite the reckless disregard of metre and rhyme, was really clever, the personal peculiarities of the company, including ourselves, being hit off with some degree of humour. The 'Poet', however, is an unconscionable drunkard and has, moreover, done several sentences for petty thefts. Having made him temporarily grateful by the bestowal of largesse, we turned in to the next house . . .

Frank the Poet was a legend in his own time, partly for the spirit with which he resisted the convict system but largely for the way in which his verse and ballads expressed the feelings of all those who laboured under it. He was probably responsible for the ballad which most poignantly evokes the Irish convict experience — 'The Convict's Arrival' (better known to some as 'Moreton Bay' or 'The Convict's Lament'), which reflects both the Irish sense of exile and the urge to rebellion. Now sung to the same tune as 'Boolavogue', it is the most powerful and one of the best known of all the convict ballads.

From contemporary accounts, it is clear that these and no doubt other ballads originating among the convicts were widely circulated within a short time of their composition. Martin Cash recorded that he received a copy of the ballad celebrating his exploits shortly after he left Hobart on 8 December 1844 for incarceration on Norfolk Island:

> On the same evening while on deck, and when all the other prisoners had been sent down below, the first mate, with a smile on his countenance, placed a paper in my hand, observing that it contained a brief account of some of my doings in Van Diemen's Land, and also that it was freely circulated through the colony, he having heard it sung a day or two after his arrival . . .

We also know from the Rev. John West's *History of Tasmania* (1852) that 'The Seizure of the Brig Cyprus' was widely circulated:

> The prisoners who waged war with society regarded the event with exultation; and long after, a song, composed by a

sympathising poet, was propagated by oral tradition, and sung in chorus around the fires in the interior.

This is not the place to pronounce on the literary qualities of Frank MacNamara's poems, except to say that he was obviously influenced by Jonathan Swift and Robert Burns; that he was well versed in the classical allusions which provided much of the vocabulary of eighteenth-century Anglo-Irish poetry; and that he was also familiar with the Irish Gaelic tradition. His magnum opus, 'A Convict's Tour to Hell', is clearly in the style of Swift's satires and burlesques, based on an idea first developed by Virgil in the *Aeneid* and imitated by many other poets, notably Dante in his *Divine Comedy*. At the same time, as a dream narrative it is very much part of the Irish Gaelic tradition best shown by Brian Merriman's 'Midnight Court'. While it sometimes verges on doggerel, the poem has a robust flow and a specific historical reference which make it the most important piece of verse by any convict author. At a political level, it is a song of revenge, expressing the conviction that in a future state of punishment and rewards, appropriate justice will be done. It was written at Stroud in 1839 at a time of relative tranquillity in Frank the Poet's life.

Nor can the foremost of the sons of men
Escape my ribald and licentious pen.
Swift

You prisoners of New South Wales,
Who frequent watchhouses and gaols,
A story to you I will tell:
'Tis a convict's tour to hell.

Whose valour had for years been tried
On the highway before he died.
At length he fell to death a prey;
To him it proved a happy day.
Downwards he bent his course, I'm told,
Like one destined for Satan's fold.
And no refreshment would he take
'Till he approached the Stygian lake.
A tent he then began to fix
Contiguous to the River Styx.
Thinking that no one could molest him,
He leaped when Charon thus addressed him.
'Stranger, I say from whence art thou,
And thy own name, pray tell me now?'
'Kind sir, I come from Sydney gaol.
My name I don't mean to conceal
And since you seem anxious to know it
On earth I was called Frank the Poet' . . .

And having travelled many days
O'er fiery hills and boiling seas
At length I found that happy place
Where all the woes of mortals cease.
And rapping loudly at the wicket
Cried Peter: 'where's your certificate?
Or if you have not one to show,
Pray, who in Heaven do you know?'
'Well, I known Brave Donohue,
Young Troy and Jenkins too,
And many others whom floggers mangled
And lastly were by Jack Ketch strangled'.
'Peter', says Jesus, 'let Frank in,
For he is thoroughly purged from sin,
And although in convict's habit dressed,
Here he shall be a welcome guest.

Isiah go with him to Job
And put on him a scarlet robe.
St Paul go to the flock straightway.
And kill the fatted calf today.
And go tell Abraham and Abel
In haste now to prepare the table,
For we shall have a grand repast
Since Frank the Poet has come at last'.
Then came Moses and Elias,
John the Baptist and Mathias,
With many saints from foreign lands,
And with the Poet they all join hands.

Thro' Heaven's Concave their rejoicing rang
And hymns of praise to God they sang;
And as they praised his glorious name
I woke and found 'twas but a dream.

REFERENCES: J. Meredith and R. Whalan, *Frank the Poet: The Life and Works of Francis Macnamara*, 1979; J. Meredith, 'Frank the Poet: A Postscript', *Overland*, June 1987; B. Reece, 'Frank the Poet', in R. Reece (ed.), *Exiles from Erin*, 1991; P. Butterss, 'James Lester Burke, Martin Cash and Frank the Poet', *Australian Literary Studies*.

Bob Reece

G

GERMAN CHARLIE, central character of a cycle of tales about the pitfalls of immigration in Australia. Many of these tales were collected by folklorist and writer Bill Scott (q.v.), who has dubbed them 'German Charlie stories'.

German folklore in South Australia The Germans in South Australia have a long and proud history. A few individual Germans had arrived even before the foundation of South Australia as a British colony in 1836, but the first organized group landed two years after the proclamation of the province. The letters they and other early arrivals sent home set in motion a chain of migration that was to bring German migrants to the shores of South Australia until the present day. From the early 1840s until 1920 Germans and their descendants constituted some 10 per cent of the population of South Australia. This large group of non-British settlers, almost twice the percentage of any other Australian colony, was to have a marked impact on the folklore of South Australia.

The Germans who first came to South Australia were 'Old Lutherans', who held to the old liturgy of the Lutheran Church and were overwhelmingly peasant farmers and artisans from Prussia. They had been forbidden to use the liturgy of their choice in their east German homeland, an area now lying in Poland, which bordered on the nineteenth-century boundaries of the provinces of Brandenburg, Silesia and Posen. Although the restrictions on worship were lifted in 1840, for the next ten years the memory of the time of repression and the fear of its return led many 'Old Lutherans' to migrate to South Australia. This large group formed the nucleus of the South Australian German traditions and it is their traditions and customs that are still most in evidence today.

However, two other groups also have had an effect on German folklore in South Australia; middle-class Germans who settled primarily in Adelaide and those who arrived after 1945 as post-Second World War migrants. This latter group has revived and enhanced many of the old customs whilst introducing others.

Perhaps the most popular pastime of the early German migrants was singing. They not only sang in Church but also formed *Liedertafel*, singing groups, the first of which was founded by the composer of the *Song of Australia*, Carl Linger, in Adelaide in 1858, followed by another in Tanunda in 1862. Both groups sang a wide range of religious and secular songs, and both *Liedertafel* still exist today with a strong following. The Adelaide *Liedertafel* now operates under the umbrella of the German Association. The German love of music also led to the founding of the Adelaide School of Music which in 1898 was to become part of the Elder Conservatorium of Music of the University of Adelaide.

During the last thirty years, prompted by post-Second World War migrants, music and dancing groups have proliferated within the German Association. This association was founded in 1886 specifically to encourage the pastimes of working people. Today within the association there are Bavarian groups, mixed choirs, children's choirs and dancing groups, none of which had been encouraged under the rather austere Prussian Lutheranism of the nineteenth century. Although for many years prior to the First World War the association was thought to be tinged with socialism and thus shunned by both the Lutheran Church and middle-class Germans, it nevertheless kept alive the customs, traditions and beliefs of the working-class Germans in Adelaide.

In country areas, in often very small communities, German music lovers developed brass bands for their own enjoyment. By the turn of the century the German communities had adopted the English custom of brass band competitions which are still a feature of country life in South Australia.

The arrival of the post-1945 German immigrants also gave the German love of festivals a great boost. These migrants did much to revive traditions and folklore that in many areas were foundering. Two world wars fought against Germany with a resulting distaste for all things German had taken a severe toll on the beliefs and traditions of the descendants of the original German settlers. The arrival of the new migrants, many refugees from a homeland that was not far distant from those of the original settlers, invigorated and made public again a folklore that had been driven underground. Because of the wars German folklore had been practised and mentioned only in the privacy of the home, or celebrated quietly in the German areas of the state. In the 1880s and the 1890s, for example, *Schützenfests* or shooting carnivals had been held in Adelaide, Hahndorf and Lobethal, to name the most important venues, but these had ceased with the First World War. The new migrants started again the *Schützenfest* at Hahndorf, enlarged it into a folk festival, made it a tourist attraction far beyond local or state borders, and gave yet another German word to South Australian English.

Similarly, on entering the Barossa Valley one is encouraged by large signs to enjoy the *Gemütlichkeit*, a message repeated in practically every restaurant in the region. There is no translation, for it is expected that the word will be automatically understood as a feeling of well-being brought about by pleasant surroundings and good company. The reaching out of German festivals into the wider community has been particularly evident in recent years in the Barossa. Again this has been greatly influenced by post-1945 migrants and the events are now enjoyed as part of the South Australian entertainment scene.

An *Oompah* Festival takes place on the Australia Day holiday weekend which with its Bavarian flavour would have had no appeal to the early, more serious-minded Prussians. The Tanunda Band holds a *Melodie Nacht* (melody night), the *Liedertafel*, a *Kaffee Abend* (coffee evening) and the Tanunda Show, an *Essen Fest* (food festival). But the greatest folk festival is the biannual vintage festival, which grew out of the church harvest festival. Originally a one-day affair which started in 1947 the event has continued to grow. Today it is a week of celebratory festivities, the high point for lovers of German food and drink being three dinners each of four courses with at least 1300 people attending each sitting.

One pastime that was and still is uniquely German is the *Kegel* Club, the Bowling Club, with its alley in Tanunda. Played with nine pins, as in the original German game, it is the forerunner of today's tenpin bowling. In the nineteenth century when the Germans in New York created such a noise carousing at their *Kegel* Clubs, bowling with specifically nine pins was banned. Another pin was promptly added, to give birth to today's ten-pin bowling.

The spiritual leanings of most of the early Germans gave rise to beliefs about the supernatural both within and without the Lutheran church. In the Barossa Valley in particular there have been in quite recent times cases of white magic exercised by people, generally women, with strange powers. The magic is usually directed against animals, although people have been involved. There is also a belief that in the nineteenth century black masses were held in the hills surrounding the Barossa Valley, although despite circumstantial evidence still in existence they cannot be authenticated.

For most Germans arriving in Australia, however, the Lutheran church provided spiritual succour in the new land as well as the companionship of like-minded friends and a familiar language. Lutheran churches of the nineteenth century with their unique design still dot the South Australian countryside and indeed, in places like the Barossa Valley, dominate it. Other evidence of the building traditions the Germans brought with them is still clearly apparent in these areas. The Germans came to South Australia so early in its settlement by the white man that they were able to plan and build towns in virgin lands to their own design. The towns were planned either on a *Hufendorf* (farmlet village) model or a *Strassendorf* (street village) model. The former still can be seen in Bethany in the Barossa Valley while the latter is evident in the layout of Hahndorf in the Adelaide Hills. Other *Hufendorf* towns such as Langmeil have had their original form absorbed by larger towns — in this case Tanunda.

The house design used in German villages and on German farms was distinct for at least thirty years after the arrival of the first migrants. The ideas of building — part of their 'cultural baggage' — understood by those who settled in the country areas caused them to build up rather than out in the traditional Australian manner, (a reflection of European land costs). Although not built into all houses, the three most obvious German features were the black kitchen, the baking oven and the attic door in the side gable. The black kitchen was found only in the larger, more expensive houses. Having a black kitchen meant that meats could be cured and the German sausages (*wursts*) smoked if desired. Black kitchens were a legal requirement at that time in Prussia where most of the houses were built of wood. In South Australia where houses were built of stone or brick they were not really necessary to reduce the risk of fire, but traditional building habits remained for many years. The baking oven was found much more often in German farm cottages. It allowed the baking of bread in the German fashion with the traditional recipes. Wood was placed in the tunnel inside the oven, the oven brought to the right heat and then the ashes scraped out and the dough put in, the heat retained by the oven bricks cooking the bread. The third feature was, and still is, ubiquitous in the South Australian landscape in German areas. The 45-degree gable of the roof enclosed an attic door which was reached by an outside ladder. This attic, which ran the length of the house, was used as a general storage area and occasionally for sleeping. Another piece of Germany until quite recently widespread throughout the state but now used primarily for display purposes or in processions was the German wagon. With its singular sloping sides and generally total lack of springing it was an all-purpose vehicle, cheap to build but immensely strong. It was arguably *the* distinguishing feature of a German farm. Clothes, with the possible exception in the very first days of settlement of the black wedding gown, have never really set the German settlers apart. Black was the colour when dressed for Sunday or best. Prussian-Lutheran sobriety was distinguished by its sombreness. More recent arrivals from other parts of Germany brought other more colourful clothing, but these are reserved for dancing and festive occasions.

Perhaps the most obvious German influence on South Australia is to be found in food. This influence, especially in the fields of pastries and sausages, continues strongly in the present day, noticeable not only in the country areas but also in Adelaide. Because a German cold sausage had no real equivalent in English it was called from the earliest times in South Australia *Fritz*, and there are innumerable stories of how this name originated. Attempts were made during the First World War to change the name to Austral, but these were unsuccessful. More successful was the attempt to change the name of the *Berliner Pfannkuchen* (a type of doughnut) to Kitchener, after the British War Minister of the period. However, since the arrival of the post-1945 German migrants and the opening of many new German bakeries the *Berliner* has become popular again, this time filled with jam, while the Kitchener bun has continued, filled with artificial cream. For well over one hundred years every Adelaide suburb and country town has had at least one delicatessen — the deli, copied from similar shops in Germany, its name being the Anglicized form of *Delikat Essen*, specialty foods. In such shops it has been possible always to buy a wide range of wholemeal, white and rye bread as well as cold meat and German-type sausages. Many delicatessens also functioned as the local sweet shop and milkbar as well as selling their more specialized goods.

While it is possible to buy most German foods at delicatessens in Adelaide, a greater variety of such food is still to be found in the country areas of South Australia, particularly in the Barossa Valley, the acknowledged centre of German influence in the state. Here every year since soon after their arrival the local women have displayed their dill gherkins, their *Sandkuchen* (German cake), *Streuselkuchen* (seed cake), *Bienenstich* (bee sting) and various other specialties. All, like *Fritz*, have undergone changes in Australia but still remain distinctively German. Competition to produce the best of any particular variety using traditional recipes, now formalized as part of the Tanunda Show, is as strong as ever. The fear of losing these old recipes, some dating from the Middle Ages, has led to publication in the Barossa and also in the other centre of German influence in South Australia, Hahndorf, of books which include not only recipes but also traditional cures for many ailments.

The early settlers from the eastern areas of nineteenth-century Germany came from a countryside that had not been industrialized, indeed which only some forty years before had thrown off the last remnants of feudalism. They lived in a way that had changed little over the centuries. These Germans brought to Australia therefore mediaeval ways not only of planning towns, building houses and cooking but also of healing. The herbal and white magic cures in the *Sixth and Seventh Book of Moses*, as well as containing such useful advice as which spells and herbs were best for ensnaring husbands or keeping husbands and wives faithful, also contained many useful and age-old remedies for a variety of ailments ranging from the common cold to menstrual pains. The book was looked upon with horror by the Lutheran Church. Many families had it secreted away from the prying eyes of the pastor in a niche behind a loose brick in a wall or a fireplace. The Church considered that trust should be put in God, not in some pagan cure brought to Australia from an era lost in the mists of time. For the same reason the Lutheran Church also disagreed with the concept of homoeopathic medicine which began in Germany and had always had a strong following. There were many adherents among the German community in South Australia. Today, although many of the various remedies have been forgotten, others that have stood the test of time are still practised.

The mediaeval influence can also be observed in beliefs approved of by the Lutheran church. The pristine cleanliness of the farms of people of German descent is due not just to the German concept of *Ordnung* (order) but to the tradition of stewardship. This tradition is a belief held by German Lutheran farmers which maintains that although they own their farm in a legal sense, in another deeper sense the good earth is only on loan from God for them to nurture and foster while they are alive. On their death it is to be passed to their sons and so on down the generations. The farmers are merely stewards of the land. By nature, therefore, they are conservationists not exploiters of the land. Land is not something to be bought and sold like a normal commodity but something held in trust for God. In the Barossa Valley and other predominantly German areas of South Australia land has been in the same family since it was originally settled in the 1840s. Even today this tradition shows little sign of weakening.

The love of the land and its trees and forests was not absent from those Germans who settled in Adelaide. A German-born parliamentarian, F. E. H. W. Krichauff, was responsible for the passing of the first Australian Woods and Forests Act (1875) to preserve, protect and replant a colony chronically short of timber. The belief in education led another MP, F. Basedow, to successfully move for the foundation of the first agricultural college in Australia, Roseworthy, in 1879. The love of nature can also be found in the paintings of Hamburg-born Sir Hans Heysen (1877–1968), who brought an Australian perspective to his work in a medium which was at that time still dominated by European images.

In Adelaide the German craftsmen showed particular skill in silver work. Again the love of nature exhibited itself in their work. The famous silver tree at Broken Hill by Henry Steiner is a fine example by one of the leading exponents of the art; other well-known naturist silver workers were Julius Schomburgk and Jochim Wendt, whose work has been thought worthy of recent exhibition. The work of other German craftsmen, particularly those who made vernacular nineteenth-century furniture in the Barossa Valley, is in high demand in Adelaide today, although the number of fakes is perhaps greater than the number of original pieces.

In conclusion it is necessary to mention one belief held passionately by nineteenth-century middle-class German immigrants to South Australia that has not survived the test of time. Middle-class Germans who arrived before World War I had an almost mystical belief in British justice and British fair play. Two world wars, the first in particular, in which anti-German feelings were swept along by the paranoia of many British Australians far from the motherland and vulnerable to internal and external enemies, brought much suffering, heartache and sense of injustice to those of German descent who had considered themselves loyal South Australians. They withdrew into their own communities, in country areas in particular, their belief in justice and fair play sorely tried. Only recently is a change of heart apparent. There has been strong demand from local communities that the German language be taught in schools in traditional German areas, and these same communities are again looking on their German heritage and folklore with pleasure and pride, and something they wish to share with all South Australians. During the First World War all things of Germanic origin in the state were suppressed and the rich German heritage of South Australia denied. One most obvious example was the changing of sixty-nine German place names on the map of the state, creating a totally erroneous impression of the origins of the nomenclature of the place names of South Australia. It is pleasing that the wounds finally have begun to heal, and the melding of the folklore and traditions of the early and later waves of German migrants into South Australia are beginning to create an even richer tradition for the German community and for Australia.

REFERENCES: Ian Harmstorf, *The Germans in Australia*, 1985; 'Some Common Misconceptions about the Germans in South Australia', *Journal of the Historical Society of South Australia* 1, 1975; Guests or Fellow-Countrymen, Ph.D. thesis, 1987; G. Young, I. Harmstorf and D. Langmead, *Barossa Survey*, 1975.

Ian Harmstorf

Gesture: see **Non-verbal communication**.

GOLDSTEIN, KENNETH (1927–), is a distinguished American academic folklorist who has visited Australia several times and is well known to Australian folklorists. He has been chairman of the Department of Folklore and Folklife at the University of Pennsylvania since 1984 and was president of the American Folklore Society in 1975 and 1976. Dr Goldstein has lectured and conducted extensive field work in the United States, England, Scotland, Newfoundland and Labrador and also in Australia in 1983, 1985, 1986 and 1988. He was the keynote speaker at Australia's first National Folklore Conference held in Melbourne in 1984. His collecting interests in Australia centred on monologue performances, bush recitation and folk song. His recordings are housed in the National Library of Australia.

Kenneth Goldstein's numerous publications include *A Guide for Field Workers in Folklore* (1964) and *Folklore: Performance and Communication* (1975) (edited with Dan Ben-Amos).

Graffiti — an Italian word meaning words or drawings scratched on a wall — are a venerable form of folk expression found in many cultures old and new. Technological progress has always been reflected in the instruments used to create graffiti, beginning with rocks or slate and ranging through blades of various kinds, chalk, pencil, ballpoint pen, felt-tipped pen and most recently the paint spray-can. Graffiti may be found on almost any public surface, including walls, billboards, sculptures, toilets, buildings and so forth. Australian historian and folklorist, Ian Turner (q.v.), once described graffiti as 'an informal communication in writing of an individual message, often of a scandalous, salacious or otherwise disapproved kind'. While contemporary folklorists would rather stress the communal and shared nature of such material, this description nevertheless succinctly encapsulates the informality, scabrousness and communicative nature and functions of graffiti. Some Australian examples of graffiti include 'Suck More Wombats', 'Free Zarb' (a reference to a Vietnam draft resister) and more recently, 'Reality is an illusion created by lack of alcohol' (or drugs). A much-quoted Melbourne example of dialogue on a graffiti wall does not need an understanding of Australian Rules football for its wit to be appreciated. 'What would you do if Jesus came to Hawthorn?' was answered with 'Get him to move Peter Hudson to full-forward'. The spray-can graffiti that have become such a feature of urban life can perhaps be considered the latest development of this folklore genre. Spray-can graffiti may seem to be little more than mindless vandalism, empty of meaning, but vandalism in itself is a clear message of alienation and defiance.

GRAINGER, PERCY (1882–1961), has a strong claim to be considered the first Australian-born musician who collected folk songs (although not Australian folk songs) and he achieved world fame as an arranger of folk music. Percy Aldridge Grainger was born in Melbourne on 8 July 1882 and educated at home by his mother Rose Grainger (née Aldridge) and private tutors. At the age of 12 he was already an acclaimed pianist, and left Australia to study at Dr Hoch's conservatory in Frankfurt-am-Main in Germany. In 1901 he moved to London and became a successful concert pianist.

Percy Grainger collected folk songs in several different countries: the United Kingdom, the Pacific Islands (with A. J. Knocks) and Denmark (with Evald Tang Kristensen). He also transcribed some early wax-cylinder and disc recordings of Australian Aboriginal, Polynesian and Indonesian music. He was one of the first to use the phonograph for field recording, and found it necessary to defend its use in the face of some conservative opposition. His definitive statement on the subject was published in the *Journal of the Folk-Song Society* in 1908. Grainger's article 'Collecting with the Phonograph' stated that 'it cannot be made too widely known that the phonograph puts valuable folk-song, sea-chanty, and morris-dance collecting within the reach of all possessed of the needful leisure and enthusiasm'.

Many of Percy Grainger's field collecting methods were in advance of his contemporaries and they are still of value to folklorists today. Grainger was particularly interested in variations in performances and in the total context of a collected item. A. L. Lloyd (q.v.) acknowledged Grainger's leadership as a collector and wrote that

> the folk singer would convey the mood of the song by a small alteration of pace, a slight change of vocal timbre, an almost imperceptible pressing or lightening of rhythm, and by nuances of ornament that our folklorists, with the exception of Percy Grainger, have consistently neglected in their transcriptions. (Lloyd, *Folk Song in England*, 1969)

Percy Grainger was also an innovative composer and arranger. He made numerous orchestral and solo arrangements of English, Danish and other folk songs, including his own collections. Some of the best-known of these arrangements are 'Mock Morris', 'Molly on the Shore', 'Shepherd's Hey' and 'Country Gardens'. According to Simon, 'Country Gardens' and Grainger 'became synonymous', and his piano score sold over 40 000 copies annually in the United States alone over a period of twenty years. Grainger had improvised on an English handkerchief dance called 'Country Gardens' as a frequent encore for his War Bond and Liberty Loan concerts given in America during the First World War. An indication of the breadth of Grainger's arrangements of folk music is given in Teresa Balough's *Complete Catalogue of the Works of Percy Grainger* (1975) which includes under 'Generic Headings' the following:

American Folk-Music Settings
British Folk-Music Settings
Early Settings of Folksongs and Popular Tunes
Scotch Folksongs from 'Songs of the North'
Unnumbered British Folk-Music Settings
Danish Folk-Music Settings
Unnumbered Danish Folk-Music Settings
Settings of Dance Folk-Songs from the Faeroe Islands
(a) Numbered Settings
(b) Unnumbered Settings
Settings of Faeroe Island Folk-Poems
Unnumbered Sea Chanty Settings

Grainger was a prolific recording artist on both discs and piano rolls, and collaborated with the physicist Burnett Cross in the construction of machines to produce 'free music', a foretaste of contemporary synthesized music. Much of his work and memorabilia is housed in the Grainger Museum at the University of Melbourne.

REFERENCE: R. Simon, *Percy Grainger: The Pictorial Biography*, 1983.

Greek-Australian folklore The stereotype of the singing, dancing Greek is all too frequently projected for external consumption both as colourful exotica in the promotion of Greece as a tourist destination and as the staple Greek contribution to picturesque ethnic festivals in multicultural Australia. The purist might dismiss both types of folkloric display as fakelore, and yet there is an important internal dimension to them which the international Zorba image threatens to decontextualize and trivialize, namely their role in the maintenance of identity, a question of self-definition and self-image as well as self-presentation to others. As the author Vasilis Vasilikos put it: 'The Greek emigré maintains his identity abroad through the language he speaks and the music he listens to', or as Steve Frangos, writing about Greek-Americans, says, 'music can be used to document and refute the assimilation process'.

Indeed, Vasilikos's statement might equally be applied to the citizens of the Greek nation–state itself, which since its inception has used the Greek language, song and other items of traditional folk culture encoded in the Greek language as fundamental points of reference for the construction and consolidation of a national identity. Thus, Greek folk song, with its perceived Homeric qualities, has been particularly useful for tracing the 'rustic road' of cultural and racial continuity all the way to ancient Hellas, and traditional folk dances have also been marshalled to this patriotic cause. In the light of all this, the performance of traditional Greek folk song and dance in Australia is a particularly highly charged assertion of ethnic identity, and hence its prominence at private and communal gatherings of Greeks in Australia, in their broadcasting and discography, and hence also the considerable emphasis placed on learning the texts and the accompanying dances in Greek-Australian education.

In seeking to establish the extent of the performance of Greek folklore in Australia one needs to go beyond its appropriation for public display in the multicultural showcase and seek to enter the 'private sphere' of its performance in a more natural context; this will entail some (perhaps specious) extrapolation from statistics and assumptions. The beginnings of Greek folklore in Australia might be assumed to date from the first settlement of Greeks in the early nineteenth century, if it is reasonable to assume that where the Greek language has been spoken, there traditional proverbs, cliché dialogues, tales, jokes, anecdotes, etc., have been uttered (if only by virtue of being embedded in daily speech and narrative), and there Greek folk songs have also been sung. It is also reasonable to assume that on stepping ashore, early Greek settlers did not immediately abjure the minor personal observances of their native culture such as veneration of small portable icons (to this day unobtrusively kept in wallets or stuck to the sun-visors of cars, trucks and taxis), traditional naming practices and celebration of (saints') name-days, evil-eye superstitions (*matiasma*), right-foot-first superstitions (*podariko*), and hospitality rituals at inaugurations, ceremonies and other special occasions (*kerasma, kolyva, vasilopita,* etc.).

It may also be assumed that performance of Greek folklore became more widespread with the massive increase in Greek immigration between 1962 and 1966, to the point where, by 1986, persons of Greek ancestry formed the second largest non-Anglo-Celtic ethnic population of Australia (after Italian) and Greek-speakers numbered 267 100. Whilst an impressively large proportion (91 per cent) of first-generation Greek immigrants have sought to integrate into Australian society to the extent of becoming naturalized, there is ample evidence of their determination to maintain and transmit their language and culture, particularly through the institution of Greek Orthodox communities which since the 1890s have been established, together with their churches and schools, all over Australia but primarily in Victoria, New South Wales and South Australia. Evidence of the success of their mission is the fact that just under half of the Greek speakers mentioned above are Australian-born and Greeks are credited with the highest level of language loyalty among the ethnic minorities of Australia. The Greek Orthodox church has reinforced the observance in Australia of calendar customs such as dyeing and cracking Easter eggs; floral decoration of the funeral bier carried in the 'Epitaphios' procession; the ritual blessing of the waters and immersion of the cross at Epiphany (the last two with due allowance for seasonal discrepancies between the hemispheres); foundation rituals (with blessings in lieu of sacrificial offerings); ritual aspersion of new buildings, cars, etc., and votive offerings of candles, silver work, etc.

The predominantly rural provenance of Greek immigrants to Australia and their modest levels of formal education, plus the fact that mass migration peaked before the extension of modern mass media throughout the provinces of Greece, might reinforce the hypothesis that of the various genres of Greek song, for instance, these immigrants would favour traditional folk song. It is interesting to note in this connection that metropolitan Greek commentators consider the emigration of actual and potential exponents from the provinces to be a major factor in the decline of folk song in Greece itself. The characteristically high residential concentrations of Greek settlement in suburbs of Australian cities and affiliation to regional associations would have offset some of the effects of their sudden urbanization on maintenance of traditional practices.

The statement by Vasilis Vasilikos quoted above further typifies the perception or expectation on the part of metropolitan Greeks that the Greeks abroad — and there are estimated to be some four million living in 'the diaspora' — will retain their cultural identity. Again historical precedent can be invoked with reference to the survival of Greek culture over centuries or whole millennia in erstwhile Hellenic dominions (Cyprus, southern Italy, Corsica, and up to 1923 Bulgaria, Anatolia and the Pontus). This expectation is echoed in Greek-Australian anti-assimilationist rhetoric where determination to observe traditional 'customs and mores' [*ithi ke ethima*] is a cliché which has permeated the consciousness of younger Greek-Australians. Recent governments of the Hellenic Republic have provided material assistance (training courses in Greek folk culture — Ministry of Culture 1989), and committees and societies have been established to manage folklore spectacles: the Lesbos bull festival was in its twelfth year in Melbourne in 1992 (albeit latterly minus a bull); a Cretan village wedding was stage-managed in Melbourne by the Greek Women's Cultural League in 1984 and a more spontaneous sequel in 1987 was acclaimed in the newspaper headline 'In Melbourne tradition takes on flesh and bones'; in November 1991 the first two-day 'Traditional festival of Second-generation Greeks' was held in Melbourne under the motto 'We hold on to the old to embellish the present and to meet the future'. Antique folklore has also been revived in the form of the 'Anthesteria' Festival of the Greek Cultural League of Melbourne, named after the ancient festival of Dionysos at Athens, transferred from its original month *anthesterion* (February–March) to October.

What the metropolis (and often the diaspora) perhaps fails to recognize is the fact that Greek language and folklife in the diaspora are not simply static clones of the metropolitan models but evolve in interaction with their local context, sometimes with disconcerting results. The report of the Inquiry into Australian Folklife draws attention to some examples of adaptation of imported traditions to local conditions — most spectacularly perhaps the fusion of bouzouki and didgeridoo music by a Melbourne Greek composer. It is at that point that the order of the ethnic adjectives should be reversed to read 'Greek-Australian' rather than 'Australian-Greek', whose balance is tilted in favour of the Helladic. This bears on a number of key questions which students of Greek folklore in Australia have to address — for example, what difference has migration made to the tradition and how is the difference expressed in the performances and texts? Is there a distinctively or even unique Australian Greek folklore? Linguists have already answered the latter question in the affirmative and proceeded to describe the version of the Greek language spoken here as an 'ethnolect' with a demonstrable two-generation tradition to date. Looking at the range of Greek music played in Australia, ethnomusicologists have both proclaimed the relative purity of the folk music tradition here (Holst 1973) and found strong evidence of its modification to meet local conditions (Tsouni 1987). Looking at calendar customs organized by the Greek communities in Australia such as processions on Greek national days (25 March and 28 October) or political anniversaries (20 July, to commemorate the Turkish invasion of Cyprus), one can point to parallel metropolitan Greek events, but the local ones have a distinct Australian colour: for example, the 25 March celebrations in Melbourne are now the culmination of a week-long festival of Greek culture, and Australia Day has become the traditional excursion day for Greek associations in Melbourne. Imported rites of passage have also been discreetly modified as the occasion demands, as when a non-Greek Melbourne undertaker was recently persuaded to drive the hearse three times round the block on the way to the cemetery, rather than have the pall-bearers carry the coffin three times round the house as Leros custom required.

Uniquely Greek-Australian folklore is perhaps best illustrated from spontaneous ephemera such as bilingual slogans, graffiti, puns and jokes. At soccer matches, where the unique configuration of local circumstances brings 'Greek' teams such as South Melbourne 'Hellas' into regular competition with Preston 'Makedonia', opprobrious taunts such as the following are heard in both Greek and English: 'You are red, you are black: where's your country on the map?'. Again the Greek section of the anti-Kerr demonstrations of November 1975 played with a bilingual pun on the late Governor-General's name [*kerato-* = horn] to produce: *Exo Kerato tou Kerata*! (Out with the devil's horn! [or cuckold's horn]). A Greek taverna in Richmond, which had taken its name '*Minore tis Avyis*' (Dawn serenade in minor key) from an imported television serial, marked a change of management in 1991 with the modified name 'Minore Now', to match the Australian TV series 'Acropolis Now'. Finally, a uniquely Greek-Australian joke: What do Greek parrots wish you at Christmas? Galah *Christouyenna*! (*Kala Christouyenna* = Happy Christmas).

However, the diaspora is not so much encouraged by the metropolis to see itself as a dynamic force for the rejuvenation or enrichment of Greek culture with a hardy, hybrid strain, rather the diaspora is admired and congratulated when it plays the heroic role of preserver of quaint folkways of yesteryear in a supposedly pure, deep-frozen form, uncontaminated by recent metropolitan concessions to Western cultural imperialism (see *Neos Kosmos* 19 September 1991 on the response in Greece to a visiting Greek dance troupe from Sydney). It would be ironic if this implied that the metropolis expected the diaspora to cringe to a model which it has itself discarded, and to act as a sort of conscience of the Greek nation.

The extent to which folklore has been adapted to the largely industrial, urban environment into which first-generation immigrants settled and transmitted to second and subsequent generations of Greek-Australians may be gauged from the evidence on folk song and theatre (see **Greek-Australian folk song**; **Rebetika**; **Karagiozis**). Folk song has been the subject of more research than other genres of traditional Greek discourse found in Australia, mainly through fieldwork projects undertaken by university students in Greek folklore courses (which have been offered at the University of Melbourne since 1975) and postgraduate dissertations. This may reflect the metropolitan Greek preoccupation with folk song as more conducive to the demonstration of links with antiquity: the etymological proximity of the very word for song in Modern Greek (*tragoudi*) to 'tragedy' has not been neglected in this regard and the avowed admiration of philhellenes of the stature of Goethe for the poetic qualities of modern Greek folk song has reinforced the importance of the verbal text. The primacy of poetry over other genres extends to modern Greek literature both in Greece, which has produced two Nobel laureates in successive decades, and in Australia, where the most celebrated Greek writer is the poet Dimitris Tsaloumas. By contrast other genres of folk literature have received scant attention, particularly those whose oriental origins or non-classical pedigree are evident (e.g. shadow theatre and fairy tales). However, it may well be that some of these genres have proved more durable than folk song in the diaspora setting: fairy tales (*paramythia*) by virtue of their international dispersion prior to Greek migration and their connection with mothers/grandmothers and children (cf. riddles, children's games and lullabies) and proverbs and traditional humorous anecdotes by virtue of being embedded in daily speech and narrative. Further research is needed to establish the validity of this supposition.

REFERENCES: G. Bottomley, *After the Odyssey. A Study of Greek Australians*, 1979; H. Gilchrist, 'Australia's First Greeks', *Canberra Historical Journal*, March 1977; Greek Ministry of Culture, *Policy on Expatriate Greeks* [in Greek], 1989; A. Grivas, 'The Greek Press in Australia and Its Mission' [in Greek], *Krikos* 79–80, n.d.; G. Holst, 'Our Greek Music World's Purest', *Australian*, 3 December 1973; A. Kapardis and A. Tamis (eds), *Greeks in Australia*, 1988; D. Tsouni, 'Music and Dance at Greek Wedding Receptions in Adelaide', *Chronico* 6–7, 1987.

Anna Chatzinikolaou and Stathis Gauntlett

Greek-Australian folk song Unlike other practices imported by Greeks, traditional music, dance and food are supposed to carry no stigma in Australia, and second-generation Greeks are said to welcome the chance to participate in the traditional music which is rarely absent from Greek private and communal celebrations, rites of passage, excursions, fundraising functions, etc., at least by dancing. Indeed, Greek traditional music would appear to provide for more than positive self-identification as Greek, to judge by the declaration of the erstwhile president of the Sydney Greek Music Lovers Club that members have a duty not only to retain this 'immortal heritage passed down to us by our proud forefathers' but also to 'share these inherited gifts with our adopted brothers'.

The degree to which Greek-Australians share the 'immortal heritage' is largely determined by the performance contexts of Greek folk song in Australia, which divide potential bearers of the tradition into broad but fairly distinct groups along the lines of their amateur or professional status and their generational status. Professional exponents tend to dominate the most public contexts, often allowing amateurs scope for little more than passive observation of folkloric display or stage-managed entertainment, whereas a living tradition might be thought to require more active involvement in relevant ritual. There are few recorded instances of amateur ritual singing by Greeks in Australia, and fieldwork is needed to establish its extent, as well as that of 'purely' oral-aural tradition of Greek folk song in Australia. The evidence of projects conducted by undergraduate and postgraduate students of modern Greek folklore at the University of Melbourne suggests that instances of 'live' oral-aural tradition of Greek folk song in Australia are rare even among the most willing and confident respondents, many of whom reported that even in Greece they learned folk songs from gramophone records, radio broadcasts or another form of professional performance. These projects further revealed that active participation in the tradition is largely confined to the first generation of immigrants who imported it from Greece; second-generation Greek-Australians have generally been reluctant or unable to sing a 'demotic song' (*dimotiko tragoudi*), and those who have obliged have often revealed an insecure or unconventional perception of what the term denotes, or insisted on first consulting the school books from which they learned such songs.

The two most extensive studies of the Greek-Australian tradition, by Chatzinikolaou and Tamis, rely entirely on first-generation amateur respondents, selected without any attempt to approximate the demographic profile of the Melbourne Greek community, but rather according to potential for confident, willing and competent performance of songs. In the event some 550 songs in total were transcribed from 'command performances' by the forty respondents, of whom two-thirds were women; a majority were aged between fifty and seventy; most had emigrated in the 1950s or 1960s; most were from northern Greece; and most had very little formal education and had hardly assimilated at all socially or linguistically into the Anglo-Australian community. The thematic range of songs recorded for both these studies and for the hands-on undergraduate folklore projects since 1977 is focused mainly on love, migration (with very little explicit mention of Australia), heroic exploits from the history of Greece, and death. This range was circumscribed by the taboo status of some types of songs (e.g. dirges) and the irrelevance of carols and ritual calendar songs out of season. Among the songs most frequently collected are 'classic' examples of the genre which have been much recorded, revived and broadcast ('Kitsos' Mother', 'Lads from Samarina', 'Forty Brave Lads from Livadia', 'Three Lads from Volos', 'When Will the Sky Clear?', 'The Willow', 'The Doe', 'Maria Dressed in Yellow', 'Kalamata Kerchief'), together with some personal compositions in folk style which were first recorded between the wars ('Don't Send Me to America, Mum', 'The Murderous Mother-in-Law'). Variation between repeated performances and different performers was minimal, suggesting that oral improvisation was restricted, but the form of the verses retained the traditional formulaic characteristics thought to be connected in origin at least with oral composition and transmission. The most striking feature of the texts in comparison with those published in collections in Greece is not, as one might expect, their use of the local Greek-Australian idiom (or 'ethnolect'), which was almost negligible, but their brevity and the high incidence of incomplete and unmetrical lines. It might be noted, however, that many standard collections of Greek folk songs are not above suspicion of unavowed selectiveness and even editorial restoration in the notorious manner of 'the Father of Greek Folklore', Professor N. G. Politis.

In performing longer narrative ballads, amateur exponents occasionally lapsed into prose paraphase or abbreviation. However, gems of narrative condensation were also uncovered, such as the following account of a cunning strategem, sung for Anna Chatzinikolaou in January 1979 by a sixty-eight-year-old woman who had emigrated from the island of Cephalonia in 1968; the contrast between the terse progress of the storyline on the one hand and the traditional tautology, triadic crescendos and 'redundant' symmetry of the individual verses on the other was further highlighted in performance by the absence of the traditional segmentation (*tsakisma*).

I've been to all the castles, all of them I've been round,
but like the Maiden's castle another I've not found.
Some Turks had been attacking for a dozen years
then younger Turks attacked it another fourteen years.
So one fine lad, a little chap, a Turkish vicar's son,
now what might he expect to gain if the castle's won?
A thousand florins hourly, a first-class stallion,
and the maiden in the castle for his own woman.
He sticks a pillow down his front, as if it was swollen,
and two unripe lemons, as though a firm bosom.
He skirts around the castle walls, sobbing, whimpering,
hurling curses at the Turks and all the while swearing.
'Open up for me, poor wretch, poor widow that I am,
for here am I, heavy with child and my time has come.'
Before the door'd half-opened, a thousand had stormed in,
and before the door'd been shut again, the castle'd been taken.

Professional performance of Greek folk song in Australia has a much higher profile and, in contrast to the amateur repertoire, extends to lengthier and somewhat more polished forms, to treatment of topical events of local Greek-Australian interest, and to use of the local linguistic idiom. A good example of this is the string of couplets composed by the Cretan *lyra* player Kostas Tsourdalakis (q.v.) in protest at the dismissal of the Whitlam government in November 1975 and urging his re-election. The text has been published with a facing translation in *Meanjin* 42, 1983, and it is to be noted that the text was originally written down (as the second couplet states) on 15 November 1975 and read over the airwaves on 16 November 1975, before being published in written form in *Neos Kosmos* on 11 December 1975 and finally sung to a ready-made traditional melody at a concert in 1982 and on an Australian-made LP in 1985. This is symptomatic of the fact that in Australia the tradition of improvising satirical, gnomic and occasional verse in Greek has found an outlet in writing, often in the form of metrical letters to the editors of Greek-Australian newspapers or magazines, and also in published collections of verse such as those of Kostas Tsourdalakis (q.v.), Stathis Raftopoulos, Katina Baloukas, Charalambos Azinos, Panayis Sotiriou and Yiannis Papadopoulos, all first-generation immigrants continuing the tradition of the broadside-rhymester (*rimadoros/piitaris*). The local practice of publishing metrical funeral and memorial notices is not practised in Greece itself.

The range of Greek regional folk music and instruments heard in Melbourne alone is broad; there are professional exponents of Cretan *lyra* (Kostas and Tony Tsourdalakis, Michael Melabiotis, Vasilis Michalakis), *lagouto* (George Tsourdalakis, George Xylouris, George Galiatsos), *outi* (Christos Bailtzidis, Manolis Galiatsos), clarinet (Charalambos Fakos, George Dragatos), *toumbeleki* (Takis Demetriou) and *sandouri* (Christos Fakos). They are supplemented for special occasions such as regional feasts and fetes (*panegyria*) by visitors from Greece, such as the Pontic *lyra* exponents Kouyoumtzidis and Tsanakalidis who visited Australia in 1985 and 1987 respectively for the Pontic feast of 'Panayia Soumela'. It has been claimed, however, that Greek-Australian musicians are the purer for having escaped the rampant commercialism of the metropolitan Greek music industry.

Greek musicians have tended to find opportunities for commercial recording in

Australia limited and prohibitively expensive. Greek-Australian discography — and broadcasting — are in themselves rich and complex research topics, but it might be noted here that while hundreds of records of a wide range of traditional and other Greek songs have been manufactured in Australia, at least since the 1950s, they have been mostly made from imported master recordings. Thus, the Greek-Australian 'Apollo' label with the catalogue prefix SY [= Sydney?] adorns at least eighty double-sided 78 rpm discs made from matrices imported by Yiannopoulos Greek Emporia (and Radio broadcasts) of Sydney, Melbourne and Adelaide, as were an unknown number of vinylite singles and EPs (on Apollo, Parlophone, Columbia, HMV labels). Only vinylite records of local artists are known to the present writers, the labels being parochial (including Antonic, Echophonic, Tsakpina, Manos, Mediterranean Melodies, Omikhli Enterprises, Sapfo, Universal [Melbourne], WG, Zorba). Imported master recordings have also been used by EMI, RCA, Image, Rainbow, and Summit to manufacture Greek LPs in Australia.

The future of Greek folk song in Australia will be to some extent determined by immigration and demographic patterns, which at first sight do not appear conducive to retention of a vibrant tradition, but beyond that the tradition appears destined to proceed along the lines of ever greater separation of its technologically sophisticated professional performers from its passive consumers, despite the endeavours of various Greek and multicultural bodies to keep the tradition alive. The loss of ritual purpose and consequent narrowing of folk song's role to a self-conscious assertion of ethnic affiliation do not constitute a foundation for anything but a museum survival in Australia — any more than in Greece itself.

The most substantial evidence of profound and creative second-generation contact with the tradition takes the form of adaptations of traditional forms by classical-trained and quasi-popular Greek-Australian composers, such as Themos Mexis, Savvas Christodoulou, Costas Tsikaderis and Constantine Koukias. Of course such use of the tradition is not confined to Greek-Australian composers, as Barry Conyngham's use of the traditional song *Samiotissa* in a bicentennial composition illustrates.

REFERENCES: R. Beaton, *Folk Poetry of Modern Greece*, 1980; A. Chatzinikolaou, Greek Folksong in Melbourne [in Greek], MA Prelim. thesis, 1979; S. Gauntlett, 'The Folksongs' in M. Carrol and S. Gauntlett, *Images of the Aegean*, 1980; M. Herzfeld, *Ours Once More*, 1986; G. Holst, 'Our Greek Music World's Purest', *Australian*, 3 December, 1973; P. Parkhill, *Cretan Traditional Music in Australia*, 1985; A. Tamis, Folksongs of Macedonia [in Greek], MA thesis, 1980.

Anna Chatzinikolaou and Stathis Gauntlett

GREENWAY, JOHN, from the University of Colorado, was a well-known American folklorist who visited Australia during the 1950s. He wrote several articles and two books based on his experiences in Australia. His articles included 'Anything like Waltzing Matilda?' (*Quadrant*, autumn 1957) and 'Folklore Scholarship in Australia' (*Journal of American Folklore* 74, 1961). Greenway's book *The Last Frontier: A Study of Cultural Imperatives in the Last Frontiers of America and Australia* was published in London and Melbourne in 1972. John Greenway was editor of the *Journal of American Folklore* from 1964 to 1968. He died in 1991.

Gumleaf playing: see **Leaf playing**.

H

HAMILTON, PETER (1924–), served with the RAAF during the Second World War. After the war he studied architecture at Sydney University. He had been interested in photography since childhood, and during his undergraduate days developed an interest in film. He began making documentary films in the 1950s, and founded a company, Wattle, to produce films and records. For many years he has lived on the north coast of New South Wales where he has been active in the alternative lifestyle movement.

HARDY, FRANK (Francis Joseph) (1917–) was reared in Victoria at Southern Cross and at Bacchus Marsh, the 'Benson's Valley' of many of his stories. His father was an itinerant, working in various dairy factories, and a compulsive gambler. Frank Hardy left school at thirteen, taking jobs as newsboy, grocer's assistant, fruit picker, factory worker and seaman. Deeply moved by the suffering of the poor in the 1930s depression, he joined the Communist Party of Australia and the Realist Writers' group in 1939. He drew caricatures, wrote concert lyrics about local identities and became interested in racehorses, betting and pub life, always deeming himself ideologically 'a rebel' (see *The Hard Way*, 1961).

In 1942 he joined the Army, and edited the *Troppo Tribune*, writing for it 'doggerel, articles, stories, reports of camp debates and Army Education lectures'. There was already an increasing pull between his revolutionary commitment and his urge to be an entertaining writer. He was encouraged then to 'write down some of the stories you tell on beer nights', and so in 1943 he put down very quickly 'A Stranger in the Camp', published in *The Man from Clinkapella* (1951). As Hardy said later of his early authorial self, 'Ross Franklyn', and of this story 'already written in his head':

> He had told it dozens of times and polished it up in the process . . . Being a bush liar is a good preliminary for writing short stories. You see a story happen . . . you repeat it in pubs until it has proven its worth, been polished up, licked into shape. Then you write it down.

In late 1944 he met Ambrose Dyson, a Bohemian concerned with the people's struggle, who then and later helped Hardy, whom he saw as 'a natural, typical Australian'. The latter's art work then was in a simple cartoon style, as in his twelve thumbnail sketches, 'Highlights of 1944', on the front cover of *Salt* for 1 January 1945. Dyson would continue illustrating and offering advice on 'Franklyn's' stories, as with those in *The Man from Clinkapella*, suggesting subtle improvements to 'The Load of Wood' (1946). Hardy's sympathetic portrait of a sustenance worker, Darky, who shows leadership and self-sacrifice, still comes alive because of the author's great skill with dialogue and handling of atmosphere. As Alan Marshall said, prophetically, in 1956, 'The tale captures something of the charm of a folk tale and could well become part of our folk lore'.

Legends from Benson's Valley (only issued in 1963) is the first fruit of Hardy the folk writer, an account of a pleasant fertile spot where the shadow of the Depression fell — a book to parallel Goldsmith's 'Deserted Village', showing a happy place blighted by economic changes and policies far away and the iniquitous 'bag' system of cheap groceries handed out to the needy each week. This highly symbolic time and place are handled nostalgically — albeit with wry humour — with great compassion for

Anzac Day march at Talwood, Qld, 1985. (Photograph, Chilla Johnston, from Peter Stanley and Michael McKernan, *Anzac Day: Seventy Years On*, Collins, 1986)

Madonna Dei Marteri during the traditional Three Novenas, prior to the Blessing of the Fleet festival, St Patrick's Church, Fremantle, WA, 1989.

Part of the annual Lesbos festival procession, which has taken place in Melbourne each February since the early 1980s.

One of Bendigo's historic Chinese dragons, Sun Loong, at the Easter Fair, 1992. Sun Loong is held to be the longest imperial Chinese dragon in the world. (Photograph, *Bendigo Advertiser*)

Loong is claimed to be the oldest imperial dragon in the world. Here the dragon passes the Charing Cross fountain during the Bendigo Easter Fair, 1950. (Photograph, Bendigo Chinese Association)

Festival of Our Lady of Terzito procession, Hawthorn, Vic., August 1986. (Photograph, Elizabeth Gilliam)

St Nicholas with two Black Peters arriving at the Dutch Club in Fairfield, Sydney, in 1959, where he presided over ceremonial speeches and the distribution of gifts. (Photograph, Australian Foreign Affairs and Trade Department)

Laura Divola in the Festival of Our Lady of Terzio, Hawthorn, Vic., August 1986. This festival is celebrated each year by families which originated in the Isle of Salina in Italy. (Photograph, Elizabeth Gilliam)

drinkers, workers, mates, battlers, and considerable sympathy for their acts of defiance, as unionists, of the system which gave them 'no work to do'. Simply sketched by typical acts and phrases, like Sparko's whose swearing was 'sheer poetry', the men are underdogs who have their moment in incidents of defiance, like cheekily 'riding the rattler', which dispel the gloom of life that, without mates, would be intolerable. While women and children all suffer, the man, as provider, fights to retain his self-respect.

The early stories of this group were written down at the same time as Hardy was researching for *Power Without Glory* (1950), a sprawling novel dealing with the years 1890–1950. By its worker readers and those further afield — it has been translated into several languages — it is deemed to be an exposure of the social and political evils of the floundering capitalist system, particularly in Victoria. In 1958 he published *The Four-Legged Lottery*, depicting the domestic tragedy that gambling brings to the two central figures, one from the ranks of the poor, the other (who tells the story) from the middle class. While both works have been branded as sensational and melodramatic, they are excellent as exemplars of the life story of personal decline told by the worker-political thinker become social commentator. Inspired by Balzac — mottoes from whom serve as mastheads — and by Upton Sinclair's prose, the tales have been popular amongst proletarian readers in many countries who have responded particularly to descriptions of the brutal and exploitative aspects of criminally organized sport and of gambling. For the writer and many of its readers, the second text, with its horrible echoes of the old convict time, does indeed speak out, as its finale has it, 'that all men shall know the tragedy of it all — the dreadful Australian tragedy of it all'.

Hardy's most significant folkloric achievement comes from anecdote collections, like *The Yarns of Billy Borker* (1965) and *Billy Borker Yarns Again* (1967), a series told by the new Hardy persona, the bar-fly Billy Borker, who has made many subsequent appearances on television. He is a larrikin, likely to pass into Australian folk tale, who tells pub anecdotes of 'pleasures and problems of ordinary people' in a style of terse irony and of throw-away Borkerisms — 'There's no more thieving goes on on the wharves . . . than in the Stock Exchange'. Together there are fifty-five yarns which their creator designed 'to capture on paper Australian folk-narrative', and which have been represented in various treasuries compiled by Bill Wannan (q.v.). Many came from Hardy's father or his own drinking mates. A key concern was always to exaggerate and yet to centre the whole on 'something typical that would throw light on the Australian character and myth, the quality of a legend'.

More significantly Hardy has brought the Lawson tradition up to date, now telling 'the yarns of modern times and cities, the man's world of the pub bar'. Borker was conceived by Hardy as a mythical character, born of the Depression, yet much older — in short, as the archetypal Australian battler, 'ironic, democratic, suspicious of authority' and so himself a fresh yet very familiar folk hero. He is always concerned at the common man's being manipulated by 'them' in positions of privilege and power. An overriding Hardy ambition of the early 1960s to write a trilogy of novels around the life of Henry Lawson did not come to fruition. Yet its narrative core, the purpose to tell the story of how 'a people and a tradition . . . a new kind of people — the Australians — came to exist' has inspired much of his work in various (literary) modes.

His other notorious folk volume, *The Outcasts of Foolgarah* (1971), which took some fifteen years to complete and get into print, was a sharp satire on local government, although also imagined by others to be 'a libel on the working class'. Although the novel caused a political sensation, its Rabelaisian and even scatological dimension rather than its comment on what can happen to the federal body politic and social has probably caused most comment among readers. Putting aside all earlier concepts of mateship Hardy mercilessly satirized federal politicians, the police and 'the effluently affluent', the purpose now being, in his words, 'to show that Australian society is a cross between a circus and a sewer', repressive and nauseous and obscene.

Clearly Hardy had already made a unique written contribution to both Australian pub folklore and to later twentieth-century Australian society's realization of its own folk memory, folk attitudes and worker conflicts and struggles. Hardy/Borker will be remembered for the abrasive and iconoclastic yarn or short story which engenders a genuine sympathy for the battler-gambler-drinker-womanizer, the otherwise irritatingly assertive folk hero. The latter is more than redeemed for us by his experience of injustice, and his splendidly wry retorts, while the good ear of Borker for slang and folk idiom is apparant in all his telling of characters and incidents 'like it was'.

In his life Hardy has won the trust and friendship of many thousands of ordinary (male) Australians as he has defied so many pressures to censor the bawdy or make less abrasive his political mockery. It is their voices, experiences and wit which he has caught in an Irish narrative style, in tales which examine, modify and yet reaffirm splendidly so many of the perdurable attitudes that make up Australia's greatest legend — the ongoing suffering, courage, humour and gregariousness of the ordinary working man.

J. S. Ryan

HOWARD, DOROTHY (1902–), was born in Texas, USA, and currently lives in Roswell, New Mexico. In 1938 she took a doctoral degree from New York University, writing a thesis on folk rhymes of American children. She has published a number of books and articles on children's traditional play, and in 1981 was given a 'Recognition Award for significant contribution to the study of play' by the Association for the Anthropological Study of Play (TAASP). Much of her American material is housed in the Dorothy Gray Mills Howard Archives at the Centre for Texas Studies/Texana Archives at the University of North Texas (Denton).

After many years of teaching in schools and colleges, Dorothy Howard received a Fulbright post-doctoral research grant sponsored by the University of Melbourne to study traditional play customs of Australian children during 1954 and 1955. On returning to the United States, Dorothy Howard published ten articles on her Australian research in major American journals such as *Folklore*, *Journal of American Folklore* and *Western Folklore*. As June Factor wrote in the *Australian Children's Folklore Newsletter* No. 4 in 1983, 'Dorothy Howard was the first person to systematically collect, collate, transcribe, annotate, and publish a comprehensive sampling of Australian children's games'. The published articles dealt with string games, marbles, counting-out customs, the toodelembuck, autograph album customs, ball-bouncing customs and rhymes, knucklebones and hopscotch. In a paper she presented to the Victorian Institute of Educational Research in February 1955, Dr Howard enumerated some of her research:

> My findings, to date (after having worked in Victoria, New South Wales, Queensland and Tasmania) include over 700 game names; descriptions of about 400 games; 175 autograph album rhymes; 50 rope skipping rhymes; 40 counting out rhymes . . . the words for about 15 singing games (with music notation for 8); a few riddles, tongue-twisters, trick rhymes; hand, finger and toe rhymes; rhymes for taunting, swearing an oath, bouncing ball; and nonsense rhymes.

The bulk of Dorothy Howard's Australian research has never been published, but in 1982 she presented all her Australian collection and some other material to Dr June Factor, the director of the Australian Children's Folklore Collection at the University of Melbourne. The Dorothy Howard Collection is now housed in the University of Melbourne Archives, and consists of traditional toys and other artefacts from Australia, Mexico and the USA, photographs and monographs plus letters and index cards. Using funding provided by the Australian National Dictionary Centre at the Australian National University, June Factor is currently preparing some of this material for publication.

A more detailed discussion of Dorothy Howard's work can be found in June Factor's *Captain Cook Chased a Chook: Children's Folklore in Australia* (1988).

HULTS, DAVID S. (1950–), a graduate of the State University of New York and of Curtin University of Technology, Perth, and also a qualified teacher and librarian, is currently head of the Social Science Department, Aranmore College, Perth. He is president of the Perth branch of the Musicians' Union of Australia, and member of the World Presiding Board of the International Organisation of Folk Art, and was joint founder and editor of *Australian Folklore*, 1986–1991; chairperson of the Australian Folk Trust (1981–84); member of Australian folklore delegations to China (1988 and 1990) and vice-president of the Australian Folklore Association, 1991–92. David Hults is the compiler of a comprehensive subject bibliography of Australian folklore and author of papers on folk craft, folklore and history.

Humour, folk Humour is an element in a great deal of Australian folk expression, being found in song, narrative, verse, some customary behaviour (see **Custom**) and folk speech forms, especially the riddle-joke and the funny story. There is almost no aspect of life, past or present, that has not been or is not a target of folk humour. However, much Australian folk humour in English revolves around a few topics and concerns. These are: drinking alcohol, together with the effects of this; the use of 'bad' language or swearing; racism and a general fear of 'others'; conflict and violence; regulation and anti-authoritarianism. Much of this humour arises at work, and especially during times of war and economic depression; immigration and ethnic and national distinctions are often a particular focus.

Humour is a constant element in the folk song, verse and yarning of nineteenth- and early twentieth-century bush life. Mostly it is characterized by a dry, understated and laconic wit that Lawson, Paterson and other writers sometimes succeeded in transmuting into more literary form. Although such characteristics are found to a greater or lesser degree in the folk expressions of most English-speaking nations, this form of humour is still frequently identified as 'typically Australian'. One particular story from this repertoire is often cited as reflecting the essence of Australian attitudes and values:

> A station-owner (or 'squatter') is driving back to his property one hot dusty day. Out in the middle of nowhere he sees a swaggie humping his drum down the road. The station-owner takes pity on the swaggie battling along, pulls up beside him and offers him a lift. The swaggie slowly turns a baleful eye on the station-owner, spits, and replies: 'You can open your own bloody gates'.

Like much Australian folklore, this story is often reprinted as well as told, and nicely encapsulates many of the themes central to that Australian self-image usually referred to as 'the Australian Legend' (q.v.). The story emphasizes the independence and anti-authoritarianism expressed by the itinerant bush worker towards the station-owner, and a dogged perseverance often associated with 'battling'. All this is represented in the archetypal figure of the nomadic, proletarian, hard-hit and hard-bitten swagman.

Many bush songs also depend for their humour on similar elements — drinking, fighting and defying authority. 'The Bullockies Ball', with its revelry and riot, 'Bluey Brink' about the shearer who was 'a demon for drink', even enjoying sulphuric acid, and 'Travelling Down the Castlereagh' ('The Bushman's Song') are only a few of many examples that might be given. While not all of these songs are uproariously funny, they typically treat serious topics, such as arson, as in the case of 'The Wallaby Brigade', in a light-hearted manner. The humour in these instances derives from situations of social and economic conflict between rural landowners and their usually itinerant labourers.

Other forms of bush folk humour also involve intergroup tensions, as in the case of the 'Jackie Bindi-i' cycle of yarns that circulate mainly in the northern half of the country, especially in Queensland. Jackie is an Aboriginal stockman or rouseabout who is usually shown replying or reacting to his white boss or other authority figure in a way that simultaneously shows an awareness of white prejudice and shows the Aborigine resisting the usually unfair demands of the boss. In one such tale, Jackie is being tried for cattle stealing and the judge asks him if he knows what will happen to him if he should tell a lie. Jackie says that he thinks he will go to hell. The judge then asks Jackie if he knows what will happen if he tells the truth. Jackie replies that he reckons he will go to gaol. Such examples suggest that folk humour is important both in maintaining and mediating racial, economic and other social conflicts.

'Bad' language — actual or implied — is a constant feature of bush humour (particularly associated with the figure of the hard-swearing 'bullocky', forerunner of the modern 'truckie'), of digger humour and of more recent folk humour forms also. What has changed is that contemporary folk humour is more likely to use — even revel in — words and images that are only hinted at in most bush humour and which is partially disguised by dashes, dots and euphemisms like 'blanky' and 'beggar' in First World War and even much Second World War soldier humour. By the time of Australian involvement in the Vietnam war, from the mid-1960s to the early 1970s, all attempts at subterfuge are largely abandoned and the appropriate terms are freely employed.

A related area of folk humour is bawdy folklore — of both males and females (see **Bawdry**). Although relatively little is known of the historical aspects of this clandestine humour, it seems unlikely that such material was as prevalent or as explicit as its contemporary expressions. Certainly, some of the classic pieces of male bawdry, like 'The Bloody Great Wheel', 'Kafoozalum' and numerous others can be traced back into the nineteenth century, but their classic status seems to largely derive from their relative scarcity, as much as from their sometimes spectacular crudity. The liberation of sexual values during and since the 1960s has no doubt influenced these developments and created a much less restrained climate of expression. Bawdry, mainly in song and verse forms, has attracted a good deal of interest from Australian folklorists, with a number of excellent collections having been published. These include Hogbottel and Ffuckes (pseuds), *Snatches and Lays* (2nd edn, 1973), Brad Tate's *The Bastard from the Bush* (1982) and Ron Edwards's *Australian Bawdy Ballads* (1973).

Children's folk expressions commonly depend for their humour upon the use of 'dirty' terms or situations, deployed in ways that parody and subvert the authority of the adult world (see **Children's folklore**).

One of the fundamental functions of folk humour — Australian or any other — is to 'make light' of the frequently burdensome realities of life. This can be clearly observed in the humour generated in situations of obvious tension and hardship, such as war and economic depression, yet is also a pivotal aspect in the folk humour of peace and plenty, as demonstrated in the humour of bush life and of contemporary urban life. Making light generally involves sardonic comment on the perceived source or sources of oppression and injustice — politicians, officers, employers, migrants. Here again can be seen the now-familiar aspects of the Australian self-image, the dislike and distrust of authority and the fear of 'others'.

Hard times always produce substantial crops of folk humour. The Australians who experienced the 'Great Depression' of the 1930s and who suffered from its effects created a body of folk humour to help them cope. Unemployment, poverty, hunger, family breakup and the 'dole' were the lot of many blue-collar workers and their families, though hardship also spread to other classes. The folk songs of this period are usually bitter-sweet reflections upon the hardship of the times and on the injustices and political failures that were seen as the main causes of depression. One song, to the tune of 'My Bonny Lies Over the Ocean', linked the experiences of the 1914–18 war with those of the depression:

We went out and fought for our country,
We went out to bleed and to die.
We thought that our country would help us,
But this was our country's reply:

Soup, soup, soup, soup,
They gave us a big plate of loop-the-loop
Soup, soup, soup, soup,
They gave us a big plate of soup . . .

Other songs commented on the political and financial aspects of the period, including those about 'old Jack Lang' who 'closed up the banks/It was one of his pranks/And sent us to the dole-office door'. Other depression folklore deals in a humorous manner with such un-funny experiences as eviction, the dole and searching for work in a depressed economy.

The Second World War ended the harsh effects of the Depression for many Australians, putting them back into full-time employment — even if often into uniform. Within the armed forces the civilian volunteers rediscovered the tradition of digger humour established in the 1914–18 war and sustained through the peace by returned servicemen. Humorous responses to war extended the tradition of digger humour, with its casual attitude towards officers and military regulation, its nonchalant treatment of valour and sacrifice and, as ever, its tradition of complaining. Not surprisingly, Australians also produced a good deal of home-front folklore during the war (see **Wartime folklore**).

Some of the items of traditional folk humour, particularly verse and tale, are still circulated by Australians in the bush and by those who rarely leave the cities. Despite their belonging largely to another time and place, stories about 'Dad and Dave' and 'The Swaggie and the Station-Owner', along with verse classics such as 'The Bastard from the Bush' and 'The Spider from the Gwydir' are a part of what many, perhaps even most, Australians would consider their folk heritage. There are also more recent forms of folklore to amuse Australians, yet many of these contemporary expressions revolve around traditional themes.

As far as it is possible to know in the current state of folklore research, the period after the Second World War is one characterized by the decline — though certainly not the demise — of some of the older forms of humour, such as the bush yarn, and the intensification of certain other forms, such as the riddle-joke. New forms of humorous lore are also developed, including reprographic (q.v.) forms and certain kinds of humorous contemporary legends (see **Modern legends**), particularly those related to the consequences of sexual philandering. The prevalence of these forms indicates a move away from the bush and the battlefield towards urban and global locations, a focus on the difficulties and tensions of contemporary life and a favouring of shorter forms. Such changes clearly reflect the perceived stresses and 'pace' of modern life and Australia's developing interaction with the rest of the world. Many of these forms are international in content and distribution and are often deployed in Australia with little change. 'Sick' jokes about topical matters of broad global interest, such as the NASA space shuttle tragedy or the mysterious or violent deaths of prominent public figures, such as John Lennon, very quickly leap the international networks of jet travel and electronic communication and go into circulation in many countries almost simultaneously. The same process can be observed in reprographic folklore, particularly that of the graphic variety which may also have the ability to transcend linguistic boundaries. In other cases such items are localized to suit Australian geography, history or other local circumstances and needs.

The themes and topics of this material

are frequently obscene, racist, sexist or in some other way derogatory. The preponderance of 'ethnic' jokes in contemporary folklore, while hardly a recent development, is one example of this trend. Another is the amalgamation of 'sick' and topical jokes to make comments upon current events and concerns. 'Blonde' jokes and AIDS jokes are cases in point. Political and religious jokes seem always to be in fashion, the same chestnuts reappearing with appropriate alterations of personalities and events.

Many of these same themes are popular in reprographic humour, with the addition of concern about communication, or its failure, together with the consequences of this. Satirical and parodic treatments of occupational hierarchies and power structures are particularly common in these forms of folklore (see **Occupational folklore**), while similar concerns often underlie much modern graffiti (q.v.).

Perhaps the most widespread articulation of humour in Australian folklore is one that we mostly take totally for granted. Australian folk speech (q.v.) is alive with humorous words, phrases, similes and comic or witty constructions of all kinds — in sayings such as 'Silly as a hatful of arseholes', 'Thick as a brick end on', or 'If brains were made of elastic you wouldn't have enough to make a garter for a canary's leg', and so on. While such forms of humorous malice tend to look simply crude on the printed page, their deployment in everyday conversation is a fundamental facet of Australian life and laughter.

Some further forms of traditional folk humour still present in contemporary Australian society include the saucy wedding telegram, humorous autograph book verse, pranks, hoaxes and initiation customs of all kinds.

Despite changes in emphasis, location and genre, the underlying themes in Australian folk humour remain basically the same. One change that is worth commenting on, though, is the apparent demise of sectarian folk humour, particularly that which used to be prevalent between Roman Catholics and Protestants until about the 1960s. Religious jokes, of course, are still told, but the virulent emotions that motivated much sectarian folklore has dissipated. While the decline of such tensions is a welcome reflection on the level of tolerance within Australian society, it could be argued that sectarianism has, to a large extent, been replaced by folk humour that concentrates on ethnic rather than religious difference.

Examples of folk humour can be found in all of Bill Wannan's anthologies of Australian folklore, especially *Come in Spinner: A Treasury of Popular Australian Humour*, (1976, first published as *Fair Go, Spinner*, 1964), in Ron Edwards's *Fred's Crab and Other Bush Yarns* (1990) and in Bill Scott's *The Long and the Short and the Tall* (1985). The humour section of most bookshops will usually contain collections of folk humour — though rarely identified as such — both past and present.

Graham Seal

J

JACKIE BINDI-I: see **Humour**.

JACOBS, JOSEPH (1854–1916), born in Sydney of Jewish parents, was educated at Sydney University and later at Cambridge. He studied Judaism in Germany, eventually becoming a noted scholar of Jewish history and culture in America, to which he emigrated in 1900. Jacobs's significance for folklore scholarship derives from his activities in England between *c.* 1888 and 1895, during which time he edited the journal of the Folk-Lore Society, *Folklore*, and published a series of his own articles on folklore theory. At this time he also published a number of children's anthologies of international folk tales. In one of these collections, for which he is still noted in the history of children's literature, Jacobs published the texts of two 'English' tales heard in his Australian childhood. These were 'Jack and the Beanstalk' and 'Henny-Penny', both included in his 1890 book, *English Fairy Tales*. The extent to which Jacobs rewrote these texts for publication is unknown and the effects of a thirty-year recollection gap between hearing and publication must also be taken into account. Nevertheless, the texts of these two tales represent some of the earliest extant evidence for the existence of 'English' folklore in Australia and Joseph Jacobs has a considerable claim to being the first Australian folklorist.

REFERENCE: Gary Alan Fine, 'Joseph Jacobs: A Sociological Folklorist', *Folklore* 98, ii, 1987.

JONES, PERCY (1914–), as described in the sub-title of the biography written by Donald Cave (1988), was simultaneously priest, musician, teacher. He lived in Geelong, Victoria, until 1930 when he went to Rome to prepare for the priesthood and continue his musical studies. In 1934 Percy Jones left Rome for Ireland and attended All Hallows' Seminary in Dublin for two years, after which he returned to the Pontifical Institute for Sacred Music. He was ordained as a priest in 1937, and returned to Melbourne in 1939 where 'he was struck by the vitality and richness he found in Catholic Melbourne', particularly among the laity. Percy Jones became chaplain to the Therry dramatic society and the Paraclete Arts Group. Another of his activities among the Catholic laity

> which was to have ... lasting effects on the Australian public at large was his work with the Ladies of the Grail, an Institute of women who ... conducted what was called the Quest ... a residential course on all aspects of Catholic life for girls, a type of highly intellectual finishing school ... Catholic culture, art, drama, plays, music, folk-dancing all formed part of the Quest. (Cave, 1988, p. 29)

Percy Jones's involvement with Australian folk music began in the early 1940s when he was asked to assist the Ladies of the Grail with their musical programme, and began to search for Australian songs for the choir. He was aided by Gerry Waite, a journalist on the Melbourne *Sun* newspaper who offered to help through his regular column. Dr Jones said to Donald Cave:

> So he ran a column in the *Sun* which he called 'Songs of Yesteryear' and as they rolled in to him he'd publish whatever he wanted and then hand them over to me. In this way I gathered my original collection of Australian folk-songs ... I decided that I would have to go around collecting the tunes ... So I went around and copied down the music as people sang to me. I didn't get around to doing very many of them but the ones that I did get were very important.

Dr Jones mentions that 'Springtime it Brings on the Shearing' was given to him by 'an old chap in Kilmore' and 'Click Go the Shears' by a man in Sydney. He nominates as his main informant Mr O'Neil from Sandringham whose mother had 'owned a hotel on the other side of Bendigo [which] was popular with the shearers and particularly with the chaps who brought down the wool, the drovers and so on'.

According to Donald Cave, Dr Jones found that many Australians lacked interest in these collected songs until Burl Ives sang and recorded them when he came out to Australia on his first tour after the war. The *Burl Ives' Folio of Australian Folk Songs* (1953) included ten songs, 'collected and arranged by Dr Percy Jones':

Botany Bay
Click! Go the Shears
The Dying Stockman
A Nautical Yarn
The Old Bullock Dray
Oh! The Springtime it Brings on the Shearing
The Stockman's Last Bed
The Station Cook
Wild Rover No More
The Wild Colonial Boy

According to Dr Jones, Burl Ives's presentation of these Australian songs marked the beginning of renewed interest in Australian folk songs.

Percy Jones provided some of the first tunes available for Australian folk songs, tunes which he obtained with great difficulty by manual notation from his informants' singing. Dr Jones said that it sometimes took 'days and weeks' to obtain a single tune, and said about Mr O'Neil, 'the trouble was I couldn't stop him — he would go so fast — I'd ask him to repeat and he'd have to go back to the beginning again!'. Dr Jones could not recall any songs collected other than the ten in the Burl Ives song book and, disastrously, all Dr Jones's papers were lost in the Ash Wednesday bush fires which destroyed his home at Airey's Inlet in Victoria in 1983. Until his retirement Dr Percy Jones was Assistant Dean in the Faculty of Music at Melbourne University.

REFERENCE: Donald Cave, *Percy Jones: Priest, Musician, Teacher*, 1988.

K

Karagiozis (Greek shadow theatre) takes its name from its comic hero, an ugly, bald, barefoot hunchback with an irrepressible sense of humour and inexhaustible capacity for bluff and cunning. Karagiozis is a caricature of the levantine Greek who always just manages to survive, usually well thrashed by his dupes, particularly his Turkish overlords (the scene being set vaguely in a period of Ottoman domination). The centenary of the 'Hellenization of Karagiozis' was celebrated in Greece in 1991, but the tradition of this type of oriental shadow theatre in the Ottoman Empire (*Karagöz* = 'black-eye') dates back at least to the seventeenth century, and there are attestations of its performance on the Greek mainland — albeit by a Jewish puppeteer performing in Turkish — as early as 1809 and of performances in Greek from the 1840s. Exponents of Karagiozis have their own myth about the genesis of the tradition.

Karagiozis interacts with a stock cast of about a dozen comic or legendary/historical characters, each with an obligatory set of peculiarities of speech, behaviour and appearance and a different signature tune. Additional to the core cast and range of plots is the prerogative of each puppeteer according to his own tastes and talents, but the plays typically revolve around Karagiozis's ingenious attempts to satisfy his hunger by faking professional skills (ranging from doctor's qualifications to those of a dictator and even an astronaut) with the help, and ultimately at the expense, of his traditional associates: the servile Hatziavatis, Uncle George the obtuse shepherd, Stavros the cowardly spiv, Nionios the pseudo-aristocrat, among others. The scope for social satire and deflation of pomposity in such plays is considerable, and the fact that the Establishment against which Karagiozis rebels has Ottoman Turks anachronistically at its pinnacle, allows him to flout authority and expose social injustice openly, even under oppressive regimes. As George Ioannou has remarked, the theme of social inequality is evident from the moment the screen lights up from behind: a stark contrast between the two traditional pieces of fixed scenery, a luxurious seraglio on the right and Karagiozis's tumbledown shack on the left, is apparent to the audience.

Behind the screen, on to which the two-dimensional puppets are pressed, a single performer controls everything: he moves the main puppets, produces all the voices, switching tone, sex and dialect as each character requires, co-ordinates the production of sound effects by assistants, and extemporizes the dialogue and the plot within the traditional framework of set storylines and formulaic speech. Many puppeteers also cut and painted their own puppets and scenery. The considerable skills needed to stage this one-man show were traditionally learned by apprenticeship, sometimes of son to father, but Karagiozis and its practitioners were never highly esteemed in Greece until quite recently, probably because of the obviously oriental provenance of the genre (see also **Rebetika**) and the comparative brevity of its tradition. The commercial development of Karagiozis into a form of children's entertainment, via broadsheets and comics since the 1920s and recently on children's television, did not serve to enhance its claims to serious attention. Foreign interest in Karagiozis, starting with the pioneering studies of Roussel, may have helped to demarginalize this traditional art form and stimulate local enthusiasm which has today grown to the extent that the dwindling ranks of puppeteers now boast national celebrities who represent Greece at international cultural events, and EC grants have been awarded for the foundation of a school of Karagiozis puppetry. Moreover, the Greek and Cypriot ministries of Culture/Education have sponsored the visits of Karagiozis puppeteers to centres of the Greek diaspora, as part of their endeavour to sustain the cultural identity of expatriate Greeks and Greek-Cypriots: thus the Cypriot puppeteer Andreas Idalias performed in Adelaide and Perth in 1991.

Karagiozis's transition to this status from the seedy cafés and makeshift children's theatres where it was traditionally performed has not met with unanimous approval in Greece, nor among Greeks in Australia, whose attitudes towards Karagiozis's migration into their midst have ranged from nostalgic affection for a childhood memory and support for anything related to popular Greek tradition, to dismissal of an unsophisticated/'anti-progressive'/un-Hellenic anachronism.

The tradition has come to Australia by various means: visits or settlement of puppeteers, and by importation of commercial audio/video recordings of performances from Greece, including those of Evgenios Spatharis and Thanasis Spyropoulos (tapes of the latter's performances being locally copied by Koala Video Productions of Melbourne).

Visits by Karagiozis puppeteers are reported to date from as early as the 1920s, but are not substantiated until 1938, when advertisements appeared in successive issues of the Melbourne newspaper *To Phos* for performances of the classic Karagiozis play *Alexander the Great* by one A. Kanaris (locally reputed to have jumped ship in Australia). Reports and documentation of performances of Karagiozis resume sporadically through the 1950s and 1960s. In 1957 Dimitris 'Mimaros' Aspiotis gave several performances in Melbourne, and around the same time Theo Poulos, an immigrant from the Peloponnese, is said to have performed Karagiozis in Sydney, Wollongong and Newcastle, and *Karagiozis the Doctor* was apparently staged by live actors in Sydney under the direction of Mandouridis at the same time. In 1967 Vangos Korfiatis visited Sydney for four months, also performing Karagiozis in Melbourne, Adelaide, Canberra, Wollongong and Newcastle, before returning prematurely to Greece, fearing for his family's safety in the wake of the April coup.

In the late 1970s, when multiculturalism was in the ascendant, two visits of particular significance occurred. Avraam Andonakos and 'Yiannaros' Mourelatos were brought out in December 1977 by a Melbourne entrepreneur to perform for two months, during which time they toured most of the eastern state capitals and provincial towns. Moreover, Andonakos disposed of his puppets to Zouganelis and Katsoulis (q.v.) before returning to Greece, thus facilitating their emergence as resident puppeteers in 1978 with a more Australian repertory. Mourelatos was to revisit Sydney twice in the 1980s, emphasizing the potential of Karagiozis for educational use. By this time Greek-Australian puppeteers began to acquire greater fame in their own right: Zouganelis toured Canada in 1983 and New Zealand in 1986, and both he and Katsoulis received Australia Council grants. Karagiozis was also well established by now on the Greek-Australian airwaves, and as a feature of Modern Greek studies: at the University of Melbourne Karagiozis was both a topic in the Greek folklore course and a stock character in the dialogues for the beginners' Greek language course; Katsoulis and Zouganelis were making school visits, educational broadcasts and publications in Melbourne and Sydney; and high-school students were writing and performing Karagiozis plays. Karagiozis had also become popular with theatre groups, both educational (such as Polyglot Puppet Theatre at the Early Childhood Development Centre, Kew, Victoria) and ethnic (such as Yefira, which staged Skourtis's *Karagiozis Almost a Vizier* during the 1977 Melbourne Greek Festival; Paranga, which acted Themelis's *The Journey* during the 1986 Melbourne Dimitria Festival; Theatro Technis Afstralias, which performed Moutafidis's *Karagiozis in Australia* in Sydney in November 1986 and in Melbourne in September 1988, and Kalamara's *Karagiozis Strikes it Rich* in Sydney and Canberra in 1987). Kalamara's *Karagiozis the Interpreter* was workshopped in English by John Saunders of the Western Australia Theatre Company and given a public preview in Perth in 1989; the same director staged Kalamara's *Karagiozis Down Under* in Perth in February 1992.

In the 1980s Karagiozis also made an impact on the creative literature published by Greek-Australians in the form of plays (e.g. Yiota Krili-Kevans, *The Time Has Come* [in Greek], 1987), verse (e.g. Elias Planguetis' *Karagiozis the Scholar* [in Greek], 1984) and prose fiction (e.g. George Papaellinas's stories in English about the Mavromatis [= 'black-eye'] family in *Ikons*, 1986); Karagiozis also inspired the Greek-Australian plastic arts (e.g. a 1990 canvas by Nikos Soulakis of Melbourne).

The symbolism of Karagiozis thus seems assured of an abiding place in multicultural Australia. As for live performance though, the acquisition of the departing Katsoulis's puppets by the Museum of Victoria may have been an ill omen for the future prospects for Karagiozis's survival in Australia, there being no obvious successors to the remaining resident puppeteer, Zouganelis (q.v.).

REFERENCES: S. Gauntlett, 'Karagiozis Almost an Aussie' [in Greek], *Chroniko Melbourne* 1, 1979; G. Kanarakis, *Greek Voices in Australia*, 1987; Y. Marinos, 'A Nation of Karagiozises?' [in Greek], *To Vema*, 29 December 1991; S. Spatharis, *Behind the White Screen*, 1976.

Anna Chatzinikolaou

KATSOULIS, DIMITRIS (1925–), a leading exponent of Karagiozis (q.v.), Greek shadow theatre in Melbourne, was born in Paramythia, Epirus, and emigrated to Australia in 1973. He lived in Melbourne from 1974 until 1991, when he returned to Greece permanently. An actor by profession, having undertaken film and drama studies in Athens in the 1950s, he toured the Greek provinces with professional theatre troupes and worked in review theatre and Katrakis' Greek Folk Theatre in Athens. He also participated in a number of films, Greek television serials and radio sketches both as an actor and as a technician. In Melbourne he co-founded the short-lived Children's Theatre and in 1974 the Professional Theatre of Greeks in Australia. During his stay in Melbourne he staged a number of plays and reviews, including his own work, *A Community without Malice*, in July 1991.

His career in shadow theatre began at the age of fifteen, when he became assistant to 'Vasilaros' Andrikopoulos in Agrinio for the summer seasons, but he learned the finer points of the art of Karagiozis from Dimitris Mollas (son of the renowned master puppeteer Andonis Mollas), whom he met in Athens in 1950; he thus claims to trace his pedigree as a puppeteer via Mollas *père* to Yiannis Roulias, one of the pioneers of the Greek shadow theatre. He began performing Karagiozis in Melbourne in October 1978 with his sons and his wife as assistants, after which he gave numerous performances all over Melbourne, in Greek schools, community celebrations and at multicultural festivals both in Melbourne and Adelaide. He also took Karagiozis to country towns, including Mildura and Renmark, but stopped performing regularly in 1986 because his puppets were worn out and he could not afford to replace them. In 1989–90 he acquired new puppets from Greece with the help of the Melbourne *Kentro Technis* and other bodies.

His repertoire mostly comprised well-known and much-recorded traditional plays such as *Karagiozis the Doctor/the Servant/the Letter-Writer, The Engagement of Karagiozis, Alexander the Great and the Accursed Serpent, The Three Questions and the Lie Told to the Pasha*, as well as *The Swing* (first performed by his master Dimitris Mollas). These plays were traditional both in content and form, and only sporadically would he insert brief references to the Greek-Australian context and colour his language with characteristically Greek-Australian idiom.

Dimitris Katsoulis confessed to seeing his target audience as comprising mainly children and their mothers, and began deliberately tailoring his Karagiozis to his perception of their needs in 1979. He started composing short radio dialogues revolving around typical experiences of Greeks in Australia and bearing little resemblance to the traditional Karagiozis, apart from use of traditional characters, no more than two or three at a time, and occasionally interspersed with non-traditional voices such as Santa Claus or an Australian called John. They were broadcast initially in the ABC educational series 'Akouste' (May 1979 to September 1981) as self-contained five-minute segments, opening and closing with the traditional Karagiozis overture and finale music. The traditionally amoral Karagiozis and his children were often transformed into unusually sententious characters, moralizing on an array of issues from proper education and upbringing to career choices, pollution, and the contrast between sound traditional Greek mores and new-fangled disco/rock/television culture rampant among Greek-Australian youth. Nineteen of these dialogues were published in Greek in 1983 by the Curriculum Branch of the Victorian Education Department, and one (*The Dowry*) in an English translation. The dialogues were revived and elaborated for a regular Friday-evening slot in the 3EA Greek programme for children (July 1986 to December 1987); students at the erstwhile Phillip Institute of Technology have translated seven of them into English with a view to publication. He also published moralistic and satirical texts and metrical letters on various topics in local Greek magazines and newspapers, often using Karagiozis as his mouthpiece.

Dimitris Katsoulis gave workshops on Karagiozis at schools and tertiary institutions all over the Melbourne metropolitan area, encouraging young Greek-Australians to make their own Karagiozis puppets and perform with them, as happened at Brunswick East High School in December 1990; but he frequently confessed that his experience over two decades persuaded him that Karagiozis could not survive as a living tradition in Australia. In recognition of his contribution to Australian multiculturalism he received Australia Council grants from 1980 to 1982, and was invited to exhibit his puppets at the National Gallery of Victoria in March 1981, in the foyer of the Melbourne Concert Hall during Piccolo Spoleto in September 1986, and in the foyer of the State Library of Victoria during National Heritage Week 1987. Finally his entire collection of puppets, scenery and stage accessories (including his working puppets of coloured acrylic, and others made in Greece by older puppeteers) was purchased by the Museum of Victoria for display in the proposed children's section of the new Southbank complex.

REFERENCE: G. Kanarakis, *Greek Voices in Australia*, 1987, pp. 400–2.

Anna Chatzinikolaou

KEESING, NANCY (1923–), was born in Sydney. She is a writer, poet, book reviewer, compiler and editor of anthologies and of books of poetry, biography, literary criticism and folklore. She has written two novels for children, *By Gravel and Gum* (1963) and *The Golden Dream* (1974). Her many literary and historical memberships include PEN, the English Association (NSW) and the Australian Society of Authors, whose journal, *The Australian Author*, she edited for several years. She is the foundation president of the ASA benevolent fund. Nancy Keesing was a member of the Literature Board of the Australia Council from 1973 to 1977, and a member of the Council from 1974 to 1977.

During the 1950s and 1960s Nancy Keesing collaborated with Douglas Stewart to produce the important anthologies *Australian Bush Ballads* (1955), *Old Bush Songs* (1957) and *Pacific Book of Bush Ballads* (1967), republished as *Bush Songs, Ballads and Other Verses* (1968) and as *Favourite Bush Ballads* (1977) (see **Stewart, Douglas**). Nancy Keesing is a member of the Royal Australian and Australian Jewish Historical Societies, and in 1978 edited *Shalom: Australian Jewish Short Stories*. Two of her books which are also of great interest to folklorists are *Lily on the Dustbin: Slang of Australian Women and Families* (1982) (see **Women and folklore**) and *Just Look Out the Window* (1985) which discusses 'superstitions, odd beliefs and possibly the truth about the weather and your fortune'. Both *Lily* and *Window* were the result of extensive independent research including the innovative use of talkback radio, and the latter book is notable for its inclusion of superstitions from Australian residents of non-English-speaking origin. Some of Nancy Keesing's other publications of folkloric interest include *Gold Fever* (1967), reissued as *A History of the Australian Gold Rushes* (1976) and *Gold Fever (again)* (1989). *The Kelly Gang* (1975, 1980) is an illustrated history text using primary source material including newspaper reports and quotations from Douglas Stewart's play *Ned Kelly*.

Nancy Keesing was appointed a Member of the Order of Australia in 1979 for her services to Australian literature.

KELLY, NED When an Australian calls someone 'as game as Ned Kelly' they offer a vernacular compliment that indicates admiration for the courage of the particular person being addressed. Australian folk speech also includes the statement 'Ned Kelly's not dead', addressed to someone thought to be robbing or in some way getting the better of the speaker. These two apparently trivial conversational items suggest the important cultural tradition surrounding the enigmatic image of Ned Kelly. They also articulate the ambivalent status of this bushranger hero in Australian society, not only during the period of the Kellys' depredations but, increasingly, ever since. Fact, fiction and folklore have circulated through Australian society for over one hundred years to create a cultural tradition that still generates heated support or condemnation in conversation, in books, films and other cultural domains.

A number of bushrangers had attracted extensive popular sympathy and active support from members of their own social groups before the Kellys. The reasons for such popularity are complex and best compared with the similar repute of some English and Irish highwaymen, and American outlaws like Jesse James. All these figures originated within certain social groups which, for a variety of reasons, felt themselves to be economically or politically disadvantaged by the government, the police, the wealthy, or a combination of all three. One consequence of this has been the widespread characterization of such outlaws as men who rob the rich to benefit the poor, Robin Hood style. In Australia, Jack Donohoe symbolized the vicarious revenge of the convicts upon their frequently brutal gaolers. And it was the dissatisfaction of the bush workers and farmers of western New South Wales with the wealthier squatters and the administration in general that led to their support of Frank Gardiner and Ben Hall. A similar dissatisfaction in north-eastern Victoria, compounded with a smouldering Irish hatred of English law, impelled the activities and the popularity of Ned Kelly over a decade later.

Edward, the first-born son of Ellen and James ('Red') Kelly, grew up in the hothouse atmosphere of a number of clan-like Irish-Australian families — the Kellys, the Quinns and the Lloyds. Each of these families had their own extensive histories of trouble with the Victorian and other police forces, surviving as they did through a mixture of legal pastoral activities and stock-stealing, or 'duffing' as this Australian version of poaching was colloquially known. The Kellys and their relations were by no means the sole transgressors in this area. In fact, this was the normal mode of existence for most free settlers at that time, the distinction between stock that had 'strayed' across usually unfenced boundaries and that which had been stolen being a fine one to make.

By 1871, at the age of sixteen, Ned Kelly already had numerous experiences with the law behind him, and had served one gaol sentence. In that year he was convicted of receiving a stolen horse and given three years in Melbourne's tough Pentridge gaol. He went into prison a high-spirited, 'flash' youth and came out a hard, bitter man in February 1874. Nevertheless, he seems to have gone straight for a time, working as a timber-getter in the Wombat Ranges and keeping out of trouble — until September 1877. Ned was arrested for drunkenness and on the way to the courthouse attempted to escape. The ensuing brawl with four policemen and a local shoemaker is notable only for the fact that two of those policemen, Constables Fitzpatrick and Lonigan, were to play small but significant roles in the coming Kelly drama.

Fitzpatrick was the first to make an entrance. Seven months after the fight with Ned, the constable, probably drunk, rode up to the Kelly homestead near Greta, alone and against orders, to arrest Ned's younger brother, Dan, on a charge of horse-stealing. The truth of what occurred then will never be known, but Fitzpatrick later claimed to have been assaulted by the Kellys, including Ned and Mrs Kelly. The family claimed that Fitzpatrick had tried to molest one of the daughters and that their actions had been justified. Six months after, Judge Redmond Barry did not agree and sent Mrs Kelly to gaol for three years, saying that he would have given Ned and Dan fifteen years apiece, if they could have been found. Of course, they could not be found and were safely hidden in the rugged Wombat Ranges, accompanied by two other young friends, Joe Byrne and Steve Hart.

In October 1878 a party of four policemen was sent into the Ranges to hunt the Kellys down. In charge was Sergeant Kennedy, a good bushman and a crack shot. He was aided by Constables Scanlon, McIntyre and Lonigan, the same Lonigan who had fought with Ned the year before. All these men had been hand-picked for their bushcraft and general police aptitude, and they intended to get the Kellys. On the night of 25 October they camped along the edge of a creek known as 'Stringybark'. The following evening the four-strong Kelly gang bailed up Lonigan and McIntyre, who were minding the camp while Kennedy and Scanlon patrolled the bush in search of the outlaws. McIntyre surrendered immediately, saving his life; but Lonigan was brave and foolish enough to clutch at his revolver. Ned Kelly shot him dead. On their return to the camp, Kennedy and Scanlon were called upon to surrender, but they resisted too, and were both killed in the ensuing gunfight. During the fighting McIntyre managed to clamber on to a stray horse and ride for his life.

The Melbourne and provincial newspapers reacted with revulsion to the shocking news that McIntyre eventually brought, enabling the Victorian parliament to rush through an adaptation of earlier New South Wales bushranging legislation called the Felon's Apprehension Act. The Victorian law, known as the Outlawry Act, rendered those persons pronounced outlaws totally outside the law. All rights and property were forfeited and the outlaw was liable to be killed on sight by any citizen. In addition, harbourers and sympathizers were liable to fifteen years' imprisonment with hard labour and the loss of all their goods.

Popular reaction, however, was somewhat different. The ballad titled 'Stringybark Creek' was sung throughout Victoria. It treats the murders with dark humour and implicit sympathy for the bushrangers.

Less than six weeks after Stringybark Creek the Kellys struck again. This time they robbed the bank at Euroa, a busy town about 100 miles north of Melbourne. The bushrangers escaped with around £2000 in gold and cash. Ned also stole deeds and mortgages held in the bank safe, an action that endeared him to the struggling selectors of north-eastern Victoria, most of whom saw the banks as 'poor-man crushers', as Ned himself was to describe them.

Acting upon false information intentionally supplied by one of the Kellys' 'bush telegraphs', or informants, the police went looking for the gang across the border in New South Wales. Meanwhile, back in 'Kelly country', the bushrangers divided up most of the Euroa loot between relatives and sympathizers. Over the next few months, many previously impoverished selectors managed to pay off their debts, usually with crisp, new banknotes.

Frustrated in their attempts to capture the outlaws, the police arrested a score of sympathizers and confined them in Beechworth gaol for periods of up to three months without trial and without evidence against them. This misguided manoeuvre made the police even more unpopular in the district, as many of the prisoners missed that year's harvest, causing severe hardship for many families.

At about the same time the reward for the Kellys was increased from £2000 to a total of £4000, an enormous sum of money at that period, but this had no effect upon the loyalty of the sympathizers either, as another of the contemporary Kelly songs points out:

Oh, Paddy dear, did you hear the news that's
going round?
On the head of bold Ned Kelly they have placed
a thousand pounds:
For Steve Hart and Dan Kelly five hundred
more they'll give,
But if the sum were doubled, sure, the Kelly
boys would live.

After the Kelly's next outrage the sum was indeed doubled.

On 5 February 1879, the gang appeared at Jerilderie, forty-six miles across the New South Wales border, where they locked the two astounded local policemen in their own cells. The Kellys spent that night and most of the next day in the town, masquerading as police officers in their stolen police uniforms. That afternoon they occupied the bar of the Royal Mail Hotel, handily adjacent to the bank, which they then robbed. Another £2000 haul was made and mortgages were burned to the accompaniment of cheers from the crowd held hostage in the hotel. Everyone was treated to drinks and a speech from Ned about the various injustices he had suffered at the hands of the police, the government and the squatters. More importantly, he left with one of the bank tellers a 10 000-word statement that came to be known as the Jerilderie Letter.

This fascinating document, now known only in a copy, catalogues the complaints and grievances of Ned Kelly and his friends, and also gives us an insight into the motives and attitudes behind their actions. Among other things, the letter complains of discrimination against free selectors and small farmers like the Kellys, claiming that the administration was working hand-in-glove with the wealthy squatters against the poor. Ned Kelly's opinion of the Victorian police is worth quoting, both for its historical value and its memorable invective. According to Ned, the police were 'a parcel of big ugly fat-necked wombat headed big bellied magpie legged narrow hipped splay footed sons of Irish Bailiffs, or English landlords'. The letter ends with a stern warning for the rich to be generous to the poor and not to oppress them:

> I give fair warning to all those who has reason to fear me to sell out and give £10 out of every hundred towards the widow and orphan fund and do not attempt to reside in Victoria but as short a time as possible after reading this notice, neglect this and abide by the consequences, which shall be worse than the rust in the wheat in Victoria or the druth of a dry season to the grasshoppers in New South Wales I do not wish to give the order full force without giving timely warning, but I am a widows son outlawed and my orders must be obeyed.

In an earlier letter, written just before the raids on Euroa, Ned had cautioned his readers to 'remember your railroads'. The full implication of this mysterious warning became apparent on Sunday 27 June 1880. The cluster of buildings and tents surrounding a railway station between Benalla and Wangaratta was known as Glenrowan. It fell to the bushrangers as easily as Euroa and Jerilderie. But this time they had not come to rob a bank, they had something more ambitious in mind.

The night before, Dan Kelly and Joe Byrne had 'executed' a one-time companion named Aaron Sherritt. Sherritt was apparently pursuing the dangerous game of double agent, playing the police off against the Kellys. But his murder also had another motive, to lure the bulk of the special district police force on to a train that would have to pass through Glenrowan on its way to the scene of the murder in Kelly country. The bushrangers planned to wreck this train and pick off the survivors, particularly the Aboriginal 'black-trackers' who had several times brought the police a little too close to the Kellys for comfort. Exactly what the gang intended to do after this massacre remains something of a mystery. It has been said that they merely aimed to rob as many unprotected banks as possible; others believe that their plans were far more enterprising, involving an insurrection to establish a Republic of North-Eastern Victoria. Whatever the Kellys had in mind, they were well prepared for a hard fight. During the months before the attack on Glenrowan, ploughshares and quantities of cast iron had been disappearing throughout Kelly country. The reason for these unusual thefts became plain when the bushrangers herded most of Glenrowan's small population into Jones's Hotel that Sunday. In the back room were four rough suits of armour, consisting of back- and breastplates and an adjustable metal apron to protect the groin of the wearer. Each suit weighed about eighty pounds and there was one metal helmet, with eye-slits and a visor, weighing about sixteen pounds. Ned Kelly was the only member of the gang strong enough to wear both armour and helmet and still manage to handle a gun.

About ten o'clock that night, after a round of singing, dancing and drinking with the crowd in the hotel, Ned allowed a few prisoners to go home because the police train had not arrived as early as expected. This blunder ensured the failure of the bushrangers' plot. One of the freed prisoners, the Glenrowan schoolmaster, Thomas Curnow, walked along the railway track and warned the police train just outside Glenrowan. Hearing the train stop outside the town, the bushrangers realized what had happened, buckled on their armour, and stood in front of the hotel to meet the police charge that very soon came. After a lengthy exchange of shots, Ned Kelly and Joe Byrne were both wounded, and the clumsiness of their armour, together with the intensely painful bruises caused whenever a bullet smashed into the metal, had become apparent. Ned lumbered into the bush to reload his revolver and fainted from loss of blood. At about the same time, Joe Byrne was killed by a stray bullet that splintered through the wooden hotel wall.

The hotel was still full of prisoners, a fact that did not discourage the police from raking the building with gunfire. A young boy and an old man were both wounded inside the hotel and a woman with a baby in her arms and her family in tow was frustrated three times in her attempts to escape by the refusal of the police to cease firing. She was finally helped to safety by the gallantry of a bystander who braved the gunfire to rescue her, though one of her children was wounded.

Shortly after this, Ned Kelly recovered consciousness and came crashing out of the bush, firing at the police from the safety of his armour. He was finally brought down by a shotgun blast to the legs and taken into custody.

The police then sent to Melbourne for a field-gun to demolish the weatherboard hotel and the two bushrangers left inside, Dan Kelly and Steve Hart. Long before the gun arrived, the prisoners were all released and the police set the hotel alight. A Catholic priest amongst the 500 sightseers who had gathered at the railway station rushed into the burning building. He found the bodies of Dan Kelly and Steve Hart, and rescued the badly wounded old man who had been forgotten by the prisoners in their rush to escape.

The sightseers waited for the blaze to subside and then hunted for souvenirs. The relatives of the dead bushrangers waited to claim the charred bodies for burial. Ned Kelly was taken to Melbourne, where he rapidly recovered from his thirty wounds and stood trial in front of the same judge who had sentenced his mother two years before. Not surprisingly, the verdict was 'guilty' and Edward Kelly was sentenced to hang. Strong campaigning to have the sentence commuted was unsuccessful, and at ten o'clock in the morning of Thursday 11 November 1880 Ned Kelly dropped through the gallows

The game of two-up being played by Australian soldiers during the First World War. (Australian War Memorial)

Improvised cricket beside the grape-drying racks, Mildura, Vic. (Photograph, Keith McKenry)

Eight-hour demonstration picnic at Ivanhoe Park, Manly Beach, NSW, from the *Illustrated Sydney News*, 19 March 1872. (National Library of Australia)

Celebrating Christmas Day on Brighton Beach, Port Phillip Bay, Vic., from the *Illustrated Australian News*, 30 December 1874. (National Library of Australia)

Only the bathing 'cossies' have changed. Bondi Beach, NSW, 1950s. (National Library of Australia)

Bocce players in the Edinburgh Gardens, Fitzroy, Vic., 1986. (Photograph, Elizabeth Gilliam)

Cricket on the banks of the Murray River near Mildura, Vic. Carmel Brown faces up to a ball from her sister Frances Fragomeni during a family picnic, December 1986. (Photograph, Keith McKenry)

trapdoor and into legend. According to folklore, his last words were 'such is life'.

After Ned's death, the stories and songs about the Kellys proliferated. One tradition had it that Dan Kelly and Steve Hart did not perish in the final conflagration at Glenrowan but escaped to South Africa where they fought in the Boer War — on the Boers' side. As late as the 1930s it was not uncommon to hear of men who claimed to be Dan Kelly, or knew someone who was. Songs were written and sung portraying the bushrangers as heroes and the friends of the poor, and vilifying the police (see **Folk songs**). One typical example is 'Kelly was Their Captain', which ends:

It was at the Glenrowan station where the conflict raged severe,
And more than fifty policemen on the scene then did appear.
No credit to their bravery, no credit to their name,
Ned Kelly terrified them all and put their blood to shame.

Books, plays, songs and even Australia's earliest feature-length film, dealt with the Kelly story in varying shades of fact, fiction and folklore, and continue to do so today. Ned Kelly has become a thriving media industry.

Even before the conflagration at Glenrowan, the Kelly gang was the subject of an early attempt at media exploitation. A Mansfield newspaper published a pamphlet entitled *The Kelly Gang, or the Outlaws of the Wombat Ranges*. This appeared in 1879 and contained an account of the bushrangers' exploits and a number of songs on the same theme, some of which are still sung today. Around about the same time a melodrama called *The Vultures of the Wombat Ranges* played to packed houses in Melbourne. Although this play was suppressed by the authorities (mainly because the bushrangers usually outwitted the police), Ned Kelly's head had hardly been severed from his body by the police surgeon when another melodrama about the bushrangers opened on the night of 11 November. Nothing is known about this piece of stage-craft other than the astounding fact that one of Ned's sisters, Kate, actually took a role in the play at its premiere. Not surprisingly, Kate was virtually ostracized by the rest of her family. She later married and moved to New South Wales where she eventually committed suicide.

Between 1880 and the turn of the century the Kelly story continued to interest writers, dramatists and songmakers, and in 1906 the makers of the country's first full-length feature film simply called their production *The Story of the Kelly Gang*. It was a box-office sensation and inspired numerous later film treatments, all of which helped to ensure Ned Kelly's eventual elevation to the status of a national figure in the public mind. And just in case the public mind proved forgetful, the publishers of popular literature provided a continual reminder in the steady stream of books and articles about the Kellys that appeared in the first two decades of the century. Reporters visited Kelly country to interview 'those who were there'. An old reprobate named Jack Bradshaw, the self-styled 'last of the bushrangers', wrote books in which he claimed to have been on intimate terms with the Kellys — as well as every other bushranger of note since the 1860s! These were just some of the varied media responses that the Kelly legend stimulated during this period.

With the development of commercial radio and the consequent importation of American country music into Australia during the late 1920s and early 1930s, Ned Kelly soon came to feature prominently in the new media of radio broadcasting and recording. Numerous songs celebrating Ned's deeds appeared on record at this time, and a few, such as 'The Ned Kelly Song' and 'Poor Ned Kelly', have caught the popular fancy and live on in oral tradition.

Since 1940 there have been more plays, novels, poems, films and even a musical about Ned Kelly. Most recently, a local television network has produced a multi-million dollar series about the bushranger for nationwide screening.

See also: **Bushrangers**.

REFERENCES: J. McQuilton, *The Kelly*

Outbreak, 1979; G. Seal, *Ned Kelly in Popular Tradition*, 1980; J. Moloney, *I am Ned Kelly*, 1980; H. Phillips, *The Trial of Ned Kelly*, 1987.

Graham Seal

Koorie languages and folk speech When one researches the history of colonization and slavery throughout the world, one finds that the first step in the process of disempowerment is to de-identify the target population by ignoring their names and applying general nominal labels such as 'indians, aborigines, blacks or natives' to those with non-white skin colouring. The British colonization of Australia was no exception — the names of our clans were neither sought nor used and ultimately the label 'Aboriginal', short for 'aboriginal native', was enshrined in legislation in 1911. Our people had no say in that matter. However, in 1988 (the bicentennial year of British colonization) many decided to discard the label, replacing it with 'Koorie' which means 'our people' and which was a password on the trade routes. It is used in the national sense and replaces only 'Aborigine/al', not the names of specific groups such as Murri, Nyunga, Yolngu, Nungga, etc.

At the time the British arrived in Australia, it is estimated that 230 distinct languages, with up to 600 dialects, were being spoken throughout the continent. It is a pity that the complexity and world view which makes them so linguistically valuable led the newcomers to make many successful attempts at their annihilation. The grounds given for these attempts at linguistic genocide were that the languages were 'primitive', 'heathen', 'gibberish' and 'rubbish' languages. The reality was that monolingual English speakers found them difficult to master, hence in their own evangelical and economic interests they deemed it necessary to supplant them with English. Inherent in the rationalization, too, was the notion of linguistic imperialism — just as the English considered themselves superior to other peoples, they also considered that the language which they spoke was superior. As British settlement was initially concentrated around the eastern and southern coastal areas because of their suitability for European-type grazing and farming, languages spoken by people who lived far from those areas, or on land considered to be arid, escaped attacks. Later, however, when English-speaking populations adapted to the warmer climates and spread further north, and as mining exploration uncovered valuable minerals in the arid parts of the country, the remaining Australian languages also became endangered. In recent years, with the advent of satellites, the greatest threat has been English-language television, a cultural/linguistic nerve gas which affects most markedly the speech of the young, the very persons needed to transmit their languages into the future.

Research to date indicates that the Australian languages belong to one family, i.e. all languages are descended from a parent or ancestral language spoken thousands of years earlier. However, over time, just as most of the languages of Europe, which are descended from Indo-European, have altered, the Australian languages have undergone considerable change, to the extent that people from widely separated speech communities are unable to understand one another. A number of attempts have been made to link these languages with languages outside the continent, but all have proved fruitless.

Language plays a unique role (when compared with European languages) in the lives of our people and communities that are still linguistically secure. Unlike other speech communities where knowledge of *all* of the language is available to each person, in Koorie societies, because knowledge is restricted to persons in certain categories and life stages, an individual has gaps in that knowledge. In most groups there are secret, taboo and special languages which are known and may only be spoken by those involved in special ceremonies, e.g., religious ceremonies. Both males and females in our communities have religious autonomy, i.e. ceremonies are conducted by each gender group, the other group not being permitted to attend or participate in special ceremonies. During these ceremonies a special form of secret-sacred language is used which, in addition to being restricted to one gender, may also be restricted to persons of certain ages. In some, but not all groups, there is a form of language or style of communication which is the only one which may be used with a relative in a taboo or avoidance relationship. The linguistic rule may prohibit certain topics, particular words, restrict communication to the special form, or limit communication to speech through a third party. In other communities the taboo may prohibit persons in certain categories from face-to-face communication even through someone else. A special form of language which is common to most speech communities is one known as 'the mother-in-law' language; this relationship can be one of total avoidance or restricted communicative interaction. In most cases, persons who are *potentially*, as well as those who are actually, in this relationship must conform to the special linguistic rules. In various groups there may be languages used by people in specific situations, such as Anitji, the language used by Pitjantjatjara mothers of adolescent sons. In contrast to the secret-sacred speech forms, styles used with persons in taboo or avoidance relationships are known to the whole community. As with other linguistic groups, speech and song styles differ for topic of narrative or song.

If all the language knowledge is to be retained in a speech community, the linguistic group must consist of persons whose life stages have placed them in the categories of speakers of all the special linguistic styles. In such a case, an elderly man and an elderly woman could, between them, hold all the knowledge. So that any language can remain viable, however, a speech community needs a number of younger people of both genders in various life stages to whom the knowledge can be passed.

An unusual linguistic practice which leads to loss of words within all Australian language speech communities is the taboo placed on the speaking of a person's name after his or her death. The period during which the taboo may last is usually several years, and then the name may only be used by someone closely associated with the deceased. In other instances there is an infinite taboo. In both cases this results in changes in the language. To overcome the problem of the word loss, most speech communities borrow a word from a neighbouring language; others fill the gap in the vocabulary by replacing it with a word with a similar meaning which was less frequently used prior to the death.

Meanings of words, or semantic distinctions, in Australian languages can be significantly different from those in European languages. This is undoubtedly due to differences in lifestyle and the emphases on what is important to the culture. For example, Bandjalang, spoken on the north coast of New South Wales, has special noun markers (small words put in front of names of people or things) for invisibility. One form is to indicate whether a person is invisible because he or she *has not yet arrived*, and another to indicate that the person is invisible because he or she *has been and gone*. These in turn are related to distance and number. Gugu Ngantjara, spoken in North Queensland, has special marking on words for 'giving', to indicate whether the item being given has left the hand of the giver, is in the process of being passed from the hand of the giver to the hand of the receiver, or has just reached the hand of the receiver.

Names carry socially significant meanings in all Koorie groups. These include clan, section, kin, personal and place names. Some clan names carry a word of the language in their title, e.g. Gabbi Gabbi (spoken in Southern Queensland) means 'the people who say *gabbi* for "no"'; similarly Yota Yota (spoken in northern Victoria) means 'the people who say *yota* for "no"'. Dialects of Gugu Ngantjara include the verb 'go' in their clan names, e.g. Gugu Uwanh means 'the language which says *uwanh* for "go"' (*gugu* means 'language').

Section names are related to lineage. In our society both one's patri-line (descent through males) and one's matri-line (descent through females) designate special roles and rights of inheritance. Children belong to different sections from their parents and must marry out of their own section. Groups may differ in the number of sections within their clan — most have four sections but others can have six or eight. The section, sub-section or moiety system operating within a community plays an important part in dictating the linguistic rules of that community. This knowledge is important, not only when one addresses people, but also in discussions about speech community members, or participation in the social life of the community. One's section name (sometimes also called 'skin' name), tells the rest of the community:

- how one must behave towards others, both within and outside the section and how they may respond;
- what land is under one's custodianship;
- what food is taboo and what is not;
- which group one may marry into, and which group one may *not* marry into;
- one's clan inherited rights (e.g. to paint certain pictures, sing and dance specific repertoires);
- which group one must support in conflicts.

As kinship is the pivot of Koorie society, it is not surprising that kin names also play an extremely important role in dictating lifestyles and language usage. One must know the names of the kin relationship and sections of *all* community members, if one is to speak the language accurately. If an outsider lives in a clan community, unless he or she is 'adopted' by someone and is given a kinship name in relation to the adopter (and therefore every other community member), that person remains a nonentity in the community and no one knows how to linguistically and socially interact with that person. Detailed knowledge of the system and those in it is important, as each person uses names for others from their own standpoint as ego. Another ego will use different names for the same people.

The English language is unable to cater for the needs of our people living in communities where lifestyle is dictated by Koorie mores. If a male were to be given English kinship terms only, he would not know who the women in the category of potential spouse were, and just as importantly, who the women in the taboo relationship were (these women are his potential mothers-in-law).

It is also usual practice for us to have several personal names by which we are addressed as well as others by which we are referred to. These may include kinship name; section name; a name describing a personal characteristic or habit; and a sacred name given at birth and which is known only to a few close relatives.

Place names in Australian languages usually embody a description of the surroundings or events which have occurred or do occur there. For example, Brewarrina, a town in mid-west New South Wales, has retained the Koorie name, which means, 'the place where acacias grow'. 'Mulli Mulli' in New South Wales means 'hills'. The naming of people and places by Creation Beings in our religious stories has not only sacred but secular significance. Names bind people to their land, which bestows upon them their identity and relationship to other clan groups.

Where marriage rules are exogamous, that is, they dictate that one must marry outside one's clan, children grow up speaking two languages — their mother's and their father's. Multilingualism is, therefore, quite common. In the Northern Territory there is evidence that our people learned Macassarese prior to the British arrival. Many Macassan words were adopted into those languages where speakers had had considerable contact with them. Our willingness to learn and use other languages is also reflected in languages spoken on the north-east coast of Queensland, where a number of Papuan words have been adopted into the languages.

Although the number of words common to linguistic groups in Australia is small (less than 100), the spread of some may have been due, on the one hand, to the interaction which the trade exchange necessitated. On the other hand, the need to communicate in trade negotiations may have aided the survival of some proto-Australian forms. Ceremonial exchange, which required people to travel considerable distances, undoubtedly also contributed to linguistic tolerance. For instance, the speakers of Bandjalang, spoken in the northern coastal area of New South Wales, walked to the Glasshouse Mountain in Queensland (a distance of approximately 216 kilometres) to join the Gabbi Gabbi peoples for the bunya-nut festivals when this fruit was ripe. These two groups did not share a linguistic boundary and, in fact, from what it known of the languages, they differed considerably from each other. Song styles as well as religious beliefs are also dissimilar. Gabbi Gabbi songs tell of the preparations which were made prior to the arrival of the visitors. A literal English translation (by the writer's mother, Ms E. Serico, who inherited the right to sing the song from Mr Warden Embry) from one of these songs is: 'On Dirijan (Bunya Mountain) I beat the bunya nuts and gave some to the visitors'.

The history of our people is passed on mainly in oral form — in stories of the events, as well as in songs to which new cycles are added as events occur. The longevity of history passed on in this form is attested in stories which obviously stem from before the last Ice Age — stories which tell of how the waters rose and cut off Tasmania from the mainland relate specifically to the melting of the ice. More recently the details of the wreck of the *Zuytdorp* in 1712, off the coast of Geraldton, was described 114 years later to the British after they had established a settlement on the Swan River.

In addition to verbal language, some groups have a complete repertoire of non-verbal communicative means — via hand signs and gestures. The need for silent communication is great when one is in the open and in danger, or in hunting shy animals, but it is also useful in communicating from afar, say from one side of a river to someone on the other side.

Frequently used written forms of communication are message sticks, which consist of symbols marked on a piece of wood for the dissemination of information from group to group and along the trade routes. Many paintings also consist of symbols which contain a history of a clan, family, or individual; tell the stories of the Creation Beings and religious laws; and in other ways are used to educate the young or transmit information. Toas are small sculptures made by Koories from wood and gypsum, often with symbolic natural objects or artefacts attached to the heads, carved natural objects, or painted designs on them. They indicate in symbols the paths taken by the Creation Beings. For messages of importance, sacred boards are sent from one group to another. After the Second World War this method was used to call a strike in twenty cattle stations across Western Australia. Another means of sending a message is the use of smoke from fires which is used to signal one's approach, warn of danger or let others know one has arrived at a particular point and is awaiting permission to enter another group's territory.

When the British arrived with linguistic terms which did not encompass our mores, gross misunderstandings between the peoples were inevitable. When they began to dismantle our languages by bans on their usage, not only did misunderstandings deepen, but the effect on the linguistically ordered world of our people was devastating.

Through contact with people from other language families, other forms of Australian languages have emerged. The first of these was Pidgin, which is the form of communication used in early contact between speakers of different languages. It is characterized by reduced syntax and simplicity of vocabulary. Use is usually limited to specific situations, e.g. trade, employment, etc. Through the increasing contact between European (and Chinese in the early days of Australian history) and our people living in northern parts of Australia, a form of Pidgin English developed, resembling other pidgins spoken in the Pacific. This resemblance is due to some of the early settlers' experience with some Pidgin before coming to Australia. The increased European contact and the fact that our people from different language backgrounds were compelled to live together had the effect that Pidgin became valuable for a while as a communication language. However, Pidgin is rarely used today, except perhaps by a few white Australians in an attempt to communicate with speakers of languages other than English. On the other hand another form of speech called Creole is widely used in the northern parts of Australia. Creoles, like Pidgins, derive from original contact situations, and are a blending of the contact languages, but unlike Pidgins, they are linguistically complete languages used by speakers whose first language usually is the Creole. Probably all languages in the world have gone through a process of creolization until they were given the official status of a language; a recent example is Afrikaans.

Australian and Torres Strait creoles exhibit the sound patterns of the Australian languages, modifying some English words to these principles. Creole languages are spoken at Roper River, Bamyili, Fitzroy Crossing, Cape York, and in some Torres Strait Islands. The estimated number of Creole speakers in Australia is approximately 34 000. Although all are subsumed under the name of Creole, the languages can differ considerably from one another. For example the type of Creole is influenced by the intensity and length of contact a speech community may have had with English. In the majority of instances many lexical items and English word orders are adopted; in most cases, however, English is adapted to the Australian phonological system. In almost all cases the Australian pronominal system, which is far more extensive than the English one, is used. Roper Creole takes its name from the Roper River, which runs into the Gulf of Carpentaria off the coast of Arnhem Land. The river forms part of the southern boundary of the Arnhem Land Reserve, which is within the boundary of the non-Pamanyungan languages. In the areas of the Roper River there were quite a number of clans, such as Jangman, Mangaarai, Jugal and Galwa, but there seems to be some disagreement about how the Roper Pidgin was established, whether by convergence of these various speech communities or by stockmen bringing the Creole from the Queensland canefields in the last century when the South Sea Island labourers were brought in.

Cape York Creole is spoken by Koories and some Torres Strait Islanders living in the northern Cape York Peninsula region of North Queensland; another creole is also spoken in the Torres Strait Islands. Some Australian languages of the area are those belonging to the Pama-Maric group (northern pama, north-eastern pama) and those of the Koko-Yao sub-group. Cape York Creole developed from Bislama which is considered a modern descendant of China-coast Pidgin English and one which may have been influenced by the Portuguese. Bislama became the lingua franca of the Northern canefields in the 1850s. It was brought into Queensland from the Solomons, New Hebrides, New Ireland, New Britain and other islands of the South Pacific and became known as the 'Kanaka Pidgin'. The language spread to the Koorie people of the north through social agencies such as church missions which brought about separation of children from parents. The prohibition on the speaking of Australian indigenous languages helped the spread of the Creole.

Comparisons between the Roper and Cape York creoles show that although they share many similarities, there also exist many differences. An example of marked difference can be seen in the third person plural pronoun, which in Roper is *alabad* and in Cape York is *dempala*. The other creoles also differ from one another.

The most widely spoken speech variety used by Koories throughout the continent is Koorie English (also known as Aboriginal English). The amount of influence from Australian languages differs regionally, from a form close to Creole along a continuum to one that is closer to English and is considered by many people to be a dialectal form of Standard Australian English (SAE).

Care must be taken however, not to make the mistake of believing that a Koorie English sentence, although containing the same words as an English one, has the same semantic interpretation. Many problems are caused for Koorie students when classroom teachers treat Koorie English as a *deficit* rather than different form of English language. In the law courts too, often Koories whose speech variety is Koorie English are disadvantaged, particularly in question–answer situations — a person pleading not guilty in Koorie English can be interpreted as having pleaded guilty in Standard Australian English.

Koorie English is characterized by the Australian language phonological influence on English words such as absence of the initial 'h' sound; stress patterns; some lexical items; a different type of non-verbal behaviour; semantic interpretations, and in forms closer to creoles by syntax. Similarly the Koorie speech communities are characterized by great linguistic diversity. The languages express the richness of a world view that is truly Australian and encompass the unique landscape and history of a people whose longevity of occupation of this land is beyond dispute. Creole and Australian language speakers are usually multilingual; however, this linguistic diversity is threatened. If language maintenance programmes are not implemented within communities, and if children are not encouraged within the school system to retain their own languages, a rich linguistic heritage will be lost to the world. It is important that the task of linguistic annihilation commenced by the English in 1788 is not successfully completed in the 2000s, or ever.

Eve Fesl

L

Larrikin Records was established in 1974 as an offshot of the Folkways Music store operated by Warren Fahey (q.v.). It had rather modest aims of releasing a handful of long-play recordings featuring traditional Australian music as the mainstream record industry had little interest in this area of music. The first recording was a collection of gold- and coal-mining songs produced by Warren Fahey (q.v.) and featuring revival folk performers: Dave De Hugard, Phyl Lobl (q.v.), Andy Saunders, Tony Suttor and Warren Fahey. Titled *Man of the Earth* the recording was assisted by a small grant from the Mining and Geology Museum of Sydney. The second album offered field recordings of Australian traditional singers and musicians as collected by Warren Fahey and was titled *Bush Traditions*. Eighteen years later the Larrikin catalogue included nearly 300 releases and covered most aspects of Australian music. Fahey now sees the label as a means of 'documenting the Australian sound, ranging from bird calls to pop music, Aboriginal music to folk song'. The company also acts as Australian distributor for many specialist folklore producers including Rounder, Folkways USA, Australian Institute of Aboriginal Studies (see **Australian Institute of Aboriginal and Torres Strait Islander Studies**), Shanachie and Real World. Through Larrikin many Australian folk artists such as Eric Bogle, The Bushwackers, Redgum, Bernard Bolan and Cathie O'Sullivan, received their first opportunity to record.

Larrikin has always seen itself as an educational vehicle and in this light has continually issued recordings of thematic folk music covering such areas as the Great Depression, sea songs, shearing songs, railway folklore, bush dance music, spoken word and field recordings. The company also operates a successful artist booking service that acts as booking agency, artist management and event management. In this area Larrikin has also co-ordinated tours of international artists including Ewan MacColl, Peggy Seeger, Joe Heany, Norman Kennedy, A. L. Lloyd (q.v.), Shirley Collins, Peter Bellamy and Stefan Grossman. It is through the efforts of Larrikin Records that the Australian record industry is both more Australian and more receptive to streams of traditional music.

Lasseter's Reef A fabulously rich reef of gold said to have been discovered somewhere near the borders of the Northern Territory (then part of South Australia) and Western Australia by Lewis Harold Bell Lasseter and a companion in 1897. Lasseter led an expedition to relocate the reef after its location was 'lost' in 1930. The expedition, as with many earlier and later attempts to locate the reef, was a failure, and Lasseter died in poorly-explained circumstances. In accordance with the folkloric processes surrounding notable and notorious figures, it was often said that Lasseter did not die and that it was another person's body that had been found. Lasseter was reportedly sighted numerous times in various places within and outside of Australia after his death. Other folkloric elements surrounding the story of the reef are similar to those found in lost treasure narratives throughout the world: a fabulously rich mine/reef/horde is found in an exotic and inaccessible location; its location is forgotten as the survivor(s) of the original party struggles back to civilization; there is a map; there are various 'mysterious' or unexplained circumstances surrounding attempts to relocate the treasure, often involving a 'curse' or similar taboo from the local indigenous people. Expeditions are still mounted regularly to search for Lasseter's Reef.

LAWSON, HENRY HERTZBERG (1867–1922), an important figure in Australian literature, is also noted for the number of his poetic works that have passed into folklore as songs or recitations or both. These include 'Ballad of the Drover', 'Freedom on the Wallaby', 'Andy's Gone With Cattle' and 'Taking His Chance', and probably the bawdy classic 'The Bastard from the Bush', to name only some. The extent to which Lawson's poems have been turned into widely circulating folk songs and verse has been demonstrated by the work of folklore collectors and also in Chris Kempster's recent compilation, *The Songs of Henry Lawson* (1989). The extent to which Lawson took already circulating items and tidied them up for literary consumption is, as with some of Paterson's work, still subject to debate among folklorists.

See also: **Folk poetry and recitation**; **Folk song**.

Leaf playing (or gumleaf playing) is not necessarily done with eucalypt leaves: the player often uses a variety of non-native leaves for ease of playing or tone, etc. Having said this it should be noted that the Australian championships are called the National Gumleaf Playing Championships, and a National Parks and Wildlife official inspects to ensure that all entrants play native leaves. The leaf was known as an instrument in many European as well as Pacific countries before the European settlement of Australia. Rumanian country people would play the pear and lilac leaf, a slip of birch bark and even a fish scale in a similar manner to that found in Australia, and it is reputed that a leaf dipped in Guinness was used to play tunes in Ireland. References to leaf playing in traditional Australian Aboriginal ceremonies are scant; some refer to spirit noises and some to children's games, yet all present references to a non-Aboriginal learning the leaf in Australia eventually trace back to an Aboriginal source. The tunes played give us no clue to the origin of leaf playing in Australia, as the main repertoire is now late nineteenth- and early twentieth-century tunes, sentimental ballads, and hymns. Leaf orchestras were frequently to be found on Aboriginal missions playing from church services, often in part or harmony.

The leaf is played in a number of ways, four of which are listed below. There are other styles, but they are less common.

- The leaf is held on the bottom lip with two fingers or a finger and a thumb. The upper part of the leaf lightly touches the upper lip. The lips are pursed, as in whistling, and air is blown gently across the leaf. The pitch is varied by varying the tension on the leaf with the fingers. Examples can be found on many audio recordings in the John Meredith Collection in the National Library of Australia. This style of playing is by far the most common and most popular.
- The leaf is folded in two with the folded edge uppermost. The leaf is fixed in place with the two fingers, but is held by the mouth. Air is blown across the top of the fold causing the lower leaf section to vibrate. Pitch is changed by varying the tension and shape of the mouth and associated facial muscles. It is possible to play this way 'no hands'. This meant the leaf could be played when on horseback while on watch with a herd of cattle, or, as was popular with many Aboriginal vaudeville acts, while playing an accompanying instrument such as banjo or guitar. This style of playing is from an Aboriginal stockman, and can be heard on a tape made by Peter Ellis, Shirley Andrews and Kevin Bradley in 1990, also held by the National Library of Australia.
- The third style is from a description and working model supplied by collector John Meredith, and was practised by his father.

A stick about as thick as a pencil is cut and the ends are trimmed squared. A slit between 25 and 50 mm long is made in the end of the stick. A leaf is cut and slid into the slit and is trimmed off around the edges with a pocket knife or other suitable instrument. Air is blown across the leaf with the split end of the stick held lightly between the lips. Pitch is varied by changed pressure on the split end of the stick.

• The fourth is from a description given to John Meredith. The leaf is folded into three and a slit is made. Air is blown gently into the slit and the inner leaf section vibrates. Pitch is varied in the same way as in the first method mentioned.

Kevin Bradley

Legends While often confused in contemporary usage on the one hand with the more conscience-shaping, or even cosmographic, term 'myth' of Aboriginal religion, and on the other with the more simple yarn or tale, 'legend' is still best defined as it was in the late nineteenth century by *The New English Dictionary* as 'an unauthentic or non-historical story, especially one handed down by tradition from early times and popularly regarded as historical'. As all societies had 'legendary periods', so did Australia have a social phase preceding more universal literacy. As in other (former colonial) territories in the twentieth century, in Australia the term 'legend' has been used more loosely for various stories about colourful or public people and their activities, in a way which may be said to be exaggerated or larger than life, and not necessarily completely credible, as well as being outside the accepted formal records of history, biography, or newspaper accounts. Legends of the settler and colonial periods tend to grow and to change, carrying into the present the (folk) traditions and perceptions of the past; they are updated or become less significant according to the evolving patterns of national life. Manning Clark, writing in *Daedalus* in 1985, in discussion of the evolving Australian concepts of 'heroes' (the core of any legend), granted that the Aboriginal inhabitants had long had mythic culture heroes (for example, Biame, Ulitarra), but said that the white settlers had ridiculed these and drawn their own popular ideals from Europe, preferring real people, such as Clive, Nelson, or Wellesley (Wellington), who had performed brave physical deeds. He then traced the evolution of white Australian heroes from the bushrangers (q.v.) and the explorers to the 'dinkum Aussie' bushmen, 'fearless and bold', then to common Australian soldiers and labourers and to the new Australian heroes emerging from politics, culture and major sports. Bill Wannan (q.v.) helps us to chart the evolution and pattern of actual Australian legends in his excellent compilation, *Legendary Australians* (1974). This volume is subtitled 'A Colonial Cavalcade of Adventurers, Eccentrics, Rogues, Ruffians, Heroines, Heroes, Hoaxers, Showmen, Pirates and Pioneers', a catalogue which makes the general point that colonial Australia was rich in colourful figures whose exploits, adventures or personalities made them talked about and notorious in the new country which seemed to bring out the best or the worst in them.

The convict period from 1788 to at least the 1820s was a time of brutal repression, the age of the lash and the gallows, a system which completely polarized the society. The earliest legends come from the victims' perspective, as in the enduring folk memory of Samuel Marsden as the 'Flogging parson', or John Price, Superintendent of Prisoners at Williamstown, Victoria, later murdered by a gang of convicts, who is supposed to have been the original of Marcus Clarke's Maurice Frere in his novel, *For the Term of His Natural Life* (1874). Others like John Larnach, overseer of 'Castle Forbes' in the upper Hunter Valley (in 1829), were not written up until later — and in southern New Zealand at that. The same is true of the explorer Eyre who, later in the West Indies, was deemed a legendary racist and murderer.

In this early phase the 'legendary qualities' highlighted were resistance, defiance, wariness of authority, and almost sole reliance on one's mates. These desirable characteristics are best found in the early ballads, as is made clear by Hugh Anderson in his *Colonial Ballads* (1965) or Russel Ward in *The Australian Legend* (1958). Their exemplars are largely anonymous, but the courage of these early heroes lives on in the resounding words attributed to 'The Wild Colonial Boy'. Thus folk yarns and balladry are the early sources of legend. As the percentage of native-born or 'currency' men and women increased, there came a dawning consciousness of the fact that they had their own bush or outback traditions, stories, attitudes and idiom.

The various patterns of folk image — many slow to emerge — may be clustered and illustrated sequentially thus: navigators, such as Dampier, Cook, Byron, Flinders; notorious convicts such as Barrington, Vaux or Wainwright; whalermen, from the convict-sympathetic Nantucket Americans to Johnny Jones or Ben Boyd; the 'bolters' or wild white men like William Buckley, Thomas Pamphlett or John Graham, or those convicts who seized such ships as *Cyprus*, *Caledonia* or *Frederick* to become temporary pirates; women 'lost' on the frontier, like Barbara Thompson or Eliza Fraser; repressive governors or convict commandants, Bligh, Price, Arthur or Darling; explorers such as Edmund Kennedy, Burke, Wills or Leichhardt; 'the ghosts' of the notorious, Dampier, Fisher, or Johnny Gilbert; those who tangled with the law in various ways, from Arthur Orton, the Tichborn claimant, to the hanged, such as Ned Kelly (q.v.), Frederick Deeming or Jimmie Governor.

Historical novels in the twentieth century have 'mythologized' the past, creating their own new legends of the hypersensitive Irish gentleman migrant in Henry Handel Richardson's *The Fortunes of Richard Mahony* (1917–1929), or of the style of family life in early Sydney in M. Barnard Eldershaw's *A House is Built* (1929), or the brutality of early pastoral development in early Queensland in Brian Penton's *Landtakers* (1934).

Traditional Aboriginal mythology, both sacred and ritualistic, has substantiated nearly all features of Aboriginal life, enforcing or explaining much of their activity. It is, in effect, a stylized version of the knowledge transmitted orally over untold generations, which lost much of its dynamic significance in the contact situation. Sacred myth has usually related to the wanderings and adventures of the ancestral beings, culture heroes, or creative beings. Modern Aboriginal legends have arisen in relation to contact encounters, especially opposition to white invasion, and massacres or violent expulsion from hunting areas and sacred sites. With the recent and increasing publication of such contact recollection, it is to be expected that Aboriginal legends of all types will be more readily known to all Australians.

Sacred or religious song-cycles of hunter-gatherers are now being written down and made more widely available in editions with translations and commentaries, so that their original religious function is likely to be replaced by one of enjoyment. In particular, some form of legend is likely to be recounted in relation to surviving stone arrangements, old mission stations and Aboriginal reserves, whether still occupied or not. More contemporary source areas are freedom fighters, artists, land rights leaders and other black role models.

While the younger non-Aboriginal Australians now reaching maturity might not seem to need many of the legends and traditionally defiant attitudes which sustained earlier generations, it is clear, not least from the post-1970 expansion of the Australian film industry, that earlier figures like Captain Starlight of *Robbery Under Arms*, Harry ('the Breaker') Morant, or 'The Man from Snowy River' still appeal to most social groups.

At mid-century the (recent) heroes were left-wing, victims of social or financial repression, swaggies, farmers, soldiers, those who had survived despite all odds, but the next generation wanted to know of the personal tragedy of refugees and migrants to this country. Thus socio-realism ceased to make new legends as much as the popular appeal of earlier history, or the determinants of religion, ethnicity or sport. At the same time Australian loneliness and the inward quest have been expressed in 'legends' of hope, doubt and fate in the work of Patrick White or Sidney Nolan. Thus, the second half of the twentieth century has been notable for the existence of two styles of legend: defiance or endurance legends of the populists who still want public affirmation that the immediate environment is under control and that destiny is of their own making; and search legends in which fate is an enigma and traditional identity has been transformed into myth. In these there is no longer the earlier schizophrenia about roots, but rather an understanding of the universal within the local, and the inner search based on both implicitly Christian and Aboriginal referents.

See also: **Folk tales**; **Heroes and heroines**; **Modern legends**.

J. S. Ryan

LLOYD, ALBERT LANCASTER (1908–82), has a considerable reputation as a translator, writer, performer and folklorist in England and in Europe. His work on British and eastern European folklore continues to be widely debated and his influence has been acknowledged by a number of British folklorists and performers. His most important folkloric works relevant to this article were *Come All Ye Bold Miners: Ballads and Songs of the Coalfields* (1952; revised and expanded 1978); *The Penguin Book of English Folk Songs* (edited with Ralph Vaughan Williams, 1959) and his major theoretical and historical statement, *Folk Song in England* (1967; revised editions 1969, 1975). As well, he published prolifically in folksong journals, magazines, and newspapers and was a broadcaster of some note. Lloyd also released a number of LP records that were extremely influential in the British and Australian folksong revival and which are also notable for their scholarly liner notes: these included *Leviathan* and *The Great Australian Legend*.

While Lloyd is a figure of acknowledged importance in British and European folklore scholarship, his significance to folklore research and collecting in Australia is controversial. A number of noted collectors of Australian folksong, dance and lore have publicly questioned the authenticity of the folksong material that Lloyd claimed to have collected while working as a farm labourer and a station rouseabout in New South Wales from 1925 to 1933. According to Dave Arthur's obituary (*Folk Song Journal* 4, 4, 1983), Lloyd claimed to have collected about 500 songs during this time, memorizing the tunes and writing the lyrics in exercise books. Some of the songs appeared on the LP *The Great Australian Legend* in rounded and refined versions that surprised most collectors of Australian folksong. Questions about the extent to which Lloyd reworked his collected materials have been persistently raised in Australian folklore journals and folk revival magazines.

Although Lloyd had a proven tendency to be less than complete in his information about the singing versions of the songs he collected, it would be unfair to suggest that he was a scholar of folksong at the time he obtained his Australian material. As he has said, he was simply interested in the songs as a singer and wrote them down for that purpose. It was not until a few years later that he began the serious study of folksong that would occupy the rest of his life. Until Lloyd's Australian manuscripts are competently examined, the extent, if any, of Lloyd's 'reworking' of his collected material must remain open to debate and conjecture. Regardless of the rights and wrongs of that debate, it must be recognized that Lloyd's Australian bush song collection is an extremely important 'missing link' in the annals of bush song, being almost the only known substantial collection of such material between the cessation of A. B. Paterson's active interest in the topic before the 1920s and the resurgence of interest and collection in the early 1950s. His evidence that such songs — in whatever versions — were being sung by bush workers in rural New South Wales during the 1920s and 1930s is of great value to folklorists and social historians interested in understanding the development and decline of the bush song.

In addition to the Topic recordings, Lloyd dealt with matters Australian in a number of contexts, including the popular journalism by which he appears to have supported himself, in a number of BBC radio broadcasts and in replies to Australian critics in the pages of *Australian Tradition* (q.v.) and *Overland*. A preliminary bibliography of Lloyd's extensive writings compiled by Dave Arthur may be found in I. Russell (ed.), *Singer, Song and Scholar* (1986), accompanied by some critically sympathetic judgements of Lloyd's folksong scholarship by a number of British folklorists.

Lloyd visited Australia again in the 1970s, this time as a lecturer and performer. His visit attracted good crowds, mainly from the Australian folk revival movement which at that time was heavily influenced by British migrants. While Lloyd's tour had an immediate impact on the performance and presentation of folksong, fuelled by the recent revised edition of *Folksong in England*, it does not appear to have had any lasting influence on folksong collection and scholarship here.

LOBL, PHYL (1937–), has been a public singer in the folk revival since 1962, but if family songs around the piano and Methodist hymns are folk music, then she has been a traditional singer since the age of four. She is well-known in Australia not only as a singer but also as a songwriter and folk activist. Between 1984 and 1987 she was a member of the Music Board of the Australia Council, its first representative of the folk arts. As a consequence of her membership of the Music Board, Phyl Lobl produced several innovative papers concerning the analysis and definition of folk music and the folk arts.

Phyl Lobl has performed with others on several records of Australian traditional material, namely *Bullockies Bushwhackers & Booze*, *Man of the Earth*, *On the Steps of the Dole Office Door*, *Game as Ned Kelly* and *Skeleton Flats* (tape). Her solo tapes and records include *Dark-Eyed Daughter* (EP), *On My Selection* (LP), *Broadmeadow Thistle* (LP), *Bass Strait Crossing* (tape) and *Hush & Joggety* (tape). The recording of *Broadmeadow Thistle* consisted entirely of songs written by Lobl. Several of her songs and/or tunes have also been published, for example in magazines such as *Tradition*, *Stringybark & Greenhide* and *SingOut*, and in books such as *The Peace Songbook* (Leigh Newton), *The Song Goes On*, *Eureka — The Songs That Made Australia*, *The Songs of Henry Lawson* and *Songs of Australian Working Life*. A collection of thirty-five original songs under the title of *Songs of a Bronzewing* was published in 1991.

Phyl Lobl has toured Australia and New Zealand and has performed in America and England. She was one of the performers who represented Australia at the Cologne Song Festival in Germany in 1982. Her particular interest as a performer is in songs which reflect 'the Australian experience', and she has engaged in much informal collection in search of performance material. She has collected more intensively material which people use with babies and very young children, an activity which has grown from her work as a music teacher in primary schools. In 1978 she published *Not Just Noise*, a music learning programme for seven- to ten-year-old children, and in 1991 she completed a folk music kit for Year 11 and 12 music students, commissioned by the Schools Commission and the Australian Music Centre. In 1991 the New South Wales Department of Education awarded Phyl Lobl the designation of Teacher of Merit in Music.

Phyl Lobl has performed and conducted numerous workshops at regional and national folk festivals on topics such as Henry Lawson — Songwriter, Parodies, Australian Songwriting, Lullabies and Dandling Songs, Aboriginal and White Views of Each Other in Song, Singing for Living, and Australian Women in Song. She has also been active in the promotion of folk music through producing and performing on radio programmes and tutoring in summer schools. She has helped to organize concerts of folk music for a variety of arts festivals and folk festivals.

LOWENSTEIN, WENDY (1927–), was born in Melbourne, and has been a proof-reader, journalist, factory worker and later a teacher–librarian. She is best-known for her seminal work in oral history in Australia, particularly through *Weevils in the Flour: An Oral Record of the 1930s Depression* *in Australia* (1978). Wendy Lowenstein was one of the founders of the Folk Lore Society of Victoria and was editor and producer of the magazine *Australian Tradition* (q.v.) for many years. She was briefly a member of the Victorian Communist Party arts committee with Ian Turner (q.v.), who introduced her to folklore and later oral history, and during 1954 they both accepted responsibility for establishing a folklore society. With her family Wendy Lowenstein spent 1969 travelling around Australia collecting folklore, some of which was published in *Australian Tradition*. The original field recordings are now part of the Lowenstein Collection in the National Library of Australia.

Wendy Lowenstein collaborated with Ian Turner (q.v.) and June Factor (q.v.) in the publication of the second edition of *Cinderella Dressed in Yella* (1978). She had also contributed substantially to the first edition (1969) and self-published her own *Shocking, Shocking, Shocking: The Improper Play Rhymes of Australian Children* (1974). Wendy Lowenstein was also a pioneer in multicultural studies, and together with Morag Loh wrote *The Immigrants* (1977), a series of oral history interviews exploring the migrant experience in Australia.

M

Magazines and other publications To deal in any satisfactory manner with the multitude of folklore and folklore-related publications is a near-impossible task because of the fragmented and sub-economic nature of much folklore publication and because of the problems of delineating a topic that relates to so many areas of social and cultural life. For reasons spelt out elsewhere (see **Aboriginal folklore**), the 'folklore' publications referred to here are not significantly engaged with Aboriginal materials, except insofar as interaction between black and white traditions has been observed and collected, as has been the case with some elements of song and music.

The earliest deliberate attempt to provide information on what we would today consider to be Australian folklore is the collection of A. B. ('Banjo') Paterson titled *Old Bush Songs* (1905) (q.v.). Paterson did not use the term 'folklore' in his collection, nor did 'Bill Bowyang' (Alexander Vennard [q.v.]) whose series of reciters published between 1933 and 1940 contains a large quantity of important song and verse texts. Collections of balladry, most notably those edited by Will Lawson (*Australian Bush Songs and Ballads*, 1944) and Vance Palmer and Margaret Sutherland (*Old Australian Bush Ballads*, 1951) continued to appear into the 1950s, a decade which saw an upsurge of interest in Australian folklore. Geoffrey Ingleton's collection of broadside ballads, *True Patriots All*, was published in 1952, followed by Bill Beatty's (q.v.) anthologies, including *Come A-Waltzing Matilda* (1954) and *A Treasury of Australian Folk Tales and Traditions* (1960). These were accompanied by the indefatigable Bill Wannan's (q.v.) anthologies of Australiana, beginning with *The Australian* in 1954 and proceeding through a great number of subsequent publications on many aspects of Australian folklore, most of which are still in print, in one form or another.

In the field of folk speech (q.v.), Sydney Baker's (q.v.) compilations of vernacular, *The Drum* (1959) and *The Australian Language* (1945), remain important classics and source works. Bill Hornadge, particularly with his *The Australian Slanguage* (1980), has continued to work in the tradition of Baker and of lexicographer Eric Partridge, as has G. A. Wilkes, compiler of an excellent *Dictionary of Australian Colloquialisms* (1978, since revised). Most recently L. Johansen has contributed an interesting volume titled *The Dinkum Dictionary: A Ripper Guide to Aussie English* (1988, since revised) which is a valuable compilation of recent Australian vernacular.

John Manifold (q.v.) and Ron Edwards (q.v.) began the broadsheet series Bandicoot Ballads in 1951 to publish traditional songs and ballads, a notion that was continued in the Black Bull Chapbooks series initiated by Hugh Anderson (q.v.) and Ron Edwards (q.v.). Both Anderson and Edwards went on to publish numerous important studies and collections of folklore under the imprint of Anderson's Red Rooster Press and Edwards's Rams Skull Press. Some other specialist publishers of folklore include the Antipodes Press, Popinjay Publications and the Carrawobbity Press.

Folk revival and folklore organizations of various kinds have been important publishers of works on Australian folklore, mainly of music, song and, to a lesser extent, dance. These organizations, such as the Australian Folk Trust (q.v.), state folk federations and related organizations, also produce newsletters and other periodical publications that frequently contain significant primary source materials and/or discussion and reports of collection and research. Generally such periodicals are simply titled the 'newsletter' of the originating organization, as is the case with the Folk Song and Dance Society of Victoria (q.v.), the Top Half Folk Federation, the Tasmanian Folk Federation, Monaro Folk Society and the Queensland Folk Federation. In some instances, and from time to time, some such publications have been otherwise titled. For instance, the newsletter of the NSW Folk Federation has often been called the *Cornstalk Gazette* (and is so at the time of writing), while the Western Australian Folk Federation's newsletter is known as the *Town Crier*, and the Folk Federation of South Australia produces the *Fedmag*. From time to time folk clubs and interested individuals have issued periodicals, usually of short duration. These have included *Rabbits' Ears* (Western Australia), *Idiom* and *Coal Dust* (both Newcastle Folk Club) and, in 1967–68, *Folksay*, issued by the Monash University Traditional Musical Society. It is true that the library accession and documentation of such publications is very patchy, even in the case of such comparatively major periodicals like *Australian Tradition* (q.v.) and *Stringybark & Greenhide* (q.v.). An attempt to document these and other folklore publications has been made in *Australian Folk Resources* (1983, 1988), published by the Australian Folk Trust, but the comprehensive bibliographical work in the field of Australian folklore has yet to be accomplished.

Periodical publications that are, or were, aimed at a broader audience than the members of individual organizations include *Australian Tradition*, *Australian Folklore*, *Stringybark & Greenhide*, *Northern Folk* and the *Australian Children's Folklore Newsletter*. All these are discussed in separate entries in this volume.

A number of book-length works that deal with Australian folklore in a primarily discursive manner deserve mention here. These include Hugh Anderson's *The Story of Australian Folksong* (1955); J. Meredith's and H. Anderson's *Folk Songs of Australia and the Men and Women Who Sang Them* (1967); John Manifold's *Who Wrote the Ballads? Notes on Australian Folksong* (1964); Russel Ward's *The Australian Legend* (1958); Ron Edwards's *The Australian Yarn* (1977); June Factor's *Captain Cook Chased a Chook: Australian Children's Folklore* (1988) and Graham Seal's *The Hidden Culture: Folklore in Australian Society* (1989).

Writers who have made greater or lesser use of folklore materials in their works, or who have anthologized folklore, include David Ireland, Henry Lawson (q.v.), A. B. Paterson (q.v.), Cyril Pearl, John Manifold (q.v.), Frank Clune, Frank Hardy (q.v.), Bill Scott (q.v.), Bill Harney, Nancy Keesing (q.v.), Wilbur Howcroft, Richard Magoffin, A. W. Reed, George Farwell, Walter Stone, A. K. MacDougal, George Blaikie, Keith Dunstan, Patsy Adam-Smith (q.v.), Hugh Anderson (q.v.) and Keith Garvey (see **Narrative literature and folklore**). As well, a multitude of anonymous, obscure and amateur writers, anthologists and editors have, since at least the late nineteenth century, been publishing and distributing usually small-scale publications, including yarns, songs, jokes and other items of folklore.

See also: **Appendix: A Bibliographic Guide to Australian Folklore**; **Recordings of folklore**.

Graham Seal

MANIFOLD, JOHN STREETER (1915–85), was born in Melbourne, into a wealthy grazier family from Victoria's Western District. He learned his first bush songs as a child on the family property, from his father, and from station hands. As a child he also came to love the work of the Australian balladists, A. B. ('Banjo') Paterson in particular. As a student at Geelong Grammar his studies included music and poetry, and he discovered the British border ballads. At Cambridge University he studied modern languages, met and married Katherine Hopwood, and joined the Communist Party. He also pursued his interest in writing poetry, especially ballads and sonnets, and in playing music. His favourite instrument was the recorder.

The outbreak of the Second World War found him in Germany as a translator for

S. T. Gill's water-colour painting of the Concert Room, Charlie Napier Hotel, Ballarat, Vic., June 1855, features Charles Thatcher's 'popular songs'. (National Library of Australia)

Ernie Dingo performing at the National Folk Festival, Alice Springs, NT, 1987. (Photograph, *Centralian Advocate*)

Alice Springs band Bloodwood toured Australia on the National Folk Touring Circuit in 1981. (Photograph, Jonah Jones)

Lay Long Yn, from Melbourne's Chinese community, sings and plays the *erhu* (traditional Chinese spike fiddle) for Oudovanh Ngakam at the Richmond Housing Ministry flats in Melbourne, April 1986. (Photograph, *Age*, Melbourne)

Traditional musician, the late Charlie Bachelor of Bingara, NSW, passed on his music to the Horton River Band, National Folk Festival, Canberra, 1984. (Photograph, Kevin Bradley)

Some recordings of traditional music and musicians. (Photograph, National Library of Australia)

an English publishing house. He escaped to England, and joined the British armed services. In 1941 his book *The Death of Ned Kelly and Other Ballads* was published. He acquired a reputation as a war poet. In 1946 his *Selected Verse* was published in the United States, and later in England.

He returned to Australia in 1949 and made his home in Brisbane, where he became active in radical politics and in the city's musical and literary life. He renewed his interest in the old bush songs, and actively sought out new material from old singers. In the early 1950s he produced, in collaboration with Ron Edwards, the *Bandicoot Ballads* (q.v.), a pioneering series of sixteen Australian songs with music. Some of these, like 'The Overlander', Manifold remembered from childhood; others were newly collected. Some, like 'Moreton Bay', 'The Ballad of Ben Hall's Gang' and 'The Drover's Dream', he pieced together from fragments remembered by several people and moulded into what he termed 'singer's texts'. On occasion he himself fitted tunes to texts.

Manifold worked hard to foster interest in bush song in Queensland, and in the late 1950s the growing body of enthusiasts formed the Queensland Federation of Bush Music Clubs under his presidency. These clubs included such notable collectors as Stan Arthur, Bill Scott and Bob Michell. In 1959 Manifold edited the Federation's *Queensland Centenary Pocket Songbook*, an important publication which contained thirty traditional bush songs collected by Manifold and other Federal members. The songs included the 'Queensland' version of 'Waltzing Matilda', which Manifold and his daughter Miranda had collected near Nambour.

Manifold's other important 'finds' included the tunes for 'The Streets of Forbes', 'The Convict Maid', and 'Freedom on the Wallaby'.

In 1964 Manifold's work on Australian bush song over the previous fifteen years culminated in the release of two major publications: *The Penguin Australian Songbook* (which contains ninety songs and remains a major reference work), and an analysis of the origins and social history of Australian folk song, entitled *Who Wrote the Ballads?*.

In his *Songbook*, each of the songs was complete, the texts scanning perfectly. This led some critics to question their purity, and to criticize him for 'silent editing'. Manifold's objective, however, was not to present a field collection, but rather to promote interest in the songs and to present them in a form for people to sing, and in this he excelled.

Despite his pioneering role in fostering interest in Australian bush song, Manifold chose to let the folk music 'boom' of the 1960s pass him by. He again became active as a poet, producing finely crafted verse, mainly ballads and sonnets. Very often his verse served to give expression to his social and political views. While his verse commanded respect from the literary establishment, he was never a fashionable poet. He remained active in the Communist Party, and was a trenchant social critic.

In 1968 his wife Katherine died of cancer, and Manifold felt her loss deeply. Five years later he suffered the first of a number of strokes, and was forced to move into a nursing care unit, where he lived out his last decade. He continued to comment on life and the human condition in verse. On 19 April 1985, two days before his seventieth birthday, he died in Brisbane. His casket was draped in the Eureka flag — he was a militant to the end.

Apart from those noted above, Manifold's publications include *The Music in English Drama: From Shakespeare to Purcell* (1956), and several books of his own verse, including *Collected Verse* (1978) and *On My Selection* (1986).

See also: Rodney Hall, *J. S. Manifold: An Introduction to the Man and His Work*, 1978.

Keith McKenry

MARSHALL, JOHN (1945–), was born in Paducah, Kentucky, USA and is currently head teacher in history at Erina High School on the central coast of New

South Wales. He is one of Australia's few holders of professional training in folklore studies, holding a Master of Arts degree in Folk Studies from Western Kentucky University. John Marshall has conducted extensive fieldwork in Georgia, Florida and Kentucky in the United States, and in New South Wales. In 1982 and 1983 he assisted Professor Kenneth Goldstein (q.v.) from the University of Pennsylvania in field collection of recitations and other verbal folklore in New South Wales, Victoria, Tasmania and the Australian Capital Territory. In 1985 John Marshall was awarded the first major fellowship by the Australian Folk Trust in order to assist a regional folklore study carried out for the National Library of Australia. One result of his folklore collection in the south-western Riverina area of New South Wales was the establishment of the River Murray Heritage Centre near Tocumwal, which collects, documents and displays the folk heritage of the area. It works closely with local schools as well as serving the local communities and the tourist trade.

Between 1985 and 1987 John Marshall taught accredited courses in Australian and world folklore to Years 11 and 12 at Finley High School in NSW. These subjects were Australia's only secondary school courses in folklore studies, and the students published *Blue Bags, Bloodholes and Boomanoomana, The Folklore of the Southwestern Riverina* in 1987. This book is one of Australia's rare studies of regional folklore, and includes children's folklore, customary folklore, personal experience narratives, recitations, weather lore, yarns, medical remedies, jokes, food preparation, superstitions and beliefs.

Medicine, folk: see **Aboriginal folklife in central Australia**; **Folk medicine**.

McDONALD, SIMON (1907–68), lived almost all his life in Creswick Victoria, a small town close to Ballarat, and worked as a woodcutter and farm labourer. He was a singer, fiddle player and writer of ballads for recitation, and is widely regarded as one of the finest traditional musicians recorded in Australia. He had a fine voice and his singing was noted for its rhythmical and emotional qualities. Simon McDonald had an extensive repertoire of English and Irish songs handed down through his family from his Irish great-grandfather. Many, such as 'The Lost Sailor', were about the sea. He was recorded during the 1960s by Norman O'Connor, and these field recordings are now held in the National Library of Australia. A Wattle recording, *Australian Traditional Singers and Musicians of Victoria*, was produced in 1963 and included Simon McDonald and four other traditional singers. Hugh Anderson produced a biography of Simon McDonald, *Time Out of Mind*, in 1974.

McKENRY, KEITH (1948–), is a senior public servant, and sometime environmental scientist, reciter, poet and folklorist. While at university his inability to sing in tune led to his reciting bush verse around rock-climbing campfires. He did not perform before concert audiences, however, until the late 1970s, when after moving to Canberra he made contact with folk music enthusiasts. Since that time he has performed frequently as a reciter in concerts, festivals and folk clubs, presenting an amalgam of history, humour and social comment that doesn't fit readily into any familiar stereotype. His festival workshops have addressed such issues as the Australian folk revival, racism in Australian folklore, the democratic tradition in Australian popular verse, and the depiction of Australian society and history in folk verse and song.

In 1983 he released his first book of modern Australian recitations, *The Spirit of the People*. In the same year he moved into the arts and cultural heritage area of the Australian public service, where he initiated the largest oral history project ever attempted in Australia, The Cultural Context of Unemployment — An Oral Record. This \$1.3 m project was administered by the National Library of Australia.

In 1984 his advice that in the heritage area governments in Australia had focused almost exclusively on Australia's *tangible* heritage — the 'things you keep' — and that measures were urgently needed to safeguard the country's *intangible* heritage, was taken up enthusiastically by the then arts and heritage Minister, the Hon. Barry Cohen. For the next four years Keith McKenry was actively involved in developing proposals for government action in this area. In 1985 he edited a guide to Commonwealth activities and resources relating to folklore, entitled *Folklife and the Australian Government*. In the same year he represented Australia in Paris on UNESCO's Committee of Government Experts on the Safeguarding of Folklore, and visited a number of folklore institutions in North America and Europe. In 1986, at his suggestion, the Commonwealth Committee of Inquiry into Folklife in Australia was established. He served as a member of the Committee, and was co-author of its Report, *Folklife, Our Living Heritage* (1987). Returning to UNESCO in 1987, he was elected Chairman of its Special Committee of Technical and Legal Experts on the Safeguarding of Folklore. The work of this Committee, and its predecessors, culminated in the formal adoption by UNESCO in 1989 of the *Recommendation on the Safeguarding of Traditional Culture and Folklore*.

In 1988 Keith McKenry left the arts and heritage area of public administration and has since worked in other areas of the Australian public service. In the same year he was elected vice-president of the new Australian Folklore Association. In 1991 he released on his own Fanged Wombat label his first album, a joint effort with his friend, bush singer and folklore collector Alan Scott (q.v.). Entitled *Battler's Ballad: Songs and Recitations of Australian Bush Life*, the album presents for the first time on compact disc Australian bush songs and recitations, performed in traditional style. The album was recorded by the National Library of Australia for its permanent oral history collection.

MEREDITH, JOHN (1920–), born and educated in rural New South Wales, has had a significant impact on the collection, study and performance of Australian folk song, dance and music. By any assessment John Meredith is a major figure in Australian folklore studies, continuing and developing a tradition of collection and scholarship that can be traced to A. B. Paterson's (q.v.) *Old Bush Songs*, the publication that originally sparked Meredith's active interest in folk song. Meredith has contributed to the collection, dissemination, documentation in photography and film and performance of Australian folklore, as well as to folklore scholarship.

John Meredith is an important collector of bush song, music and dance, both because of the amount of material he has obtained and published and for the repertoires of a number of his chief informants, particularly Sally Sloane (q.v.) and 'Duke' Tritton (q.v.). A good deal of his early fieldwork has been published in his and Hugh Anderson's (q.v.) *Folk Songs of Australia and the Men and Women Who Sang Them* (1967), the latter part of the title indicating the importance that Meredith attached to the performers themselves. This was a distinct and welcome development of Paterson's approach, which is almost totally devoid of informant attribution and social context. Volume 2 of *Folk Songs of Australia* (with Patricia Brown and Roger Covell) has since been published. These two works form the core of Meredith's published collections of song, music and dance.

John Meredith has also collected and published other forms of folklore, particularly folk speech, in *Learn to Talk the Old Jack Lang* (republished in 1991 as *Dinkum Aussie Slang*), and has published works on local and social history, including *Ned Kelly: After a Century of Acrimony* (1980) with Bill Scott, and also works on aspects of Aboriginal culture. He has published a number of more specialized folklore studies, most notably *The Wild Colonial Boy* (1982), *Duke of the Outback* (1983) and, with Rex Whalan, *Frank the Poet* (1979), as well as a number of smaller anthologies of bush songs and ballads, often under the auspices of the Sydney Bush Music Club (q.v.), of which he is a founding member, editing the journal

Singabout (q.v.) during its early years.

As a member of the original Bushwhackers band, Meredith has also been influential as a performer of folk music and has continued to be active in this area, having written and co-written a number of musical dramatic works on folklore themes.

In 1989 the National Library of Australia mounted an extensive exhibition of Meredith's work, particularly of his photography, in association with the third National Folklore Conference. John Meredith continues to record folk song, music and dance in many parts of Australia, and has become increasingly involved in filming traditional performers. His folklore collections are mostly with the National Library of Australia and with the National Film and Sound Archive. The bulk of his work remains unpublished in these archives, providing a rich resource for future study and an incomparable record of important aspects of Australian folk song, music and dance as it was performed from the 1950s onwards.

John Meredith was awarded the Medal of the Order of Australia in 1988 for his contribution to the collection and preservation of Australian folk traditions, and in 1992 was made a Member of the Order of Australia.

See also: **Bush dance**; **Folk revival**; **Folk song**; **National Library of Australia**.

Modern Legends The terms 'modern legends' or 'contemporary legends' have largely replaced the earlier and more widespread term 'urban legend', since their content is not confined to the cities, nor are their tellers. Some modern legends in fact require rural settings as part of their atmosphere of isolation and fear. The Australian film-makers Nadia Tass and David Parker used one such legend — tongue in cheek — in their prize-winning film *Malcolm*. The story is set on a lonely road where a father and his child run out of petrol, and the father leaves the child to sleep in the car while he walks back to a distant garage. All night the child hears a terrifying 'thump, thump' on the roof of the car, and later is rescued by the police who warn the child not to look back. The child disobeys this instruction and sees his father's headless body thumping on the car roof, having been hung on a tree by an escaped maniac from a nearby criminal asylum. It is not hard to see the derivation of the instruction 'not to look back' from the Biblical story of Lot's wife and the classic Greek legend of Orpheus and Eurydice. In all cases, the disobedient one suffers terrible consequences. Modern legends incorporate the same themes as those in ancient mythology, themes such as death, violence, sex, obedience to or transgression of society's norms, although the modern stories are firmly set in the present day, and are international in their circulation.

Two of the world's best-known researchers into modern legends are Paul Smith, editor and co-editor of popular collections and academic publications on the subject, and Jan Brunvand from the University of Utah. Brunvand has produced a number of books such as *The Vanishing Hitchhiker* (1981) and *The Choking Doberman* (1984). Brunvand has identified about 500 different legends with innumerable variations, and has classified them into ten major categories:

Legends about automobiles
Legends about animals
Horror legends
Accident legends
Sex and scandal legends
Crime legends
Business and professional legends
Legends about governments
Celebrity rumours and legends
Academic legends

Modern legends reflect society's preoccupations, fears and anxieties, including fears about technology, such as stories about razor blades in water slides, cement mixers dumping their load in an alleged sexual transgressor's new car, and about babies in microwave ovens. These legends are different from yarns, tall stories or jokes (see **Folk tales**) in that they are told literally, rather than simply to entertain, by people who believe them to be true and (usually) to have happened to someone-they-know-who-knows-someone, the 'friend of a friend' or FOAF as some folklorists have entitled this elusive source.

It is hard to identify a specifically Australian contemporary legend. Amanda Bishop published the story of a clothed kangaroo, entitled 'The Gucci Kangaroo' in her collection of stories of that name (1988). The kangaroo is knocked down by some drivers and thought to be dead, whereupon the drivers decide to dress the animal in a sports coat and cap and take photographs. The animal revives and hops away, still wearing the driver's expensive coat — and wallet. Although this legend has been identified in nineteenth-century Australia, there are also international variations, and in fact Amanda Bishop refused to nominate any stories as being specifically Australian, as is suggested by the 'Gucci' kangaroo. Similarly, Australia's best-known collector of modern legends, Bill Scott, has declined to nominate 'uniquely Australian' legends in his numerous publications, including *The Long and the Short and the Tall* (1953).

Folklorists today are less willing than formerly to reject out of hand any suggestion that a particular legend may have originated in a true event. Given that manufacturers are still prosecuted for adulterated food, it is not surprising that food adulteration legends such as the 'Kentucky fried rat' or 'The mouse in the Coca Cola' proliferate. Like many legends and other folk tales, they involve an element of imaginary revenge or reclamation of power by the powerless as regards these gigantic multinational corporations.

Like most folklore, modern legends are of unknown origin, and contemporary folklorists are more interested in the dynamics of telling such legends. The noted Hungarian-American folklorist Linda Degh discussed tellers and transmitters of modern legends at the 1990 conference of the American Folklore Society in Oakland, California. She found children and adolescents were rarely sources; among the more common sources were scholars, fiction writers, docu-drama producers, clairvoyants and journalists. Certainly Australian newspapers are a rich source of modern legends presented as 'news', such as the following from the Melbourne *Age* 'Odd Spot' of 22 September 1988 which has appeared in many other places:

> A burglar in Treviso, Italy, stole a family's car but returned it the next day with a note of apology and invited the family of four to dine at his expense at a smart restaurant. While they were at dinner, he burgled their home.

Given the 'cautionary tale' element in many modern legends, it is not surprising that they should be essentially adult folklore, sometimes told to children or adolescents, rather than an aspect of children's folklore itself. A classic example would be the many legends told about hitchhiking, such as that about the youthful driver who picks up a 'little old lady', luckily identifying her in time as a male murderer with an axe (or knife) by 'her' unusually hairy wrists. Linda Degh also spoke to the 1990 AFS conference about Halloween legends, particularly about 'poisoned candy' tales. The police department in Bloomington, Indiana was issuing printed warnings to parents about such Halloween dangers, although Linda Degh checked with the department and found that they had no reported cases. Such cases show the similarity between the transmission of rumour and that of modern legends.

Gwenda Beed Davey

MOYLE, ALICE (1908–), was born at Bloemfontein, South Africa, and has lived in Melbourne, Sydney and Canberra since 1913. She is one of Australia's most noted researchers into Aboriginal music, and was made a Member of the Order of Australia in 1977 for academic service, particularly in the study of Aboriginal music.

In 1959 Alice Moyle wrote the article 'Sir Baldwin Spencer's Recordings of Australian Aboriginal Singing' in the *Memoirs of the National Museum of Victoria* and subsequently published extensively concerning early sound recordings of Aboriginal music and her own field recordings. She has contributed numerous articles and reviews to Australian and overseas journals, principally on Aboriginal, Torres Strait Islander and Papua New Guinean music. She contributed 'Aboriginal Music and Dance: Reflections and Projections' to J. Kassler and J. Stubington (eds), *Problems and Solutions. Occasional Essays in Musicology Presented to Alice M. Moyle* (1984). Alice Moyle has made frequent contributions to encyclopaedias and dictionaries such as *The New Grove Dictionary of Music and Musicians* (1980) and to the forthcoming *Companion to Music in Australia*. She has compiled eleven discs/cassettes from music she recorded on Aboriginal settlements between 1959 and 1969. Most have been published by the Australian Institute of Aboriginal Studies (see **Australian Institute of Aboriginal and Torres Strait Islander Studies**) of which she was a foundation member. She is currently Honorary Visiting Fellow at the Institute. She is a member of numerous Australian and international organizations in musicology and ethnomusicology and is a member of the International Council for Traditional Music (ICTM). Her main interests lie in the development of sound archival methods in Australia, especially with regard to early recordings of Aboriginal music. She is the founder, convener and chairwoman (her preferred term) of the Sound Heritage Association, which is currently producing a series of sound recordings entitled Australia's Heritage in Sound to celebrate the advent of sound recording in Australia.

Mulga Wire has been the internal publication of Sydney's Bush Music Club (q.v.) since 1967, before which date the publication was titled *Singabout*, first published in 1956.

MURRAY-SMITH, STEPHEN (1922–88), was a literary editor, educator, historian, radical nationalist and campaigner. He is perhaps best known as founder (in 1954) of the magazine *Overland*, the motto of which — 'Temper democratic, bias Australian' — he adapted from Furphy, and which he continued to edit until his death. Educated at Geelong Grammar and at Melbourne University, he was a commando in New Guinea during the Second World War and spent a decade (until the Hungarian uprising) as a committed Communist while working as a student teacher, journalist, peace activist and trade union official. During his post-war student days at the University of Melbourne he co-compiled collections of radical and bawdy songs, which appeared as *Rebel Songs* and *Snatches and Lays*. The latter was not 'officially' published until 1973, but circulated from the late 1950s. It was regarded as both a valuable record of an underground oral tradition, and a blow against the rigid censorship practised at the time. In 1966 Murray-Smith completed a Ph.D. on the history of technical education in Australia up to 1914, and worked thereafter in the Faculty of Education at the University of Melbourne, retiring as Reader in 1987. At the time of his death, he was an Associate of the History Department.

Stephen Murray-Smith's publications include *The Tech* (1987, with A. J. Dare), a centenary history of the Royal Melbourne Institute of Technology which drew on his extensive primary research in this area. His great passion as a historian and traveller was for isolated communities, particularly the islands of Bass Strait; he journeyed to settlements as remote as St Kilda (UK), Tristan da Cunha and the Antarctic. Significant publications in this area include *Sitting on Penguins, People and Politics in Australian Antarctica* (1988); *Mission to the Islands, The Missionary Voyages in Bass Strait of Canon Marcus Brownrigg 1872–1885* (1979); *Bass Strait: Australia's Last Frontier* (1969, revised 1987); 'Beyond the Pale' (1973) a paper on the evolution of the mixed-race community on Cape Barren Island; and two Bass Strait bibliographies (1973, 1981). He collected extensive oral records of the long-term inhabitants of the Bass Strait islands. Murray-Smith edited several anthologies of Australian short stories (the earliest, *The Tracks we Travel*, in 1953) and published a monograph on Henry Lawson in 1962 (revised 1975). He delighted in a project in which he collected reminiscences of nineteenth-century Australian childhoods through an appeal in the *Australian Women's Weekly*. This appeared as 'Messages from Far Away' in *The Colonial Child* (Royal Historical Society of Victoria, 1981). He compiled the first authoritative *Dictionary of Australian Quotations* (1984), as well as *Right Words* (1987), a guide to Australian usage of the English language; both are standard references. He was made a Member of the Order of Australia in 1981 for services to literature and education.

Stephen Murray-Smith campaigned on many social issues, especially historic or environmental concerns, such as saving cattlemen's huts in the Alps, the Percy Grainger Museum in Melbourne, the continued manning of Australia's lighthouses, or the view from Oliver's Hill near his home on Victoria's Mornington Peninsula, which had been marred by newly-installed telegraph poles. He was active in the successful debate against the death penalty and in the unsuccessful debate against metrication, amongst many other long-fought causes.

Musical and poetic characteristics of children's folklore Children's playground rhymes, chants and songs are profoundly musical. 'Music' does not just mean singing in the traditional sense, but the intoning of texts at varying levels between speech and song. The way this music is intoned, the number of children participating, the accompanying text and the nature of the related activities are all integral to the performance. The kind of counting-out rhyme used for determining who will be 'he' for 'tiggy' ('There's a party on the hill', 'Eeny meeny' and so on) is a good example. Besides keeping the basic pulse of the rhyme, the children are required to recite the words (thus internalizing both the rhythms and the intonation patterns), and remember the foot-counting ritual.

Child lore is modelled on adult lore. Young people borrow a great deal from adults, taking what they can remember into their own private domain, manipulating, parodying then claiming ownership of it. Sometimes a whole text or melody is borrowed; at other times only fragments of the text are remembered, which can become the basis for a new creation. Nevertheless, we can find out a great deal about what children think and how they learn from their playground rhymes and songs. Not only do they learn social skills, but they also explore the basic elements of language, music, poetry, drama, and dance. Mathematical skills such as counting, grouping, estimating and spatial awareness are also part of the playground 'curriculum'.

Nine interrelated musical and poetic characteristics of child folklore can be identified: forms of intonation; tonal repetition; vocal and choral characteristics; rhythms; melodic motifs; tonal clusters; parodies; rhetorical devices; and thematic material.

Forms of intonation

There are basically two kinds of intonation: spoken intonation (with unstable pitches) and sung intonation (with stable pitches). Within a continuum which begins with speech and ends with song are a number of intermediate forms, which include recitation, monotone, *sprechstimme*, calls and chant.

The least pitch stability is found in casual speech, with its many interconnected tones. If the voice is projected, as in recitation, the intonation is heightened and a few pitches tend to dominate. Through constant repetition of sounds, words and phrases, these pitches increase in stability.

In monotone, where there is no melodic interest, the rhythms are often very well defined. The rhyme shown in example 1 (all examples are listed at the end of this entry) was created by a Year 5 boy. The performance remains firmly anchored on the one tone until the end, where the performer derives great satisfaction from the final pun. It's a rude rhyme, but cleverly put together.

At times children use a kind of *sprechstimme*, which is also half spoken and half sung, but instead of being stable, the pitches will be vague and uncertain. It often happens when children try to but can't transfer to sung intonation. In example 2 the performer has an idea of what should be sung, but is finding difficulty in expressing it in melodic terms.

There are several forms of sung intonation used by the young. One is 'the call', frequently used to call another person, which consists of two stable pitches, the second tending to fall a minor third below the first (*so–me*). The repetition of this interval has been claimed by the Hungarian composer and musicologist Zoltan Kodaly and others to be a universal children's chant formula. Evidence so far suggests that Australian children seldom use the call for chanting without the addition of one or more other pitches.

Children's chanting is, of course, another form of sung intonation. Chants tend to be repeated many times and are limited to a few stable pitches — usually two, three or four.

Vocal and choral characteristics

Some children make music singly (vocally), but most prefer to perform in a group (chorally). If a rhyme is spoken, and so has no known tune, the way it is intoned will often relate directly to the number of performers. For instance, a single child is likely to recite a rhyme, but if a group begins to recite the same text, the pitches tend to become more stable, and sung intonation (chanting) may result (see example 3).

Tonal repetition

Because children tend to use a limited number of stable pitches, certain tones occur more frequently than others, and are often repeated consecutively. If the words are sung, these repeated tones are most likely to correspond with the first or fifth degree of the major scale (*do* or *la*). The most common form of tonal repetition tends to be tonic 'hammering' (*do*, *do*, *do*, *do*, etc). The five-note song shown in example 4 is a good illustration of this, as seen in the 'be, be, be' and 'me, me, me'.

Where the music is not sung but recited, a main tone is likely to emerge, which rises or falls slightly during the intoning of the rhyme. In the rhyme used for school excursions shown in example 5, the students are far more interested in the words than in the melodic organization, so the rhythms dominate the recitation. Note how the voices rise as the children stress the rude word!

Rhythms

In 1954 the French researcher Constantin Brailoiu hypothesized that regardless of culture, children's songs share the same distinctive qualities of rhythm, metre and form. He claimed that children's texts are scanned in trochees, that is, the rhythms are in groups of two syllables, one stressed and the other unstressed (binary rhythms). Brailoiu put forward four models and their variants (see example 6), which he believed were universal manifestations of 'childlike rhythm'. All these rhythms are found in Australian children's folklore.

Mary Brunton, in an unpublished Master's degree dissertation (1976), has studied the way South Australian primary school children use music patterns, and sets out rhythmic hierarchies found in playground rhymes and songs. She found that

- the opening poetic feet tend to correlate with the presence or absence of a 'leading in' tone (anacrusis);
- most examples tend to start on a strong rather than a weak beat;
- nearly all songs begin on the first beat of the bar;
- unstressed opening syllables tend to be lengthened to adapt the text to a melody beginning with a strong beat.

Brunton puts forward six basic rhythmic patterns, all of which can be adapted, varied, 'swung' (syncopated) and made into 'composites'. She was aware of, but unable to obtain, Brailoiu's study, so it is interesting to note that the most preferred pattern in her study is a stretched out (augmented) version of Brailoiu's first model (see example 7).

Most Australian researchers into children's folk music agree that Brailoiu's schema is applicable to their studies. For example, Margaret Kartomi relates Brailoiu's first three models to her work on Aboriginal children's playsongs at Yalata in South Australia (published in *Miscellanea Musicologica* II, 1969). Another ethnomusicologist, Cheryl Romet has found that Brailoiu's first model and its variants apply to her work in West Javanese children's folklore (see *Australian Journal of Music Education*, October 1980).

Syncopation in children's folklore is rare, and tends to be limited to a particular rhythmic motif within a rhyme or song. However, young people may borrow a syncopated song or jingle from the adult repertoire to use as the basis of a parody.

Melodic motifs

By disregarding tonal repetition, Brunton has identified seven basic melody patterns or motifs. These are different from what she describes as 'core chant' melodies, which are repeated over and over again:

1 mediant-supertonic-tonic (*me-re-do*);
2 tonic-submediant-dominant (*do-fa-so*);
3 submediant-dominant-mediant (*la-so-me*);
4 tonic chord (*do-me-so*);
5 dominant-upper tonic (*so-do*);
6 dominant-subdominant-mediant (*so-fa-me*).

Tonal clusters

A number of chants and songs are based on small 'clusters' of six notes or less. While most of these clusters relate to the pentatonic (five-note) mode (*do, re, me, so, la*), others appear to be fragments of the heptatonic (seven-note) scale (*do, re, me, fa, so, la, ti*).

The tones *me, so, la* form the most common cluster, in a three-note chant formula which is peculiar to Australian children and is especially common in skipping rhymes (see example 8). Another common three-note chant formula is derived from the call, which is used developmentally in the Kodaly method of music teaching. Lois Choksy (*The Kodaly Method*, 1974) believes that this formula is common in American children's folklore, but in Australia it has not been widely recorded. It is based on the same cluster as the first chant, but in different order (*so, me, la*), and with different stresses. Example 9 is an instance where Australian children have used this 'American' children's chant formula with a uniquely Australian text.

Common in both classrooms and schoolyards is the 'Mary Mac' formula based on the four-note cluster *so, la, ti, do* (see example 10). The melody rises stepwise with a final hammering of *do*, and is then repeated. Children continue to enjoy creating new verses to this chant. In this version there is a swing on 'Mac, Mac, Mac' (*do, so, do*), but another version has also been recorded where the same words are hammered out on *do, do, do*.

A five-note cluster frequently used by children is based on the pentatonic scale (*me, so, la, do*), with *ti* used as a passing note. It forms the basis for at least eleven rhymes, nine of which are used for clapping. As the melody has a 'lead-in' (anacrusis), and all the rhymes are scanned in iambic feet, that is, in groups of one unstressed and one stressed syllable, the poetic and rhythmic match is perfect. Handclapping rhymes tend to be scanned in iambic feet, allowing the performers time to co-ordinate the handclapping on the first stressed beat. This is why the unstressed initial beat is often augmented by the performers (see example 11).

One six-note cluster is based on the stepwise pentatonic progression of *so, la, do, re, me* with *ti* used as a passing note. It forms the basis of the popular skipping song 'Vote, vote, vote'. This lively melody is related to both the evangelical chorus 'Jesus loves the little children' and the song 'Tramp tramp tramp the boys are marching'. It begins by skipping through the pentatonic scale in retrograde (see example 12).

Another cluster is based on the major scale (*do, re, me, fa, so, la*). It is the nucleus of the tunes 'Twinkle twinkle little star', and 'This old man, he played one' which are used by children for a variety of parodies. Both tunes involve movement either up or down the cluster, as in the popular children's song about school shown in example 13.

Parodies

Australian children tend to be conservative about the kinds of tunes that they use for parodies. The tunes can be divided into four categories:

- Popular songs and advertising jingles
  Once children learn a catchy song or jingle, it's open slather! Consider this parody on an Australian Rules football advertisement of the seventies:

  Up there Cazaly, meatballs and sauce,
  Fly like elastic, eat like a horse.

- Traditional songs, including folk songs
  The sea shanty 'What shall we do with a drunken sailor' forms the basis for this parody, where young males explore violent and sexist language. The last lines are a common addendum to many other children's rhymes:

  We are the boys of the red triangle,
  Every girl we meet we mangle,
  Ooh, aah, off with the bra,
  Down with the knickers and in to the car.

- Christmas carols
  This parody, recorded in 1980, can be heard in many versions, of which children and adults alike have claimed ownership!

On the first day of Christmas my true love gave to me
A partridge in a pear tree.

On the second day of Christmas my true love gave to me
Two Chiko rolls, and a partridge in a pear tree.

On the third day of Christmas my true love gave to me
Three meat pies (etc.)

On the fourth day of Christmas my true love gave to me
Four sexy sheilas (etc.)

On the fifth day of Christmas my true love gave to me
Five rubber thongs (etc.)

On the sixth day of Christmas my true love gave to me
Six Eskies icey (etc.)

On the seventh day of Christmas my true love gave to me
Seven skinned koalas (etc.)

On the eighth day of Christmas my true love gave to me
Eight Ron Barassis (etc.)

On the ninth day of Christmas my true love gave to me
Nine billies boiling (etc.)

On the tenth day of Christmas my true love gave to me
Ten Foster's Lagers (etc.)

On the eleventh day of Christmas my true love gave to me
Eleven Aerogard cans (etc.)

On the twelfth day of Christmas my true love gave to me
Twelve football teams (etc.)

- Nursery rhymes
  These simple melodies lend themselves to all kinds of parody:

  Oh dear, what can the matter be,
  Four old ladies stuck in the lavatory,
  Stuck there from Monday to Saturday,
  Nobody knew they were there.

When rhymes or songs are parodied, we can be certain that the new texts will ridicule something or somebody, using rude language wherever possible! This is part of a youthful stage of exploration, where children are discovering language, poetry and music and using these discoveries to invent their own creations.

Rhetorical devices

In their exciting study *Speech Play* (1976), the American researchers Mary Sanches and Barbara Kirshenblatt-Gimblett have demonstrated that both adult and child traditions share the common poetic language. They append fifteen different examples of rhetorical structures in children's rhymes to illustrate their discussion.

Australian children also use verbal patterns similar to adult rhetoric, and these patterns are often emphasized by 'musical rhetoric'. Most children use rhetorical devices simply because they are part of a known text or song, or are useful tools for spontaneous creativity, not through any wish to be deliberately persuasive, witty or to create a work of art. Since part of the function of repetition is to assist the memory, rhetorical devices emerge very early in children's folklore. Teachers and parents can use these structures as poetic and musical building blocks to further develop creativity in young people (see example 14).

The most common rhetorical devices in children's folklore are repetition of sounds (assonance, alliteration, onomatopoeia and rhyming), repetition of words, phrases, lines and verses, and concatenations.

All sound repetition is profoundly musical. Therefore, to say that a child is 'not musical' is simply not true! Fascination with sound, together with the urgent desire to explore it, is a fundamental characteristic of children of all ages. Here onomatopoeia is used in a parody of a well-known song. Notice the repetition in the third line, echoing the first. It's a very basic stage of poetic and musical exploration:

Row row row your boat
Gently down the stream,
Putt putt putt putt,
Out of gasoline.

It is common to find the repetition of words or phrases repeated in identical positions in consecutive poetic lines (parallelism). In the clapping rhyme 'My boyfriend gave me an apple' (see example 11), the parallel movement occurs at the beginning of each line. In this next song the parallel repetition occurs at the end of each line and is combined with homophonic word play:

A sailor went to sea sea sea
To see what he could see see see,
But all that he could see see see
Was the bottom of the deep blue sea sea sea.

The concatenation is a fascinating area which has inspired considerable overseas research. Extra words, phrases, lists, or whole rhymes may be appended to an existing rhyme like links in a chain. These concatenations are most frequently affixed to skipping, counting out and handclapping rhymes requiring some physical challenge. Other rhymes are more likely to contain verses and choruses. Short texts (two-three-and four-liners) provide the most common basis for concatenations:

Doctor, doctor over the hill,
Melissa Green is very ill,
Will she live or will she die!
Live, die, live, die . . . .

'Floating' texts, or 'floaters' are longer concatenations, usually consisting of a complete new rhyme. One function of the floating text is to allow the skipper or clapper extended time in the game, providing that she or he can overcome the challenge set by the words. Some floating texts are likely to continue the rhythmic flow set by the first rhyme, and to be linked by some kind of verbal formula, for example 'This is what (s)he said'. Alternatively, floating texts may be linked by association of ideas. In this clapping rhyme a new, faster rhythm provides the challenge to the clappers:

Under the bamboo,
Under the tree, cha cha cha,
True love for me, my darling,
True love for me, cha cha cha,
When we get married we'll raise a family
Of sixty thousand children in a . . . (*linking idea*)
Row row row your boat (*floating text*)
Gently down the stream,
Chuck y'r teacher overboard
And listen to her scream, Aaaah! (*tag*)

Thematic material

Many studies show that children have not lost their love for make-believe and fantasy. Children also have an irresistible urge to jeer at authority, but as the Opies have explained, 'The teacher can, however, rest assured that no matter how personal the lyric may appear, he or she is unlikely to be the original of its sentiments'.

Improper rhymes are an important stage of development (see example 15). They help children to satisfy their natural curiosity about body and sexual functions. It seems that where rhymes are exclusively owned by girls (skipping rhymes, clapping rhymes and the like) reference to sexuality or scatology is less common than in rhymes used only for amusement and shared by both sexes. Clapping rhymes are mostly owned by girls and still tend to reinforce traditional sex roles. A number of rhymes equate life-long happiness with the rearing of as many children as possible. 'Under the bamboo', where one can raise a family of 'sixty thousand children in a row' without a qualm, is one such rhyme.

Children's playground rhymes and songs involve a great deal of learning, yet, as Davey and Factor have suggested, neither they nor their elders make the connections between schoolyard and schoolroom learning. Perhaps one reason for this is the way some adults still regard children's play: as something trivial, and therefore unworthy of serious consideration. The value of folklore in the creative development of children can be seen as young people explore through sound and movement the fundamental areas of music, language, poetry, dance and drama. In doing this, children are developing skills and knowledge in important areas of sensory communication: verbal, aural, visual, kinaesthetic and tactile ways of thinking and knowing. Through this exploration they come to better understand themselves, their schoolyard society and ultimately the world they live in.

Example 1

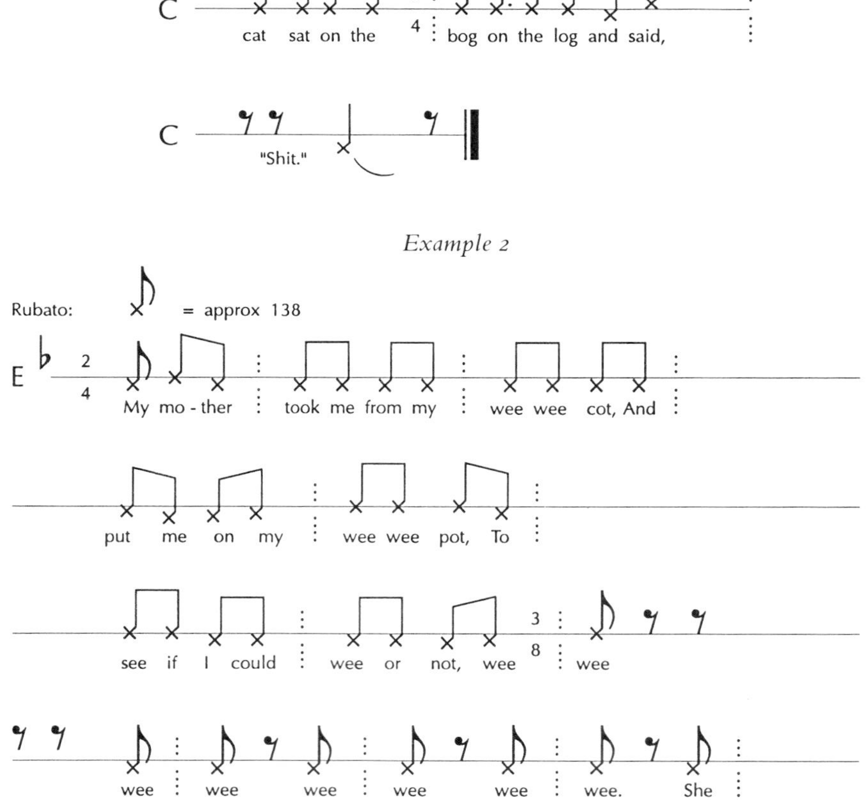

Example 2

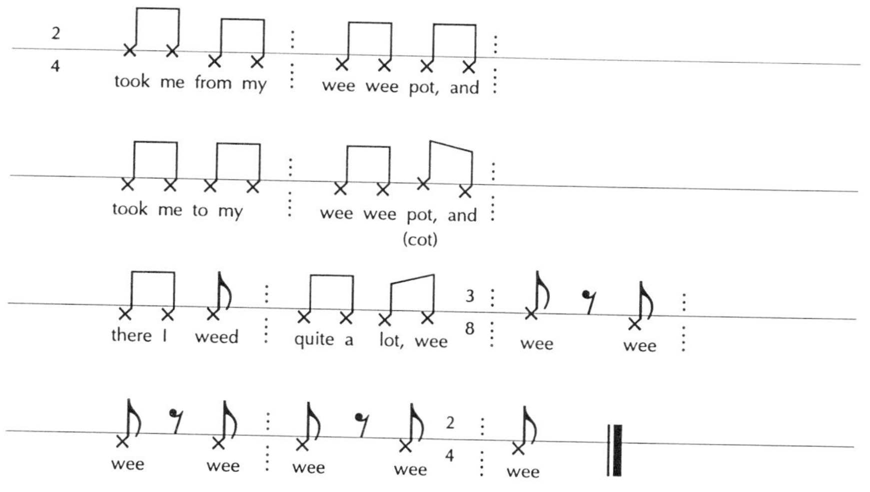

Tentative: tonal cluster:

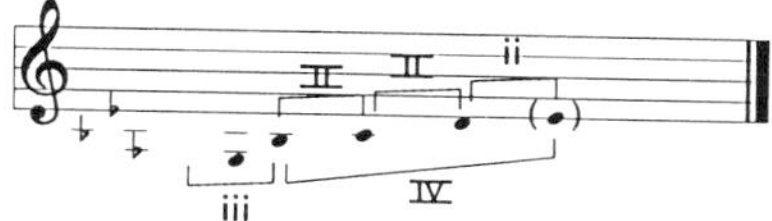

Example 3

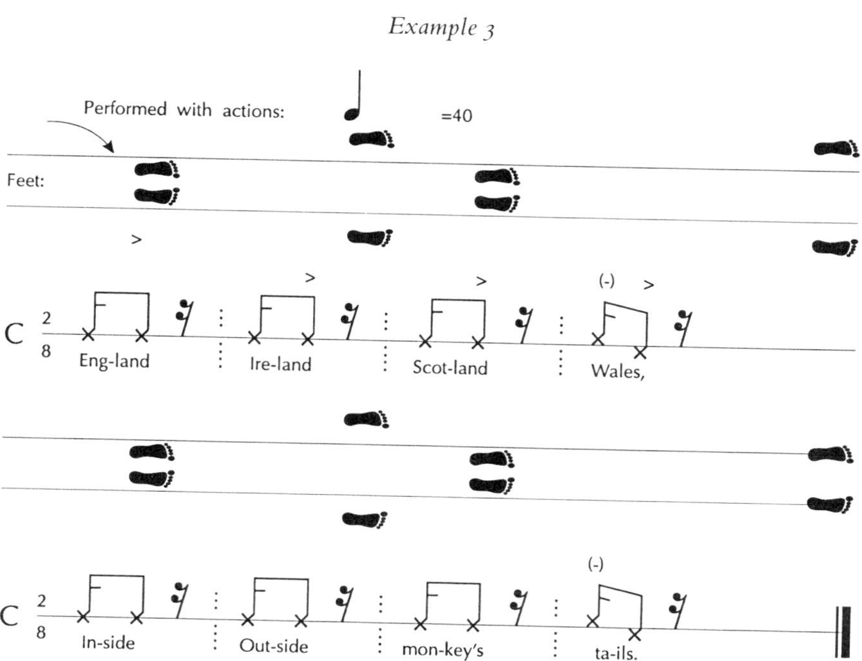

Example 4

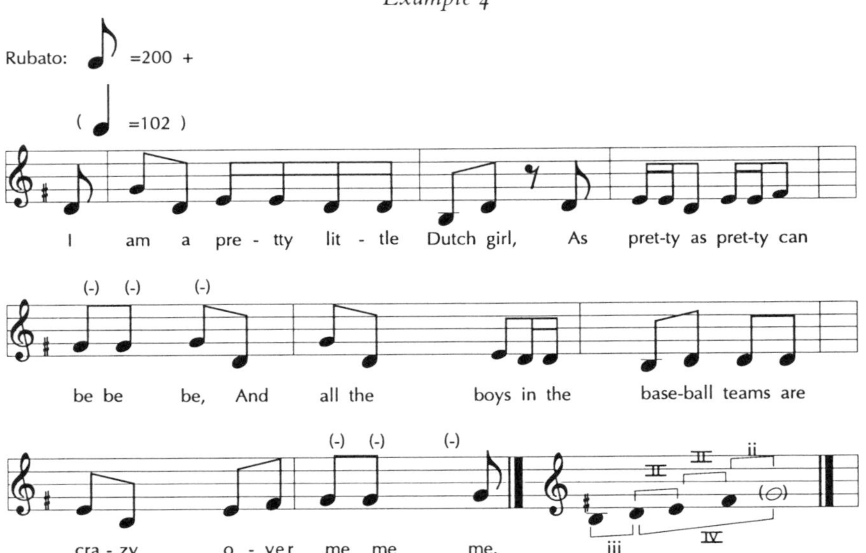
Rubato: ♪ =200 +
(♩ =102)
I am a pre - tty lit - tle Dutch girl, As pret-ty as pret-ty can
(-) (-) (-)
be be be, And all the boys in the base-ball teams are
(-) (-) (-)
cra - zy o - ver me me me.
II
II
ii
iii
IV

Example 5

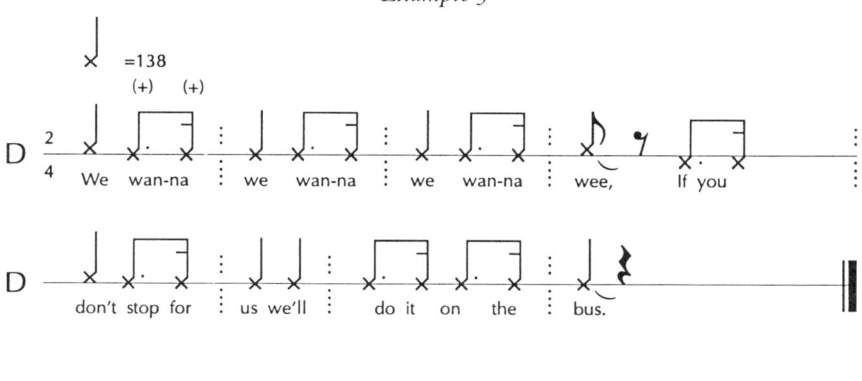
=138
(+) (+)
D
2
4
We wan-na we wan-na we wan-na wee, If you
D
don't stop for us we'll do it on the bus.

D

Example 6

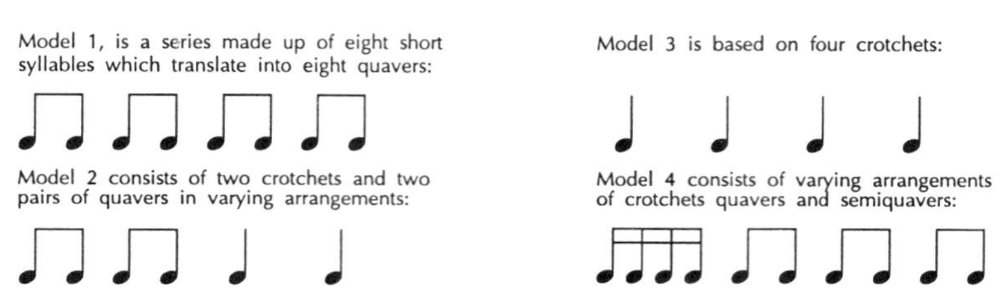
Model 1, is a series made up of eight short syllables which translate into eight quavers:
Model 2 consists of two crotchets and two pairs of quavers in varying arrangements:
Model 3 is based on four crotchets:
Model 4 consists of varying arrangements of crotchets quavers and semiquavers:

Example 7

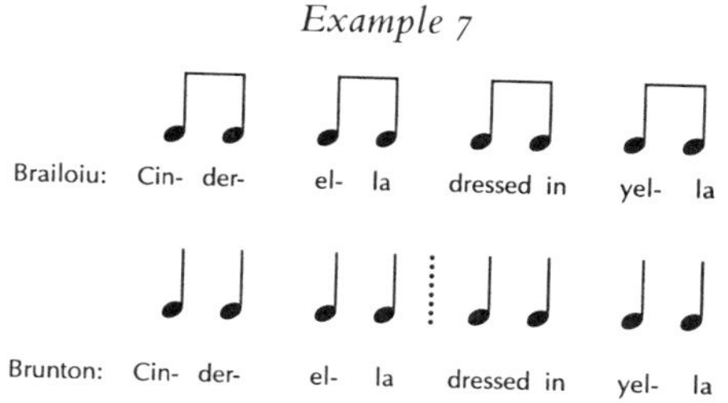

Example 8

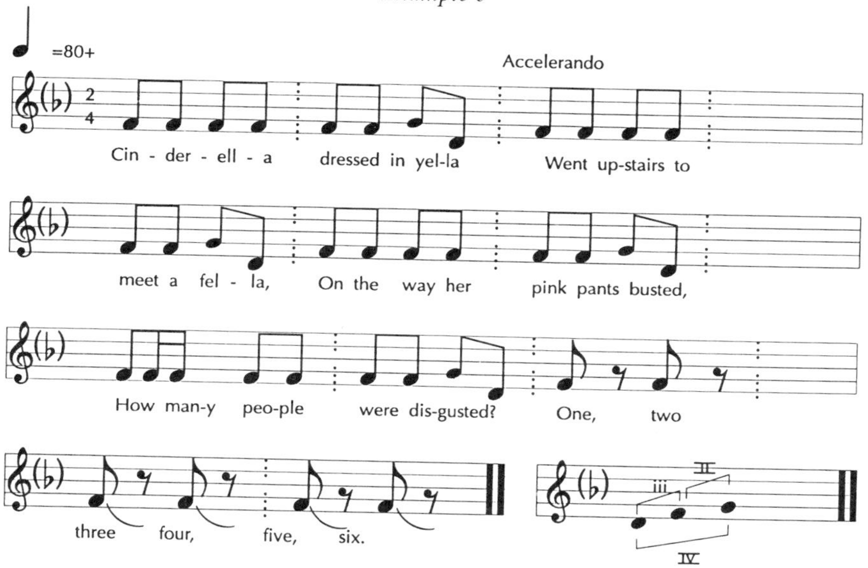

♩ =80+

Accelerando

Cin - der - ell - a dressed in yel-la Went up-stairs to

meet a fel - la, On the way her pink pants busted,

How man-y peo-ple were dis-gusted? One, two

three four, five, six.

iii II IV

Example 9

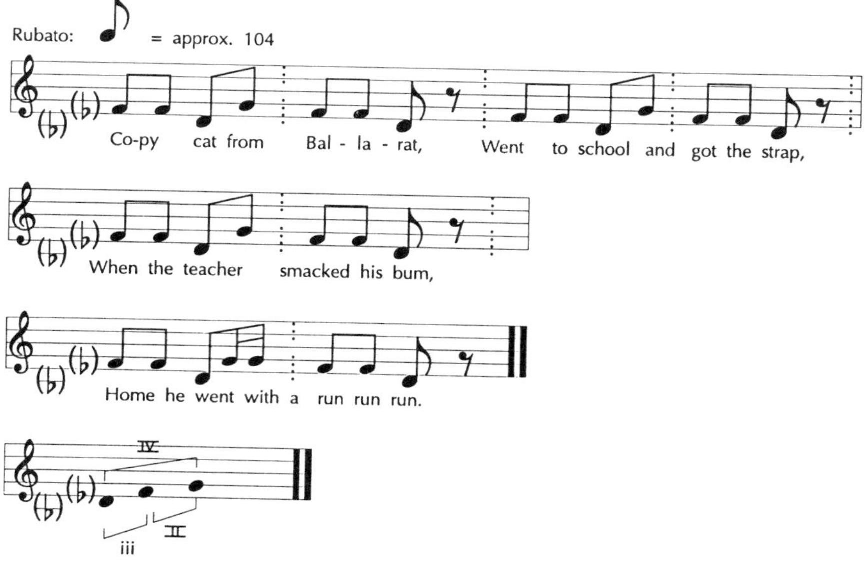

Rubato: ♪ = approx. 104

Co-py cat from Bal - la - rat, Went to school and got the strap,

When the teacher smacked his bum,

Home he went with a run run run.

IV iii II

Example 10

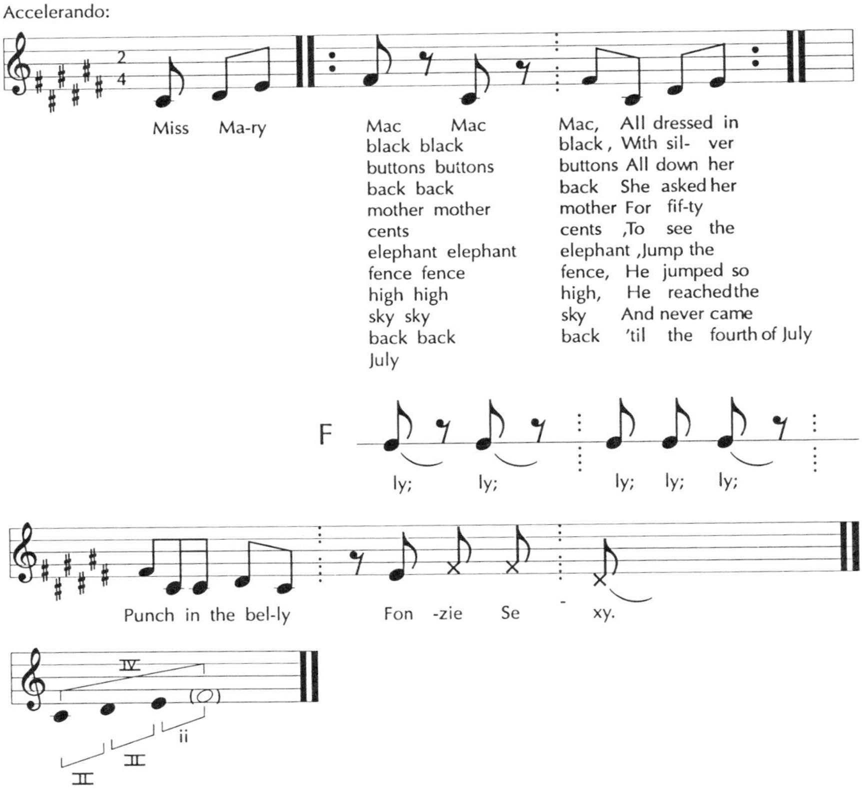

Accelerando:

2/4

Miss Ma-ry

Mac Mac
black black
buttons buttons
back back
mother mother
cents
elephant elephant
fence fence
high high
sky sky
back back
July

Mac, All dressed in
black, With sil- ver
buttons All down her
back She asked her
mother For fif-ty
cents ,To see the
elephant ,Jump the
fence, He jumped so
high, He reached the
sky And never came
back 'til the fourth of July

F

ly; ly; ly; ly; ly;

Punch in the bel-ly Fon -zie Se - xy.

IV

ii

III

III

Example 11

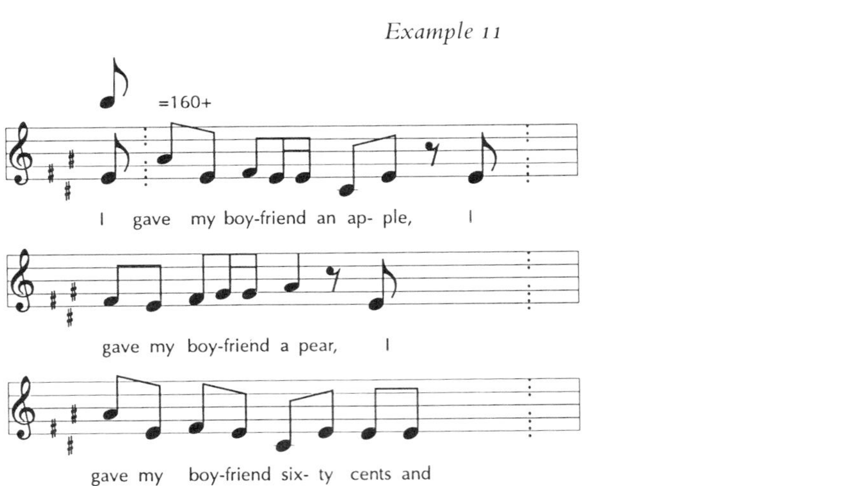
=160+
I gave my boy-friend an ap- ple, I
gave my boy-friend a pear, I
gave my boy-friend six- ty cents and

tore his un- der wear. I kicked him o-ver Aus- tra- lia I
(-)
kicked him o- ver France I kicked him o-ver Ha- wa- ii, And
3
done a hu- la dance da'l da da da.... d d.
iii
II
II
ii
IV

Example 12

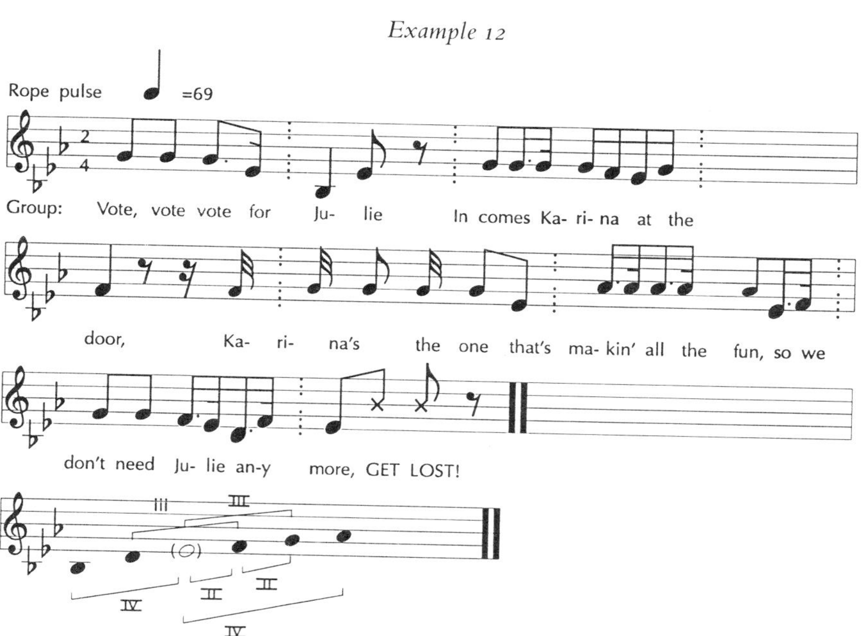
Rope pulse =69
2
4
Group: Vote, vote vote for Ju- lie In comes Ka- ri- na at the
door, Ka- ri- na's the one that's ma- kin' all the fun, so we
don't need Ju- lie an-y more, GET LOST!
III
III
IV
II
II
IV

Example 13

Rubato: ♩ =116+
We break up, we break down We don't care if the
school burns down, No more Eng-lish, no more French,
(dewn)
(+) (+)
No more sit-tin' on the old school fence.
Accelerando
(-)
If the tea-cher in- ter- feres, knock her down and
(dewn)
(++) (+) (+) (+) (+)
box her ears, If that does not serve her right,
(+) (+) (+)
Blow her up with Veg- e- mite.
V
iii
III
II
II ii II
IV

Example 14

(-)
I'm Pop-eye the sai-lor man, full stop, I live in a car-a-
van, full stop, I op-ened the door and
fell through the floor I'm Pop-eye the sai-lor
man, full stop, full stop, com-ma com-ma com-ma
full stop, full stop.
III
II
iii
II
IV
V
II
ii

Example 15

=200+

C ♯

I wan-na be a bra, bu bu - du bu, A cross your heart

bra, bu bu - du bu, My boobs are pum-pin'

C

PO-WER du du du du du - dl du - dl du - dl

du du du - du du du - dl du - dl du

I hate be-in at the back, I like be - in on the

track, I wan - na be a bra, bu bu - du

C ♯

bu, A B - M - X bu bu bra!

(Get your B M X bra at toy stores everywhere.)

REFERENCES: G. Davey and J. Factor, 'Cinderella and Friends. Folklore as Discourse', paper presented at the ANZAAS conference, 1980; L. Geller, 'Linguistic Consciousness-raising: Child's Play'. *Language Arts* 59, 2, 1982; H. Hall, 'Children's Playlore: Implications for the Teaching of Poetic Language'. *Educational Magazine* 40, 3, 1983; A Study of the Relationship between Speech and Song in the Playground Rhymes of Primary School Children, Doctoral dissertation, 1984; 'When All's Sung and Done: Music in the Primary School Playground', paper presented at the XVIII International ISME conference, 1988; 'From Playground to Schoolroom: The Joy of Rhyming'. *Proceedings from the 18th National Conference of the Australian Early Childhood Association*, 1988; G. List, 'The Boundaries of Speech and Song' *Ethnomusicology*, 7: 1–16, 1963; Wendy Lowenstein, Field Recordings from Western Australia, State Library, Melbourne, 1960; I. Nash, 'Reciting the Real Benefits of Rhyme'. *Times Educational Supplement*, 1987; Opie, I. and P., *The Lore and Language of Children*, St Albans, Paladin, 1959; Moira Robinson, 'Children's Playground Rhymes and Traditional Verse'. *Pen* 47, 1984.

Hazel S. Hall

Musical instruments After the first European settlement of Australia and up until 1910 or so, our population was largely rural, and people had to entertain themselves, often in remote situations and primitive conditions. The musical instruments played were simple, portable ones like the fiddle, the concertina, tin whistle and so on; quite a number were home made. After the Second World War modern transport enabled people to travel to centres with more sophisticated professional music outlets, and radio, juke box, picture theatre and later television provided commercial entertainment, augmented more recently by the video and the disco. Musical instruments in Australian folklore were in some danger of becoming museum pieces.

A strong folk revival since the late 1950s has resulted in the formation of a number of specialist groups in the cities, and many of the instruments described here as 'bush instruments' are again being played in the city. However, the specialist clubs and modern city 'bush bands' are playing the instruments differently (and using a different repertoire), following the Anglo-Celtic style, a demonstration of the obsession with revived British folk music which did *not* hold a major place in traditional Australian folklore.

The early Bush Music Club (q.v.), formed in Sydney in 1954, at first concentrated on collecting and promoting bush songs, and its original Bushwhackers' Band recorded some of these on the 78 rpm records of the day. In turn, a number of traditional fiddle, accordion and concertina players, originally from the country, were attracted to the club, and so today there is an inner circle of dedicated people perpetuating the authentic Australian songs, dance tunes and style of playing the instruments of the bush. Pioneers in this field of collection include John Meredith, Alan Scott, Bill Scott, Ron Edwards, Warren Fahey, Hugh Anderson, David De Hugard, Chris Sullivan, Mark Rummery, the late Joy Durst and the late John Manifold. Many of their collections have been lodged with the National Library of Australia's oral history section and this material will become available to be played by future generations.

Following the success of the Bushwhackers' Band, particularly with the play *Reedy River* (q.v.) of 1954, city-based bush bands suddenly formed everywhere with lagerphone, bush (tea-chest) bass, spoons and the ubiquitous guitar (which previously had not belonged to the bush) as the major support instruments; and this obviously was not representative of a typical get-together of musicians in the bush. John Manifold (*Australian Tradition* 33, September 1973) suggests that these types of groups were more reminiscent of the 'Foo Foo' bands of the ships (skiffle groups). At least these early bush bands included a selection of bush songs, recitations and tall stories among their repertoire.

Subsequent generations of bush bands have become more professional, with the trained violinist and inclusion of instruments such as the piano accordion, perhaps the English-style concertina or the multi-row button accordion played in the English melodeon or Anglo-Celtic style. It was then possible for the repertoire to include — and in due course be dominated by — racing Irish jigs and reels in a variety of keys. The piano accordion, English concertina and multi-row button accordion all allow the hands to travel quickly about the keyboard, minimizing bellows action, and play relatively lightly and smoothly, relying on modern amplification to project the sound, and on rhythmic support by instruments such as guitar and *bohdran* (Irish hand drum). Except possibly for the Irish fiddler who played for solo items and step-dancing at country dances and at concerts, this style of music, which is now promoted by modern bush bands and the specialist folk clubs, was not a major part of the tradition in the Australian bush. John Meredith said in *Folk Songs of Australia* volume 2:

> Sadly, these — for the most part very talented — young [folk revival] musicians turned their backs upon the very tradition they imagine they are keeping alive. Almost without exception their music has been learned from Irish fiddle-tune books, and their repertoires are exclusively jigs and reels. Most of the dances they perform have come from the same source or have been recently made up. Yet they call themselves 'bush bands' and identify themselves by names having a strong Australian flavour.

People in the bush made use of whatever was available (home-made instruments such as kerosene-tin dulcimer, saucepan banjo and cigar-box fiddle were quite common), and there could be unusual combinations. An illustration in the *Bendigonian*, 14 January 1897, depicts a Bendigo miners' camp on the goldfields at Mulgarrie, Western Australia, with three musicians. The lineup is tin whistle, violin and sax-horn. Another photograph of the Colemane brothers at Brawlin near Cootamundra, New South Wales, around the turn of the century (*Singabout* 6, 1, 1966) shows a combination of clarinet, concertina, and violin. A further photograph (courtesy Ray Grieve) taken somewhere in Victoria about 1905 shows a bush scene with a group of musicians sitting in front of two tents playing mandolins and a pedal harmonium. There is at least one recorded instance (Yandoit House, Yandoit, near Daylesford, Victoria — 1920s to 1950s) where dances were held to the music of a harmonium, sometimes in combination with button accordion. Another unusual instance is that of dances held near Euroa (1930s) solely to the music of a one-string fiddle, home-made from blackwood, a gramophone horn and pick-up, and piano wire. Family dances were held in the kitchen of the Cusack family at Moyreisk near Avoca, Victoria, during the 1920s and there were no musical instruments; Mrs Cusack 'lilted' tunes and they would dance a set such as the Lancers without a break between figures, or the call. Gumleaf playing (see **Leaf playing**) was at times utilized. Mr Arthur Lumsden has an account of a 'shearers' band' which consisted of a mouth organ and several Jew's harps tuned to C, E, G for chordal accompaniment.

These are one-off instances, however; although sometimes a bush community could muster up several musicians for special occasions — in which case the lineup would generally include some or all of fiddle, concertina, button accordion, tin whistle, bones, piano, mouth organ, and possibly a bass drum or one made from a kerosene tin and kangaroo skin — it was more common to have a lone fiddler or 'squeezebox' player. It was the German single-row button accordion played on the knee and the German-style concertina above all that predominated in the bush. The accordion, even more than its concertina cousin, was wheezy and sluggish in comparison to its modern counterparts, but easy to play and ideally suited to the dances that were most popular in Australia. These were not Celtic, but dances of European origin (see **Bush dance**), well established by the days of the gold rushes of last century and concurrent with the rise of popularity of the squeezeboxes. The waltz, Schottische, polka, Varsoviana, barn dance, two-step and simple set tunes and single jigs and reels in major keys were perfect for the 'oom pah' button accordion and German concertina. Good dance music must have internal rhythm or lift to make the music bounce and to lift the dancers' feet. This can be provided by having some notes short and some long. The single-row button accordion played on the knee and the German-style concertina swooping down at the ends of phrases fully utilized dotted-note rhythms with extra emphasis from the critical push/draw bellows action. The method of playing maximized use of chordal vamp on the beat. The fiddler used a short up-and-down bowing technique similarly to make very danceable music, and this can be particularly noticed in the Charlie Batchelor style on recordings held by the National Library of Australia. Dance pianists, in a similar fashion to the button accordionist, played vamp — all fingers down on the chord notes under the melody note, coupled with alternate bass vamp in the left hand.

These techniques developed to the full because of the need to 'fill a hall' with music in the absence of amplification and rhythm instruments such as guitar and *bohdran*, and because there was often only a lone player. Only a certain type of music suited these players and the tunes were modified because of the action of the instruments, particularly the critical push/draw bellows action of the squeezeboxes. Tunes were simplified and stripped down to a basic form so that a strong rhythm was maintained, and then dressed up again with triplets, dotted notes, chordal coupling and downbeat vamp automatically following the instrument's action.

Where did the tunes come from? Again there is enormous variety in origin. The role of the piano has been overlooked as one of the contributing sources of initial dance tune material; dance sheet music was widespread, and there were special editions for children learning piano. Then there were tunes from the music hall, brass band, broadsheets, etc., plus a core of original dance material both from Britain and Europe. All of this at some stage could become part of the bush musician's repertoire as he or she was quite able to pick up tunes, or snippets of tunes, by ear. Changes might well occur along the way. The late Clem O'Neal explained that new tunes were learned when people were away on shearing and droving trips. In some cases they would remember the whole tune when they returned, but quite often they would have forgotten part of it. In these cases they would either combine parts of several tunes or else they would make up a piece to fill in for the bit that had been forgotten.

> The only way things was, was that someone would go away on a shearing trip and he'd remember part of a music, part of something. He'd have to keep it in his head; when he came back perhaps he'd remember only part of it. So to make up a dance tune, he'd probably remember parts of three bits of different things which someone had played in a town or somewhere or other, and he'd combine them together. Someone else would hear him play that, and eventually new tunes got created from one listening to the other and these seemed to go right up and down some twenty or thirty miles along the river. (*Concertina Magazine* 1, winter 1982)

Note that this is precisely how 'folk music' developed in the days before notation.

Stringed instruments
The *violin* or *fiddle* was one of the earliest instruments to be brought out and played by the first European settlers and it held top place for entertainment along with the piano. Later in the nineteenth century the button accordion, concertina and mouth organ became equally prominent. It was not until the 1920s that modern combinations of piano accordion, saxophone and the like competed for pride of place with the violin. 'Here, as in Europe, the violin was heard as a "folk instrument" before it reached the concert hall', according to John Manifold in *The Violin, the Banjo and the Bones* (1957). It is the style of

playing and repertoire that differentiates the ear-playing fiddler from the trained violinist. The fiddler will rest the violin under the chin, the instrument pointing out or down in a relaxed manner in contrast to the strict upper hold of the trained violinist. Many fiddlers use even more unconventional holds such as resting the violin in the crook of the arm or even on the hip or chest. The bow also may not be held in the classical manner; some fiddlers hold it from at least a third of its length from the base. Hence, the fiddler plays lightly with short, sometimes rapid, up-and-down bowing, 'jigging' out lengthy notes, whereas the violinist would sustain with strong vibrato and long bowing. Fiddlers use only occasional vibrato, if at all, to highlight an end-of-phrase note. The slower tunes such as waltzes and Schottisches are played in a bright danceable rhythm — almost jig style. Usually fiddlers play only in first position and make use of open strings. Some players were in the habit of tuning a tone or more below concert pitch. A few who were used to playing alongside a fixed-key instrument such as a button accordion or concertina would retune the fiddle so as to be able to follow usual finger positioning when confronted with another squeezebox tuned in a different key. Fiddlers tend to favour jigs, reels, hornpipes, and the more traditional set tunes, Schottisches, waltzes, polkas, mazurkas, Varsovianas and barn dance tunes from earlier times.

References to early dances held in Australia, particularly in the bush, highlight the community importance of the sole fiddler. He was often an Irishman and equally adept as an MC, dance instructor and player for solos and step-dancing. Some players had both musical training and the ability to fiddle by ear. Many old-time dance-band violinists made use of strong bowing on the upper strings, particularly the E string, so as to be heard above the squeezebox and piano, and were equally at home with the slow waltzes, or fiddling jigs and lively set tunes for the quadrille and Lancers.

Home-made fiddles were quite common. Materials used included satin ash, yellow wood, orange box and even wood from the apricot tree. Bush ingenuity prevailed in remoter districts and one amazing example is a violin made from galvanized iron in perfect shape with scrolled neckend. Another, held by the National Museum of Australia, is made from a coconut shell. The cigar-box fiddle (one string) was also quite a common innovation of the bush musician.

The traditional Australian fiddler has almost disappeared. There is a certain revival in the city folk clubs, but the revived Irish Celtic tradition often draws classically trained violinists. John Meredith et al., *Folk Songs of Australia*, volumes 1 and 2, has a good representation of Australian fiddle music collected from elderly traditional players. This material is gaining interest from a core of young enthusiasts in the bush music clubs and at National Folk Festivals.

The *Stroviol* (Stroh violin) takes its name from manufacturer Charles Stroh, and appeared around 1901. It has a type of megaphone for directionally projecting the sound, plus its own 'miniature fold-back', a smaller funnel which directs the sound back to the player's ear. It is tuned and played in the same way as a violin and is reputed to have three times the volume. In this instrument the usual body of the violin (sound box) has been replaced by an ingenious mechanism and membrane connected to the amplifying horn. It was invented primarily to improve the recording of violin music above other orchestral instruments for making Edison rolls and 78 rpm records.

The *one-string fiddle* (one-string cello), possibly a derivative of the Japanese fiddle, was a popular busker's instrument during the depression years of the 1930s. It also has a history of use in the bush, at dances, and like the musical saw was a popular entertainment instrument at concerts, masonic lodge functions and so on. It would be strung with either piano wire or the gut used for cello strings. Manufactured by the British Stroviol Company, it had an amplifying funnel like that of the

Stroviol attached at the base of the instrument, with a membrane to transmit the sound from the string to the horn. It was held like the cello over the shoulder, clasped between the knees, and bowed with an ordinary violin bow. Home-made examples were also known, one being constructed from Australian blackwood, with the funnel of a gramophone and membrane pick-up.

The role of the *piano* in Australian folk music and particularly dance music has been overlooked. Presumably this has been partly because it is not an easily portable instrument, but perhaps more because its role is considered the domain of the classically trained musician. However, the piano was used for dance accompaniment when available. By the 1920s, under the influence of the foxtrot and ragtime music, the piano had become an essential requirement.

In *Australia's Music* (1967) Roger Covell details the importance of the piano in the bush, and notes that in fact 700 000 were imported during the first 100 years. On grinding bullock drays they were transported to remote places to be housed often in wattle-and-daub huts, and were deemed more important than essential furniture. Pianos were also common in Mechanics' Institute halls, pubs, schools and other venues. In the absence of radio, not to mention television and video, the piano was the mainstay for family entertainment, singalongs and keeping up with the latest tunes and music of the fashionable dances.

Sometimes no end of trouble would be taken to get a piano for a dance, as described in *Then the Water Wheel Turned*, a history of Lockington, Victoria (1967):

> There were many dances in those days in barns or woolsheds where we danced until morning to the music of violin, accordion and sometimes a piano . . . No obstacles were let come in the way of our fun in those days. The McIntyre boys would bring along the piano for dances in Rankin's barn. The instrument was conveyed in a four wheel carry-all, horse drawn of course, then loaded again for the trip home after the dance. Everyone would lend a hand, and nobody worried about getting home in the early hours of the morning.

Even pianos that were badly out of tune would be usable. Many pianists (the untrained variety) were able to automatically transpose the key so as to get the piano over the 'cracks' and tune in with the squeezebox, e.g. play in A flat while the accordion or concertina was in G. It was not unusual for them to carry a tuning key and 'touch up' the main piano notes of the chords of the accordion or concertina key, if things were really tough. Dance pianists had an instinctive style of playing and improvising vamp and rhythm accompaniment. A piece of music played by a classical sight reader could be 'undanceable', yet the same piece of music played by the dance pianist (whether a reader or an ear player) would be rhythmic and enjoyable to dance to. Most dance pianists played by ear; some could read music as well. The bass accompaniment was strong and by the turn of the century alternating single notes and chords. The melody was also played vamp style, whenever notes sounded by the little finger (or third) could be sustained in suitable passages, and the lower fingers strike lower chord notes below the melody note on the off beat.

The *mandolin* is a simplified lute with eight strings tuned in four pairs, as for the violin. It is a fretted instrument which developed from the mandola. It is the rapid striking back and forth across the double strings that gives the mandolin its beautiful characteristic ringing sound and tremolo. Fingering is the same as for violin but aided by fretwork positioning and a plectrum. Although the mandolin can be used for rhythmic chord strumming, it is more usually deployed to pick the tune. Advanced players will utilize the instrument to its fullest by using the fine tremolo action for which the double strings were intended, and also by using all the lower strings for chord harmony while playing the melody on the upper string(s). It is commonly regarded as the national instrument of Italy. The Neapolitan (bellied) model was the typical instrument used in Australia and was popular in the bush around the turn of the century. It was, however, probably more a concert and self-entertainment instrument than a dance-band instrument. The flat-bodied model of the mandolin with steel strings has enjoyed a healthy revival in the cities, possibly influenced by the revival of American bluegrass music. The banjo mandolin, in contrast, seems to be shunned by the folk enthusiast, yet was a popular acoustic dance-band instrument earlier this century. It looks and sounds like a banjo, but is strung and played the same as a mandolin.

Presumed to be of African origin, the *banjo* was in use amongst slaves on the plantations of the southern states of the USA and became the minstrel instrument of the nineteenth century. The popularity of wandering 'Ethiopian' minstrel troupes (really American Negroes) was high in Australia by the middle of the nineteenth century and there is an account of the singing of plantation songs to the strumming of banjos by American Negroes (presumably runaway slaves or ship deserters) on the goldfields in the centenary issue of the *Bendigo Advertiser* (1951). The early fretless banjos had gut strings and were most likely played with simple chord accompaniment, since a 'picked' tune would not have been heard, at least in a dance-band situation. By the turn of the century and certainly later with the age of jazz, and aided by the newer steel strings, picking the tune became equally popular. By the 1940s banjo clubs had formed all over the country, and it was not until then that they were known in rural areas, except on rare occasions. Strangely, the banjo, which is an excellent rhythm instrument, takes a low profile in the contemporary folk revival; and it seems quite contradictory that it is almost absent from the bush band scene in contrast to the guitar, which never enjoyed popularity in the bush until far more recently following the 1950s American country and western craze.

Two styles of banjo at least had developed earlier last century — English five-string and American plectrum style. It is in fact the English who perfected the five-string banjo, used more chords, and also first developed the claw hammer mode of playing ('thimble playing'). A thimble was wrapped around the forefinger and using it with the thumb to vibrate the fifth string made it resonate with the other strings. This technique was used for playing quick marches, jigs, reels, hornpipes, etc. The American plectrum style was used on four strings and a certain amount of tone loss resulted because of the removal of the fifth string. There is a certain mutual lack of regard between players of the two styles.

The English five-string banjo had strings tuned in perfect fifths and minor thirds, making it a unique chord instrument. Playing chords across strings 3, 2, 1 at any fret produces a major triad with the root note on the third string, thus making the learning of pages of chords unnecessary. This cannot be achieved on any other fretted instrument. Players of banjo could alternate between chord accompaniment and 'breaks' where the melody would be picked with great effect.

Despite early records of banjo in Australia, particularly as a minstrel and performer's instrument, it was barely known in rural areas until well into this century. It was an expensive 'middle-class' instrument and well beyond the resources of the average 'bushie'. Meredith and Anderson note in *Folk Songs of Australia*, volume 1 (1968), that 'Of the hundreds of items forming the basis of this book not one tune was played upon nor a single song accompanied by a fretted instrument'.

The *guitar* was never part of the Australian traditional music scene until the 1950s and 1960s folk revival in the cities. It seems to have followed in the path of American hillbilly music of the 1930s and more recently the country and western influence of the 1950s coupled with contemporary folk singing trends. John Manifold in *The Violin, the Banjo, and the Bones* makes the point that it may have enjoyed some popularity amongst the German settlements of the Queensland coast, but most nineteenth-century references to guitar in advertisements of the day promote it as a genteel classical parlour instrument for upper-class ladies.

In the contemporary bush band scene, judicious use of the guitar provides accompaniment once provided by the piano, and is most useful generally for singers accompanying themselves. It is portability, conveniently provided by modern transport, that has helped the guitar to gain popularity to the extent that was once enjoyed by the button accordion, fiddle and concertina. Don Topley in an article in the Bendigo *Advertiser*, 17 August 1988, on the Wedderburn Old Timers made a point about the contemporary city bush bands:

> The guitar of course, predominated, yet it is certain that no shearers' shed dance or drovers' camp sing-song had a guitar — it was too bulky, too sensitive to heat, cold, and damp; and too quiet.

Whistle and flute group

Piccolos were popular in the bush, perhaps more so than flutes, and the *fife* was a common instrument in school bands and bands for drum and fife. Most likely the fife was one of the first instruments to be brought out by our settlers along with the whistle, owing to its portability and low cost. Fifes were also the instruments of the early regimental bands, and later the Volunteer Rifle Corps which were established throughout the country by the 1860s had drum and fife bands. Dances were held to the music of the drum and fife band during this period. The fife and *tin whistle* have the same fingering so it is understandable that a large proportion of the population of the day changed from one instrument to the other, although the fife is a side-blown flute whereas the tin whistle is an end-blown or fipple flute belonging to the flageolet family (although the tin whistle has all six holes at the front, whereas the flageolet has two thumb holes at the back). The tin whistle or 'penny whistle' (so-called because a penny was given to the player for each tune) is no longer made of tin (except for the Clarke brand) but of brass, which is sometimes nickel-plated and now costs several dollars or more. Some older instruments were also made of wood, bone or celluloid, and sometimes bushmen whittled their own instruments out of a hollow stem or bamboo. Because of its low cost and portability, the tin whistle, along with the mouth organ and Jew's harp (see 'Percussion', p. 288), circulated widely to all parts of the country. It also had the advantage of being unaffected by sun, water or dust. Like mouth organs and button accordions, tin whistles are fixed-key (diatonic) and come in a range of keys determined by the diameter and length of the instrument. To play in another key it is normal to change whistles and play again with the same fingering. For the next octave repeat the fingering but blow harder. It is a very easy layman's and ear player's instrument. It is easy to get some accidentals by 'half noting' (half covering the hole) or by certain finger combinations, and therefore it is not hard to play key of G on a D instrument or F on a C instrument. The whistle is an excellent instrument for solo performances and effective for playing various music including hornpipes, reels, waltzes, jigs and barn dances. The simple technique (fingers off one by one to go up the scale) allows the performer to slide ('smear') from one note to the next, which is a characteristic feature of the tin whistle sound, accompanied byn appropriate triplets and trilling. Because of this simple but characteristic style, renowned flautist James Galway considers the tin whistle to be far more versatile than the recorder, and an excellent lead-in to the classical flute. Perhaps the simplest of all instruments, the tin whistle, in the hands of an experienced player, can make the finest music.

Because the whistle was so widely played in the bush in days gone by it was considered an essential component of any old-time bush band, along with the fiddle, squeezebox and the bones. In the right hands it could be the sweetest of all whistles and flutes, 'floating' and trilling above the other instruments of the dance band. An example of the old-time Australian style of whistle playing in a bush band can be heard on the first EP recording put out by the Wedderburn Old Timers Orchestra, and subsequent LPs. There has been a resurgence of whistle playing in the folk club and modern bush band circuit, but this tends to be in the revived Anglo-Celtic style of Britain and Ireland.

Free-reed instruments

Almost every Australian country home or bush shack had a 'squeezebox' (either a German-style concertina or a button accordion) or a mouth organ. They were as much the layman's instrument as the guitar is today, and are the instruments most of our elderly folk think of as the fundamental component of an old-time bush band, along with the fiddle, tin whistle and bones.

Free reeds were first developed in China about 2500 BC in the form of the Sheng, a type of mouth organ with a gourd as a wind chamber and a number of bamboo wind pipes.

The modern free reed is an independent metal tongue fastened into a slit so that it can vibrate to produce a note when activated by a flow of air from the lungs (mouth organ) or bellows (concertina and accordion). The length and thickness of the reed determines the pitch, which can be sharpened by filing metal from the tip, or flattened by filing metal from the base. There are two reeds per space with an open/shut valve (plastic or leather strip) system so that the first note of the scale is sounded by blow or push and the next note of the scale by suck or draw. As an example, consider the key of C. The first note of the scale is C on a blow or push, then suck or draw for D, and so on right up the scale with each successive mouth organ space or squeezebox button. However, if this were followed through, the next C, an octave higher, would be a draw. The manufacturers of these instruments made the B a draw note and the C a blow note. By this slight rearrangement of the basic blow/suck scale, all notes blown together will be in tune, sounding in tonic chords, thirds, fifths or octaves; likewise on the draw (dominant). It is an inherent quality, therefore, that these instruments are very rhythmic because of the critical push/draw system. It will be immediately apparent that these instruments are fixed-key, but instruments in other keys are available, pretuned and arranged in exactly the same pattern and played as that described above for key of C (e.g. in the key of D, D is blow/push, and E suck/draw). No sharps or flats can be played, except those that are an inherent part of that key.

The *mouth organ* or *harmonica* (also called the harp) was played in Australia from the 1830s. Christian Buschmann of Berlin, a piano tuner possibly inspired by the 'Jew's harp' (a simple free-reed instrument), developed the first mouth organ, called 'mouth aeolian', as a simple piano tuning aid in 1821.

The harmonica took on worldwide popularity when in 1857 clockmaker Matthias Hohner founded the first production line factory in Drossinger, southern Germany.

A range of mouth organs in most major keys is available; the simplest is a ten-hole single-reed instrument covering a scale of two octaves from the fourth hole (the lower three holes do not form a complete scale, but are used for chords and effects played under lower melody notes by 'tonguing'). Many unique effects can be achieved on a mouth organ. Two ways of sounding notes are commonly used. In 'lipping' the lips are pursed only over the space of the note to be sounded (in a manner comparable to sipping through a straw). With 'tongue blocking', the mouth covers four holes with the lower three blocked by pressing the tongue to the side across them. The top note is sounded by blowing from the corner of the mouth, and the lower notes are produced in vamp-style rhythm by the tongue lifting on and off to the beat. Most bush musicians played in this style. The hands are cupped around the instrument and the vamp is further enhanced by lifting one hand on and off in time. Wavering the hand creates a pleasant tremolo. Slowly altering the cavity size of the cupped hand, coupled with sudden draw or expired lung capacity, gives a 'wah wah' effect. Lipping can be used to the tune in a staccato fashion, and glissades can be produced by rapidly running the lips up or down the instrument. Bush musicians had a trick of holding a tumbler or pannikin in cupped hands below the instrument to improve resonance, and until recently special 'echophone' mouth organs could be purchased with a cup-shaped receptacle (the instrument attached to its side) so that a warble could be achieved with ease while playing.

Two-holed double-reeded instruments were also popular. One row of reeds would be slightly out of tune with the other so that when they were played together a pleasant French accordion tremolo sound (termed 'beating') would be heard. These instruments are louder than the single-row mouth organ, but cannot achieve all special effects such as 'wah wah'.

A 'blues harp' is a ten-hole single-reed mouth organ, but the reeds are thinner to aid 'bending' of the pitch. With blues music the instrument is played in another key (for example one might play in the key of G on a C instrument) so that the normal blow/draw relationship is reversed. The music (particularly the end-of-phrase sections) uses mostly draw notes, which, with sudden added lung capacity thrown in, increases the 'wah wah' effect and makes it easier to achieve 'dragging down' of notes ('bending') to give semi-tone intervals. Several keys, including minor keys, can be played on a C harp in this way. This unique sound is characteristic of the blues mouth organ style.

The chromatic harmonica was produced in the 1930s by adding a C sharp row alongside the C row. A slide closes off air to the C sharp row until the slide is depressed, when the corresponding accidental above the natural note is sounded. All semi-tones are therefore available to the player. An average player would still play in the 'home key', simply using the slide for occasional accidentals to the tune, whereas an advanced musician could theoretically play in any key, although the reading of music might be necessary.

The mouth organ was one of the most widely used instruments played in the traditional style for entertainment, concerts and dances in the days of our grandparents and their forebears. There were, and still are, mouth organ bands such as the Yarraville Mouth Organ Band in Victoria (established 1935). It was also often the instrument that led on to button accordion or concertina playing. Although the sale of mouth organs remains high today, traditional-style entertainment is rare, even in the folk scene, and most modern performers play in the American blues style.

Development of the *button accordion* (sometimes now known as the 'melodeon', although this is a name which rightly belonged to the suction-operated American reed organ, and is now sometimes used for the modern 'melodica') dates back to Buschmann of Berlin, 1822, with his construction of the Handaoline — a bellows-operated mouth organ designed as a piano tuning aid. Buschmann's Handaoline was described in further developments as a 'conzertina'. Damian of Vienna improved on Buschmann's instrument and in 1829 the 'accordion' (Akkordeon, Hand Harmonika) was patented; Wheatstone's English concertina was patented also in 1829 as an improvement on the accordion. The word 'accordion' derives from the German word *akkord*, meaning chord or harmony, as its scale is arranged to chord (arpeggio) progression.

The accordion has on its right-hand side the same scale as described generally for the free-reed instruments and comparable to the mouth organ. It is operated by button keys and bellows with the push/draw system, two notes of the scale per button. As with the mouth organ, the button accordion is pretuned to a diatonic fixed key, i.e. the scale is seven-tone with no chromatic intervals. The accordion's range is usually about two and a half octaves, commencing from the third button. The left-hand side has a plain 'oom pah' bass which in the simplest form is operated by two spoons or buttons. On the push the lower button gives the bass tonic note (the keynote), and the upper button, the octave tonic chord. Similarly on the draw it will be the equivalent dominant bass note (a fifth above the bass tonic) and chord; thus for a key of C instrument the push chord will be C and the draw chord G. Generally, therefore, the bass automatically tunes with the melody. The right-hand scale is arranged mouth-organ style so that push notes run in order of a C arpeggio, C, E, G, C, and the draw notes are D, F, A, B.

Accordions were advertised for sale and were in use in Australia at least by the 1850s, along with flutinas, concertinas and harmoniums. However, the early instruments were small, not well-made, brass-reeded and therefore had a short life span. The brass or German silver reeds were on a single mounted unit, similar to a mouth organ reed bank; if a reed went out of tune or broke (a common occurrence with brass reeds), it was difficult to replace. Although the early German concertinas were not particularly durable either, there were at least some well-made and reliable English brands in the German style, so these were the preferred dance instruments (where extra volume might be required) from the 1870s through to about 1920. Other reasons for the early concertina preference and the rise of popularity of the accordion after 1900 are discussed in *Concertina Magazine*, issues 20, 21, 22. Probably the early accordions were more popular for home-made entertainment rather than as a dance-hall instrument. They might lend themselves to a Sunday session in the vestibule or on the veranda, the family dance in the kitchen, and as a learning instrument for aspiring young musicians. In fact there were special children's models such as the 'Peter Pan' and 'Little Lord Fauntleroy'. Some had curious built-in effects of special interest. The small 'Ajax' model had its own peculiar 'vox humanum'. The air passed through a wooden chamber between the bellows and the reeds, via a wooden lid hinged with chamois, and weighted with a small lead ball. Operation of a lever with the thumb while playing raised the lid a little so that it fluttered, setting up a vibrato. The instrument would literally shake while in operation, and the sound was curious if not pleasant.

After the turn of the century good-quality accordions were manufactured in larger numbers. They now had reliable steel reeds mounted separately in blocks and sealed with beeswax. Reeds could be individually replaced and the beeswax resealed with a soldering iron. The modern accordion retains this type of reed bank, and, experienced players know not to leave the instrument in the summer sun or in a car on hot days, as the wax can melt and the reeds fall out.

The majority of the early accordions were of German manufacture and this remains so today. Such was the popularity of the button accordion by or after the turn of the century in Australia that it was dubbed the 'bush accordion' (there was also a brand by that name), presumably partly to differentiate it from the modern piano accordion which would have been catching the eye of trained musicians by the 1920s, and certainly by the 1940s (although it was invented as early as 1852, by Bouton of Paris). Button accordions suit an ear player whilst piano accordions are better for those that need to be trained in playing or reading music.

The bush accordion was a 'knee accordion' and it was usually a stop accordion, having several banks of reeds which could be brought in and out of action by operating knobs at the top of the accordion. These were connected to wooden slides with slots that opened air passages to the respective reed banks. This type of accordion could have two, three or four such stops, one which could couple one row of reeds an octave higher, one giving an octave lower, and one which would give tremolo or 'beating' because it was slightly out of tune with the main reed bank. Obviously operation of the coupling stops altered the tone and volume as well. Popular German accordions of the day included the Ludwig, Imperial, Melba, Monarch and so on. Such was their popularity in Australia that the German factories stamped special metal work to make the instruments seem Australian models and advertised as 'specially designed for the Australian climate', sometimes with Australian animals and names.

The most highly regarded knee or stop accordion was the 'Mezon Grand Organ', which was made in Saxony. It was extremely well made, with high-quality steel reeds, and possessed its own unique haunting tone. Regarded as the Rolls Royce of the bush accordions, the Mezon could fill a dance hall with its sound, and the instrument was seldom known to go out of tune, fracture a reed or break down. The Second World War saw the end of production of the Mezon, and the most reliable to take its place and find acceptance in the bush was the four-stop Hohner, and the 'Vienna' model.

Perhaps because of competition from the modern piano accordion, which was fully chromatic (i.e. has a piano scale with all the sharps and flats), produced the same note push or draw and had every chord facility in the left hand, more sophisticated models of button accordion were brought into production. Two- and three-row Hohner 'Vienna' models and the 'Erica' and 'Corso' were manufactured, so that choice of two or three keys usually a fourth apart with extra chording facility, including some minor chords, were available.

Within the folk revival scene of the cities the two-row D/G Vienna accordion is most popular, together with the A/D/G three-row instrument. This allows the accordion player to follow the whims of the string players (violin, mandolin, guitar) who in these keys maximize use of open string and first position fingering. There are also other button accordions which can be played in any key, such as the C/C sharp and the B/C instruments, and the chromatic 'button row accordion' favoured for Scottish country dance music. With these more sophisticated instruments many complex jigs and reels that were once the realm of the Celtic fiddler can be played. In Australia this style has only developed in the Anglo-Celtic revival scene of the city folk clubs where the piano accordion, English system concertina and melodeon style of playing is to the fore. These instruments play relatively smoothly because they are heavy, requiring shoulder straps and/or have the same note whether one plays push or draw, thereby minimizing bellows action.

The typical style of the old-time bush accordionists was quite different and distinctive. Generally they preferred the single-row knee accordion, such as the four-stop Hohner or a two-row Vienna model which gives a choice of shoulder-strap or knee-playing style. Because of the push/draw bellows action, and because the instruments were light and could be played on the knee where extra anchorage and arm leverage was maximized, the music was punched out with beat and end-of-phrase emphasis which the modern sophisticated instrument players generally lack. The old bush accordion was a superb dance music instrument and its style is seldom reproduced by the modern accordion players.

Apart from the inherent rhythm of the button accordion, the style and effects are quite individual. The simplest style is the one-note melody, but most players add some lower notes as well. Some play in octaves; some accentuate the rhythm in a similar effect to the 'tongue blocking' of the mouth organist, by bringing all fingers down to add extra chord vamp below the melody note, on the beat and/or off-beat. Other unusual effects can be given on held notes by rolling the fingers over lower buttons and, in the case of the knee accordions, a waver given by shaking the leg in the style of a musical saw player. Because of the need to maintain air in the bellows, passages with many long notes are generally modified with a few push/draw intervals, a little in the style of a fiddler 'jigging' out a minim.

As he was often the sole player, the old-time squeezebox player made full use of the bellows and bass rhythm which was essential for dance music. The bush musician was happy to play in one key all night, and would not have been able to carry more than one accordion to a dance anyway. The Nariel musicians play in C this way for a complete dance, whilst some generations back their forebears played the Mezons in A. Harry McQueen preferred the C/F accordion, playing F on the inner row which allowed him extra chording facility. The Wedderburn Old Timers Orchestra play in G, C, or F, swapping at times between two-row accordions, the key having been selected earlier to provide a suitable singing range.

Whilst the simple button accordion might lack the versatility of its modern counterparts, it more than makes this up with character and is the 'dinkum' folk instrument. Australian button accordion music can be heard on recordings by the Wedderburn Old Timers, and on a cassette entitled *Bush Fest* from Mudgee, available from the National Library of Australia. Many recordings of various accordion players (e.g. Harry McQueen) have been lodged with the National Library.

Concertinas are of three main types — those using the English Wheatstone system, the German and Anglo-German type and the Duet type. Only the German-style concertina constituted any major part in Australian tradition. Like the button accordion, it was the squeezebox of the bush. A feature of the hexagonally shaped concertina is the location of each reed on an individual metal shoe housed separately in individual wooden compartments. It is this construction that gives the concertina its particular tone and 'ringing' sound, enhanced when the instrument is gently swayed while playing.

As mentioned earlier, Wheatstone developed the first concertina which in its earliest 'symphonium' form was patented in 1829. In its final form this instrument was double-action — i.e. sounded the same note on the push and the draw of the bellows — and it was fully chromatic, having the same range as a violin, and was therefore playable in any key. To play a scale, the player works up two inside rows of buttons, alternating from row to row and from hand to hand. The additional two outside rows of buttons are the accidentals (sharps and flats).

Bush musicians generally found the English concertina impossible to fathom if they were ever 'unfortunate' enough to find one. The Wheatstone instrument lent itself to playing from written music and was accepted or tolerated in classical circles. It became a stage and concert instrument, but is relatively unknown in these circles today. It is within the contemporary folk scene in England and the Anglo-Celtic folk club circles in the cities of Australia that the Wheatstone system and the Duet system/s concertinas find the greatest revival of interest (the Duet concertinas have a treble scale on the right hand and a bass scale on the left hand, with buttons sounding the same notes each way). They formed the basis of the concertina bands and Salvation Army bands in England.

The German concertina (Angloconcertina) or 'konzertina' was brought into being in 1834 by Carl Uhlig. The German concertinas had long colourful bellows and were often referred to as 'Chinese lanterns'; they were not particularly durable instruments, having a mouth-organ-type reed bank like the early accordions, making playing with any volume and reed repairs difficult. It was English manufacturers who produced the good-quality reliable German-style concertina with individually fitted reeds in compartments and who also provided the Anglo-German model (1843) or simply 'Anglo' concertina. These are not to be confused with the English-style or Wheatstone-system concertina described above. The German-style concertina had the push/draw diatonic scale of the mouth organ or button accordion. It could be compared with the right-hand keyboard of the button accordion cut in half, so that the player played the first octave of a scale on the left-hand side of the concertina and transferred to the right hand for the next octave. The Anglo concertina combines the German style with refined English manufacture and has an extra row of accidentals added. Bush musicians generally did not know what these were for. The term 'Anglo' concertina is usually used when referring to the two-row German-style concertina of English manufacture, but technically applies to the three-row model which has the accidentals available.

The Anglo concertina shared many of the characteristics of the button accordion,

Mateship: *In the men's hut*, from *Illustrated Australian News*, 1 April 1891. (National Library of Australia)

A cluster of icons. John Williamson, Tamworth, *c.* 1984. (Photograph, *Northern Daily Leader*)

The giant guitar, Tamworth Country Music Festival. (Photograph, *Northern Daily Leader*)

Ted Egan's concert at the Longyard Hotel is a popular feature of the Tamworth Country Music Festival. (Photograph, *Northern Daily Leader*)

Musicians take shelter from the tropical sun beneath one of Darwin's traditional elevated buildings, Top Half Folk Festival, 1985. (Photograph, Keith McKenry)

Eric Bogle at the Hands of Fame Ceremony, Tamworth Country Music Festival. (Photograph, *Northern Daily Leader*)

Yothu Yindi, who combine Aboriginal traditions with modern musical styles, have been acclaimed in Australia and overseas. (Photograph, *Age*, Melbourne)

although the concertina was lighter to play and quicker in action than its 'wheezy' cousin. Because of its split keyboard, it was possible (depending on the tune) to play melody in both hands at once, an octave apart, or to use the left hand to play a counter-melody or arpeggios, or to simply vamp like the accordion. The shorter range in one hand of the concertina may have been its only disadvantage in comparison to the accordion, but experienced players were adept at 'jumping octaves' and turning the tune to keep it within the range of the five buttons in the right hand, and if needed, had the facility to momentarily carry the tune into the left hand or to cross-row to carry the scale higher on to the right-hand side.

The German concertina had two rows of buttons, i.e. it could be played in a choice of two keys a fifth apart — say, C and G. Some players played straight on the C row as if for mouth organ or button accordion. More typically many played the scale of C so far on the C row and then continued across to the right-hand G row to extend the range and to allow additional cross-coupling and chording with the left hand. Some bush musicians preferred G, as the higher pitch carried further; this would also be preferred by the fiddler. It was all a matter of personal technique and choice. Some players performed in a fairly quiet relaxed style, only gently swaying the instrument as the tune required. Many were real showmen and swung the instrument overhead, in circles and figure eights, dramatically swooping the instrument down at the end of phrases. This gave an unusual ring to the music, making use of the Doppler effect and could be compared to the revolving speakers of an organ. Some concertina players insisted on standing on a chair on stage so their style would not be impaired by overcrowding with other musicians. A few might perform the 'Cobblers' dance or 'Frogdance', playing the concertina while performing a type of vigorous Cossack dance on their haunches, kicking forward and hopping from one foot to the other. Some could do a circular waltz while playing the concertina behind their partner's back.

Along with the fiddle, the concertina was the principal dance instrument of the bush until the 1920s, when the button accordion found more favour. There are probably a number of reasons for this. The concertina was a complex instrument to manufacture and repair, and perhaps too expensive in comparison to the well-made accordions available after the turn of the century. It was more versatile too. Until the turn of the century, set dances featured as prominently on the programme as the waltzes and barn dances. The concertina could handle all, and its quick action could outstride and outlast the wheezy accordion on some set tunes, jigs and reels. After 1900 more and more waltzes and couples dances featured on the programmes; the accordion was better for these. Moreover, with modern transport people could travel to larger regional and town dances where the greater volume of the newer-model accordions, particularly the Mezon and Hohner, would be valuable. The accordion had multiple banks of reeds to produce a choice of several voices, and extra volume and pleasant tremolo.

Today the German-style Anglo concertina enjoys a healthy revival in the folk club scene, along with the English-style instrument. Recordings of traditional concertina players collected by John Meredith and Chris Sullivan are held by the National Library. There is also the *Concertina Magazine*. A recording by John Meredith of an Australian player, Fred Holland (playing an Australian-made Stanley instrument), is featured on one track of the cassette *Bush Fest*, obtainable from the National Library of Australia. Good Australian-style concertina playing by David De Hugard can be heard on his LP recording *The Magpie in the Wattle*.

The *flutina* is a type of accordion which had a reverse bellows action (i.e. draw/push) and which had an inside row of accidentals (thus, it was comparable to a C/C sharp instrument.) Its reeds were separately housed, as in a concertina, so that it had a specially sweet sound; a shortness of air from the bellows appeared to be its main disability. Although widely advertised for sale in the 1850s, it is a rare museum-piece instrument today. The flutina brought out from Essen in Germany in the 1850s by the original Conrad Klippel of Nariel had the date 1835 stamped inside it. Conrad Klippel disappeared en route to a dance at Mitta Mitta; he was believed to have drowned, as his horse and flutina were found on the banks of the river. The flutina is still in the possession of the Klippel family, and Mrs Beat Klippel can demonstrate the 'Bells of St Mary's' on the instrument, as can her grandson Jason. Both have been recorded and filmed by John Meredith and tapes lodged with the National Library oral history archives.

Percussion

No bush music group or old-time dance ensemble was complete without *bones* or *spoons* (generally soup spoons). In the bush they were often the only percussion instrument backing the piano, squeezebox or fiddle. Pairs of rib bones made excellent rhythm instruments and were most likely introduced into Australia by the itinerant minstrel shows. The use of bones as rhythm instruments dates back to antiquity and an interesting history is given in *How To Play Nearly Everything* (1977) by Dallas Cline. Good sets of bones were collected from bullock carcasses in a paddock, where they would have been bleached by the sun and hollowed out by ants. Some people purchased rib bones from the local butcher and boiled them down in the stockpot to remove excess flesh before placing them on an ants' nest to be cleaned. The bones would be cut back to an appropriate length and balance, and would be touched up on a wheel if necessary, or polished and buffed for a smoother finish. Some sets were made from a hardwood such as blackwood or redgum, and ebony sets could be purchased from a music dealer. The pair of bones was held curved inwards either side of the middle or index fingers. One was clasped tightly and the other loosely so that with a particular 'throw' and flick of the wrist and elbow, a 'clickety click' rhythm would roll out. Expert players could work a set in each hand and sometimes play cross-rhythms.

The *bush bass* (tea-chest bass) is commonly made from a tea chest (or oil drum or tin wash-tub), a piece of cord such as heavy string or hay band, and a broom handle or sapling. The cord is strung from near the corner of the tea chest to the top of the stick, which can be held in place by driving a nail into the bottom and placing this in a hole in the tea chest. The traditional way of playing the bush bass is to sound the different bass notes purely by straining the cord with appropriate leverages and plucking it. Some modern groups slide a ring over the broom stick to vary the length of cord and hence the pitch when plucked; however this is not the traditional manner in which the tea chest was played. The original Bushwhackers Band of Sydney (1952) introduced the bush bass along with the lagerphone to the revived bush band scene.

An ancient instrument found throughout Europe and Asia, the *Jew's harp* was popular in Australia in the days of our great-grandfathers and their forebears. Smaller than an adult's hand, it consists of a flat, pear-shaped metal frame with a metal tongue. The term 'Jew's harp' is probably a corruption of 'jaw harp', and it was also known as 'juice harp' because of the saliva and blood from beginners' tongues. Around the turn of the century it must have been one of the most common instruments in the bush, along with the mouth organ and penny whistle. Obviously the small size, portability, low cost and simple method of playing contributed to its widespread popularity. Literally thousands of Jew's harps were imported into Australia each year by the middle of the nineteenth century. The instrument is played by pressing it against the teeth, which must be held wide enough to allow the metal tongue to vibrate between them when 'twanged'. The metal tongue is plucked in the time of the tune to be played and the note produced by altering the volume of the mouth cavity, and sighing at the same time to produce volume. It was more likely a self-entertainment instrument, but may have been played along with the mouth organ for small family dances held in the vestibule or kitchen, in the old days in the bush.

It is strange that the modern groups which rely on amplification have not introduced the Jew's harp at times. Perhaps it is because of a country and western association that it has been avoided. Although it has a range of only a few notes, simple tunes can be played and certainly as a bass accompaniment is effective. Whilst John Williamson's song 'Old Man Emu' is not traditional, his original rendition does highlight the interesting effect that can be rendered by Jew's harp accompaniment. Arthur Lumsden, in an article in *Australian Tradition* 2, 3 (1965) refers to a 'shearers' band' which consisted of a mouth organ and several Jew's harps. Jew's harps could be purchased in a range of notes. Fine tuning could be achieved by filing the metal tongue so that three different instruments could play C, E or G for chordal accompaniment to the mouth organ. The tongue had to be removed for tuning and then riveted back into place. Other tunings to different keys could offer greater musical effect. In Australia it has been a much neglected instrument of the folk revival period.

A forerunner of the *lagerphone* could be the 'Jingling Johnny' of Napoleonic times, a stick covered in bells or shells and sometimes used by shepherds. Islanders of the Torres Straits also have similar shell-covered rhythm instruments. However, the history of the lagerphone is obscure. There do not appear to be any early references to such an instrument in use in the bush, and it could not have existed in its modern form prior to the introduction of crown seals as bottle tops about 1912. In the 1950s lagerphones suddenly appeared all over the country, becoming the signature of the revival bush bands. When the original Bushwhackers Band was formed in about 1952 by John Meredith and Brian Loughlin, a talent quest/concert was being held in Meredith's home town of Holbrook. A rabbit poisoner entered an instrument made from a broom handle covered with beer-bottle tops, played it with a stick — and won the contest. John's brother, Claude, was impressed with the instrument and made one for himself. John Meredith dubbed it 'lagerphone' and introduced it to the Bushwhackers Band, which needed a backing instrument as at that time the principal and only band instrument was the button accordion. By the mid-1950s the Bushwhackers Band had formed the Bush Music Club and was also performing in the enormously popular play *Reedy River*, and the lagerphone was before the public eye. The *Australian Post* magazine at this time featured an article on how to make and play a lagerphone, and suddenly they sprang up all over the country with innovations such as cow bells, drummer's bells, car-lamp reflectors and so on. Later, one of the most brilliant performers ever of the lagerphone was Liz Eager of the former Mulga Bill's Bicycle Band. It is interesting to note that the Gay Charmers' Old Time Band of Lake Charm district between Boort and Kerang in Victoria used a lagerphone until they could afford a drum set, but they dubbed the percussion instrument 'boozaphone'.

The *musical saw* (carpenter's hand-saw) was quite a popular entertaining instrument, particularly at bush concerts, and also occasionally at dances. It is played by holding the handle between the knees and straining (bending) the other end with the left hand to alter the pitch, while bowing (with a violin bow) across the saw to play the tune. Vibrato and sustain were achieved by shaking the right knee in a short rapid up and down motion, with the heel well off the ground. The sound is not unlike that of a swannee whistle (see below), and good players would be perfectly in tune, a few having the exceptional skill of playing chord notes by the clever adaption of harmonics. Some saws were better tempered, working well, whilst others were difficult to play with a constant sweet tone. A special brand of musical saw of reliable quality and tone used to be available from musical instrument dealers. There are also several accounts of the cross-cut saw being played. The knuckles or two sticks across the middle finger were used to 'rattle' out a resemblance of a tune.

The *swannee whistle* (slide whistle) is comparable to a bicycle pump, consisting of a tube with a fipple on one end and a washer attached to a rod at the other end. The notes of a tune are simply achieved by sliding the washer along the tube to alter pitch, somewhat in the manner of playing a trombone. The slide is usually finely wavered to create a vibrato, hovering or oscillating either side of the range of the note, as exact tuning is difficult. A novelty instrument that found favour and gave amusement in the ragtime bands of the 1920s Charleston days, it later found a vogue in some of the old-time dance bands in the country, although is seldom seen now.

Various *home-made instruments* were popular in the bush, including the kerosene-tin dulcimer, bow harp, saucepan banjo, wall fiddle, the 'Barcoo dog' and various drums and rattles. Details of the kerosene-tin dulcimer are given in *Folk Songs of Australia*, Vol. 1, and the other above-mentioned instruments are described in *Skills of the Australian Bushman* (1979) by Ron Edwards.

The modern *drum kit* did not come into vogue until the early jazz bands appeared around 1912, and more particularly by the 1920s. Traditional groupings of musicians relied on the bones and spoons for rhythm, also keeping and emphasizing time by thumping the floor with their feet while playing the squeezeboxes; sometimes a home-made drum made by stretching a kangaroo skin across a kerosene tin would be used, or a bass drum might be available. Most old-time bands couldn't afford a modern drum kit until well into this century.

One accessory from the modern drum kit that found popularity with the country old-time bands was the now old-fashioned and antique frog mouth and skull 'glocks'. These appeared during the 1930s and allowed the drummer special glockenspiel runs of notes to be played during a tune, and particularly to emphasize and embellish end-of-phrase passages.

See also: **Folk music in Australia: the debate**.

Peter Ellis

N

Names Real names, both for places and persons, are pivotal in many legends. Colloquial or familiar names, and their popular meaning or association, have an important role in Australian folklore, even though many possess what may be deemed a limited historical or regional/familial currency. This whole general field of study, onomastics, has in folkloric forms the customary two divisions — names of places, and periphrases, nicknames or popular labels for (groups of) people. Most of Australia's European-style place names or toponyms are in some degree 'signposts to the past', each containing one or all of the following types of information: geographic/topographic description; historical association in mind of the often unknown namer; or some linguistic/ethnic detail, if a name should be in form Irish, Gaelic (e.g. Inverell = 'river confluence — swans'), or from any language other than English. Aboriginal place names are equally pointers to past thought, although an actual associated incident may be impossible to retrieve now, and so they will seem more purely descriptive if a plausible etymology can be arrived at (see J. S. Ryan, *Papers on Australian Place-Names*, 1963).

Australia the land mass or general area has for itself many popular/general names, such as: Terra Australis (by early navigators); Botany Bay (early in the convict period); Down Under (when referred to by the British); the Lucky Country (a self-designation in the 1960s); the Farm (when doubters are discussing overseas investment in the nation); or the Clever Country (when pondering on education or technology here in the 1990s). A like modern and recent name, from the comic strip concerned with the extrovert Barry McKenzie, is Bazzaland, while a more enigmatic general label (phonetically a loose abbreviation) is Oz. The wellsprings of this ongoing linguistic mode of jocular derisiveness have often been said to be the naming jests found in the street talk and (scurrilous) ballads of London, Dublin or Glasgow in the latter part of the eighteenth century.

Examples of this continuing wit, seldom purely descriptive, are the following state names: Bananaland (for Queensland); Cabbage Garden (for Victoria); or the Apple Isle (for Tasmania). The state capitals too have their sobriquets, e.g. the Coathanger (Sydney, from the shape of the Harbour Bridge); or the City of Churches (Adelaide). Some city or town toponyms give syncopated forms like Brizzie (for Brisbane), or they are used of their residents in forms like 'a Bankie' (someone from Bankstown, NSW). Yet other (similar) regional names of origin have related or still relate to persons from the particular state, e.g. Croweater (from South Australia); Cornstalk (from New South Wales); Bananalander (from Queensland), etc. Many place names are contained in phrases referring to specific styles of behaviour, or to individuals of note. In the former category is the 1970s phrase 'Balmain boys' (tough working-class folk from this Sydney suburb); in the latter, 'the boy from Bowral' (where Donald Bradman first played cricket); 'the colt from Kooyong' (of a promising Liberal Party parliamentarian); 'the Emmaville express' (a girl sprinter from that small New South Wales town); or 'the Lithgow flash' (another New South Wales champion woman sprinter). Yet other phrases refer to the place or its products, e.g. Banana Curtain (of Queensland's southern state boundary), or 'a Coolgardie' (i.e. a home-made food safe, as devised at that Western Australian place), or 'fit as a Mallee bull' (like one from that Victorian district).

Mythical Australian place and personal names are continually being added to colloquial speech, as in the following designations for imaginary locations of various kinds. Thus, while '(the) Never Never' is a general phrase for unsettled areas in northern Australia, other more general (southern) notional primitive spots are Woop Woop, Snake Gully (the home of the most rustic of folk), the remote Nar Nar Goon, Oodnagalahbi (an imaginary and backward community), as also is the more recent (1953–) Bullamakanka. Speewah has long been a legendary pastoral station, the setting for the greatest lies and exaggerated outback feats, while Booligal in western New South Wales is a symbol of heat, dust and flies, even as 'the Alice' (Alice Springs) symbolizes traditional hard drinking, reckless dares, etc. Isolated places are also the locales for ghostly figures — e.g. the Nullarbor Nymph, a semi-naked wild white girl reported seen near Eucla; or various reputed savage predators of sheep, such as 'the Emmaville panther', or 'the Tantanoola tiger' (q.v.). Yet other (distant) names are used in rhyming catch phrases for a generally adverse situation, e.g. 'Things are crook in Tallarook', or 'Got the arse at Bulli Pass'. 'Fredsville' is the home of the ordinary, unimaginative modern Australian, 'Fred' (the name may be modelled on New South Wales' Fredricktown). An ordinary uncultured Australian is styled 'Bruce' in the United Kingdom.

Place names can be used in various (figurative) ways, as in: 'went to Callan Park' (i.e. 'had a breakdown'), from the name of a Sydney psychiatric hospital; 'follow the Bridle Track' (i.e. 'ride across country'), from a Northern Territory phrase for the Milky Way; weather contexts, where the 'Albany' or 'Fremantle doctor' refers to a cool wind later in a broiling day in southern Western Australia; while 'Bogan, Darling or Wilcannia showers' (all referring to places far inland) describe (local) duststorms, as does the Sydney term, '(a) brickfielder'; and 'Beliando spew' and 'Barcoo sickness' refer to retching illnesses contracted in such searingly hot and socially isolated places. Periphrases are used: of various industrial towns, as with the Silver City or the Steel (for Broken Hill), or Coal City (for Newcastle); in such political terms as Cocky Corner (for the National Country Party's bloc in the federal parliament); or in such a common noun from a place name, as 'a gympie' (for a hammer or drill made at Gympie), or 'some Parramatta' (once a common phrase), for the cloth made at that New South Wales gaol. Many other place names are used in hypocoristic form, especially ending in *-o*, as in Darlo (Darlinghurst); Flemo (Flemington); Kenso (Kensington); and Paddo (Paddington). (Such abbreviations are also used for religious denominations, as in Bappo [Baptist], Congo [Congregational], Metho [Methodist], Salvo [member of the Salvation Army].)

Names for persons have many colloquial forms, most being vehicles for folk speculation, prejudice or cultural reference far beyond the mere designatory function. Thus, many ethnic terms have a derogatory aspect, as in: 'Limey' for an Englishman (nineteenth-century); Herman (a pre-1914 German settler); Balt (a post-1945 'Baltic' immigrant); 'chink', 'Chinkie', 'Chow' (earlier terms for those of Chinese descent); Ding (of Southern European antecedents); or Yugo (a Yugoslavian immigrant). More semantically neutral is the phrase 'at the Greeks' (merely referring to the local milk bar or café run by Greeks), while 'Jacko' was an approving generic term for the Turkish soldier in the First World War.

Folk names for various places, persons or activities have been generated since white settlement. Early convict terms were 'the Female Factory' at Parramatta; 'Ocean Hell' for Norfolk Island; or 'Tench' (from Penitentiary) for the convict barracks in Hobart. Other names so generated — perhaps even 'Buckley's chance' — were for bolters or wild white men (i.e. convict escapees who lived with Aborigines), or for the various (later) bushrangers of fame, Starlight, Moonlight, Red Cap, Thunderbolt, etc. Terms with a name element are legion, as in 'Barcoo (pastoral) challenge' (in shearing generally), or 'a Jacky Howe' (sleeveless flannel shirt) from an early shearer so named. Outback slang gave 'the Old Bloke' and 'Hughie' as terms for God. Derisive attitudes to British migrants were to be found from 1912 in 'Pom', 'Pommification', 'Pommyland' or 'Bombay bloomers' (for men's long shorts similar to those worn by the British in India).

Football of the various codes has generated a considerable onomastic folklore for club emblems, colours, etc. Thus, the Australian Rules code produces the Bays (from Glenelg) or the Blacks and Whites (from the Swan District), while the Victorian Football League generated the Hawks (for Hawthorn) and the Bulldogs (Footscray). Sydney Rugby League produces Bears, Sharks, Berries, Magpies and Sea-Eagles, and Brisbane gives us Broncos and Diehards. Cricket's popular names are rather for grounds, as with the Gabba (Brisbane) or the Wacca (Perth), or sections thereof (as with Sydney's 'The Hill').

Storytelling has generated many names for (mythical) raconteurs, from 'Frank the Poet' (Frank MacNamara, q.v.) in the 1820s to 'the Colonial Minstrel' and goldfields balladist Charles Thatcher (q.v.); from 'Tom Collins' — the pseudonym was adopted by Joseph Furphy, from a mythical bush character who was reputed to start all the idle rumours in the Riverina — to clancy/Clancy (from A. B. Paterson's poem, 'Clancy of the Overflow'), for 'a (newspaper) story'; or the bemused 'bloke' from C. J. Dennis's 'Sentimental Bloke'. Since 1918 folk humour has created such symbolic figures as: Ginger Meggs, image of unaffected Australian boyhood; Bill Bowyang, a yarn-spinner together with his mates from the legendary Gunn's Gully; Alex Gurney's (Second World War and later) diggers Bluey and Curley; Frank Hardy's Billy Borker, and so on. One and all illumine the growth of such national characteristics as laconic humour, the ability to survive, and gritty earthiness. Unlike the others, Borker was a city figure, now widely accepted, even as the 1950s 'Nino Culotta' (John O'Grady) has been for his Italian migrant humour and self-mockery. Asian or Pacific folk hero names have yet to be widely recognized in general Australian society.

The world of theatre and popular entertainment has produced various (allusive) proper name phrases, such as: 'do a Nellie Melba' (i.e. make a further 'last appearance', after the singer who made several 'last appearances'); '(to be) a Dave and Mabel' (i.e. like A. H. Davis's simple characters of the early twentieth century who, through radio, became universally identified as unsophisticated rural Australians); 'the Firm' — J. C. Williamson Ltd, the theatrical entrepreneurs, known for spectacular musicals; or La Stupenda, the popular way of referring to soprano Joan Sutherland. State lotteries with some philanthropic dimension gave their titles to the lexis — 'the Casket' in Queensland, or 'the Charities' (now obsolete) in Western Australia.

Many seeming personal names are used for products/institutions of various sorts, e.g. 'Grannies' for Granny Smith apples; 'Gregory's' or 'Melways' for a street directory (especially for capital cities); 'the Jesus Hilton' for Sydney's expensive St Vincent's Private Hospital run by nuns; 'a Douglas' or 'a Kelly' (brands of axe); and 'Barrett's twist' for any chewing or pipe tobacco; or Cappo (a Capstan cigarette). Derisive terms for pretend pastoralists with little personal commitment are 'Pitt Street farmer' (from Sydney) or 'Collins Street farmer' (from Melbourne). Zambuck, a commercial form of inflamation-reducing ointment, has become 'zambuck', a first-aid person.

Nicknames, often not in upper case, have always been widespread, particularly for politicians. R. G. Menzies was known as 'pig iron Bob', while R. J. Hawke was styled 'Yellow-cake Bob' (for his policy on mining uranium), or the 'silver Bodgie' from his groomed hair. 'Hawkespeak' was used of his confusing political utterances, while Joh Bjelke-Petersen, once Queensland premier, was called 'the flying peanut', since he grew peanuts and actually flew his own plane. Nicknames have always existed in very large numbers, although it is difficult to accept that the thousand or more listed by Taffy Davies in his *Australian Nicknames* (1977) have wider currency than a particular bar or sporting team. More generally encountered are: Blue and Bluey for a red-haired person; Chiller for

Charles; and such abbreviated forms with the *-o* suffix, as Jimmo, Tommo, Johnno, Freddo, or Sallo and Betto. Today any male Kelly may be called 'Ned' after the bushranger; any male Paterson 'Banjo' after the poet; and any male Hall may well have the nickname 'Ben' after the bushranger. More generally nicknaming has developed into a fascinating branch of linguistic and folkloric inventiveness, embellishing social history, embracing character assessment and assassination, combining physical description and achievement, and always employing humour and ingenuity. Arguably the main source of the most widely used nicknames is popular journalism, as discussing politicians, e.g. 'Nifty' (Nev) (Wran); music, e.g. 'Slim' Dusty; or sport, e.g. 'the great White Shark' (the fair-haired golfer, Greg Norman). Yet other names are applied from outside, as in the British term, 'the (Dirty) Digger' for Rupert Murdoch, the aggressive Australian-born international newspaper tycoon. It has been well said that the major social areas for generating modern nicknames/popular name designations are: boxing; cricket; the football codes; radio personalities; criminal activity; gossip columns; pop music and entertainment personalities/combinations; and racism or jingoism of various threatening/despised ethnic groups.

In all periods there have been many proper names in rhyming slang, as in these still current terms: Al Capone telephone; Arthur (Murray) a hurry; Hedy (Lamarr) car; Lady (Godiva) driver (golf); Oscar (Asche) cash; Slapsi (Maxie) taxi; or Werris (Creek) a Greek.

Nomenclature is undoubtedly one of the most exuberant, inventive, cruelly shrewd and humorous aspects of both folk memory and folk imagination in this country.

J. S. Ryan

Narrative, folk: see **Folk narrative**.

Narrative literature and folklore Traditionally, tertiary literary studies have included Homer's great epics and perhaps some early English ballads, but apart from this, academics in Australia have largely ignored both oral traditions and the influences of contemporary oral forms of expression. Walter Ong, in *Orality and Literacy* (1982), points out that the concept of 'study' is only possible in literate cultures which are dependent on acts of reading and writing; this has tended to exclude orality from scholarly attention, as has the primacy given to authorship and the individual creative act. Also, literary scholarship has been confined to texts which are considered to have aesthetic value. The privileged position of print within our culture and the circulation of particular elitist forms of written expression within schools and universities led Roland Barthes to remark that literature is that which is taught.

In contrast, folkloric narratives have always been thought of as traditional rather than original. The collectors of such narratives in print form have, in the past, emphasized their traditional content, giving little attention to aspects of individual performance. Structural analyses of folkloric tales have also emphasized their content rather than the style in which the story is told or the specifics of the situation in which they are narrated. Some folklorists wished to claim the same aesthetic values for folk narrative as were claimed for the modern novel, and referred to folk forms as oral literature (an expression abominated by Ong) in an attempt to gain some status for this field of study. Elizabeth Fine notes that in the past folklorists, although working with non-literary texts, 'have tended to impose a literary or ethnolinguistic model of the text on orally composed and transmitted discourse ...' (*The Folklore Text*, 1984). Frequently, collections of folkloric material have been treated as if they had a written source. Judgements have therefore been *literary* judgements. As folkloric narrative was not deemed to meet with the aesthetic requirements of literary value, folklore was marginalized in narrative studies. This lack of aesthetic value assigned to the folkloric text was, Fine argues, largely due to the text appearing 'stylistically barren and flat' when transposed to print. The formulaic, repetitive and cliched aspects of oral narrative are necessary qualities for its transmission in oral form, and part of the pleasure it gives listeners; but these same attributes are considered to be faults in a printed text. Some folklorists would now argue that the folkloric event cannot adequately be translated from oral performance into print. The quality and significance of much oral tale-telling, they would argue, depends on the particulars of the performance; the text only *lives* in performance. Fine suggests that, while recognizing that any forms of recording will alter the nature of the text in question, folklorists do the performance greatest justice when details of the performance situation, gesture, tone and audience response are all noted in the transcription. Context is as important as content.

Current schools of literary criticism now also place considerable emphasis upon the social context which produces the text and which the text reproduces. Scholars and general readers are increasingly interested in texts previously excluded from the canon on aesthetic grounds — indeed, the concept of aesthetics is itself under greater scrutiny. Such battles for territory have been encountered before. Late last century literature in the vernacular began to be perceived as a threat to classics and philosophy within the universities. Margo Culley points out that the arguments then used against the study of vernacular literature were that

> it was too easy — it had no substance, no organizing body of knowledge, no rules, no theory, in short nothing to promote rigorous mental training, the discipline that was the justification of an education. ('Women's Vernacular Literature', in Hoffman et al. [eds], *Women's Personal Narratives*, 1985)

The literary canon was formulated 'to organize and establish the discipline' of English studies. It is in quite recent history that regional studies, with their own vernaculars and socio-political concerns — the area of Australian literature, for example — have been involved in a struggle for recognition as valid and significant areas for scholarly research. That battle largely won, ironically mostly through the establishment of a new canon, now it is the turn of non-literary texts to claim a hearing. The impulses for women's studies, Aboriginal studies, multicultural studies and studies in popular culture all produce their own textual emphases.

As is to be expected, the producers of these texts draw upon traditions other than those laid down for the literary canon. These traditions are frequently oral, folkloric forms. The broadening of the concept of text (to include any type of communication in any medium), and the greater emphasis on interdisciplinary approaches mean that many forms may now be included under the rubric of cultural studies. Practices, games, rituals, songs, sayings, jokes, and oral narratives, as well as previously marginalized literary forms such as the diary and the autobiography, are now of interest to academics.

Previously, the distinctions made between oral forms and print literature emphasized the differences rather than the similarities between them. In fact, there is much cross-fertilization between the two. Sandra Dolby Stahl suggests that folkloric forms can provide

> the basis for plot or subplots . . . the basis for narrative form . . . the source of 'local color' in the stylization of each work . . . the traditions that are used throughout the works as common allusions.

Folklore is a rich plundering ground for the creative writer.

It is important to note that tradition and innovation are active in both literature and folklore. However, the transformative power of print is in danger of 'fixing', or seeming to fix, a traditional form which has always been fluid and adaptive. The appropriation of folkloric forms into print may alter the function of that form in its usual medium forever. The Aboriginal traditional tale is a case in point. Any discussion of folklore and literature in terms of indigenous people without a print culture raises some problems. In a society in which almost all narrative traditions are oral (there is no scope in this article to consider the narrative role of Aboriginal art), it may be difficult to differentiate between the folkloric and the formal; perhaps the distinction is one that only applies to print cultures. Distinctions have been made, largely by linguists and anthropologists rather than folklorists, between formal or religious stories and those tales which are freely exchanged and appear to play no part in ritual or taboo, which are deemed to be folkloric. In the past such tales were frequently collected, translated, reproduced and promoted as stories for white Australian children. Tales of origin — explanations for certain geological features of the landscape and the distinguishing characteristics of particular animals — are very common in this canon. These popularizations for children, from which the bawdy and other material considered indecent has been excised, have frequently been expressed in literary forms familiar to those from a Western European culture. The models are the European fairy story, Rudyard Kipling's *Just So Stories* and the Bible. Such adaptions of form may make these stories more accessible to a European readership, but they inevitably alter their meaning and their social function. To the reader such stories may have a charm and quaintness which appeals, but they are not likely to bear the majesty and truths of experience that they may do for the Aboriginal listener. (It is important to note, however, that these stories may also have begun to be modified orally as their narrators were introduced to the European models of the folk and fairy tale; (see Mudrooroo Narogin, 'Aboriginal Responses to the "Folk Tale"', *Southerly* 4, 1988). When these narratives are defined as 'children's stories' there is a danger, too, of perceiving all Aboriginal folklore as functioning at the level of a child's understanding.

Such adaptions may be seen as a form of theft, a new colonialism in which Aboriginal culture is appropriated and reshaped in terms of European mythology. Collectors of such material frequently express their awareness of the dual possibilities of their role: as preservers of a culture; or as pallbearers at its funeral rite. Many early collectors were aware of the sensitivity of the material. As an example, among the tales first collected in 1896 by K. Langloh Parker and later published as *Australian Legendary Tales* (1954) is the story of 'Dinewan the Emu and Goomble-Gubbon the Turkey'. This story is available in the Appendices in the Euahlayi language. Guided by a glossary, the reader is led to believe that the story is translated very closely in terms of content, if not in the syntax of the original — though even here some concessions have been made. It is interesting to note, however, that the English version of 'Dinewan' concludes with a coda which is not present in the original: 'And ever since that time a Dinewan, or emu, has had no wings, and a Goomble-gubbon, or turkey of the plains, has laid only two eggs in a season'. This coda is illustrative of the impulse to closure to be found in print texts; the closure summarizes, completes and closes off potential meanings. This collection of stories is presented in conjunction with art works in Aboriginal 'style' by a European artist, Mary Durack, and the implied audience is Euro-Australian. There are no indicators of the text's original method of transmission, the style, tone and gesture of the speaker, or the nature and response of the audience. But for its time the collection appears to be truer to the oral text than some Europeanized versions that are to be found. Rarely do collections of Aboriginal oral narrative acknowledge the active bearer of such material. An exception to this is Catherine H. Berndt's collection *Land of the Rainbow Snake* (1979) in which each storyteller is named and the source from which Berndt first heard the story is identified. (In this collection all the storytellers are women, each of whom, with one exception, first heard the story from a man — usually her father or uncle.) An attempt to achieve an authentic Aboriginal voice in narrative can be seen in *Gularabulu* (1983), a collection of Aboriginal stories from the West Kimberleys, told by Paddy Roe, an Aboriginal outback worker, and transcribed by Stephen Muecke, a Euro-Australian academic. Not all the stories are traditional; some are quite recent, demonstrating the evolving quality of contemporary Aboriginal folk narrative and its responses to European culture. Muecke believes his choice of Aboriginal English rather than formal literary English opens up communicative opportunities between white and black readers, but he recognizes, too, that such texts, removed from their original social and oral contexts, are open to many differing responses. The orality of the stories' narration is emphasized, with pauses, emphases, extended vowel sounds, occasional gestures, all indicated in the text. Such texts are, of course, very difficult to read. Even if one struggles to overcome the absence of European literary conventions, the reader cannot learn all the conventions of Aboriginal oral narrative by reading a few stories. Missing, too, is a good deal of other cultural knowledge which would be available to the Aboriginal listener. An Aboriginal story in another collection states: 'Those who ask, however, must have the right sympathy or they will hear nothing'. (C. W. Peck [ed.], *Australian Legends*, 1933).

Modern Aboriginal writers may now assume a literate audience from both black and white communities. They also draw on folkloric forms and beliefs. Mudrooroo Narogin (formerly Colin Johnson) has drawn upon a variety of traditions in his narratives. His novel *Sandawara* (1979) incorporates the true story of the Aboriginal rebel or freedom fighter Sandawara within a contemporary work of fiction, drawing upon a mixture of literary and oral forms: history, fiction and oral legend. Sally Morgan's autobiographical family history, *My Place* (1987), is a contemporary work in which oral forms and folkloric attributes are happily united in a text which has met with enormous popular success. Morgan's narrative links transcribed interviews with the literary forms of the autobiography and the mystery. While a one-off personal narrative may not strictly speaking be considered folkloric in form, this narrative abounds in folkloric beliefs and practices. Several of these have as their source traditional Aboriginal belief, others derive from Christianity, and many are familiar to women, in particular, or various cultural backgrounds within the Australian community. They are to be found in Morgan's narration of her grandmother's beliefs, particularly her home remedies for various ailments. And there are many examples in Gladys's story (Morgan's mother) relating to pregnancy and childbirth. A repeated structuring device in the narrative is the call of a bird which is understood to have special significance to the hearer. Sally's hearing of the bird call becomes a metaphor for her awakening to a source of knowledge that is linked to her Aboriginality. At the conclusion of the narrative the bird call heralds the approaching death of Sally's grandmother. Narratives concerning precursors of death are commonly collected by folklorists. The most common precursor is the call of the 'death bird', heard in the vicinity of the person who is about to die. The belief is not peculiarly Aboriginal, but widespread amongst many cultures.

A folkloric form which is often thought to be peculiarly Australian is the yarn (see **Folk tales**), an oral form of tale-telling which is frequently adapted by creative writers. The yarn and its first cousin, the tall story, have come to epitomize, for many, the Australian character and the Australian sense of humour. Nevertheless, Bill Wannan (*Tales from Back o' Bourke*, 1984) points out that much that we think of as typical of an Australian yarn can be found elsewhere, particularly in countries with a similar recent frontier experience, such as the American West. Wannan remarks that the content of such yarns may differ very little from one culture to another; the distinctiveness of the Australian yarn lies in its manner of telling, 'vocabulary, traditions, accents and environment that subtly influence the way they're told'.

We tend to think of the yarn as the discourse of the bush worker in Australia at the turn of the century. One of the causes of this historical 'fixing' is probably the work of Henry Lawson. Lawson might well be regarded as the greatest of Australia's literary yarn-spinners. Lawson's stories, largely centred on bush life, frequently depict its bleaker aspects, but his more humorous narratives such as 'The Loaded

Dog' and 'Bill the Ventriloquial Rooster' are closely modelled on the yarn. Lawson was commonly regarded as a 'folk writer' who captured the concerns and idiom of the rural working class, but the colloquial style disguises careful literary craftsmanship, and especially his use of irony, a quality largely conveyed by tone of voice and facial expression in oral yarn-telling but requiring considerable literary skill to convey in print.

A recent popular yarn-spinner is Albert Facey in *A Fortunate Life* (1984). Facey, an unskilled writer, has not the technique to convey understatement and irony in written form, so that the effect is to make the events narrated seem like fact or history, rather than exemplary tales. Print gives authority to authorship: the tales told by Facey to his family and friends come to us via the written word as if with 'official approval'. Facey's text includes a variety of stories which seem to draw on folkloric traditions — although one has to rely largely on shrewd guesswork here. This is particularly true of the section on Gallipoli. The incident in which Facey and a friend blow out a candle that has been burning for 'over one thousand years' is a folkloric yarn which has been recorded elsewhere (Joan Newman in *Australian Folklore* 4, 1990).

Another particularly plentiful source of material for authors of fiction is the urban legend (see **Modern legends**). These folk narratives that are often repeated as if they were real-life experiences are a significant aspect of contemporary urban folklore. Such narratives frequently reflect major concerns within the culture and may serve as a warning to avoid certain kinds of behaviour. Jan Harold Brunvand, in his study of urban legends, recognizes that each one has 'a strong basic story-appeal' (*The Vanishing Hitchhiker*, 1981). Occasionally such stories appear in the mass media, repeated as news. Such an authoritative rendition of the story may confirm the belief that it is true, but may temporarily alter its variant propagations within the community. A regional legend which circulates in the south-west of Western Australia has been taken up and utilized for the basis of a masterly gothic tale by Tim Winton in *In the Winter Dark* (1988). Folklore has it that a mysterious beast roams the forests of this corner of the state and savagely attacks livestock and pets. The animal is variously considered to be an escaped circus leopard or tiger, a Tasmanian tiger, or a huge feral cat. Winton dedicates his novel to 'the Nannup Tiger wherever you are'.

In more humorous vein, Peter Carey takes up a well-worn legend of the performing elephant, who, having been trained to sit on a large red tub, mistakenly crushes a small red car. In *Bliss*, Carey playfully acknowledges the folkloric source of his story, but reclaims it as a true event within his narrative. Blackly comic, too, are David Ireland's use of jokes, legends and folkloric customs in the barroom world of his characters in *The Glass Canoe* and *The Unknown Industrial Prisoner*. Graham Seal, in *Australian Folklore* 2, 1988, has noted Ireland's deployment of nicknames (see **Names**) largely derived from a character's occupational behaviour and his use of repographic lore to illustrate class and racial conflicts of Australian social life.

An interesting literary experiment combining folklore and fiction was conducted by Glen Tomasetti in her novel *Thoroughly Decent People*. Subtitled 'a folktale', this narrative is replete with folkloric practices, colloquialisms and rituals of an Australian suburban family in the 1930s. Gardening habits, games, recipes, superstitions, wise sayings and children's rhymes convey a detailed, believable and recognizable social world. Autobiography, too, is a rich source of domestic folklore. Among the best is Hal Porter's *The Watcher on the Cast-Iron Balcony* with its massive detail of material life, interior decoration, clothing and food. It abounds in the songs and sayings of children, folkloric belief expressed in children's games, chants and practices, as well as the domestic world of women. Porter is born 'Thursday's child with far to go, brought forth under the sign of Aquarius . . .'. Thus are established the beliefs which inform his childhood. The abundance and warmth of this world signifies a sense of belonging — to a family, to a social group. Porter's mother shows him the images in the patterns on flowers; simple handcrafts from objects in nature; paper dolls, houses of cards. Her domestic tasks are conducted using time-old methods and divided traditionally, allocating specific tasks to particular days of the week. There are not only these warm supportive aspects of folklore, but its more xenophobic aspects — racist beliefs and taunts, sectarian hostility expressed in Catholic/Protestant chants. These chants, with their many variations, appear regularly in the recorded memories of Australian autobiographers.

Folkloric practices and expression pervade nearly all contemporary narratives. It is frequently those small incidents and expressions which give rise to feelings of recognition on the part of the reader. Such moments might be the simple gesture mentioned in a scene in Helen Garner's *Honour* in which a woman remembers her mother testing if the washing was dry by putting the garment to her lips; or in Dorothy Hewitt's recollection of the tale told as a dire warning in the account of the copulating couple who became stuck in the act of coitus and had to call for medical assistance; or in the title of a book by James Simmonds about Azaria Chamberlain, subtitled *Wednesday's Child*.

As has been shown, Australian writers of narrative frequently employ folkloric forms and include folk beliefs in their work. The effects upon the practice of folklore are mostly limiting, restrictive ones — the 'fixing' of print, for example, relegates folkloric expression to a nostalgic past or alters its function through the challenging authority of print. Most of the evidence here suggests that the bulk of the appropriation is by writers; but the traffic is almost certainly two-way. Just as other forms of expression are appropriated by folklore, as in parodies of advertising jingles, so too this is likely to be the case with written narrative. For folklore is a vital, innovative force which adapts and changes according to circumstance and to cultural needs. It will continue to provide material for creative writers, but it will also continue to thrive in its own right.

REFERENCES: Enid Neal, 'Oral Narrative and Audience Response', in S. Gunew and I. Reid (eds), *Not the Whole Story*, 1984; Sandra Dolby Stahl, *Literary Folkloristics and the Personal Narrative*, 1989.

Joan Newman

National Film and Sound Archive, The The National Film and Sound Archive was established in 1984 to increase awareness, appreciation and understanding of the nation's heritage of moving image and recorded sound. It does this by acquiring, preserving and presenting relevant material of national significance. The Archive currently holds more than two million items, including films, discs, audio and video tapes, piano rolls, photographs and artefacts. It is now recognized as the national centre for preserving film and sound recordings.

In collaboration with other major collecting institutions, the Archive collects and preserves the nation's traditional folklife heritage. The Folklife Program has acquired, particularly in the past few years, many unique film and sound recordings. The Program gathers material through collaboration with collectors; by donation; and through its modest recording activities. It seeks to cover early European settlement, post-war migration, folk revival and Aboriginal folklore.

The Archive's collection includes commercial releases ranging from 78 rpm records such as the Wattle and Carinia recordings to LPs from the 1960s to 1980s and more recent CDs. However, a large component of the Archive's folklore collection is unpublished. It encompasses recordings of several national and regional folk festivals, the films of John Meredith including *The Man from Cookamidgera, Songs of Granny Cotter* and *The Real Folk* series, and collections such as the series compiled by Rob Willis (for example, *Ethnic Folklore in Australia* and *Women in the Bush*).

The Folklore Collection also has a significant multicultural component: it includes performances by and interviews with Australian artists from a variety of cultures. Complementing this is ethnographic material such as the Fanshawe Collection from the Australasian region.

The Archive holds the Australian Dance Video Collection which contains: the Bellbrook Lancers; the Forbes Lancers; the Shirley Andrews Collection; the Phillip Ashton Collection; the Schuster Collection; Finnish Dances in Australia; the Marshall Mount Dance; Tom Gollogilly; dances from the Eureka Youth League; Dorothy and Peter Taylor of Tralee; dances of the Barossa Valley.

The Archive has taken advantage of local collecting activity by acquiring much significant material from the Canberra and Monaro area. Three of the more prominent of these collections are the Buchtman, the Monaro Folk Society and the Terry McGee collections.

Other material presently being detailed for the collection include the Queensland Folk Federation Collection, the Pibrochead Collection, the National Pipers Convention Series, the Australian National Gallery Gamelan recordings, copies from the ABC's 'Music Deli' recordings (courtesy of 'Music Deli'), the Folklife Centre Launch and the Geoff Wooff Collection.

National Folk: see ***Northern Folk***.

National Folk Festival: see **Australian Folk Trust**; **Festivals**.

National Library of Australia, The, traces its history back to 1901 and the formation of the Commonwealth Parliamentary Library which began early in its life a commitment to documenting Australian national life and achievement. In 1960 the National Library Act was passed by the Commonwealth parliament, thus separating the national library and parliamentary library functions which had been developed in tandem over the previous sixty years. Amongst other functions, the Act imposed on the National Library the responsibility of developing a comprehensive national collection of library material — defined broadly to encompass books, manuscripts, pictures and photographs, maps, music, film and oral history — on Australia and the Australian people. Within several of these collecting areas, but especially through the oral history collections, Australian folkloric traditions are strongly documented.

The oral history programme which began during the 1960s was based upon American experience and in particular the programme at Columbia University in New York. Local advisers, especially Dr Edgar Waters (q.v.), provided recommendations for the overall development of both the oral history and folkloric collections. While in the early years the Library placed greatest emphasis on the collection of oral reminiscences of well-known Australian men and women, it did not neglect a commitment to Australian folklore. Several notable achievements were made which together laid the foundation for the higher profile which the Library developed in the field of Australian folklore during the 1980s.

In 1963 the Library acquired the John Meredith Collection of Australian traditional music, songs and recitation. This pioneering collection remains arguably the most notable collection of such material extant, though other good early material has since been added to it. During the 1970s the Library built gradually on the resources acquired through John Meredith, with acquisitions including field recordings of traditional performances by Colin McJannett and Warren Fahey, recordings of concert performances of folk song and music donated by the Monaro Folk Music Society and others, and interviews dealing with rural life, the personal experiences of Aboriginal Australians and Scottish traditions in Australia recorded by John Blay, Kevin Gilbert and Emily Lyle.

During the 1980s, with the return to active fieldwork by John Meredith (q.v.), the National Library entered a vigorous phase of folkloric recording and collecting. By 1991 John Meredith had added more than four hundred new reels to the original thirty-two which dated from the 1950s and 1960s. With financial assistance provided by the Library, Meredith completed a series of journeys through New South Wales, Tasmania and Queensland, much of it virtually virgin territory as far as folklife collecting was concerned. He is currently recording on Digital Audio Tape with DAT equipment supplied by the National Library. In addition, the Library initiated recording agreements or purchase arrangements with several other respected collectors, including Alan Scott (q.v.), John Marshall (q.v.), Chris Sullivan, Mark Rummery, Peter Parkhill and Wendy Lowenstein (q.v.). Edgar Waters has continued to provide the National Library with valuable advice and guidance in the shaping and developing of its folkloric collections. At the same time, through the award of two research Fellowships, the Library has provided opportunities for the American and Australian folklorists Sandra Dolby Stahl and Gwenda Beed Davey (q.v.) to carry out detailed research on the Library's collections. The acquisition of an extensive collection of personal and literary papers of the historian and writer Hugh Anderson (q.v.) who carried out pioneer research into Australian folklore has added a valuable documentary dimension to the oral history holdings.

The Australian folklore collections which have been built up in the National Library are a rich but as yet generally little known or appreciated community resource. The Library's folklore programme has been well supported over the past twenty years by prominent collectors. Even so, much remains to be done. The long-term survival of original fieldwork recordings is considered to be a prime archival responsibility (see **Audio preservation**), and it is only if this essential task is recognized and supported that archival agencies can move confidently to the next stage, the production and release of commercial recordings, cassettes or compact discs which will bring the resources of great research and archival institutions such as the National Library into ready circulation within the community. The National Library of Australia works closely with the Australian Folk Trust and with other collecting institutions and individuals, both to continue the enrichment and development of its collections and to ensure the better availability of materials.

Nicknames: see **Names**.

Non-verbal communication is an aspect of folklife of particular importance in Australia's multicultural society. International research, such as that of Michael Argyle and Desmond Morris, has indicated that cultural differences in non-verbal communication can lead to misunderstandings or hostility between people of different ethnic origin. Anglo-Australians regard avoidance of eye contact, for example, as a sign of untrustworthiness; impatience in a queue or queue-jumping as anti-social, and standing 'too close' in a conversation or in an elevator is seen as intrusive or offensive and an invasion of personal space. Yet all these behaviours may have different meanings in some Asian and Middle Eastern cultures, where, for example, avoiding eye contact indicates deference or modesty.

It is hard to identify aspects of non-verbal behaviour which are unique to Australia, apart from the so-called Australian salute which involves waving a hand in front of the face to discourage the ever-present flies. Australians might cross their index and middle finger either literally or in words ('I had my fingers crossed') for luck, in a sticky situation or to enable them to evade the truth. Similarly, adults might say 'cross my heart (and hope to die)', an evocation from childhood, as an indication of good faith. Australians of Italian, Greek or other Mediterranean origin might make the Devil's horns with the index and little finger for a variety of purposes, including avoiding tempting fate or the 'evil eye'. A raised thumb can indicate either encouragement ('good onyer, mate!'), confirmation ('she's right') or offence. Insulting hand gestures have grades of offensiveness, ranging from the relatively innocuous thumb, through two upraised fingers, to one (the third especially rude). A variety of gestures of sexual appreciation or interest are usually self-explanatory; their use by Greek-Australians is satirized in the television series 'Acropolis Now'.

Many such gestures are of course situation-specific, and used only among peers or in extremely informal situations. The complex, informal social rules governing body contact among men are also highly situation-specific. Normally, Anglo-Australian men do not touch each other in public, unless in situations of heightened emotion such as on the football field, in battle, on greeting or farewelling a travelling friend, or when offering congratulations, as on the birth of a child. In such situations, bear hugs are acceptable. Given the subtleties of these aspects of Australian folklife, and the likely consequences of transgression, some instruction for immigrants would seem warranted.

REFERENCES: M. Argyle, *Bodily Communication*, 1975; M. Argyle, A. Furnham and J. A. Graham, *Social Situations*, 1981; D. Morris et al. (eds), *Gestures: Their Origins and Distribution*, 1979.

Northern Folk, edited by Ron Edwards (q.v.), was published by the Cairns Folk and Jazz Club from April 1966 until June 1968, from which time it continued as *National Folk* until ceasing publication in 1970. The publication catered for the interests of folk revival performers as well as publishing the results of fieldwork and discussions of various aspects of Australian folklore.

Nursery rhymes Despite gloomy statements made by some Australians that television, commerce and 'the pace of modern life' have killed old traditions, particularly in family life, the nursery rhyme is alive and well in Australia, in English and a host of other languages. Most English-language nursery rhymes have come relatively intact from the British Isles and are documented in Australia well before the twentieth century. Banjo Paterson's compilation *Old Bush Songs*, for example, was first published in 1905 and included the songs 'Immigration' and 'John Gilbert, Bushranger'. Both were to be sung to the air 'Four and twenty blackbirds baked in a pie'. This traditional English nursery rhyme was well known in Australia in the nineteenth century, as it is today:

Sing a song of sixpence,
A pocket full of rye;
Four and twenty blackbirds
Baked in a pie:
When the pie was opened,
The birds began to sing;
Oh wasn't that a dainty dish
To set before the King?

What is a nursery rhyme? A nursery rhyme is a traditional short poem, song, lullaby or game which parents, grandparents and other adults use with infants and young children. For most people a nursery rhyme is their first experience of poetry, music and formalized humour. It may be ancient, it may be funny, and it may teach the child something, such as counting or telling the difference between left and right:

One two three four five,
Once I caught a fish alive;
Why did you let him go?
Because he bit my finger so.
Which finger did he bite?
The little finger on the right.

A nursery rhyme can also be pure nonsense, although scholars might argue from now to kingdom come about its meaning:

Half a pound of tuppeny rice,
Half a pound of treacle;
That's the way the money goes,
Pop goes the weasel!

Most nursery rhymes are learned by word of mouth, although adults sometimes use books to jog their memory or to enjoy the illustrator's interpretations of the ancient poems. Every year hundreds of Australian adults buy books of traditional nursery rhymes for young children. They are one of the most popular types of children's literature, and range from cheap mass-produced editions and paperbacks to expensive art books.

Publication of nursery rhyme books is almost as old as printing (see Baring-Gould,

The wedding of Amy-Helen Sonter and Robert Weekes was held on 4 May 1987 at Christ Church, Queanbeyan, NSW. The bride made her own dress and those of the bridesmaids. (Photograph, *Queanbeyan Age*)

Dimitris Katsoulis, assisted by his wife and Nick Zarkadas, behind the white screen during a performance at Brunswick High School, Melbourne, 1990.

Rose Dichiera with a loaf ready for baking in a traditional Calabrese wood-burning bread kiln, Mildura, *c.* 1980. (Photograph, Keith McKenry)

ing Matilda' informally from other members of their family. The power of oral tradition is such that every Australian, new or old, knows the Great Australian Folk Song, and its use with very young children gives Banjo Paterson's poem a claim to be, perhaps, the first Australian nursery rhyme.

The fact that a song or rhyme has a known author is no barrier to its passing into the folk tradition, if 'the folk' take it to their hearts. 'Twinkle Twinkle Little Star', 'Wee Willie Winkie' and 'Three Little Kittens that lost their Mittens' all have acknowledged authors, and the poet Longfellow is commonly credited with

There was a little girl, and she had a little curl
Right in the middle of her forehead;
When she was good she was very, very good,
But when she was bad, she was horrid.

An Australian rhyme which is usually regarded as anonymous is 'Kookaburra Sits in an Old Gum Tree'. 'Kookaburra' was once exclusively a school or Guides' and Scouts' song, often sung as a round, but nowadays it is increasingly part of the family repertoire for very young children, together with old favourites such as 'Little Jack Horner' and 'See Saw Margery Daw'. As a song which belongs more exclusively to children than 'Waltzing Matilda', 'Kookaburra' has perhaps a better claim to be regarded as 'the' Australian nursery rhyme:

Kookaburra sits in an old gum tree,
Merry merry king of the bush is he;
Laugh, kookaburra, laugh, kookaburra,
Gay your life must be.

'Kookaburra' is not anonymous; both words and music were written in the 1920s by Miss Marion Sinclair, who died in Adelaide in February 1988.

There are other songs and rhymes which are moving into the Australian nursery rhyme repertoire from a variety of other sources, some traditional and some not. Iona and Peter Opie collected 'Michael Finnegan' in school playgrounds in Britain, but it is often used in Australia as a nursery rhyme, told to children by adults. One version reads:

There was a man called Michael Finnegan,
He grew whiskers on his chin-e-gan;
The wind came out and blew them in again,
Poor old Michael Finnie-innie-inegan.

Music-hall and popular songs have also moved into the nursery, and the Australian favourite is probably 'How much is that Doggie in the Window?'. In addition, 'Daisy, Daisy', 'K-K-K-Katie' and even 'When Father Papered the Parlour' are still sung by some adults to their infant and young children.

The borrowing process goes both ways; the Beatles used many fragments of nursery rhymes such as the phrase 'See how they run' from 'Three Blind Mice' in their song 'Lady Madonna', and the popular Second World War song 'I've Got Sixpence' was an almost literal adaptation of 'I love sixpence' from the eighteenth-century publication *Gammer Gurton's Garland.* The wartime version runs thus:

I've got sixpence,
A jolly jolly sixpence,
I've got sixpence
To last me all my life;
I've got tuppence to lend
And tuppence to spend
And tuppence to send home to my wife.

In 1784 the version printed in *Gammer Gurton's Garland* was as follows:

I love sixpence, a jolly jolly sixpence,
I love sixpence as my life;
I spent a penny of it, I spent a penny of it,
I took a penny home to my wife.

Nursery rhymes provide the adult world with more than a repertoire of reuseable song material. They also provide a continual supply of symbols, images and poetic phrases, often used in the titles of books, plays and television productions such as *My Fair Lady*, *Bells on her Toes*, and 'Upstairs, Downstairs', or for trade names such as Tom Piper or Four 'n Twenty. They also become incorporated into idiomatic speech, so that a foolish person may be described as a Simple Simon or a disgraced public figure as a Humpty Dumpty.

There are a number of kindergarten or primary school songs and rhymes which are popular in families and seem to be entering into oral tradition. 'I'm a Little Teapot Short and Stout' was a popular novelty song in the 1940s and was published by Clarence Kelley and George H. Sanders. It has survived by oral transmission and is thus a strong contender for folkloric status. Other popular titles are 'Bananas in Pyjamas', 'Little Peter Rabbit' and 'Five Little Ducks', but only time will tell whether they survive to become true folk nursery rhymes. Australian parents have also appropriated to the nursery the folk songs 'Old Macdonald had a Farm' and 'This Old Man', the latter receiving a fillip from its use in Ingrid Bergman's popular film *The Inn of the Sixth Happiness* (1958). 'This Old Man (he had one, etc.)' is an interesting example of the enormous amount of education involved in these inspired pieces of traditional nonsense:

This old man, he had one,
He played knick-knack on my drum:
Knick-knack paddy whack, give a dog a bone;
This old man came rolling home.

'This Old Man' teaches counting, practice with rhymes and with sounds such as 'knick-knack, paddy whack'. Learning the words involves quite a feat of memory, surpassed only by even longer cumulative rhymes such as 'The Old Woman and her Pig' or 'This is the House that Jack Built'.

Iona and Peter Opie wrote that children 'become joyful and wise listening to the same traditional verses . . . wherever the English word is spoken'. Nursery rhymes are one of the strongest pieces of evidence for the persistence of British traditions in Australia, although the rhymes are told by royalists and republicans alike, and their contents are often delightfully irreverent or anarchic, as in 'The Grand Old Duke of York':

The grand old Duke of York,
He had ten thousand men;
He marched them up to the top of the hill
And he marched them down again.
And when they were up they were up,
And when they were down they were down,
And when they were only half way up
They were neither up nor down.

It is not only adults from English-speaking backgrounds who pass traditional nursery rhymes on to their Australian children. This Italian nursery rhyme was collected in Melbourne in 1976:

Signorina Patatina
Con le gambe di gallina,
Con la vesta di veluto;
Signorina, ti saluto!

(Little Miss Potato
With her little chicken legs
And her velvet dress;
Little Miss, hello!)

Nursery rhymes, in many different languages, pose a hundred different questions. Why do parents amuse their children with unlikely juxtapositions like a cat and a fiddle or a cow which jumped over the moon, or in the Greek '*O Pontikos*', with a mouse and a monastery? This rhyme comes from Footscray, in Victoria, 1986:

The mouse jumped from the window
And his mother asked him,
'Where are you going, my skipper?'
'I'm going to get marble
To build a monastery
To put my children in,
So that the fleas won't eat them.'

Which rhymes do immigrant parents still tell to their children in Australia, and which ones (if any) have they discarded? Are there more similarities or more differences between English-language and other nursery rhymes?

There are certainly some culturally specific features in nursery rhymes told to young children. Most of the rhymes in the core English-language repertoire are secular, whereas religious sentiment is common in Italian and Spanish rhymes. '*Agua, San Marcos*' was collected in Melbourne in 1975:

St Mark, please send us water:
For the wheat that is already very beautiful,
For the barley that already has grain,
For my melon that already has a little flower,
For my water-melon which also has a flower,
For my olive tree that already has one olive
on it.

Nevertheless, it seems that there are more similarities than differences in the nursery rhymes still in use in Australia (see Davey, Folklore and the Enculturation of Young Immigrant Children in Melbourne, M.Ed. thesis 1983). The use of nonsense humour is particularly widespread, humour which Thompson's *Motif Index of Folk Literature* (1955) designates as 'humour of lies and exaggerations'. The Spanish '*Vamos a cantar mentires*' literally says 'Let's go out and tell (sing about) lies':

Now we are going out, and we're going to tell lies.
The first lie is that the rabbits are in the sea
And the sardines are in the mountains . . .

Why do parents use so many 'lies and exaggerations' to their children, given that adults place so much emphasis in their child-rearing practices on children's telling the truth ('Tell the truth and shame the devil')? The issue is not simply one of morality, since one of the basic rules of conversational discourse is the assumption that the partners in the exchange will tell the truth, that you say what you believe to be true. As Catherine Garvey (*Play*, 1977) points out, the deliberate violation of this discourse rule is a source of humour for both children and adults. Children may misname others ('you're a crocodile'), parents employ nonsense rhymes (Simple Simon fishing in his mother's pail), and adults engage in tall stories. The latter may be improvised, as in fishermen's tales, or traditional, as in the stories told about the Turkish folk character Nasreddin Hodja. One story about Nasreddin tells of his boiling a cauldron over a candle, as a lesson to a mean friend who would only put one candle in his window at night to show Nasreddin the path.

In his 1925 classic *From Two to Five*, Kornei Chukovsky discussed 'the tendency to violate the established order of things' which is found in folklore told to young children. He gave as a Russian example

Listen, my children,
And I'll sing you a fiddle-faddle:
'The cow sat on a birch tree
And nibbled on a pea'.

Chukovsky emphasized particularly the reality-testing function of nonsense rhymes, the *pereviortyshi* or 'topsy-turvies'. In order for the child to realize that humour is involved in the performance, the child must have a knowledge of the real order of things. Nonsense rhymes play with notions of causality, with the functions and relations of objects, and with the classification and ordering of things in the world, all processes which aid the child's developing 'construction of reality' as Piaget called it. The nonsense rhyme demonstrates power over reality, a reality which can be manipulated and turned topsy-turvy, because the knowledge is there which can turn it right again.

No discussion of nursery rhymes would be complete without some comment on that hardy perennial, the 'political origins' of nursery rhymes. Almost everyone discussing 'Ring-a-Rosy' will describe in great detail the origins of the rhyme in the Great Plague of London, and say that 'a-tishoo' refers to sneezing brought about from the 'posies' or nosegays carried to ward off infection, and 'all fall down' indicates people dying.

Ring-a-ring-a-rosy,
A pocket full of posies,
A-tishoo! A-tishoo!
We all fall down.

Yet in their introduction to *The Oxford Dictionary of Nursery Rhymes* Iona and Peter Opie reject the popular belief that most nursery rhymes were originally political satire. They condemn most of the theories as 'political squibs' lacking in real evidence, and describe Katherine Elwes Thomas's book *The Real Personages of Mother Goose* (1930) as 'a curious mixture of fact and fable'. Although they believe that there is some evidence for a few political origins, such as that 'Charlie over the water' was in fact Bonny Prince Charlie, they state that 'the bulk of these speculations are worthless. Fortunately the theories are so numerous they tend to cancel each other out'. The also note that 'Ring-a-Rosy' was not published in any form until the nineteenth century, despite the proliferation of earlier books of nursery rhymes. Yet 'Ring-a-Rosy' remains intriguing, not least because versions using the same tune occur in several different languages. The Italian '*Girotondo*' has no apparent plague references, but refers to the world turning and the earth falling down. 'Ring-a-Rosy' is only one example of the endless fascination and persistence of the nursery rhyme, in Australia and elsewhere.

Gwenda Beed Davey

O

Occupational folklore used to be thought of as the verbal, material and behavioural practices of well-defined task-specific groups, usually male and usually engaged in manual labour in locations away from urban centres. Shearers, overlanders and gold-diggers are Australian examples of such groups. In other countries attention focused on cowboys, lumberjacks, agricultural labourers, peasants, seamen, and the like. These occupational groups certainly have a wealth of folklore and in the Australian context such lore has also been extremely influential beyond the occupational folk group itself, contributing to images of 'Australian-ness' and national character. However, such occupational groups, while having a strong air of romanticism attached to them, are not typical of industrialized societies and in Australia have been the exception rather than the rule for much of the last two hundred years.

Since the 1960s folkloristic attention has moved to occupational groups that are more common in modern life. American folklorists and cultural anthropologists have been particularly, though not uniquely, active in this area, providing ethnographic studies and analyses of numerous contemporary occupations, including air hostesses, firefighters, and waitresses. Very little such study of particular occupational groups, rural or urban, has been carried out in Australia. However, enough general work exists to complement the few specifically occupational studies and to allow some useful observations about this important area of folklore. The enormous range of occupations, even within a country with as small a population as Australia, means that the following observations are intended as a guide to occupational folklore in general rather than a definitive statement about the folklore of any particular occupational group.

Work sites have four main types of folklore circulating within them. The first is the folklore brought into the workplace by workers by virtue of their participation in the broader society. This folklore includes topical jokes, ethnic jokes, bawdry (q.v.), rumours, legends (q.v.) and folk beliefs (q.v.). The second is occupation-specific lore. This is the lore of workers with a particular expertise or skill and/or industry, such as draftsmen or women, or upholsterers, or machinists. As members of an occupational group, such workers will use jargon terms for tools of the trade (for example, coal miners know that 'joggling' a piece of wood involves cutting it to a required size and shape) and for processes used in their occupation, and they will have in-group knowledge of personalities and events peculiar to their calling. Such lore is frequently used to define the particular occupational group against other occupational groups within the same industry or workplace — scientists against engineers, plumbers against carpenters and so on. The third main type is industry-specific lore. People with particular skills may be found working in a variety of industries — draftsmen or women may work in the building industry, in the mechanical engineering industry, or in a government bureaucracy. Such industries may be scattered across regions, nations and indeed the globe, yet folkloric communication is still maintained through a variety of methods — by face-to-face meetings at conferences, seminars, and other meetings; by mail, by computer and facsimile transmission, or by internal company or bureaucracy communication channels, such as newsletters and bulletins. Common folklore forms circulating amongst such groups include reprographic folklore (q.v.), jokes and narratives of various kinds related to typical occupational situations. They are often critical of occupational hierarchies, power flows, communication and human resources aspects of the occupational group. Fourth, there is workplace folklore which is peculiar to the group of workers in a particular office, factory, school or other work site. Such lore typically includes allusions to events, personalities and concerns known, or at least fully known, only to those people present at the workplace. Because of the markedly esoteric nature of such lore, comprehensible examples are difficult to provide in brief; however, readers need only reflect upon the nature of their own work experiences to appreciate the existence of this aspect of occupational folklore and its importance in knitting together those in a particular workplace.

Certain forms are particularly characteristic of occupational folklore: the anecdote ('the worst boss I ever had', 'the biggest stuff-up we ever made'), jokes, graffiti (q.v.) and reprographic lore. Although we have come to think of songs as a characteristic expression of occupational folklore, song is unusual in most modern urban occupational groups. Even musicians typically tell nasty jokes about each other rather than sing occupational songs.

Behavioural forms frequently found in occupational contexts include observance of the Melbourne Cup, April Fool's Day, office parties (particularly the Christmas party), farewells, initiation rituals and pranks, hoaxes, general 'stirring' (which may involve practices such as rumour-mongering, practical jokes or simply refusal to co-operate), 'foreign orders' (unsanctioned use of time, skills, equipment or materials for personal, communal and extra-occupational benefit), as well as customs specific to particular occupations and/or industries, such as 'topping out' of large new buildings by hoisting a tree or other flora to the top and assembling all those involved with the construction for a drink and a final gathering. Further examples of occupational custom include student 'muck-up' days and commemorative customs and revels, picnic days, informal sporting contests, quiz nights and other such occasional and periodic events.

The tone, intent and content of much occupational folklore is humorous; it is also often crude, and frequently critical of certain aspects of working life, particularly of the distribution and exercise of power (the workers against the bosses) and the absence or failure of communication within the industry or workplace.

As with most folklore, a primary function of occupational lore is the development and maintenance of group coherence and identity. In the occupational situation this involves such things as worker pride in the products and processes of his or her occupation. It also involves worker prejudices of various kinds, especially those involved in the inevitable tension-points of power, authority, control and communication, or worker prejudices against client groups with whom they come into contact. An example of this may be found in the folklore of public servants and workers in private bureaucracies, such as insurance companies, who are in continual contact with members of the public over such potentially difficult matters as personal and business finance, social security provisions or the provision of health care, to name only a few such areas. Much of the rationale of occupational folklore, then, is to allow workers to cope with the tensions, conflicts, aggressions and frustrations that inevitably arise in any work situation. We can also speculate that folklore may be used to generate such tensions, perhaps as a form of release from the possible monotony of a particular task. Either or both of these possibilities may operate in any particular situation.

Occupational folklore is continually evolving and adapting to changes in the nature of work, of technology and society at large. The virtual disappearance of the previously widespread ballad song and verse forms from occupational folklore is one devolutionary example of this, while the rise of the reprographic or photocopy-lore form provides an excellent example of the evolution of new folklore forms, both as a response to changing social, economic and occupational circumstances and as the adaptation of new technologies to the perennial needs that folkloric expression appears to satisfy.

REFERENCES: P. Adam-Smith, *Folklore of the Australian Railwaymen*, 1969; R. Byington (ed.), *Working Americans: Contemporary Approaches to Occupational Folklife*, 1978; T. Radic (ed.), *Songs of Australian Working Life*, 1989; G. Seal, 'Worktime and Folktime: Folklore in a Modern Office', in R. Edwards (ed.), *Proceedings of the 3rd National Folklore Conference*, 1988; G. Seal, *The Hidden Culture: Folklore in Australian Society*, 1989, Chapter 6, 'The Folklore of Work'.

Graham Seal

O'CONNOR, NORMAN (1923–), was a foundation member of the Folklore Society of Victoria (q.v.). During the 1960s Norman O'Connor, Pat O'Connor and Maryjean Officer travelled throughout Victoria seeking and recording traditional songs, recitations and dance music. Norman O'Connor's recordings, including those of the traditional singer Simon McDonald (q.v.), are now housed in the National Library of Australia.

'On the Wallaby Track': see **Radio and television programmes**.

P

Patchwork quilts Australians loosely use the word 'quilt' to describe a bed covering, regardless of its construction: a quilt may consist of one layer of cloth, or else of two or three layers stitched together, and it may consist of patchwork in some form. In North America the word is used more strictly to describe a bedcover consisting of several layers which are usually decoratively hand-stitched together through the process called quilting. In contrast Australians appear to follow usage from Britain and Ireland, which allows a freer interpretation of the word. Quilts in Australia can be whole pieces of cloth (such as the industrially produced Marcella quilts popular early in the twentieth century), hand-knitted or crocheted, or made from some form of patchwork. The processes of patchwork can be further described as either 'pieced' (an American term describing patchwork in which geometrically shaped pieces of fabric are seamed together) or 'appliquéd' (a word of French origin used to describe the process of sewing pieces of fabric on top of a background fabric). These words were not usually those that Australians used; generally it would appear that quiltmaking was simply called 'patchwork', and the finished result called a quilt, or sometimes a counterpane or a rug.

Patchwork quilts are to be found throughout Australia (Rolfe, *Patchwork Quilts in Australia*, 1987), and examples exist from the first half of the nineteenth century through until the present when patchwork and quilting have become popular leisure crafts for women. However, while many Australian women in the past made quilts, the tradition of patchwork and quilting was not as strong in Australia as it was in North America, where quiltmaking was an almost obligatory skill for girls and women in the nineteenth century. Since a majority of Australians originally descended from emigrants from Great Britain and Ireland, quiltmaking in Australia during the nineteenth century largely followed traditions emanating from these regions. Many settlers coming to Australia brought quilts with them. Unless a quilt has either family provenance or overt Australian content it is usually impossible to say with confidence whether a quilt was made in Australia or was brought here from overseas.

Quiltmaking was mostly, but not exclusively, a female occupation. Some quilts are known to have been made by men, usually in situations where either isolation or ill-health gave them time, which they filled with making patchwork. There is also a nineteenth-century tradition of quiltmaking by military men, in which intricate quilts or cloths were made from thousands of tiny pieces of fabrics from coloured woollen uniforms. Several of these exist in Australia, one being attributed to a veteran of the Boer War.

In the past, quiltmaking existed as a folk tradition because quilts were made as practical or decorative additions to the home. They were not made to be sold, although some were made to raise money for good causes, such as the Red Cross or the Country Women's Association. It is meaningless to describe women's work in terms of amateur or professional, as until recently, much of women's time was fully occupied with domestic work (a situation which has now changed because of reliable birth control and the technological advances affecting housekeeping). Training for this domestic role occurred largely within the home. While domestic work had a great economic importance to the survival and well-being of the family, it was not 'paid' work in the sense of either wages being received or goods sold. Needlework, both utilitarian and decorative, was highly regarded as an important skill for women and girls, as it made a major contribution to the creation and maintenance of the clothing, linen and bedding of a family. In fact, when many nineteenth-century people used the word 'work' in relation to women, needlework was implied (a 'work-basket', for instance, was a basket containing needlework tools and requirements). Women generally pursued quiltmaking from two differing, but often overlapping, motivations: first, to create a beautiful object of needlework to decorate a bed or room; second, to economically use materials at hand to make a utilitarian bed cover. The overlap of the two intentions occurs because many finely worked quilts also used materials left over from dress-making or other sewing, and conversely, many scrapbag creations show that even while the materials used may have been humble, attention was still paid to design and decorative effect.

Several major styles of quiltmaking can be distinguished and, although the popularity of these styles may have waxed and waned at different times, each style persisted for many decades, some even spanning centuries. The existence of many examples of the same style from different times and places indicates that a widespread tradition was being followed. Quilts, being textiles, are most vulnerable to decay. If, as generally intended by the maker, a quilt was used as an everyday bed cover, it was subject to the ravages of light, soil, the removal of that soil by washing, and the general wear and tear of handling and use. The survival of quilts, therefore, was generally not assured. Quilts from the past which do still endure are thus random remnants, the existence of which hints at greater numbers of their ilk that have been worn out and thrown away. The fact that many quilts of similar styles remain would seem to indicate how widespread and persistent traditions were.

While it is often imagined that patchwork was a craft which was born of necessity, in fact the opposite appears to be true. Originally, patchwork quilts in England were made of expensive materials, notably imported Indian chintzes and their European-made derivatives. Hand-sewing small pieces of fabric together is a time-consuming process, indicating both leisure and the availability of fabrics beyond necessity. However, the increased industrialization of the production of fabrics meant that throughout the nineteenth century fabrics became cheaper and more abundant. Also, rapid mechanized sewing became an option after 1860 when sewing machines began to be widely distributed. While at the beginning of the century patchwork was a 'lady's occupation', by the end of the century it had become associated with thrift and making do. Changing fashions also contributed: in the early part of the century clothes were generally made from rectilinear shapes, leaving little or no waste fabric, whereas later in the century fashion dictated fitted clothing (assisted by the invention of paper patterns for home dressmaking) which left odd-shaped waste scraps from the cutting.

Consistent with the idea of patchwork being a lady's occupation in the early part of the century, the few existing quilts in Australia from this era are attributed to wealthier women colonists. One instance is the patchwork quilt top (collection of the National Trust of Australia, New South Wales) believed to have been created by Elizabeth Macarthur (1766–1850), wife of John Macarthur. The patchwork was made of hexagons in the English style of patchwork over papers, and the quilt may have been unfinished, because it has no backing.

Quilts are known to have been made by convict women on board ships coming to Australia. Although sewn by the convicts, the work was instigated by ladies under the direction of the Quaker philanthropist and prison reformer Elizabeth Fry. Making patchwork quilts fulfilled several aims: it kept the women busy, and therefore, their benefactors hoped, would keep them out of trouble; it taught the women sewing skills (which by implication indicated some absence of such skills); and as the women were permitted to sell the quilt on completion, it possibly gave them some little capital against their new life in the colonies. The sewing requirements for the patchwork were all supplied by the committee of ladies superintending the work, and each convict was given haberdashery (needles, scissors, thread, etc.) on embarkation, as well as two pounds of patchwork pieces. The scheme was first set up by Elizabeth Fry in 1818 with the approval of government officials who were happy for it to continue. In the 1838 report on transportation, the haberdashery and patchwork pieces are listed among the requisites for the convict women. The operation of the scheme was described in the report:

> what is the effect . . . of the voyage upon a number of females of indifferent character? They have some occupation during the voyage; they are allowed certain patchwork, which keeps them in some degree occupied; but still while they are at their work they will converse. As far as I have been able to collect from surgeons who have attended convict vessels, and from those persons who have attended female convicts on their arrival, the giving them this employment has had the effect of occupying their minds to a certain degree, but it does not last to the end, they get over their work by the time they have gone two-thirds of the voyage. But their conversation is very obscene, and the results very bad. (Select Committee on Transportation, *British Parliamentary Papers*, 1837)

The patchwork may have employed their fingers, but it did not control the convict women's tongues.

Sanctioned by officialdom and organized by the well-intentioned committee of ladies, the quilts made by convicts would not thus be considered a folk tradition. However, some of the convicts may well have learned skills and personally profited from the exercise. One at least wrote back to the Fry family describing her success in Australia, and saying that 'the patchwork quilt which now covers her bed, was made of the pieces given her by the ladies when she embarked'. The only known surviving quilt was made by the women of the convict ship *Rajah*, and it bears an embroidered inscription:

> To the ladies of the convict ship committee
>
> This quilt worked by the Convicts of the ship Rajah during their voyage to Van Diemans Land is presented as a testimony of the gratitude with which they remember their exertions for their welfare while in England and during their passage and also as a proof that they have not neglected the Ladies kind admonition of being industrious
>
> June 1841

The large quilt (collection of the Australian National Gallery) is made in the medallion style (i.e. a style in which a centre design is surrounded by successive borders). The centre is of *broderie perse*, a technique in which motifs from a chintz print are cut out and appliquéd to a background. Surrounding this centre are rows of simple patchwork, with a final border of *broderie perse*. The quilt is unlined.

While patchwork may have been the province of ladies at the beginning of the nineteenth century, by the end of the century it had become widespread, and as such, had become a folk tradition. A number of major styles of quiltmaking emerge: oversewn patchwork over papers, the log cabin, crazy patchwork, simple shapes made into medallion-style quilts, embroidered and appliquéd quilts. While in the early half of the nineteenth century quilts were made of cotton fabrics, in the latter part of the century it was fashionable (and recommended by women's books and magazines) to create patchwork from silks and velvets. Use of these fabrics in patchwork reflects their increased use in clothing at this time, when industrialization had made them cheaper and more widely available. Silk fabrics can be seen in all styles of patchwork until the end of the century, after which this fashion declined.

One predominant style of patchwork is the English form (so called today to distinguish it from American patchwork), in which shapes are cut from paper, covered with fabric, then oversewn together. The most frequently found shape is the hexagon, followed by a diamond based on a

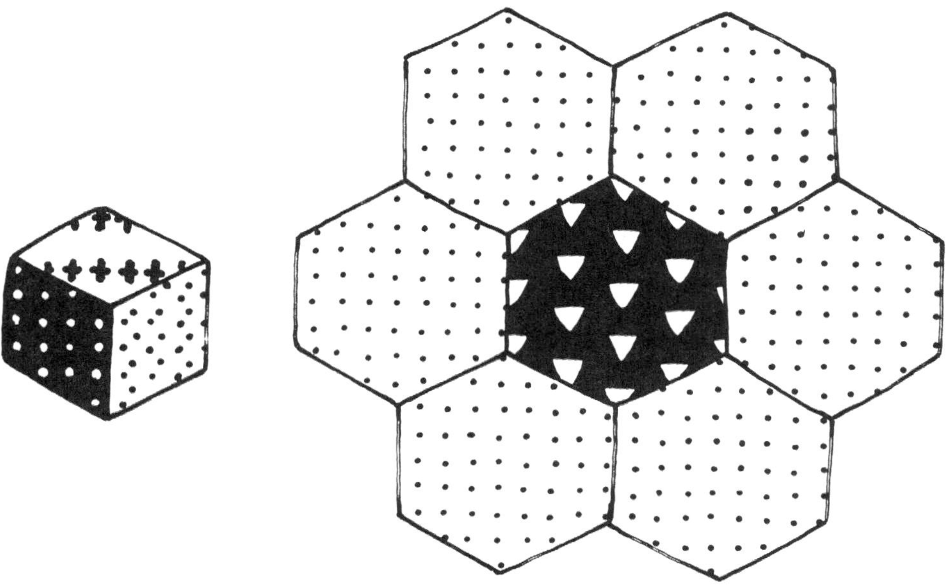

English patchwork style

division of the hexagon, although other shapes were also used. Fine examples of quilts using the English technique are seen in three surviving quilts made by Sophia Wilbow (1829–1924). One quilt is made of hexagons, and the other two of diamonds (one is in the collection of the Yass Museum, New South Wales). Sophia made seven quilts, one for each of her daughters. Her husband, who had been a farmer and publican in the Windsor district of New South Wales, was an invalid for the last years of his life before his death in 1896, and it is believed that she worked on the quilts by his bedside.

Log cabin quilts are also found widely throughout Australia, most having been made from the latter part of the nineteenth century through to the middle of the twentieth century. Quilt researchers believe that the log cabin is a style that probably arose in the mid-nineteenth century, but its origins are obscure. A log cabin quilt is usually made of square units of design (called blocks by the Americans), which contain a small centre square surrounded by successive narrow strips of fabric, with half of the strips light in colour, and half dark. The overall pattern of the quilt is created through the arrangement of the light and dark tones into patterns such as diamonds, stripes, steps, or variations of these. The process of making the blocks involved stitching the centre square and strips onto a square of foundation fabric; this foundation ensured that the quilt was strong, and gave an extra layer between the top and the backing of the quilt. The sewing could be done by hand or machine. Because the log cabin design uses small squares and narrow strips, it was an ideal way of using leftover scraps of fabric. For example, in the 1980s ninety-year-old Elvena Russell (née Smith, 1888–1985) described how her mother and sisters made a log cabin quilt some seventy years earlier in the village of Hall (now in the Australian Capital Territory). She remembered which of the fabrics came from the leftovers of her own or her sisters' dresses and petticoats and the occasions on which some of the dresses were worn. Her mother, Jane Smith (1859–1928), set her daughters to making the log cabin blocks on the machine, then arranged the finished blocks into a pattern to make the quilt. Sarah Monument (*c.* 1862–1952), from Stawell in Victoria, stitched some nine thousand tiny strips into her quilt (collection of the Australian National Gallery), which was completed over eighteen years between 1910 and 1928. The quilt was just called 'patchwork' by the family, and the word 'log cabin' was not used.

Log cabin patchwork style

Crazy patchwork was a style which became widespread in the English-speaking world during the 1880s and 1890s. Begun in America, it was enthusiastically promoted through magazines and needlework books such as *Weldon's Practical Patchwork* (published in England). As many crazy quilts and other items of crazy patchwork (such as tea cosies, sachets and cushions) are found in Australia dating from around 1890, it is clear that the style was very popular. Crazy patchwork was made by covering a foundation block of fabric with random-shaped patches of rich fabrics such as velvet, silk and brocade. The edges of the patches were covered with a variety of embroidery stitches, and often extra embroidery or (occasionally) painting was added to the centre of the patches. Decorative and impractical, these early crazy quilts existed more as a showcase of the makers' skills with the needle than as useful bedcovers. While it could be argued that such quilts may be more popular art than folk art, in fact women used crazy patchwork as a vehicle to express their own ideas. For instance, many crazy patchwork quilts contain Australian motifs, such as wattle flowers or Australian animals. One quilt in the collection of the Wangaratta Historical Society, Victoria, has a spray of Sturt's desert pea in the centre, and several king parrots on other blocks. The quilt was worked by Marianne Gibson (1837–1911) in about 1891. Such Australian motifs were the maker's own addition to the style, and certainly were no copy of the published patterns which came from America or England. Other women used the style as a sentimental record of family life and linkages. For instance, Marion Gibson (1824–1908) of Hay, New South Wales, called her crazy quilt a 'Friendship quilt' because she consciously collected scraps of fabrics for it from all her friends and relations, both in Australia and in Scotland (where she had grown up and married). Marion later wrote that she 'went in for "Federation" on that "quilt"', because she included fabric from a tie which she collected from a man whom she connected with Federation. Another quilt, whose maker is unknown, is covered with brooches, beads, buckles, buttons (including several with small photographs on them) and badges (including a Melbourne cricket club badge for the 1903–04 season). Most certainly this quilt was a collection of particular family memorabilia.

Crazy patchwork style

Australian women absorbed the technique of crazy patchwork into their repertoire of needlework so that it became a traditional method of quiltmaking. The use of crazy patchwork changed over time, becoming more utilitarian during the twentieth century. As the fashion for silk patchwork declined after the end of the nineteenth century, crazy patchwork was made from printed cotton or sometimes woollen fabrics, and these quilts were made for practical rather than decorative purposes. The style became less elaborate, with embroidery only being added around the patches rather than in the centre as well, or else the embroidery was omitted altogether. Examples of such quilts are found from the early part of the century through to the 1950s.

Another style of patchwork involves simple geometric shapes seamed together (generally without papers) to make what is now termed a medallion-style quilt in which a centre pattern is surrounded by successive borders. Sometimes quilts had a special decorative centre, such as a late nineteenth-century quilt which had a red and white unofficial nineteenth-century Australian coat of arms appliquéd to a centre panel. This centre was surrounded by rows of squares and triangles. As some of the patches are themselves patched, and the backing entirely made of small pieces of a variety of white fabrics stitched together,

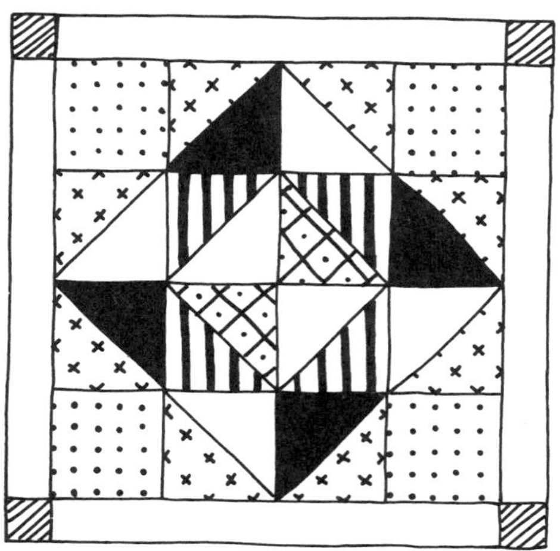

Medallion patchwork style

it would seem that the maker, Mrs Brown of Bowning, New South Wales, was being very thrifty with her materials. Unlike most Australian quilts, it was flat-quilted (i.e. quilted without padding between the patchwork and backing) in a pattern of hearts and crescents. However, this style of patchwork was more in need of quilting than other styles, as the patchwork did not lie flat and, if hand-sewn, was not as strong as the oversewn patchwork using the English piecing method. A medallion-style quilt was made by Elizabeth Hawkey (1829–1909) of Clunes, Victoria. The patches were hand-stitched together, and patchwork was used on both the front and back of the quilt. The patchwork layers were machine-quilted to a centre layer of old blanket, but it is not known at what date this quilting was done.

In the twentieth century many thrifty women made utilitarian quilts from samples of men's suiting fabric which were obtainable from country shops or tailors. When the sample books of fabrics were outdated, women collected them and sewed together the swatches to make hard-wearing warm quilts. These quilts are widely found throughout rural areas. While the rectangular shapes of the samples were often retained, sometimes the samples were cut into other shapes, such as triangles or strips. A fine example is a quilt in the collection of the Powerhouse Museum, Sydney, made by Caroline West (1872–1947) of Trundle, New South Wales, in about 1930. The rectangles of fabric were cut into triangles, and these were arranged into a strong pattern of lights and darks. Warm, dark and sturdy, such quilts were often used on boys' beds.

Another style of quiltmaking features embroidery, a style which probably derives from the popular nineteenth-century fashion for doing outline embroidery using red or (less commonly) blue thread on a white background. Printed patterns were published for this kind of embroidery, and often the designs derived from Kate Greenaway drawings of children wearing old-fashioned clothing. An example of this fashion being transmuted into an unsur-

The Misses Hampson, Tasmania. Detail from The Westbury quilt. 1900–03. Cotton. 224.5 × 188.8 cm.

passed example of folk art exists in a quilt in the collection of the Australian National Gallery made by the Misses Hampson of Westbury, Tasmania, from 1900 to 1903. Rather than red embroidery on a white background, the quilt has white embroidery on a red background. The motifs and inscriptions are drawn from either the makers' surroundings or else represent their values and beliefs. Central on the quilt is Queen Victoria, and other motifs emphasize the makers' allegiance to the Empire (such as a fine 'British lion'). Virtues such as Faith, Hope and Charity are featured, and industry is encouraged ('Work is the best antidote for worry'). Houses and shops from Westbury are embroidered, as well as a variety of animals, all with their names beneath them (Boz the dog, Barney the rooster, and Polly and Kitty the cows, for instance). The quilt may have been made for a raffle, as it also has the inscription 'Good luck to the winner of this'. This quilt illustrates the essential tension in folk art described by Michael Bird, in *Canadian Folk Art*, 1983, being at once both within a tradition (ethnosyncratic), but very much an individual statement (idiosyncratic).

Linked with the popularity of outline-embroidered motifs was a vogue for making collections of embroidered signatures. These were sometimes made for sentimental purposes (such as signatures from a group of people on board a ship, for instance) or else a fee was charged for each signature to raise money for some good cause. Such collections of embroidered signatures were made into quilts, especially during both world wars when signature quilts were made to raise money for the Red Cross. For example, in 1918 Mrs J. B. Ross (1854–1927) of Tarwin, Victoria, made a quilt now in the collection of the Red Cross Society in Victoria, which raised £6/1/6 at 6d per signature. Mrs Ross had a particular interest in raising money for the Red Cross, as three of her four sons enlisted and one son became a POW in Germany. Her youngest son was killed at Fromelles, France, in 1916.

Few appliqué quilts exist in Australia, and generally these quilts were the work of individuals using the traditional technique of appliqué in an idiosyncratic and expressive manner. This is unlike North America where there were many widely-shared appliqué block designs. In Australia sometimes appliqué and embroidery were combined, such as in a large Tasmanian quilt (collection of the Tasmanian Museum and Art Gallery) made by an unknown quiltmaker. Constructed in the medallion fashion, the quilt has a central panel surrounded by successive borders, all of which are appliquéd or embroidered. A careful study of the motifs used (which include a hand of cards, many beautiful fans, flags of the Empire, contemporary cartoons of two dragoons, a cricket wicket, tennis racquets and tea things) suggests a humorous, though patriotic, comment on life in a colonial outpost of the Empire. A date of 1887 and the words 'Jubilee of Queen Victoria' are embroidered on the quilt, as well as a letter bearing a Tasmanian address. Other examples of appliqué being used in a very individual manner are the quilts (or perhaps wall-hangings) made by Mary Jane Hannaford. Three of her quilts exist in the collection of the Australian National Gallery: one with an Australian theme, one with a nativity and creation theme, and the third with the theme of time passing and remembrance. This latter has the inscription 'M. J. H. aged 84 years. 1924' worked on it. Each quilt is in the same naive style, with strips of joined patches surrounding the pictorial motifs, all of which are crudely stitched to a background fabric without any lining added.

The style of patchwork called Suffolk puffs (known in America as yo-yo's) is found in Australian quilts from around the turn of the century until the middle of the twentieth century. These quilts were made by joining circles of gathered fabric together. Many were made of all-white fabrics (often with matching pillow shams), but quilts of print fabrics also exist. An example is a quilt of printed cotton fabrics made by Olive Bibb (1881–1923), from Sydney, in about 1910. Fabrics showing several different colourways indicate she was probably using fabric samples.

Quilting (that is, the patterned stitching

joining the layers of a quilt together) is generally absent from most Australian quilts. Two of the most common styles of quiltmaking, the log cabin and crazy patchwork, did not need to be quilted because of the way they were constructed onto foundation fabrics. The other major style, oversewn English-style piecing, sits firm and flat and does not require quilting to strengthen it, although quilting is possible (some English examples are quilted.) Where quilting has occurred in Australian quilts, it is often partial, such as a layer of padding quilted to a backing. However, there are some quilted quilts in existence, indicating a limited carrying-through of English and Irish quilting skills to Australia. English quilting involves the elaborate patterned stitching of three layers: a top of either plain fabric (termed whole cloth) or simple patchwork, a layer of padding (termed wadding) and a backing. Elaborate patterned quilting is very rare in Australian-made quilts. Flat quilting is typical of many Irish quilts, in which a simple pattern (such as a chevron pattern of zigzag lines) is stitched through only two layers, the patchwork top and quilt backing, the latter frequently being recycled cotton bags. Flat quilting is found in some Australian quilts, such as the medallion quilt made by Mrs Brown (see above). Isabella Cooke of Alstonville, New South Wales, some of whose family came from Ireland, made nine quilts for her children during the 1920s and 1930s. One surviving quilt is made of plain white fabric, and is flat quilted with a chevron pattern.

How patchwork and quiltmaking skills were transmitted in the past is difficult to determine, as information about the transmission of domestic skills has not been considered important enough to collect, although some inklings can be gleaned through oral history, and also through women's diaries and letters. It is most likely that the skills were passed informally, from one woman to another. Probably the women were often of different generations in one family, such as a mother or grandmother teaching a daughter or granddaughter. One small unfinished piece of patchwork, made in about 1916 by Beryl Kilby (1904–1988), illustrates this process. Beryl was a young girl of twelve or thirteen learning to use a sewing machine. Under her mother's tutelage, she made a small piece of patchwork in the medallion style, and her mother explained to her that this was how patchwork was done. Fabrics included pieces from her own dresses and fabrics from her mother's youth, as well as a piece which came from one of her grandmother's dresses. Alternatively, many women probably shared skills and patterns with each other. An example of this is noted in Annabella Boswell's diary of 1844:

> Marion was busy finishing a patchwork table cover of silk . . . She kindly gave me the patterns, and we at once looked out all our pieces of silk and began to make one for my aunt. (*Annabella Boswell's Journal*, ed. Herman Morton, 1981)

English books and magazines published patterns for patchwork during the second half of the nineteenth century, and women were influenced by these. However, just as it is usually the women who already knit who are the buyers of patterns for knitting, it is arguable that published patchwork patterns may have been perused more to extend already existing skills and knowledge rather than to impart totally new skills. For instance, a booklet about patchwork (a section reprinted from Caulfeild's *Dictionary of Needlework* of 1882) was discovered amongst Sophia Wilbow's (see above) belongings when she died.

During the 1950s and 1960s, patchwork and quiltmaking largely fell from favour because they were perceived to hark back to days of deprivation during the wars and the depression. Some hexagon work continued to be done, but log cabin patchwork and crazy patchwork almost disappeared altogether. In the mid-1970s there was a wide resurgence of interest in craft, and many women began exploring crafts such as pottery, macramé and needlework. Women who travelled overseas were influenced by the quilt revival which was happening in the United States, and these women returned full of admiration for traditional

American techniques and designs. The enthusiasm generated was such that quilting groups and organizations spontaneously formed throughout the country; teachers began classes both privately and through leisure education facilities; exhibitions were held; quilt shops opened; and books containing Australian designs were published. The paid teaching of skills was a break from the informal transmission of skills of the past, but was consistent with the contemporary trend to have all skills taught more formally through educational institutions, and also with the entry of a majority of women into the paid workforce with a consequent re-evaluation of the time/money relationship. The majority of quilts made after the mid-1970s followed American trends and ideas, but some women made quilts with original and Australian themes, especially at the time of the Bicentennial in 1988. While quilts continued to be made for beds, quilts were also made to hang on walls, as purely decorative objects. The extent to which the current enthusiasm for quilts will result in a new folk tradition of quiltmaking has yet to be seen, but the signs are there indicating that a new generation of women have learnt skills which will continue to be practised for many years. For example, quilts are now frequently being made to raise money for charity. Quilts are also made as presentation gifts, such as a group of women making a quilt for a friend who is having a baby.

See: **Wagga rugs**.

REFERENCES: Marion Fletcher, *Needlework in Australia: A History of the Development of Embroidery*, 1989; Wendy Hucker, 'Bush Quilts', *Textile Fibre Forum* 8, 3, 1989; Jenny Isaacs, *The Gentle Arts: 200 Years of Australian Women's Domestic and Decorative Arts*, 1984; The Quilter's Guild, *Quilt Australia*, 1988; Margaret Rolfe, *Patchwork Quilts in Australia*, 1987.

Margaret Rolfe

PATERSON, ANDREW BARTON (1864–1941), who wrote under the pen-name of 'Banjo', is Australia's most popular poet — witness the sales of his numerous works during the last century or so. For Australian folklore studies, Paterson's significance is twofold. First, a number of Paterson's poems, including 'Waltzing Matilda' (q.v.), 'A Bushman's Song' and 'The Billygoat Overland', have become folk recitations or songs. Second, Paterson compiled the best-known and most influential collection of folk songs, *Old Bush Songs (composed and sung in the bushranging, digging and overlanding days)*, published in 1905. Going through numerous subsequent and frequently amended editions and reprints, *Old Bush Songs* was for decades the definitive collection of Australian folk songs.

An authoritative study of Paterson's methods of collecting the material that made up *Old Bush Songs* has not yet been published, but it is already clear from subsequent collection and recent research that Paterson was only able to retrieve a smattering of the large repertoire of traditional song that was, and still is, circulating in rural and urban communities. It has also been shown by Philip Butterss in *Songs of the Bush: The First Collection of Australian Folk Song* (1991) that Paterson used without acknowledgement a substantial number of texts published in the Brisbane newspaper the *Queenslander* eleven years before the publication of *Old Bush Songs*.

See also: **Collectors and collections of Australian folklore**; **Folk poetry and recitation**; **Folk song**.

Phar Lap: See **Sporting folklore**.

Photocopy lore: see **Reprographic folklore**.

Political songs It would be fair to say that at least half of Australia's traditional song repertoire could be described as political. Even if one does not subscribe to the theory that all folksongs are political songs, one certainly has to acknowledge the fact that the 'folk' have never seemed particularly interested in composing and circulating songs in praise of work or even offering a few complimentary words to the boss. Similarly, there are very few real folk songs that celebrate the law-makers, the politicians and the bureaucracy. Historically, Australia grew out of a very turbulent past and a chronological survey of our folk songs shows the paths of struggle. The socially aware will see that this struggle continues even today.

Convict songs bemoaned the fate of the transports, the gold-rush songs are full of despair and warnings about the possible arrival of the troopers, many of the bushranging ballads make statements about police harassment and the shearers seem to have sung of little else than their exploitative conditions. After the turn of the century the songwriters became concerned with an antipodean look at the international confusion over labour solidarity, the continuing threat of international war and the role of the working class in the new society. Our political evolution can be traced from songs of the colonial period through to songs describing the latest incident in the 1990s. Australians have always composed songs and sung them loudly, to state their political and social responsibility and to express their frustration. Historically these political songs were often the only way to retaliate against an unjust system or law. Open rebellion was dangerous and the composition and transmission of a good song could, at least, relieve some of the anguish and provide a safety valve hissing away at the enemy. They were also a valuable vehicle to promote solidarity; a chorus of singers is a strong barrier to break. It has been suggested that the 'Wild Colonial Boy' was considered a 'treasonous' song in colonial Australia and that the convict chain-gang work crews would quietly sing the song to antagonize their soldier-keepers. This may simply be a folklorist's romantic view of history, but few could deny the vehement power carried by such a song as 'Moreton Bay' with lines that shout:

My back with flogging was lacerated,
And oft-times painted with my crimson gore.
And many a man from down-right starvation
Lies mouldering now underneath the clay.

The 'folk' were more than happy to borrow popular tunes for parody and just as happy to have accepted poets to do their flag-waving. Charles Thatcher (q.v.), that prolific songster of the goldfields, found a ready paying audience for his songs that ridiculed the law and, in particular, the licence-hunting troopers in songs such as 'Where's Your Licence?' and 'The Song of the Trap'. Thatcher took other political issues such as immigration and the 'Chinese Question' and delivered them to his audiences in song. No doubt the antagonistic sentiments expressed in 'The Fine Fat Saucy Chinaman' were typical of the day and frustrated diggers found particular satisfaction in hearing songs that warned of the dangers of being overrun by the Chinese. Later poets, especially Henry Lawson (q.v.) and 'Banjo' Paterson (q.v.) followed similar paths by delivering poems chock-full of the political anger of their times. Lawson, never one to shy away from a working-class issue, was ever-ready to fire a line of warning to 'Texas Jack' or bite with lines such as:

So we must fly a rebel flag as others did before us,
And we must sing a rebel song and join in rebel chorus.
We'll make the tyrants feel the sting o' those that they would throttle;
They needn't say the fault is ours if blood should stain the wattle.

These classic words from 'Freedom on the Wallaby' were Lawson's parody on an earlier bush song, 'Australia's on the Wallaby', and have been quoted and sung at political demonstrations ever since. Lawson would have liked that! Paterson, although a man of some wealth, also wrote of the injustices done to the Australian worker and it is interesting to track a song like his 'A Bushman's Song' (also known as 'Travelling Down the Castlereagh') and see how the 'Banjo's' original text has changed. The original text ran:

I asked a cove for shearing once along the Marthaguy:
'We shear non-union here,' says he. 'I call it scab' says I.
I looked along the shearing floor before I turned to go —
There were eight or ten non-union men a-shearing in a row!

A version of this song sung by Joe Watson of Caringbah, New South Wales, in 1973, like several other collected versions, offered the last line as:

> There were eight or ten damned Chinamen
> a-shearing in a row!

This is a perfect example of the folk using a song for their own political ends, and a purpose which, one suspects, Paterson would have endorsed. Racism, of course, is only one ingredient of political songs, but one that is sure to produce the most hatred and usually the worst doggerel. The old songs against Chinese emigration are not so far removed from the songs written over the past ten years to attack the arrival of the Vietnamese and Cambodian boat people. In talking to Joe Watson it was clear that, despite his use of words such as 'Chows', 'Chinks' and 'Coolies', he did not harbour any real ill-feeling against the Chinese people; he was simply reflecting the popular attitudes of one hundred years ago. The common habit of calling the local Chinese vegetable delivery man 'John the Chinaman' or some generic name implies a similar, if less overtly racist, lack of cultural awareness.

Political songs come in all shapes and sizes. However, they seem happiest when riding along on a popular tune. Parodies offer a ready vehicle and that's exactly what is needed when the song has been written as a political solidarity-builder. A good example is the 1940s anti-budget song 'On Top of No Smoking', which is an obvious parody of 'On Top of Old Smokey' and starts off with:

> On top of no smoking
> Because of the cost
> They're taxing our drinking,
> Our pleasures are lost.

Another good parody is from the 1930s Depression; with a few quick changes 'It's a Long Way to Tipperary' becomes 'It's a Long Way Down the Soup-Line'. It's interesting to see which songs are most frequently parodied for political song; the following would certainly make the top of the list: 'Tramp Tramp Tramp', 'Waltzing Matilda', 'Click go the Shears', 'Rum and Coca Cola', 'Marching through Georgia'.

University songbooks have always offered a healthy selection of political songs, as have bushwalking songbooks, union newspapers such as the *Worker* and *Common Cause*, and the *Bulletin* magazine. Among the most fruitful areas are the factory noticeboard, oral transmission and the publications of small action groups. Particularly fruitful historical sources are turn-of-the-century Australian music-hall songbooks, for they offered songs lampooning contemporary political leaders and commenting satirically on Australian nation-building. One example is 'the Gumtree with Six Branches'. The story finds a young man dreaming and all of a sudden a native tree starts to talk to him, sing to him, about the challenge of building Australia into a strong nation. The song has an extraordinary chorus that goes:

> One branch is called Victoria and one is
> New South Wales,
> then South and West Australia, each
> gallantly prevails,
> with Queensland and Tasmania, all rich
> in mines and ranches,
> that's Federal Australia, the gumtree
> with six branches!

The great key to folksong research is to discover that these songs, the long and the tall, the short and the small, are part of a living tradition. People still sing the old songs and, just as importantly, they are still creating new songs in this same tradition. The song about Bob Hawke, 'The Mild Colonial Boy', or 'Nothing could be Crazier than Malcolm Fraser Grazier', are not so far removed from the songs written by those anonymous convicts some two hundred years back along the track.

REFERENCE: W. Fahey, *The Balls of Bob Menzies*, 1989.

See also: **Folksong**.

Warren Fahey

Popular culture and folklore Though the categories of 'folk', 'popular' and 'high' culture can be criticized, they emerged from the desire of particular social groups as well as scholars to account for differences in the social uses, meanings and contexts of cultural acts. If we are to continue to use these categories fruitfully to account for contemporary social uses of cultural acts, we must understand aspects of the historical, social and scholastic construction of these categories.

The distinctions between high culture, folklore and popular culture may emphasize the technological means employed, or the relationship of an expressive practice to social or ideological structures, or the alleged formal structure of the art forms. Thus, high culture is sometimes assumed to be of particular complexity and hence of expressive potential, or to be supported by social groups of high status and social power, or perhaps to be dependent upon official patronage and so on. Folklorists distinguish their realm of study from popular culture, which they generally regard to be commercially mediated culture produced by the organizations of industrial society. Folklore is seen as in some ways independent of these institutions, whether as the continued existence of pre-industrial cultural forms, or as the continuation of modes of communication or knowledge not dependent upon industrially based institutions. Folklore, whether seen as oral tradition, or as unofficial culture, or as small-group communication and expression, is being distinguished from those expressive genres which use industrial technology and media, are cultural commodities, or are generally available in a market-place.

Scholars of popular culture are less likely to categorically distinguish folklore from the rest of popular culture. Often they assume that the archaic survivals which would be labelled as folklore are a marginal part of contemporary popular culture, and of relatively little significance. Some historians of popular culture would insist that before the formation of industrial communications industries, no distinction could be made between folk and popular culture (see, for example, C. W. E. Bigsby, 'The Politics of Popular Culture', in Bigsby [ed.], *Approaches to Popular Culture*, 1976). Others insist that the historical existence of a folk culture separate from popular culture was merely the reflection of a transitory stage when the market had only partly penetrated popular cultural forms. Other critics, such as Dave Harker, may point to the origins of folklore studies in discredited intellectual or social movements, and maintain that the value of folklore as an analytical concept is compromised by its intellectual heritage (Harker, *Fakesong*, 1985).

Many of the distinctions between folklore and popular culture result from differences in perspective of scholars. There are few items studied by folklorists which would not be of interest to those who regard themselves as studying popular culture, and conversely most popular culture carries or contains aspects which folklorists would identify as folklore. Harold Schechter shows countless examples of folklore-derived motifs in popular literature, and argues that folklorists' techniques of analysis are essential to the understanding of such works (Schechter, *The Bosom Serpent: Folklore and Popular Art*, 1988), but students of folklore and popular culture each bring different perspectives and limitations to their subject matter which are embedded in broader social attitudes to culture, the society, the individual and aesthetics.

The categories of folklore and popular culture emerged in Western European culture in the nineteenth century. As Laurence Levine has argued, it was only then that social elites began vigorously to nominate particular cultural forms as high culture in order to distinguish them from the mere 'entertainment' indulged in by lower orders and increasingly being made widely available by new technologies and industries of communication and by the social structures of the new industrial cities (Levine, *Highbrow/Lowbrow*, 1988). The idea of folklore played a role in the delineation of a high culture from popular culture. When folklore became a well-defined field of study, it was thought of as the sum of survivals from pre-industrial social conditions; in many cases the surviving customs were thought to carry evidence of great antiquity. Within this perspective folklore was an ally of high culture against its main adversary: those entertainments which threatened hegemonic descriptions of the place of high culture. For the historical and human depth of folklore, its creativity founded in the collectivity of society, and its supposed relationship to the national spirit, all were stressed to expose popular culture as lacking these positive characteristics.

An important body of folklore scholarship emerged in Britain in the second half of the nineteenth century. Its main figures were cultivated and erudite gentlemen amateurs, who established a remarkable tradition of scholarship which formed the basis of much twentieth-century anthropology. However, it was founded on theories of cultural evolution, which have generally been rejected by twentieth-century anthropology, as scholars became more concerned to ground their investigations in detailed studies of particular social structures. Folklore scholars emphasized comparing and recording particular social traits or actions through time or between separate social groups. British folklore scholarship failed to secure itself a place in academic institutions within a general intellectual culture comparable with its co-disciplines of anthropology and sociology (see Kenneth Thompson, 'Folklore and Sociology', *Sociological Review* 28, 2, 1980).

In America folklore studies were more successful in transforming themselves into an accepted academic tradition. This was due to the continuing close association with the schools of cultural anthropology and with literary studies of oral 'texts'. This has led to the continued vitality of American folklore studies, and finally to its recent attack on the apparent marginalization of its subject matter in the modern world. Around the early 1970s this confrontation stimulated debate as to whether folklore is a type of text defined by its genre or a particular kind of social interaction. The 'text–context' controversy, although finally unresolvable, has since stimulated much of the most interesting work in folklore studies (Zumwalt, *American Folklore Scholarship*, 1988).

The idea of folklore has grown alongside, and to a large extent independent of, these developments and declines in academic discourse, and has been of considerable public and political influence. German romanticism brought forth the idealized people *das volk* and this construction underwrote most of the European nationalist movements of the nineteenth century. It was widely assumed that the potential citizens of a nation in some way owned a historic folk culture which would both inform their unified political action and justify the existence of the state which could express this. This conception of folklore took shape in the nineteenth century as part of the cultural project of the European national bourgeoisie to legitimate their role within the nation states which they were forming. By the twentieth century this historic role had become somewhat dispensable, and the political inspiration which founded the pursuit of national art forms abated in many European societies.

A new political perspective on folklore emerged from the 1930s to the 1950s, which has had considerable influence on the way in which folklore and popular culture have subsequently been seen. It was not until the 1930s that the intellectual left took over some of the cultural constructions of the nineteenth-century bourgeois folkloric activists to define a new populist nationalist folklore. This shift of interest was activated by the Popular Front policy of the Comintern in 1936, which, in conjunction with a move of socialist activists towards political co-operation with less radical groups, espoused a new promotion of nationally distinct aspects of culture by socialist intellectuals, particularly through the movement known as social realism. This was enthusiastically followed in the United States, where intellectual groupings like the 'people's songs' movement gave the social vigour of folklore as evidence of the right of lower-class groups to greater political power. The movement stimulated an increased interest in folklore, and moved its place away from the conservative ideologies of 'the invention of tradition' to making it part of nationally based radical populist movements (see Smith, 'Making

Folk Music', *Meanjin* 44, 1984; Hobsbawm and Ranger, *The Invention of Tradition*, 1983).

In Australia there has been virtually no official academic study of folklore until very recently, but a significant and publicly influential body of writing and discussion has come out of the radical populist definition of folklore. A serious concern with Australian folklore commenced in the early 1950s through the activities of Australian radical nationalists, who saw folklore as part of the project of defining a national popular character. This stimulated a programme of collection of folk songs in rural areas, and a few of the activists were drawn towards broader notions of folklore. Thus, Australian folklore was the cultural legacy of white rural Australia, as seen within the 'Australian Legend' (q.v.) of the male rural worker. Australian folklore has overwhelmingly been seen by these activist folklorists as the traditional understandings and ways of life of nominated 'typical' Australian groups, subordinate groups who are asserted to have defined this 'folklore' which is, or should be, understood in a unique way by other Australians. In most cases activists in the field have been interested in their subject matter from a historical and a nationalist perspective, and the folk revival has remained the major social and intellectual centre for involvement in folklore studies.

Popular culture studies in the English-speaking world have developed along a different path, and this has contributed to the different approaches now taken by scholars from this tradition and folklorists. The study of popular culture, in the sense of commercially mediated culture, has a shorter history than that of folklore. Until relatively recently the dominant mode was critical. Before the Second World War most scholars from all parts of the political and cultural spectrum were determined to show how so destructive a phenomenon as popular culture could be produced, and where its power resided. In the English-speaking world the first commentaries on popular culture, which appeared after 1945, were dominated by conservative ideas emanating from the influential literary criticism of the Cambridge academic F. R. Leavis, and from the cultural pessimism of the German 'Frankfurt school' of Marxism (for example Leavis, *Mass Civilisation and Minority Culture*, 1930, and Adorno, *Prisms*, 1967). In both cases, from very different political perspectives, mass culture, as it was termed, was seen as inauthentic, dehumanizing and stupefying, and ultimately a tool of social control. Though some radical critics in the 1940s and 1950s lionized a certain number of popular genres, such as jazz, these were accepted because of the possibility of seeing them as folk musics in the sense espoused by the radical populists (for example see Finkelstein, *Jazz: A People's Music*, 1948). The cultural limitations of these enthusiasms can be seen in the selection as authentic of near-extinct genres and styles and the rejection of any recent developments. However, in the 1950s and 1960s a certain tolerance of popular culture as such started to emerge in some academic disciplines.

The most influential work on popular culture in the English-speaking world today emanates from, or has been influenced by, the group of sociologists of culture associated in the 1970s and 1980s with the Birmingham Centre for Contemporary Cultural Studies. In general these writers have worked within a Marxist tradition which has sought to reinstate the importance of ideological and cultural elements in the creation and maintenance of political power. They have drawn heavily on the writings of the Italian Marxist Antonio Gramsci and on the French Marxist philosopher Louis Althusser. Central to their work is the idea that the dominant classes maintain their position through ideological and cultural means, but that their position is never won outright, and subordinate groups continually resist incorporation within dominant hegemony. Thus, for these writers, popular culture is seen as a site where political struggle occurs. The literature on popular culture that has emerged from the Birmingham school, often identified as Cultural Studies, has frequently studied specific 'sub-cultural' groups, the way in which they use expressive genres to assert their identity and to resist the dominant culture and ideology. But the analysis usually makes little distinction between deliberately expressive or communicative acts and interpretable social structures or practices. Some representative works are Dick Hebdige, *Subculture: The Meaning of Style* (1979), Paul Willis, *Profane Culture* (1978) and Stuart Hall's influential essay 'On Deconstructing the Popular' in R. Samuel (ed.), *People's History and Socialist Theory* (1981). There is a good deal of convergence between the discourse of cultural studies and that of some folklorists. In general, folklorists tend to place greater emphasis on textual analysis and comparison, whereas cultural studies emphasizes ideological and contextual interpretation. A folklorist will generally be highly sensitive to the history of the item, a cultural studies scholar to the historical construction of the practice within which it now exists.

In the USA much of the study of popular culture developed within a post-war cultural pluralism, in which the American dominance of popular culture made the debate about the nature of mass culture one about American political culture generally. Liberal pluralist sociology rejected the model of the monolithic culture industry, and emphasized the idea of taste groups, which, though influenced by their social background, could choose to use the cultural products of the media produced in the democratic economy. Herbert Gans, in *Popular Culture and High Culture* (1975), develops this approach. George Lewis identifies three approaches in American popular culture studies ('The Sociology of Popular Culture', *Current Sociology* 26, 1978). The social-indicators approach attempts to show the way in which the content of popular culture mirrors dominant cultural patterns in belief or attitudes. A second branch of research uses systems analysis to investigate the way in which social and economic institutions produce and distribute popular culture. A third stream is in communications research, investigating audience reactions and social composition of audiences, which has developed out of positivist behavioural sciences.

In Australia, sociological study of popular culture in the American style has been relatively rare. Of much greater importance has been the development of interest in social history. Cultural historians have been central to the radical nationalist project in Australia in which much folklore study has originated. This has ensured that scholars of popular culture have generally accepted the historical conceptions of folklore as that form of popular culture which has been superseded by the mass forms of production and distribution. Thus, the 1890s, the key site for the definition of Australian culture in the eyes of the radical nationalists, were seen as the last site for a post-colonial yet pre-modern society, where 'a shift from folk-art to an essentially imported culture designed for mass consumption' occurred (Introduction to D. Walker and P. Spearritt [eds], *Australian Popular Culture*, 1979). This ideal of folk art as a pre-contaminated form of popular culture is of little use for attempts to locate such forms in modern society, except as remnants of largely defunct cultural forms, and the association of folklore studies with social history has not helped the independent establishment of folklore studies.

From the early 1980s cultural studies influenced by the Birmingham school has had considerable impact on cultural criticism in Australia. However, those studies which have appeared, though addressing areas which command the interest of folklorists, have shown little interest in the perspectives of this discipline. In particular, they often fail to locate the practices they comment on in a historical perspective, and the interpretative analysis sometimes does little more than reiterate generalized stereotypes, sketches of national character (for example, Fiske et al., *Myths of Oz*, 1987). An active folklore studies tradition could suggest much more closely observed and historically informed methods of enquiry. Recent developments in Western

Australia, where cultural studies and folklore studies are being carried on side by side at Curtin University, suggest the possibility of cross-fertilization of this type.

There are a number of areas to which both disciplines can contribute techniques of analysis, especially those cultural phenomena which exist in a range of social contexts so that the face-to-face forms and commercially mediated forms are inseparable, as well as those genres which are in transition from folk to popular. The best of recent folklore scholarship is well informed by popular cultural studies, and has creatively absorbed its concerns. Folklore studies, as a less influential and fashionable field, has had little reciprocal effect, and some studies of popular culture are the poorer for their lack of awareness of both the historical depth of cultural genres and motifs which folklorists identify as 'tradition' and the techniques of close analysis of performance context which modern folklore scholars have developed.

A transition from folk to popular can reflect basic economic transformations, or may be activated by social movements, or (more usually) both.

Many, if not most, genres and commodities of popular culture have their historic origins in smaller-scale, more limited cultural forms and contexts. For example, the development of the commercial genre of country and western music from the singing and instrumental styles of poor white Americans in the south-eastern states has been well documented (Bill Malone, *Country Music USA*, 1985). This form developed when recording companies in the 1920s recognized that a cultural form — though they would not have so referred to 'hillbilly' music — which was hitherto regarded as valueless and limited had an exchange value as a commodity. Thus, the move from folk to popular necessarily involves the creation of markets which did not previously exist.

Another example of this can be seen in the recent shifts in the relationship between autonomous children's culture (see also **Children's folklore**) and commercially produced and distributed versions of it. Children's play culture has been seen as one of the most stable and enduring clusters of folklore in the Western world, and within this the verbal culture of the play rhyme has been of central importance. Although children in Western society are surrounded by official educational culture and specially created commercial popular culture, and the genres of adult folklore and popular culture produced for children from lullabies to fairy tales and nursery rhymes, these are not in themselves part of the play culture of children, except as subject matter for commentary and ridicule. One remarkable exception to this may be a recent series of books of children's rhymes edited by the Australian folklorist June Factor (q.v.) over the period 1985–90, which appears to have renegotiated some of the boundaries of the cultural sphere of children (Factor, *Far Out, Brussel Sprout!*, 1985, and its successors). In these books unaltered children's playground rhymes, as collected through the folklorists associated with the Institute of Early Childhood Development in Melbourne, are published in a form readable by primary school children. They sold extremely well and appear to have been independently chosen by children. Many collections of such material have been issued before for a scholarly adult market, with no impact upon the users of genre itself, but these newer publications stand in a much closer relationship to their audience. Their long-term effect is yet to be seen. It may be that the two genres of oral children's rhymes and printed versions of these will continue to exist side by side and, in spite of their parallel use by children, exert little cross-influence. On the other hand, printed texts may influence the oral tradition of children and encourage standard versions. The investigation of this issue would involve considerable research.

Why, then, has this folk genre now gained a 'popular' form? The most important factor is the enormous expansion in the publishing industry's output of children's literature which has taken place over the last twenty years. A new book market has opened up within which the child, although not economically autonomous, is encouraged to participate. Following moves in educational literature, much more literature aimed at autonomous reading has been issued, and the child is now seen within this market as an independent consumer. Much of this adult literature for children has taken the lead from children's culture to establish its mood and preoccupations. For example, the works of children's authors Paul Jennings and Roald Dahl show a continual fascination with the scatological and with the primitive individualism of the child which would be quite foreign to the moralistic and improving tone which until recently pervaded children's literature. It would seem that a much more tolerant attitude to rebelliousness in children's behaviour has emerged since the 1970s, and adults are much more likely to collude in the child's critique of the adult world than to dismiss or suppress it.

A transition from a folk context to a popular one always involves these kinds of economic and cultural transformations. Folklorists often use terminology of loss, standardization and decline to describe such transformations, but these judgements can cut short the possibility of analysis and understanding of the social processes involved. Nor does the tracing of persistence in content when folk items are transformed into popular forms offer much understanding of these processes.

Revivalism is another type of transformation of a folk genre into a popular one. Although usually underwritten by the sort of economic shifts described above, revivals are initiated less by shifts in the economics of culture than by ideologies of culture. Revivalism sees itself as maintaining or reactivating folk genres, but its central task is the creation of new social contexts and meanings for these genres. In revivalism, items of folklore are identified and promoted, and so become part of a general popular culture. They are taken into the public domain carrying an aesthetic aura of folkness, whether this 'folkness' is taken to be a national spirit, a class consciousness or a regional identity.

In general, revivalism has been associated with socially influential urban groups who, driven by an image of the 'folk' significance of a genre, attempt to give it a wider currency. This involves moving cultural objects and actions from face-to-face contexts into the public arena. Revivals differ in the political and social aims which they foster, but in almost all cases they involve the preservation or dissemination of cultural objects which are seen as threatened by some social or historical forces. The most prominent of these movements have been revivals of traditional music and dance, which have dominated public consciousness of the nature of folk art in Australia (see **Folk music in Australia: the debate**). The folk music revival was clearly part of the explosion of popular music produced by and aimed at young people from the 1950s onward, though it tended to define itself against mass-produced popular music. Thus, folk music stood for authenticity of expression, independence of thought and political freedom, whereas commercial popular music was deemed to be mindless, conservative, and American-imposed.

The other main revivalist trend in Australia has been in the ethnic arts arising from massive immigration in the post-war period. National and regional organizations formed, sometimes as little more than formalized friendship groupings, but sometimes extending to become large institutions. During the 1970s the prominence of these organizations increased with the policies of multiculturalism and with the success of larger groups as political lobbyists. A number of government-sponsored and semi-autonomous organizations fostered a broadened public role for performers who could be seen as speaking for an ethnic community, and the folkloric concert presented to a broader public became the cultural arm of the policy of multiculturalism. The ethnic artist became a new category of public performer, usually operating in semi-public and local performance contexts and sponsored by official organizations. Even so, few ethnic artists could move far from their original audiences.

Some with experience or awareness of professionalism in their country of origin were tantalized and frustrated at their inability to extend this initial success. The acceptance of non-Anglo cultural genres in mainstream culture was rather partial and tepid, and even for its political enthusiasts, a temporary dedication. Also, most artists had a relatively conservative view of the relationship between their genres and their audiences. As entertainers of a particular group they were confident and skilled, but in front of other audiences with little awareness of the social meanings of their genre, even obvious virtuosity was insufficient. Performers frequently were unable or unwilling to turn the foreignness of their styles into an exploitable exoticism. Thus, the determination of the meaning of ethnic art was often dominated by sympathetic Anglo-Australian entrepreneurs, whose broadly liberal political perspective of cultural tolerance and musical eclecticism seldom promoted the specific qualities and skills of performers.

Revivals and allied promotions of folk-like material have stimulated a number of responses from folklorists. Dorson coined the term 'fakelore' to describe the deliberate fabrication of cultural material which exploited an aura of folkness. Others have described the new public uses of folk genres as 'second existence'. Another approach suggests the term 'folklorismus' as 'the conscious commercialisation of folk-festivals, dances, costumes and crafts', which is the way in which these genres most frequently appear today (Dorson, *Folklore in the Modern World*, 1978). The general term 'folkloric' might be used to describe the presentation of cultural actions in a public arena using an aura of folkness to establish the conditions of their reception (G. Smith, 'Folklore, Fakelore and Folkloric', in *Proceedings of the 3rd National Folklore Conference*, 1988). Recent studies by folklorists have often taken a tolerant line on contemporary urban re-creations of folk genres, and stressed the need to

> eschew the . . . value judgements involved in distinguishing the real and the fake and rather become concerned with a description of the change in performer-audience relationship and in the setting . . . By doing this we will in effect describe the changeover from the folk to the popular, but this change does not need to carry with it any feeling of debasement. (Abrahams and Kalčik, 'Folklore and Cultural Pluralism' in Dorson [ed.], *Folklore in the Modern World*, 1978)

The categories of folklore and popular culture are limited and sometimes distorting, but they can tie an investigation into a rich history of scholarship. They need not be abandoned if they are subjected to continual critical examination in use. For example, how have the ideas of 'folk' and 'popular' in relation to any specific example emerged? Are there significant differences between the understandings of scholars and participants in this regard? Do, for example, participants think of their actions as folk or as culture at all? If so, what do they understand by these terms, and if not, how do they categorize their actions? Are there other groups with attitudes on these questions?

When informed by a critical awareness of their own cultural context, the analytical techniques of folklorists and scholars of popular culture will remain powerful tools in the understanding of non-elite cultures.

Graeme Smith

See also: **Folk Club Music**.

Q

Quadrilles were one of the most popular dance forms in the nineteenth century. They are danced by four couples facing inwards (the name derives from an Italian word for soldiers formed up into a square), a form which was also common in folk dance. The first one introduced into French ballrooms was called simply 'the Quadrille', and it was based on earlier dances (known as contredanses and cotillons) which had been adapted from French folk dances; five of these performed to lively tunes in 2/4 and 6/8 tempi were combined to make up the five figures of the Quadrille. It was introduced to London society in 1815 and arrived in Australia in the 1820s. Along with other quadrilles of the period, it quickly became popular here and combined with the closed-couple dances to oust the folk dances from the social dance scene. The Quadrille (otherwise known as the First Set) would often be danced several times in the evening, using different music each time. As the opposite couples danced together, the heads or tops first, and then the sides, the quadrilles retained much of the sociability of folk dances but with more variety of steps and music. Other early quadrilles were the Parisian and the Caledonians (danced to Scottish music). Later the Lancers (or Second Set), which had taken longer to be accepted in England, became very popular in Australia. Others very popular here were the Alberts, Royal Irish, Prince Imperials, Waltz Cotillon, Fitzroys and Exions. The latter two are made up of favourite figures from other quadrilles and appear to be Australian in origin. Many of the quadrilles developed many versions, such as the Alberts with two waltz figures. Not only were there differences between states, with some especially interesting versions in Queensland, but often places quite close together had quite different versions. One variation common in country districts was in the last figure of the First Set; this was known as the Stockyards or the Bull Ring and in it all the sets joined together in a big circle all around the hall.

Some quadrilles enjoyed a certain period of popularity and then disappeared; examples are the Kent, Lurline and Coulon's quadrilles in the 1870s and the National quadrille in the 1890s, as well as some made up locally, such as the Metropolitan quadrille (Perth) and the Triplet (Melbourne). The others listed above continued to be danced everywhere up until early this century. Everyone knew them, a caller was not necessary, or if used, gave only a brief reminder of the next step.

Styles would have varied somewhat in different social circles, but all historical accounts emphasize that dancing was always vigorous. The more sedate 'set turn partners' was replaced by 'swing partners' and many country versions included extra swinging.

The most popular quadrilles were still danced in many country districts until the 1950s and in a few places right up to the present day. Now, nearly all those mentioned here have been revived, and are being danced again by those enthusiasts who enjoy our traditional Australian dances.

See also: **Colonial or heritage dance**; **Dance**.

Shirley Andrews

Quilter's Guild Inc., The was founded in March 1982, and has the following aims:

- to promote the art and craft of patchwork and quilting;
- to bring quilters together and encourage the establishment of quilt groups;
- to encourage and maintain high standards of craftsmanship and design in both traditional and contemporary quilting;

- to foster interest in the history of patchwork and quilting;
- to organize exhibitions;
- to publish a magazine.

Membership in the Quilters' Guild is open to anyone interested in quilts and quilting. Junior membership is available to those under eighteen.

General meetings include guest speakers, slides, films, etc., and workshops are organized regularly; these are open to Guild members, and to non-Guild members on payment of a surcharge. Workshop tutors are both local, interstate and international. Exhibitions are held each year, and members and groups are encouraged to exhibit their work.

The Guild publishes a newsletter, *The Template*, and has also produced two books: *Quilt Australia* was produced in conjunction with an exhibition at Centrepoint, Sydney in 1988; *Australian Quilts. The People and Their Art* (1989) draws together the forty-three quilters commissioned by the Guild to create small quilts that could tour even the remotest parts of Australia — 'the Suitcase Exhibitions'.

A library is also available to members and the Guild sells patterns (of raffle quilts), badges, postcards and T-shirts.

Quilts: see **Patchwork quilts**.

R

RADIC, THÉRÈSE (1935–), was born Thérèse O'Halloran in Seddon in Melbourne's western suburbs. Her ancestry is illustrious, and includes Jonathan Swift (*Gulliver's Travels*) and the convict Isaac Wyse, transported in 1814 on the same ship, the *General Hewett*, as the architect Francis Greenway and for the same offence — forgery. Thérèse Radic was educated at the Star of the Sea Presentation Convent at Gardenvale in Melbourne, specializing in music, and studied in the Bachelor of Music course at Melbourne University, taking piano under Lindsay Biggins and Waldemar Seidel and harmony with the composer Dorian le Gallienne. During a two-year period in England with her husband Leonard Radic, Thérèse Radic attended Cecil Sharp House, absorbing as much as possible of folk music studies. When she returned to Australia she accepted the suggestion of Dr Percy Jones that she specialize in Australian music history. Her Master of Music was taken out in 1969 and her Doctor of Philosophy in 1978, both from the University of Melbourne. Her Ph.D. thesis was on the topic of Melbourne music to 1915.

Dr Radic began her career as a playwright with *Some of My Best Friends Are Women* (1976), co-authored with Leonard Radic and produced by the Melbourne Theatre Company, and has written numerous plays for theatre and radio. She drew on her work in the Grainger Museum at Melbourne University to write *A Whip Round for Percy Grainger* (1982), and in 1986 *Madame Mao* was produced by the Playbox Theatre Company and Circus Oz. This play has also been performed in both England and the United States. *Peach Melba* was first produced by Playbox Theatre Company in 1990.

In 1986 Thérèse Radic published two biographies, *Bernard Heinze* and *Melba: The Voice of Australia*, as the result of her interest in the mythic quality of what had been written and said about both of them — and how it contrasted with the hard evidence. She has also produced two songbooks, *A Treasury of Favourite Australian Song* in 1983 and *Songs of Australian Working Life* in 1989. At the time, Dr Radic was one of the few Australian musicologists interested in Australian folk songs. She writes:

> I believe 'folklife' is the matrix of cultural life — all cultural life. It is not primitive, simple or a stage in our development as a people. It is complex, continually in a state of change and renewal and is ignored by the scholar at the scholar's peril. In music it is demonstrably the first source. Look at traditional Aboriginal music. Once thought to be the ULTIMATE primitive sound of savages it is now used to show claims to land rights are based, or can be based, on information encoded in orally transmitted songs — the songlines, the maps to places and the past which these people move into 'folksong' that has survived all we have done to them.
>
> But folklore, for me, extends even to subjects like Melba, Heinze and Percy Grainger. Legend, myth — lies surround them. Why? Why do Australians treat their one-time heroes in this way, creating myths in order to destroy? And that's what we *do*.

Apart from her books and plays, Thérèse Radic has produced numerous articles and radio programmes. From 1984 to 1987 she was a member of the executive of the Australia Council and is currently a member of the Music Advisory Panel of the Victorian Ministry for the Arts, and of the Australian Writers' Guild's National Stage Committee, and the State General Committee. Folk music is only one of her interests, but one of central concern, and she intends to develop its link with Australian working life further in her current writing project, a history of music in Australia. Dr Radic is currently an Australian Research Council fellow writing a history of music in Australia, and is attached to Monash University's music department.

Radio and television programmes Folklore is, and has been, represented in the Australian broadcast media mainly on ABC radio, in public and community radio and, to a far lesser extent, on commercial radio. Programmes have taken a variety of formats, though the magazine style has predominated, followed by the single-programme documentary and with the very occasional documentary series. As well, some productions of the ABC's Social History Unit, such as 'Talking History', have treated folkloric topics. Given the aural nature of the radio medium it is not surprising to find that attention has been mainly on song, music and, increasingly, on narrative folklore forms such as the tale and legend. While these programmes are too numerous to detail individually here, they contribute to the persistent grass-roots interest and activity in folk music, song and other forms of folklore that adapt well to the aural medium. A number of radio series have reached a national audience, and some of these are mentioned below.

- 'Australia All Over' is a very popular series broadcast nationally on ABC radio and hosted by Ian 'Macca' McNamara. The programme combines folk and country music, yarns and almost everything else, and is noted for its listener participation, from Cape York to Antarctica. A number of recordings have been produced as an offshoot of the programme.
- 'The Coodabeen Champions', an ABC Radio National weekly programme featuring Tony Leonard, Ian Cover, Greg Champion, Simon Whelan and Jeff Richardson, evolved from a football chat show on Melbourne community radio station, 3RRR. The Coodabeens' irreverent discussions cover almost every aspect of Australian popular culture and folklife, such as sport, family customs, humour, music, language (and misunderstandings thereof), children's games, pastimes and hobbies. Their distinctive speech style is itself a parody of media conventions and includes a strong sense of the ridiculous elements in everyday working, cultural and family life. 'The Coodabeen Champions' have an estimated 300 000 listeners, many of whom contribute to the programme. The Coodabeens also present a Saturday morning football show in Melbourne.
- 'Music Deli' is an ABC Radio National programme produced and presented weekly by Stephen Snelleman and Paul Petran. Both traditional and contemporary folk music from all cultures are presented, with over 70 per cent Australian performance. Before 'Music Deli' the folk musics of non-English-speaking peoples were rarely heard on mainstream radio. Many recordings used on the programme are made specially by the 'Music Deli' team in concerts and studios around the country, and the programme is frequently broadcast live from concert venues and festivals, including recent national folk festivals. 'Music Deli' has also produced over fifteen of its own concerts and a number of commercial recordings.
- 'On the Wallaby Track' was an ABC-FM stereo radio series produced and presented by David Mulhallen and Murray Jennings between 1987 and 1990. In magazine format, 'On the Wallaby Track' concentrated on distinctly Australian material and performers.
- 'The Songs and Stories of Australia', an ABC-FM stereo radio series produced and presented by David Mulhallen since 1991, has focused on the diversity of Australia's folk heritage, with documentary-style thematic productions forming the basis of the series.
- 'Sunday Folk' was an ABC-FM stereo radio series produced and presented by David Mulhallen between 1979 and 1987. Designed to reflect the best of Australian and international folk performers, this long-running programme was a mixture of interviews, reviews,

Galvanised iron dingo decoy, SA, *c.* 1930. (Australian National Gallery)

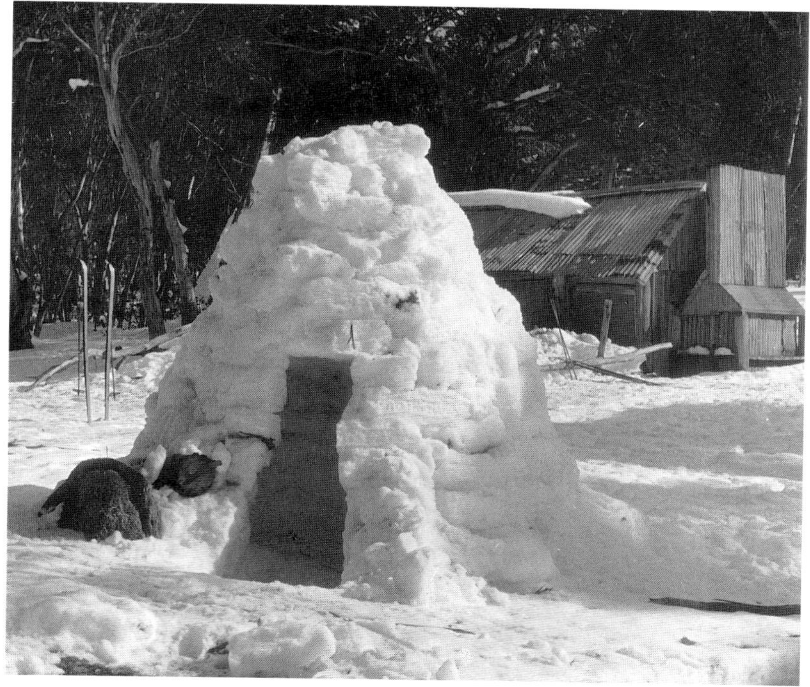

Two kinds of folk buildings used by ski-tourers: a snow igloo and Tin Mine Hut, south of Dead Horse Gap in the Snowy Mountains, July 1981. (Photograph, Keith McKenry)

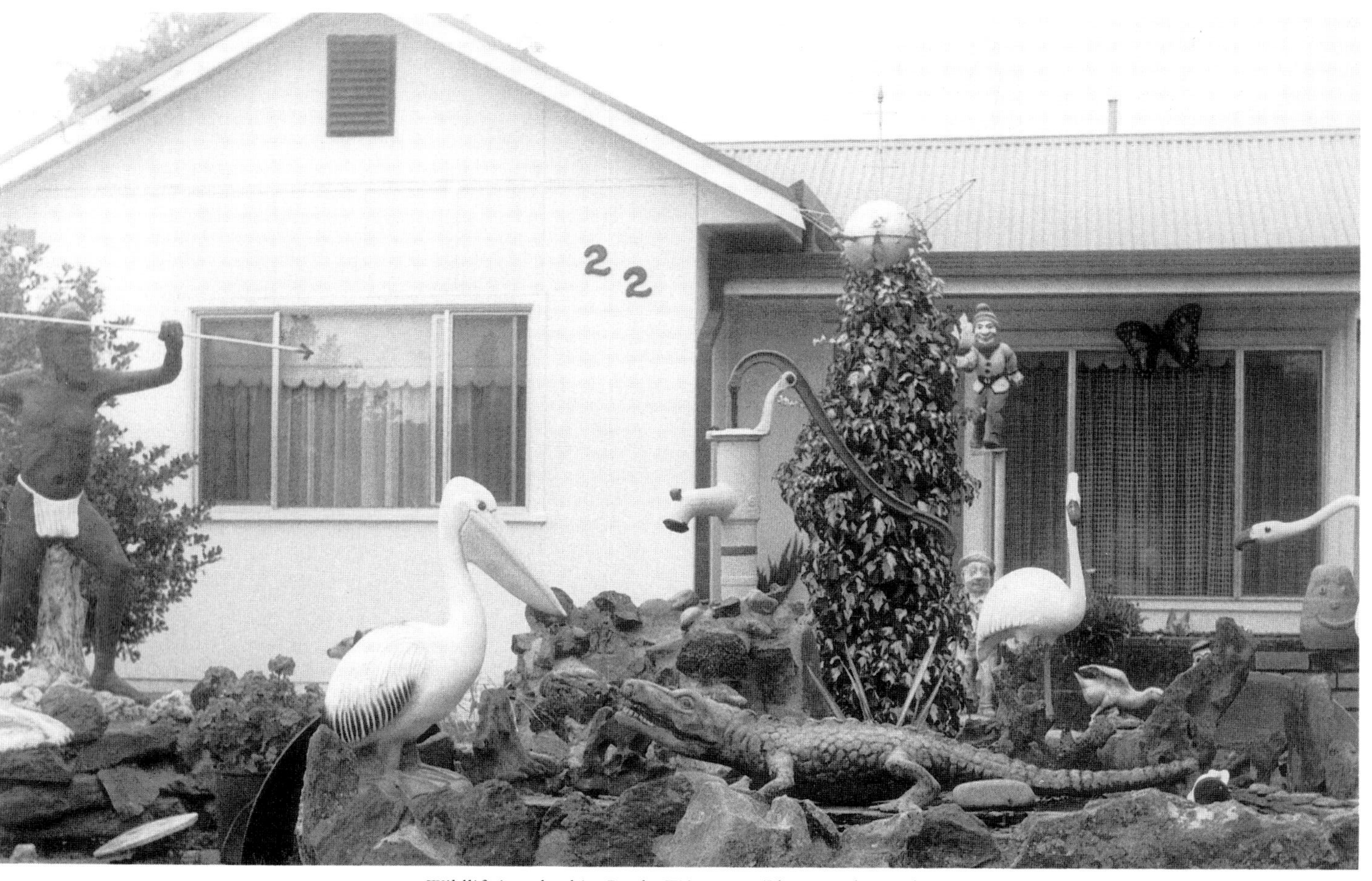

Wildlife in suburbia, Perth, WA, 1986. (Photograph, Keith McKenry)

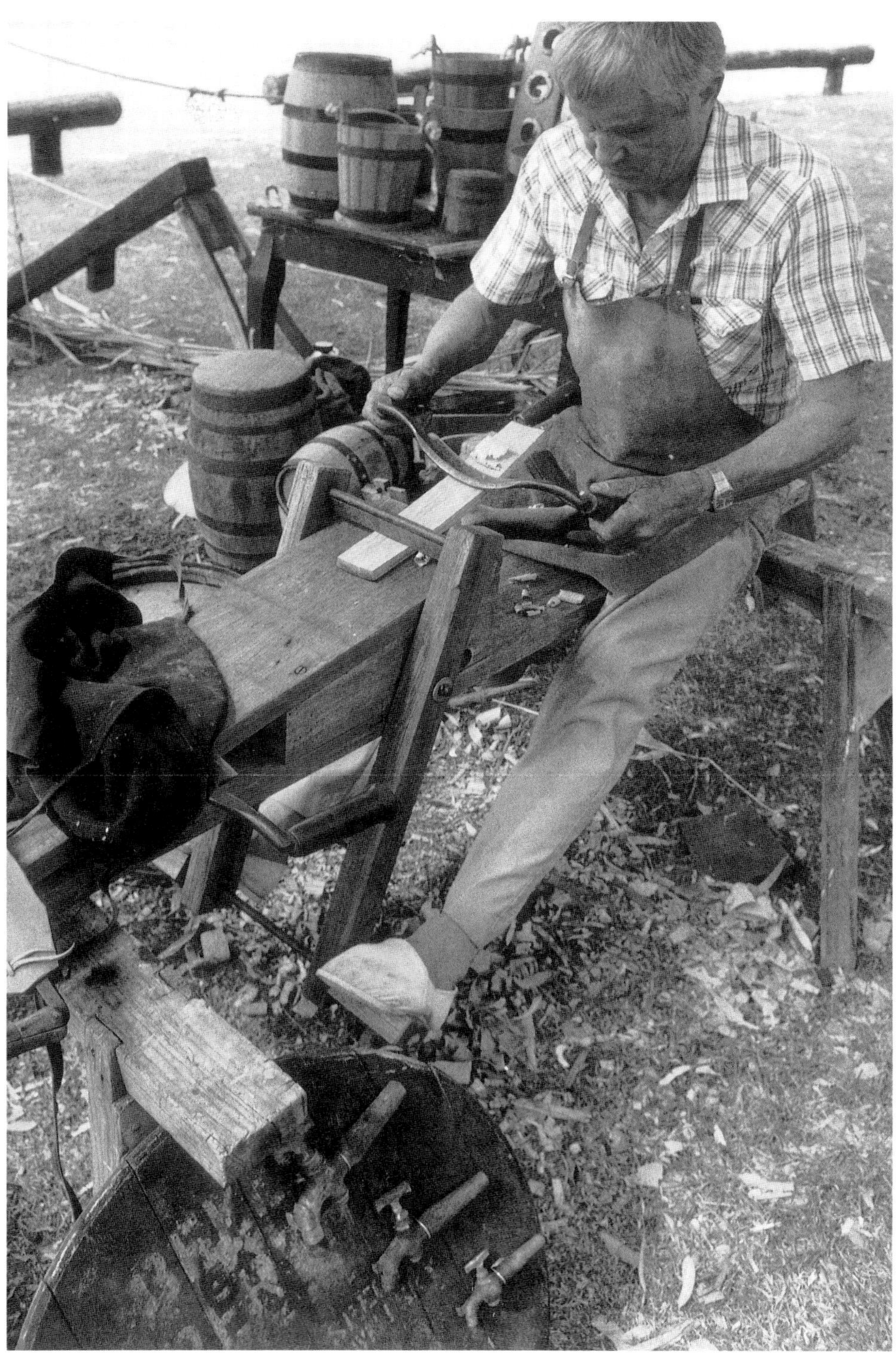

A cooper practising the traditional craft of barrel-making, Mudgee, NSW, 1985. (Photograph, Keith McKenry)

'Making do' with kerosene-tin furniture. A chest of drawers, *c.* 1920, made from painted tin, pine, hardwood and brass, by a member of the Ross family of Muckleford, Vic. (Collection and photograph, Australian National Gallery)

Jimmy Possum chair made from blackwood, *c.* 1925, Deloraine, Tas. (Australian National Gallery)

profiles, documentaries and special features of all kinds.

Only very few folk-related programmes have yet appeared on Australian television. At the time of writing, the most recent was 'That's Australia', hosted by John Derum and produced by Warren Fahey (q.v.) for the ABC in the late 1980s. The straightforwardness of this series made it popular with viewers and its cheapness made it popular enough with the network to ensure its survival beyond one series. A number of musical and variety programmes have also featured some aspects of folklore, as has the occasional documentary, particularly some of those screened by the Special Broadcasting Service (SBS) and some episodes of the ABC's 'A Big Country' series.

Rebetika, 'Greek blues', also known in Australia as 'Piraeus Blues', 'Zorba music', 'Bouzouki songs', and even 'Greek bodgie songs', have been one of the most popular and commercially resilient varieties of popular music in twentieth-century Greece — certainly the most exportable and the most notorious. Expatriate Greeks have been very influential in the development of rebetika; this is particularly true of Greek-Americans, whose ready and uncensored access to modern technology and commercial distribution in the early days of recorded sound had a catalytic effect on the gramophone recording of rebetika in Greece itself.

Australia can also justifiably claim to have contributed significantly to the flowering of the genre in recent times, not merely as a lucrative destination for touring rebetika musicians from Greece or an export market for recordings, films, television serials, and related books and magazines, but also as a centre of production and recording of rebetika (see **Stathoulopoulos**, **Thymios** and **Apodimi Compania**), and of stimulation of interest in the genre on a global scale. It was an Australian musicologist who produced the best-selling tourist guide to rebetika (G. Holst, *Road to Rembetika*, 1975 — unfortunately disseminating the deviant pronunciation 're*m*betika' in the process), and the University of Melbourne was among the first in the world to offer undergraduate courses and foster scholarly research on the genre. The Australian-produced documentary 'The Blues of Greece' (1982, narrated by Anthony Quinn) declared Melbourne to be one of the world's rebetika capitals on account of the local popularity of the genre and its local exponents, a claim confirmed by the memorial concerts for Vasilis Tsitsanis in March 1984, which brought together half a dozen local rebetika ensembles for repeated performances to packed audiences at the National Theatre before touring interstate. Even the Anglo-Australian press recorded this 'outburst of plinking' or 'bouzouki bonanza', defining rebetika as 'the music Greeks like to hum when they have had a bad day at the races or the wife has run off with the refrigerator mechanic' (*Age*, 10 March 1984).

The entertainment guides of Greek-Australian newspapers attest that rebetika have been standard musical fare in Greek *tavernes* and at Greek festivals and multicultural concerts since the mid-1970s. Several cafés, restaurants, clubs and pubs have advertised rebetika as a speciality. More recently, rebetika have been widely broadcast on Australian radio and television, mainly by SBS TV (which has repeatedly screened both series of 'Piraeus Blues', Kostas Ferris's feature film *Rebetiko* and two documentaries on the genre), Melbourne's Radio 3CR (which has a regular rebetika slot from 1.00 to 6.00 am on Sundays) and Sydney's 2SER FM (rebetikalaika from 9.30 to 11.30 pm on Sundays).

To a large extent popular interest in rebetika has been sustained in Australia by importation of records and films from Greece, by duplication of Athenian gossip-columns in Greek-Australian newspapers and magazines, and by touring exponents of rebetika, some of whom have been tempted to stay in spite of the limited scope for a career in Greek song in Australia (see **Stathoulopoulos**, **Thymios**, for example). The earliest such tours may have been those claimed for 1925 and 1926 by the Greek-American guitarist/vocalist George Katsaros, but he seems to have visited as a member of a theatre troupe rather than as an exponent of rebetika, and it was not until the post-war wave of Greek immigration was well under way that Greek musical *tavernes* started to operate on a regular basis. By the late 1950s the notoriety of rebetika was a major issue in Greece, and echoes of the sanctimonious condemnations of the songs as decadent and anti-Hellenic can be found in the Greek-Australian press, but significantly without reference to a local tradition. The arrival of bouzoukia in Melbourne in 1960 was announced in *Neos Kosmos* as 'incredible but true', and while no details are given of the provenance and repertoire of the musicians, their venue, 'MU Hall' at 158 Hoddle Street, is claimed to have the style and menu of an Athenian *taverna*. The proliferation of musical establishments in the metropolitan Greek style was stymied by Australian licensing laws, a concept which some would-be proprietors seem to have had some difficulty accepting. However, by 1962 the bouzouki was perceived to be sufficiently acclimatized in Australia to sustain the first of a long sequence of visits by leading exponents of rebetika and rebetika-derivatives. Stelios Kazantzidis and his ensemble gave two successful performances at the Palais theatre in Melbourne, but following a dispute between artists and organizers the tour was prematurely terminated in Sydney amid scenes of bottle-throwing and pandemonium at Paddington Town Hall and subsequent threats of legal action. Among those to follow this inauspicious start were Skordilis (1968), Manolis 'The Gipsy' Angelopoulos (1971), Theodorakis (1972 — negotiations having started in 1964), Bithikotsis (1974), Kety Grey (1976), Vangelis Perpiniadis (1979), Sotiria Bellou (1979), Dalaras (1980), Alexandra (1989), and the rebetika revival band Palio Spiti (1989).

Promotion of rebetika to the broader Australian public has involved some connivance in the Zorba stereotype: an advertisement for the Greek Inn in South Melbourne in the *Age* (30 October 1984) proclaimed 'Greek is great. Smashing Christmas parties up to 150 people. Zorba music, dancing, plate smashing and all that'. It has also featured some reductive cross-cultural equations: reviewing the first episode of 'Piraeus Blues' for the *Age* (18 May 1985), Denis Pryor summarized the plot as 'Nana Mouskouri meets Arthur Daley'. But sustained efforts at explaining the genre to the uninitiated have been undertaken by the Sydney/Melbourne *Boîte* (established in 1978 with funding from the Arts Council and state bodies) through its annotated concerts and resource kits; the Brunswick Community Arts Office through concerts and discography (see **Apodimi Compania**); ABC-FM's multicultural music programme 'Music Deli'; and even the Greek government, which included an exhibition of 'Music in the Aegean' from pre-classical antiquity to the rebetika of Markos Vamvakaris in the Greek pavilion at the Australian Bicentennial Expo '88. The exhibition guidebook claims that rebetika are a 'marvellous tradition' and 'work wonders'.

In the course of all this, rebetika have tended to be romanticized as the timeless *cri de coeur* of the rebel, which, given the embourgeoisement of its exponents in Greece and its political patronage by Papandreou's PaSoK government, is a gross oversimplification. The accommodation of rebetika within Australia's official multiculturism is another form of institutionalization of this erstwhile persecuted genre and every bit as paradoxical as that attested in metropolitan Greece.

REFERENCES: S. Gauntlett, 'Folklore and Populism: The "Greening" of the Greek Blues', in *Proceedings of the Fourth National Folklore Conference*, 1991; 'Orpheus in the Criminal Underworld. Myth in and About Rebetika', *Mandatoforos*, 1991; K. Karayiannis, 'Australia Condemns Artists to "Death"' [in Greek], *Neos Kosmos*, 11 November 1982.

Stathis Gauntlett

Recordings of folklore A comprehensive discography of Australian folk recordings available commercially on disc, cassette or compact disc has not yet been published. The difficulties involved in such a project are comparable to those encountered in discussing folklore publications and are related mainly to the frequent 'underground' or sub-economic nature of such releases. Consequently, many folk recordings are only sporadically documented. As commercially available recordings often contain accompanying notes of scholarly value, frequently unavailable elsewhere, this situation is particularly unfortunate. This article provides only the broadest overview of commercial (as opposed to archival) recordings. It encompasses both fieldwork recordings issued commercially, as well as revival performers of what is usually considered to be 'traditional' material.

The earliest known attempt to provide recordings of folk music is generally said to be the efforts of Wattle Recordings (q.v.), a small independent label, issuing both revival performances and fieldwork recordings. Wattle was followed by Peter Mann's Discurio (q.v.) label in Melbourne and, in turn, by Sydney's Larrikin Records (q.v.) and Sandstock Music (q.v.), to name the most prominent only. There have been smaller companies issuing folk recordings in most regions of Australia, though they are too numerous to mention individually here.

As access to the technologies of recording became greater for non-specialists in the 1960s and 1970s, there was a gradual increase of 'homegrown' or 'do-it-yourself' releases, particularly as cassette quality and affordability improved. Solo artists, duos, bands and reciters of all kinds have committed their performances to some form of sound recording medium and continue to do so in the era of the compact disc, despite the heavier financial investment required for such productions. Quality of sound, performance and material naturally varies, but there is a busy network of small-scale, kitchen-table recording of folk and folk-related materials in Australia waiting to be properly documented. These recordings include not only 'bush bands' and other revival performers, but also traditional performers, either recorded in the field or in a studio or concert situation.

Other sources of recorded folk performance include the oral history sections of the National Library of Australia and the Australian Broadcasting Corporation (almost exclusively radio). Occasionally the large record companies have found it worthwhile to issue revival performers of one kind or another, though no field recording has ever been released by such a company. The major repository for commercially recorded sound (and visual) folk materials is the National Film and Sound Archive (q.v.) in Canberra.

See also: **Magazines and other publications**.

Reedy River is a musical play, variously said to have been written or 'improvised' by Dick Diamond and first produced in 1954 (published 1970). Set during the shearers' strike of 1891, the play includes a number of traditional bush songs, together with some material composed more or less in that idiom, such as 'The Ballad of 1891'. It was performed with considerable success throughout Australia by the amateur left-wing New Theatre movement and is revived from time to time.

Reprographic folklore encompasses the variety of folklore forms produced and reproduced by means of reprographic machinery, including the duplicator, photocopier, computer and facsimile machine. Early examinations of this form of folklore were primarily American, where the term 'Xerox-lore', after the brand name of the most common photocopier, was and often still is used. Recent work, including that done elsewhere in the world, has tended to use the less-restrictive term 'photocopylore', though this is clearly inadequate to describe the range of materials and processes involved in the production and transmission of such items.

Reprographic folklore often parodies the more official forms of modern communication, such as the letter (see **Chain letters**), the memorandum, the notice, the examination paper, the application form, instructions, and so on. Collections of jokes — usually obscene, racist, sexist or any combination of these — are another staple of reprographic lore, usually compiled on computer data bases and updated, printed out and distributed as necessary.

Graphic renditions of racist, sexist, political, bawdy, occupational or other topics are also common in cartoons by anonymous folk artists or through the reworking of formally produced cartoon strips, advertisements or other graphic materials. Games, amusements and pastimes of various types are also a feature of this form of folklore.

Reprographic folklore is an extremely important modern folklore form and is encountered throughout the industrialized world. It tends strongly towards the vulgar (particularly with regard to bodily functions), the scabrous and the satirical in tone and content, frequently criticizing aspects of the formal institutions of modern life, such as occupational hierarchies and power relations. Gender relations, interethnic tensions, technology, communication breakdown and problems of modern life such as stress are other common topics of reprographic folklore. These may be rendered in pictorial form, in print or typescript, or a combination of these.

Recent research on reprographic and other contemporary folklore forms suggests that there is a strong relationship between some reprographica and two other central folklore forms, the riddle-joke and the legend, with certain themes and topics — such as the AIDS epidemic — being treated similarly across all three genres. While this may be of particular interest to the folklorist rather than the general reader, it does suggest that certain serious contemporary social concerns are being dealt with through the processes of folklore.

RYAN, JOHN SPROTT, was born in Dunedin, New Zealand, and was educated at the universities of Otago (MA), Oxford (MA), Cambridge (Ph.D.) and New England (Ph.D., Dip.Cont.Educ.). He was originally trained as a classicist and mediaevalist, but has spent most of his academic life teaching (English) language and mediaeval and regional culture. He is currently in the Department of English, University of New England. His related professional affiliations include a life membership of the Folklore Society (London), Fellow of the Royal Geographical Society, Fellow of the Society of Antiquaries (Scotland) and Honorary Fellow of the Commonwealth Academy of Biography (Cambridge). He has been a member of the Australian Institute of Aboriginal and Torres Strait Islander Studies (q.v.) since 1965.

John Ryan has a strong commitment to folk/oral culture, especially in western Europe and the South Pacific. He edited the *Armidale and District Historical Society Journal* (1966–78) and is presently editor of *Australian Folklore* (q.v.) and of the Yearbook of the Centre for Australian Language and Literature (Armidale). Relevant books by him include *Papers on Australian Place Names* (1963), *Charles Dickens and New Zealand: A Colonial Image* (1965), *Australian Fantasy and Folklore* and *The Land of Ulitarra: Early Records of the Aborigines of the Mid-North Coast of New South Wales* (1964, revised edition 1988).

S

Sandstock Music was founded in Newcastle, New South Wales, in the late 1970s by Jill Gartrell and Lee Baillie as concert promoters for folk musicians, both local and overseas. Sandstock soon moved into importing records produced by the musicians they represented and also into distribution of a number of Australian labels. In the 1980s Sandstock Music began releasing recordings on its own label, firstly by licensing overseas recordings and eventually by producing original recordings. By 1992 Sandstock had produced over thirty recordings, and was a major focus for production and distribution of folk music in Australia. Right from its inception Sandstock Music has recognized the importance of cultural diversity in folk music, and some of its first concert promotions in Newcastle included the Latin ensemble Papalote as well as the Bushwackers and Eric Bogle. Sandstock currently distributes about eighty labels encompassing both traditional and contemporary folk music and what they call roots music from the *a capella* of Sweet Honey in the Rock, through the 'techno pop' of Najma, whose music draws upon the Ghazals tradition, and the popular ballad style of Ireland's Mary Black, to zydeco from Louisiana.

SCOTT, ALAN (1930–), was born at Caboolture, Queensland, and has had an interest in Australian folklore since 1950. In 1954 he joined the original Bushwhackers' Band in Sydney; he was the first secretary of the Bush Music Club (q.v.) when it was established in October 1954 and was associated with the BMC's *Singabout* magazine as both contributor and editor. Alan Scott assisted John Meredith (q.v.) in some of his earliest field recording and began independent field recording in New South Wales in 1955. He made some field trips for the oral history section of the National Library of Australia during the 1980s, and his recordings are deposited with the National Library. Some of Alan Scott's collected material was published as *A Collector's Songbook* in 1971 by the Bush Music Club. Over the years he contributed articles to many magazines, including *Australian Tradition*, *Stringybark & Greenhide* and *Cornstalk Gazette*. He represented the NSW Folk Federation on the Australian Folk Trust between 1987 and 1991 and is a foundation member of the Australian Folklife Centre's Program Advisory Committee.

Recordings made by Alan Scott include *Drover's Dream* (1955) with the Bushwhackers, *Bush Music Club* (1958), *Dinky Di* (1964) with the Bush Music Club, *Native Mate* for UNESCO Folk Songs of Asia and the Pacific (1988) and *Battler's Ballad* (1991). Alan Scott is a regular performer of songs with concertina accompaniment at folk clubs, concerts and festivals.

SCOTT, BILL (W. N. SCOTT) (1923–), was born in Bundaberg, Queensland. He left school at fourteen years of age, served in the Royal Australian Navy between 1942 and 1946 and worked at many trades until 1956, since which time he has been associated with the book trade as bookseller, editor, publisher and, since 1974, writer. He lives at Warwick on the Darling Downs in Queensland.

Since 1956 Bill Scott has been engaged in the collection, performance and publication of European folklore in Australia, and assisted John Manifold in compiling *Folksongs from Queensland* (1959) and *The Penguin Book of Australian Folksong* (1960). He has produced numerous books dealing with Australian folklore, notably *The Complete Book of Australian Folk Lore* (1976), *Bushranger Ballads* (1976) (with Pro Hart), *The Second Penguin Australian Songbook* (1980), *Australian Humorous Verse* (1983) and *The*

Long and the Short and the Tall (1985).

In addition to his work in Australian folklore, Bill Scott has a number of original publications to his credit, including novels, children's novels, poems and historical and biographical material. He has also produced school texts, magazine articles and radio programmes. He was highly commended by the Children's Book Council for two novels for young people, *Boori* (1978) and *Darkness Under the Hills* (1980), and received the Mary Gilmore Award for his short story 'One is Enough' (1964). In 1978 he spent six months as artist-in-residence at Goulburn College of Advanced Education.

Together with Stan Arthur, Bill Scott made the first commercial recording of Queensland folksong, *The Billygoat Overland* (1958), and he appeared on the recording *Folksongs from Queensland* (1959). Both were issued by Wattle Recordings (q.v.). Bill Scott's current interest as a folklorist is in collecting urban legends and spoken entertainments generally. He was awarded the Medal of the Order of Australia in 1992 for his services to folklore in Australia.

SEAL, GRAHAM (1950–), has studied and collected folklore in Australia and Britain since 1970 and has taught and lectured at universities in Australia, Britain, China and the United States. He has published numerous articles in British and Australian folklore journals and, with David S. Hults, co-founded *Australian Folklore* in 1987. Publications and recordings include *On the Steps of the Dole-Office Door: Oral Images of the Great Depression in Australia* (Larrikin Records, 1977); *Ned Kelly in Popular Tradition* (Hyland House, 1980); *Australian Folk Resources* (Australian Folk Trust, 1983, 1988); *A Broken Song: The Voice of Australians at War* (ABC, 1984); *Banjo Paterson's Old Bush Songs* (Angus & Robertson, 1984, 1985, 1986) and *The Hidden Culture: Folklore in Australian Society* (Oxford University Press, 1989). He has been Vice-Chairman of the Australian Folk Trust, Vice-President of the Australian Studies Association and President of the Australian Folklore Association, and was awarded a National Folklore Fellowship in 1986. Seal founded a specialist Australian folklore publishing house, The Antipodes Press, in 1989. He currently lectures in Folklore and Australian Studies at Curtin University, Perth, where he established the Western Australian Folklore Archive in 1985. Seal is co-editor, with Gwenda Beed Davey, of *The Oxford Companion to Australian Folklore* (1993).

Singabout: see ***Mulga Wire***; **Bush Music Club**.

SLOANE, SALLY (1894–1982), was one of the most outstanding of many traditional singers and musicians who contributed to the field recordings made by John Meredith (q.v.) and others in the 1950s and 1960s. She played fiddle, button accordion and mouth organ, and had an extensive knowledge of songs and tunes. Many of these were learned from her grandmother who emigrated from County Kerry in Ireland in the 1840s, or from her mother's family repertoire, and are both Irish and English in origin. She was born in Parkes, New South Wales, and lived for most of her life in New South Wales country towns. Two of her songs are included on the Wattle Recording disc *Australian Traditional Singers and Musicians* (1957), namely the bushranger ballad 'Ben Hall' and the Irish song 'The Wee One'. The Larrikin recording *A Garland for Sally* (1983) gives a larger selection of her music, songs and unique traditional singing style. Her songs and music have been widely discussed and are featured in John Meredith and Hugh Anderson, *Folk Songs of Australia and the Men and Women Who Sang Them* (1968).

Speewah: see **Crooked Mick**.

SPOONER, DANNY: see **Folk performers**.

Sporting folklore Like all social groups, the practitioners and the followers of particular sports have their own in-group or esoteric folklore. Names, events, terms and various references and allusions involved in this folklore will be mysterious, even cryptic to the uninitiated. Surfboard riders, for example, have an elaborate set of descriptions for various types of waves, board manoeuvres and, indeed, for aspects of life outside of (though related to) surfing itself, such as drinking, sexual activity and cars. In the cases of some other sports, the practices and expressions involved are more widespread and public. This is true of sports such as boxing, football, cricket and horse racing, all of which have a wide following in Australian society. These sports have *national* resonances due to their relationship to central myths of Australian society — masculinity, toughness, the suspicion and resentment of 'others', itself related to the 'colonial cringe' and especially seen in cricket games with English sides. Other less popular or glamorous sports (e.g. women's hockey), while having their own folklore, do not have the broader national appeal of those sports mentioned above.

Some of the most characteristic elements of sporting folklore are the colourful terms of speech associated with particular aspects of the sport: jargon terms for various parts of play, such as 'square leg' in cricket or the 'dead man's corner' in Australian Rules football or, most colourful of all, the 'barracking' of spectators at football matches, which includes disparaging of the opposing side and, if necessary, the umpire and other officials. An umpire in Australian Rules, for instance, may be referred to as, among other things, 'a white maggot', a 'white dog' or a 'white mug', usages derived from the traditional white uniform of umpires.

Folk costume of various kinds is also a feature of sporting attendance. Apart from the colours and uniform worn by the players or participants in a sport, followers may well adopt a particular form of attire. The often elaborately formal dress of turf followers is one example of this, though some Australian 'punters' satirize these dress conventions by wearing outlandish attire, particularly at such occasions as the Melbourne Cup. Supporters of a club in team sports also habitually wear the colours of their team in one or other (sometimes bizarre) form of dress. Football grand finals are the customary moment for such display.

There are many other customs associated with different sports. These include gambling customs, not those regulated by the Totalisator Agency Board, or 'TAB', but the more informal and illegal betting practices that take place in public houses and workplaces across the country. 'SP' (starting price) or off-course betting is still widely practised in Australia, despite having been outlawed for decades. The Melbourne Cup event, of course, is perhaps the pinnacle of Australian sporting and betting interests. Work ceases in many places while the Cup is being run and many people who would not normally bother betting on horses either 'put something on' at the TAB or participate in informally organized 'sweeps', often combining the Cup viewing with a 'liquid lunch'. In recent years it has become the custom for women to wear a particularly formal or outrageous hat to such lunches, in folk mimicry of the fashionable headgear displayed at the Melbourne Cup event itself.

The relationship between watching or listening to the more popular sports and drinking alcohol is a close one that also involves a range of folkloric practices and expressions located in hotels. Races and matches are watched and listened to in the pub, bets are laid there, yarns and jokes are exchanged there and, naturally, alcohol is consumed there. While it is not the purpose of this article to deal with the extensive folklore of the pub, this cannot be ignored in any discussion of sporting folklore. Bill Wannan's *Folklore of the Australian Pub* (1972) is an excellent introduction to this topic.

Many folk beliefs connected to sport have developed over time. Some of the more widespread of these include the belief that in 1917 the Australian pugilist, Les Darcy, was given 'dope' on the night of a big fight in Memphis, Tennessee, by jealous Americans fearful that he would beat their favourite. This belief (untrue — Darcy died as a long-term result of blood poisoning combined with over-exertion during training after a period of relative inactivity in America) is extremely persistent and is even encapsulated in at least one folk song. In the same category is the belief that champion racehorse Phar Lap was similarly nobbled fatally by the Americans in the early 1930s.

Perhaps one of the most important folk elements of sports are their capacity to generate heroes and, less frequently, heroines. The outstanding nature of a particular sportsperson's achievement is usually the basis for such celebration, though tragedy is also likely to be commemorated. Boxing has numerous songs, recitations and stories about marathon bouts. Horse racing has its share of young jockeys killed in racecourse accidents, such as Alec Robertson, whose death in the Caulfield Cup in the 1890s is commemorated in a ballad, as is that of Willie Stone of Brisbane, who met a similar fate. The various football codes each have their pantheons, headed perhaps by Australian Rules's great Roy Cazaly (at least in Victoria), as does cricket, where Sir Donald Bradman — 'the Don' — reigns supreme. While aviation is not generally considered to be a sport, nevertheless Sir Charles Kingsford-Smith ('Smithy') has something of the aura of folk hero. More recently, marathon runners have become popular heroes, particularly Robert De Castella ('Deek') and the venerable Cliff Young.

The degree of popular admiration attained by sporting personalities is usually reflected in a nickname (see **Names**) of some sort being accorded them. Some of those mentioned above have nicknames, but there are many other examples of this widespread folk practice. A few are: 'Boy' Charlton the swimmer, 'Demon' Fred Spofforth the cricketer, 'Gelignite Jack' Murray the motor racer, 'the Great White Shark' for Greg Norman the golfer, 'the Lithgow Flash' for Marjorie Jackson the athlete, while boxer Jeff Fenech is known as 'the Marrickville Mauler' and tennis player John Newcombe is referred to as 'Newk'.

Graham Seal

STATHOULOPOULOS, THYMIOS (1934–92), Melbourne bouzouki-player, composer and exponent of rebetika (q.v.) songs. Born in Diavolitsi, Messinia, southern Greece, he was a permanent resident of Australia from 1964. His claim that he is the first and most authentic rebetika musician in Melbourne features prominently both in his compositions and in his local publicity. He recalled that his father had learned to play *saz* (a precursor of the Greek bouzouki) in prison, but died before passing on these skills. In 1948 Thymios decided to take up the bouzouki after hearing a band from Athens at the village fair of Meligala and eventually managed to buy a six-string bouzouki from Kalamata. He learned to play rebetika songs by ear from gramophone records and started to play them at village fairs in his native province. In 1951 he ran away and 'lived rough' on the outskirts of Piraeus, busking in taverns, before enlisting for national service. Soon after returning to civilian life he married Yiota Panayou from northern Greece, who sang several folk songs and popular songs on Columbia records in the early 1960s.

The couple was brought to Melbourne in October 1964 to perform at a Greek club, but finding that it had closed down before they arrived, they started at the Olympus Club in Chapel Street, Prahran, accompanied by drums, guitar and electric organ. The band's repertoire was somewhat international: the male vocalist, an Egyptian Greek, would sing in seven languages, while Thymios would play rebetika and traditional folksongs on his bouzouki. He was persuaded to extend his visa, even though he thereby incurred a large debt in respect of his outward passage. In 1965 the couple released their first record in Melbourne, a double-sided 45 rpm single, largely at the instigation of the lyricist Yiorgos Papadopoulos, who also met the production costs. One of the two songs was a *tsifteteli* (belly-dance rhythm) with alternating Greek and Turkish lyrics; both were sung by Yiota, accompanied by Thymios on bouzouki, an accordion, guitar and bongos. The record was frequently aired in the then fledgling Greek radio broadcasts (typically between advertisements for purveyors of Greek baptismal and trousseau ware).

Within the year 'Olympus' burnt down, and Thymios became joint proprietor of the Monte Carlo Cabaret at St Kilda Junction, renaming it Nostalgia — Greek Club. It opened every night and was well patronized, particularly by single Greek males, a species in abundance in Melbourne in the mid-1960s. Unable to afford a liquor licence, the joint owners of the club frequently had their stock of alcohol confiscated by police. After two years the premises were demolished to make room for the highway, and Thymios resorted to plastering in order to meet his increasing domestic expenses. He continued to play bouzouki at wedding receptions, christening parties, and Greek social events in hotels, halls and various other venues and with numerous other musicians, maintaining a diverse but traditional Greek repertoire.

In 1967–68 he appeared in a bouzouki floor show as part of an international cabaret at the Zorba Club, Prahran, playing 'Zorba's dance', 'Never on Sunday' and other 'cosmopolitan' Greek songs. At the same time he also opened a café of his own nearby, where he would entertain the clientele with live music on Sundays, but the absence of a liquor licence restricted his career as a proprietor. Around 1975 he changed from a six- to an eight-string bouzouki for greater ease of playing. He used the same instrument, made in Greece by the renowned Zozef Terzivasian of Piraeus, for over fifteen years, having it refretted locally when necessary. In 1979 Thymios appeared on Friday and Saturday nights at the Aristotle Tavern, Fitzroy, with the Cypriot singer Varnavas, and also participated in rebetika soirées at the Melbourne branch of the mulicultural *Boîte* and at the Tsakpina Café in Melbourne, in which setting the Stathoulopouloi were filmed for the Melbourne segment of the documentary *Rembetika, the Blues of Greece* (1982). The proprietor of the café, Kevin Zervos — self-styled 'koala-*mungas*' [*sic*], *mangas* being a synonym of *rebetis* — also produced Thymios's second record in 1980, an EP titled *Tsingana Mou Satrapissa* ('My Haughty Gipsy Girl'), featuring four songs composed and played by Thymios, with vocals by Varnavas and Douloudis on baglamas.

In 1980 Thymios abandoned plastering and went to the Kalyva *taverna* in Richmond, where he played until his death twelve years later, apart from an interlude at Happy Bistro, Springvale, in 1985. Thymios provided the whole night's entertainment at Kalyva, from early evening until dawn, accompanied by various guitarists and a *baglamas*. The clientele was diverse, with non-Greek customers often forming the majority, but Thymios's repertoire remained firmly focused on rebetika by Tsitsanis, Vamvakaris, Yenitsaris and Mitsakis, and traditional folk songs; he preferred to add older songs to his repertoire rather than modern, non-traditional compositions. He was conscious of making a contribution to Australian multiculturalism and enjoyed the accolades of all his enthnically diverse clients, but missed the immediacy of rapport with large, culturally attuned audiences such as he experienced at village fairs in Greece. He had long since abandoned his dream of returning to Greece with enough money to open his own rebetika club in Tzitzifies, the citadel of Athenian bouzouki music.

In 1985, with financial assistance from the Music Board of the Australia Council, he produced an LP titled *The Rembetika of Melbourne*, featuring twelve of his own compositions (including the four from the earlier EP). They were inspired by his sad experience of expatriation, but also emphasize his pride in his pedigree as an authentic rebetika musician, announce his abomination of pseudo-American/European Greek songs, and celebrate his rebetika instruments (bouzouki and baglamas) and their famous exponents — such as Tsitsanis, whose passing he commemorated with a rebetika-style dirge at the Melbourne memorial concert in March 1984. The Australian colour of his compositions is mainly implied in the traditional Greek song-theme of exile, or is conjured up incidentally — e.g. in the Australian-Greek phrase *ime sportis* ('I'm a sport'). His song 'The Bouzouki in Melbourne' is a rare instance where the Australian context explicitly dominates. A free translation runs:

In Melbourne's lovely evenings now,/in clubs and taverns the bouzouki plays.
And people dance and live it up in the purely Greek way,
getting merry on Aussie wine and breaking all the plates.
Beautiful Greek girls, full of grace,/proudly dance zeibekiko.
And so do fine blonde Aussie girls,/they dance away their passions.

Stathis Gauntlett

STEWART, DOUGLAS (1913–85), was a notable Australian literary figure as a poet, editor, critic, publisher and prose writer. In his major biography *Douglas Stewart* (1974), Clement Semmler quoted Nancy Keesing's earlier description of Douglas Stewart as 'the most versatile writer who ever lived in this country'. In the early 1950s Douglas Stewart became convinced that an anthology of bush ballads should be compiled and published, and because of his extensive critical and journalistic commitments (including editing the Red Page of the *Bulletin*), he invited Nancy Keesing to be his collaborator. Their research drew extensively on early copies of the *Bulletin* and other publications in the Mitchell Library and elsewhere, and was published as *Australian Bush Ballads* (1955) and *Old Bush Songs* (1957). Both volumes were pioneering collections of Australian ballads and were followed by numerous other publications, both academic and popular. After these two initial collections, Douglas Stewart did not take an active part in collecting folk materials, although he was a publicist in the folk revival (q.v.) of the 1950s and 1960s; but evident in his later writing is the inspiration of folklife and legend, not only from Australia and New Zealand, but from Scottish balladry (*Glencoe* 1947), and from Pacific and Antarctic exploration (*Terra Australis* (1952) and *Fire on the Snow* (1944)). Douglas Stewart drew on many other aspects of Australian folklife in his comedy drama *Fisher's Ghost* (1960) and his Birdsville Track poems of 1954, which included much folklore about place names, people and incidents. One other poem, 'Arthur Stace', celebrated the Sydney resident of that name, who over many years chalked the word ETERNITY on footpaths throughout the city.

Stockman's Hall of Fame: see **Australian Stockman's Hall of Fame**.

STONE, WALTER: see **Magazines and other publications**.

Storytelling guilds of Australia Formed in Melbourne, in 1978, the Storytelling Guild has since expanded to include storytelling guilds in New South Wales (1980), South Australia (1983), Tasmania (1984), Australian Capital Territory and Queensland (both 1987), and in 1988 Western Australia. Each guild has its own constitution, newsletter and activities programme. There is no central co-ordinating body, though national conferences are held from time to time.

Membership of storytelling guilds is open to anyone interested in storytelling. The aims of the guilds are to promote storytelling in the community by bringing together people from all sections of the community to share experiences and stories; to encourage people to tell stories and to develop their skills in the art; to produce a newsletter (the *Harper* has been issued since 1979); to produce a directory of storytellers; and to organize storytelling activities.

Stringybark & Greenhide, founded and edited in Newcastle, New South Wales, by Lester and Cecily Grace, was published mostly quarterly from 1979 to 1986. Produced in glossy magazine format, *Stringybark & Greenhide* contained news, reviews, collected materials, newly composed materials and research articles on various aspects of Australian folklore.

STUBINGTON, JILL (1944–), was born in Brisbane and graduated BA from the University of Queensland in 1967. From 1967 to 1973 she worked as research assistant to Dr Alice Moyle, a Research Fellow in Australian Aboriginal music for the Australian Institute of Aboriginal Studies, now Australian Institute of Aboriginal and Torres Strait Islander Studies (q.v.). In 1978 Jill Stubington graduated Ph.D. from Monash University with a thesis entitled *Yolngu Manikay: Modern Performances of Australian Aboriginal Clan Songs* based on fieldwork in north-east Arnhem Land during 1973–75. She began research into non-Aboriginal Australian traditional music in 1985 when she was appointed Lecturer (now Senior Lecturer) in the Department of Music at the University of New South Wales. Her lectureship at the University was initially partly funded by the Australia Council specifically to encourage research and teaching in Australian traditional music, and is today one of Australia's rare academic posts teaching any aspect of Australian folklore. The lectureship is now permanently funded by the University.

Dr Stubington has produced numerous publications, mainly concerning Aboriginal music. In 1984 she edited (together with Jamie C. Kassler) *Problems and Solutions: Occasional Essays in Musicology Presented to Alice M. Moyle*. In 1987 Jill Stubington organized a forum for collectors of folk music, and its proceedings were published in 1989 as *Collecting Folk Music in Australia: Report of a Forum Held at the University of New South Wales, 4–6 December 1987*.

Superstitions: see **Folk belief**.

T

Tantanoola Tiger One of a number of mysterious animals, usually referred to as 'tigers', and said to terrorize various rural areas. Tantanoola is in South Australia, but other similar animals are thought to be at Nannup (WA), in Tasmania, in Victoria (the 'Yowie') and elsewhere. Legends surrounding the origins of such beasts are similar throughout Australia and elsewhere in the world, often involving animals that have escaped from travelling circuses at some usually indeterminate time in the past.

TATE, BRAD (1943–), was born in Newcastle, New South Wales and grew up in an isolated farming community near Tuena on the New England tablelands, where his first musical influences were local woolshed dances. He was employed as a secondary school teacher in New South Wales between 1964 and 1987. Brad Tate has made substantial collections of Australian folklore, principally songs, recitations and dance music. An early collection of bawdy songs and recitations was published as *The Bastard from the Bush* in 1982. Subsequent fieldwork in the Hunter Valley and the New England area in NSW and in Victoria was published as *Down and Outback* (1988). Brad Tate also engaged in field collecting in Britain during 1974. *Which Side Are You On?* (1989) included musical settings and annotations to an extensive collection of political and Labor song texts compiled by the Australian miners' poet, the late Jock Graham.

THATCHER, CHARLES (1831–78), was a popular entertainer on the Victorian goldfields in the 1850s and 1860s. He composed his own texts, usually to existing tunes, and both described and satirized many aspects of life on the goldfields, including the experiences of the 'new chum', Chinese immigration, bogus land sales and electioneering. Like S. T. Gill's drawings, Thatcher's ballads are a rich source of information about everyday folklife in colonial times. Charles Thatcher published compilations of his own songs, such as *Thatcher's Colonial Songster* (1857) and *Thatcher's Colonial Minstrel* (1859). Hugh Andersen (q.v.) published a selection of Thatcher's songs in *Goldrush Songster* (1958) and a biography of Charles Thatcher, *The Colonial Minstrel*, in 1960.

TOMASETTI, GLEN (1929–), was born in Melbourne and is a singer, collector, lecturer and radio and television scriptwriter using folk music and social history. She has always sought out songs, stories and practices describing our experience in a specially Australian way, but also in commonality with other peoples of the world. Her activities in the post-war song revival include introducing bush ballads and folk singers to ABC radio (1964–70); directing the concerts held at Emerald Hill on Sundays in 1963; and co-founding a folk music coffee lounge, known as Traynor's, at Frank Traynor's Jazz Agency after it opened in February 1964. Glen Tomasetti has toured as a singer in China (1958) and Germany (1980). From 1964 she wrote many topical songs on themes such as peace, war, conservation and equal opportunity for women. After the High Court case on equal pay in 1969, she wrote 'Don't Be Too Polite, Girls!'. Her records of topical and traditional songs include *Glen Tomasetti Sings* (1961), *Will Ye Go Lassie, Go?* with Brian Mooney and Martin Wyndham-Read (1964), *The Ballad of Bill White* (1966) and *Goldrush Songs* (1975). She has also recorded folk songs in languages other than English, principally French.

Glen Tomasetti's first novel *Thoroughly Decent People — A Folktale* (1976) was the culmination of postgraduate research from

1954 on folklore in Australia, and extended the image of 'typical' white Australians to include married city people of the 1930s.

TRITTON, DUKE (1886–1965), spent many years as a shearer and outback worker, and wrote of his life in *Time Means Tucker*, published in Sydney in 1959. He was a foundation member of Sydney's Bush Music Club (q.v.) and assisted John Meredith in collecting a number of songs. He frequently performed songs that he had learned in the bush, as well as those that he wrote himself. Of his own songs, 'Shearing in a bar' is one of the best-known. Duke Tritton played an important part in the Australian folk revival of the 1950s and 1960s. Writing an obituary in the *Australian*, Edgar Waters stated that Duke Tritton would be remembered 'because many folk song enthusiasts suddenly understood for the first time . . . that folk songs are not created by Peter, Paul and Mary, but by men like Duke Tritton'.

TSOURDALAKIS, KOSTAS (1927–) **and GEORGE** (1939–), are a Cretan folk-music duo playing in Melbourne. George emigrated to Australia in 1963 from the small town of Melabes, southern Crete, and Kostas joined him the following year after spending two years as a *Gastarbeiter* in West Germany. Coming from a large and indigent family, both brothers left school young, but continued their musical education in their family setting; their father and his brothers between them played mandolin, fiddle and the Cretan *lyra* (a short-necked bowed lute) and all enjoyed singing. Kostas started playing his uncle's *lyra* in 1946, initially at small gatherings of friends and kinfolk, but in Germany performed for larger and increasingly diverse audiences and also began composing occasional verse in traditional couplet form. George took up the *lagouto* (long-necked fretted lute) *c.* 1952, against his father's wishes, in order to accompany his youthful friends who played other traditional instruments.

Since 1964 the Tsourdalakis duo has played at all sorts of Greek community and private functions and celebrations in Melbourne and beyond, appearing at the *Boîte* (q.v.) (a showcase for ethnic arts), contributing to the Piccolo Spoleto festival in 1987 and to multicultural concerts in other states of Australia. They have also appeared in radio and television broadcasts, including the SBS TV co-production 'Flowers of Rethymnon', commemorating the 1941 Battle of Crete, and an episode of 'The Sullivans', in which George accompanied the *lyra* of Kostas's son Tony, who continues the family line of traditional musicians.

In 1984 they recorded an LP titled 'Cretan Traditional Music in Australia' at the instigation of the ethnomusicologist Peter Parkhill, who transcribed, transliterated, translated and annotated the recorded songs for the accompanying booklet. The record includes examples of the occasional verse which Kostas had been composing for over twenty years, not only for singing, but also for public recitation or publication in printed form in newspapers (of Australia and Crete) — or indeed for direct dispatch to eminent public figures such as ex-President Kyprianou of Cyprus and Mikhail Gorbachev. His heroic, political, gnomic and satirical couplets clearly belong to the tradition of the Cretan rhymester (*rimadoros*), infused both with Cretan dialect and with occasional Australian-Greek words (such as *yiounio* = trade unions; *thri dambloo zet* = 3ZZ). Some of his verses are somewhat out of sympathy with prevailing social mores:

Society is ruined, mankind is in a mess.
— How does the deity put up with all this wickedness? . . .

Women wearing trousers are now a common sight,
their mini-skirts get shorter and our minds can't work right . . .

A curse on him who granted so much liberty,
infecting and destroying all of society.

Equality is what she seeks and tries to cast the male
into the darkest depths of life, and even God turns pale.

Kostas Tsourdalakis is particularly keen to share his traditional values and skills with younger generations of Greek-Australians, and he has regularly demonstrated his musicianship and skills in versification at schools, tertiary institutions, teachers' gatherings, and on educational radio.

See also: **Greek-Australian folklore**; **Greek-Australian folk song**.

REFERENCES: R. Beaton, *Modern Greek Folk Poetry*, 1980; P. Parkhill, 'Two Folk Epics from Melbourne' 3, in *Meanjin* 42, 1985; P. Parkhill, *Cretan Traditional Music in Australia*, 1985; K. Tsourdalakis, *The Cretan Poet and Lyra Player Kostas Tsourdalakis* [in Greek], 1983.

Anna Chatzinikolaou

TURNER, IAN (1922–78), born in Victoria, was one of the founders of the Folk Lore Society of Victoria in the 1950s. Following military service in the Second World War he became a leading student activist at Melbourne University. After graduating he worked for organizations such as the Victorian Peace Council, Australian Railways Union and Australasian Book Society. He was a leading member of the Communist Party arts committee until his outspoken opposition to the Soviet invasion of Hungary caused his expulsion. Returning to study he wrote his doctoral thesis on the Industrial Workers of the World and at the time of his death he was Associate Professor of History at Monash University. He published several books on Australian culture and labour history.

Ian Turner's *Cinderella Dressed in Yella* (1969), a compilation of Australian children's playground rhymes, was the first Australian scholarly publication on children's folklore (the work of Dorothy Howard [q.v.] in the 1950s was published in the United States). The book achieved notoriety when it was briefly banned from transmission through the post on the ground of obscenity. The second edition of *Cinderella* (1978) was compiled by Ian Turner in collaboration with June Factor (q.v.) and Wendy Lowenstein (q.v.). Ian Turner had a deep interest in Australian traditional and popular culture (see **Graffiti**), and collaborated with Leonie Sandercock to write a history of Australian Rules football, *Up Where, Cazaly?* (1981).

Two-Up, also known as 'swy', is a gambling game involving the tossing of two coins. It was very popular among Australian troops in the First World War. Often claimed as a uniquely Australian gambling game, two-up is in fact a close variant of similar gambling games played elsewhere in the world, including Scotland and Germany. This has not obstructed the passage of two-up into popular nationalist myth, aided by its illegality and the fact that it is nevertheless tolerated on Anzac Day (q.v.).

U

Urban legends: see **Modern legends**.

V

VAUX, JAMES HARDY (1782–?), was born in England and transported to Australia for theft, arriving in Sydney in 1801. He was resentenced numerous times for theft and forgery and while imprisoned at Newcastle, NSW, compiled a dictionary of slang. Vaux's *Vocabulary of the Flash Language* was included in his two-volume *Memoirs*, which was published in 1819, and includes some 700 examples of both colonial and London 'flash language'. At least fifty of Vaux's examples, such as 'bolt', 'pinch', 'racket', 'sting', and 'yarn', are still in widespread use in Australia in the 1990s at all levels of society. Vaux's compilation is almost certainly the first dictionary of Australian colloquial English or folk speech.

VENNARD, ALEXANDER VINDEX ('BILL BOWYANG') (1884–1947) Journalist, publisher and occasional swagman, Vennard took 'Bill Bowyang' as a pen-name around 1915 and also published as 'Frank Reid'. His main claim to fame is as the editor of a column in the *North Queensland Register* called 'On the Track', which ran from 1922 until 1947. The column published verse, song and other items of bush lore, mostly contributed by readers. The popularity of the column led Vennard to publish selections from his column in at least nine volumes between 1933 and 1940. These were variously titled *Old Bush Recitations, Australian Bush Recitations* and *Bill Bowyang's Bush Recitations*, but were more generally known as Bill Bowyang Reciters. Despite the titles, many of the items also exist as songs, and overall Vennard's collection is a most valuable documentation of the traditional material in circulation during the first half of this century. Folklorist Hugh Andersen (q.v.) has researched Vennard's life and work for publication in a six-volume series titled *On the Track With Bill Bowyang*.

See also: **Folk poetry and recitation**.

Vernacular architecture: see **Australian house, The**; **Building and folk architecture**.

Victorian Folk Music Club, The, The Victorian Folk Music Club was founded in 1959 with these aims: to preserve and popularize Australia's folklore heritage of traditional songs, poetry, and dance; and to encourage the composition of contemporary work in the folk idiom. In line with these aims it has published, or contributed to the publication of, a book of songs (*Joy Durst Memorial Australian Song Collection*) and dance music from 1830 to the present (Peter Ellis, *Collector's Choice*, volumes 1, 2, and 3), plus cassettes of music for dancing. In co-operation with the Folk Lore Society of Victoria it also put out *Australian Tradition* (Wendy Lowenstein [ed.]) from 1963 to 1974. The Club publishes a regular newsletter. Members have researched and subsequently taught colonial dancing and run workshops and singabouts to encourage participation in the various folk arts.

Among the highlights of the Club's year are the woolshed balls held twice a year in May and October and the Nariel Creek black and white folk festivals at Christmas-New Year and on the Labour Day weekend in March.

The first National Folk Festival was run by a group consisting of Club members who formed the Port Phillip Folk Festival Committee (later to become the Folk Song and Dance Society of Victoria).

Victorian Folklife Association Inc., The, is a non-profit organization funded by the Victorian government through the Victorian Ministry for the Arts and managed by a committee representing a broad spectrum of the folk arts. It aims to promote the activities and cultural programmes of all the various peoples comprising the Victorian folk community and to assist cross-cultural exchange. The Association was established to raise both public and private support for folkloric activities and events in Australia and especially Victoria, and to encourage the collection, preservation and presentation of folkloric materials and activities. The Victorian Folklife Association seeks to establish and manage the operations of an Australian Heritage and Folklife Centre in Victoria.

Wagga rugs The Wagga rug has been defined as 'An improvised covering, usually of sacking' (*Australian National Dictionary*, 1988). Used extensively throughout rural Australia, the Wagga rug was constructed in two different ways. Men in Australia made Wagga rugs from jute bags (wheat bags are most often spoken of) which were crudely stitched together, and used as a rough outdoor sleeping cover, especially for bush work. Early references to masculine Waggas can be found in some of Henry Lawson's stories, such as 'The Darling River', in which he wrote 'a spark would burn through the Wagga rug of a sleeping shearer . . .'. Sometimes the sacks were padded with spare wool, or tea-tree bark. Women, on the other hand, created bedcovers which they called Wagga rugs by roughly sewing old pieces of clothing and blankets to a base fabric, frequently a bag of some kind, then covering the whole with cheap bright cotton fabric or some kind of simple patchwork. Many older Australians will automatically call even a modern quilt a rug because of the association with this kind of crude patchwork. Hessian chaff bags were prized for their use in creating Waggas because they were soft, in contrast to the thick rough texture of wheat bags. The use of some kind of bag is probably the common element between the men's Wagga and the women's domestic Wagga, although not all Waggas made by women contain bags. The concept was basically to use materials at hand, such as bags, old blankets and worn-out clothing, to create a warm bedcover. 'Necessity is the mother of invention', quoted one woman describing Wagga construction. These recycled textiles, however, were usually hidden by the covering of either a cheap cotton (usually cretonne) or some kind of simple patchwork. The cover was often stitched loosely to the inner layers by hand (so that the cover could be removed for washing) or else the layers were stitched together by machine. Many people commenting on the use of Wagga rugs state how heavy they were to sleep under. The general value placed on thrift meant that Waggas were made not only by poorer people, but were generally part of a 'make do and mend' attitude toward life.

The name Wagga rug apparently derives from the town of Wagga Wagga, which was in an important wheat-growing area as well as being situated on major stock routes. Men's Wagga rugs are also sometimes called 'Murrumbidgee blankets'. Sometimes Wagga is spelt 'wogga' or 'wogger', but these would appear to be phonetic corruptions for a word which may have only been heard rather than seen. The construction of bedcovers using old clothing as padding is known both in England and Ireland (Averil Colby, *Quilting*, 1978), so 'Wagga' would appear to have been an Australian addition to an existing imported tradition.

Margaret Rolfe

WALKER, MARGARET EVELYN (née Frey) (1920–), was born in Bendigo, Victoria, and her childhood was spent in various Victorian country towns. She matriculated from MacRobertson Girls High School, Melbourne, as head prefect in 1938. Between 1939 and 1942 she studied at and graduated from Melbourne College of Pharmacy. In 1942 she began dance training with Xenia and Eduard Borovansky. She was a founding member of the Melbourne Ballet Club and spent seven years studying classical ballet, character dance and choreography. She danced in major city seasons of the early Borovansky Ballet Company. In 1944 she joined the Melbourne New

Theatre, creating dances and ballets on contemporary Australian themes for theatre productions. She formed the Unity Dance Group to take dance to factories and community audiences. At this time she also learnt recreational national dances from Melbourne youth groups. In 1946 the first recreational dance group for children was formed in the Junior Eureka League.

In 1949 Margaret married John Walker, a Melbourne architect and town planner. In 1951 she was one of nine members of the Unity Dance Group sent to the third World Congress of Youth and Students for Peace in Berlin. The group, with other artists, formed the nucleus of a programme produced by the Walkers which highlighted Australia's cultural history. Seventy-eight members of the 133-strong delegation took part. Aboriginal songs and dances, bush songs, town ballads, mimes and ballets on modern themes showed Australian folklore to large audiences. After the Festival, dance delegates visited Hungary and the Soviet Union collecting dances which they performed on returning to Australia.

In 1952 the Unity Dance Group was involved in the National Youth Carnival for Peace and Friendship in Sydney. Many dance groups were formed and Margaret established the Association of Australian Dancers (AAD), a national body which initiated a wide range of dance activities focusing on creating Australian dances, summer schools and international dance exchanges. The Unity Dance Group also participated in the New Theatre production of the first Australian musical, *Reedy River* (q.v.), in Melbourne. This led to research and promotion of bush songs and dances.

The Walkers moved to Sydney and in 1958 they adopted the first of four children. Margaret continued to further the work of the AAD with branches in several states. She also set up the Roseville Dance Centre with classes for children and housewives. Senior dance students from other ballet schools were to join the Centre, attracted by folkloric character dance as an alternative dance form to classical ballet. Today her son Kim works professionally as a dancer and choreographer.

When ensembles from Russia and other areas which had a strong folklore background visited Australia, Margaret was helped by the Edgley organization to arrange classes for dancers and teachers. In 1967 two specialists came from Moscow with the Osipov Orchestra and they gave the impetus to work towards establishing a professional folk character dance group in Australia. This was called Dance Concert. In Melbourne, the Kolobok Dance Company was formed by Eva Segal and Marina Berezovsky. Marina was one of many talented and professional people with whom Margaret worked who brought traditions of significance from other countries to enrich Australia after the Second World War.

In 1967 Margaret Walker produced the first multinational programme with Dance Concert and guest dancers from ethnic communities. In 1973 she evolved a one-hour programme for schools and community events to encourage interest in the rich and beautiful dances, costumes and music. This programme involved both performance and participation and was very popular. The Australia Council, New South Wales government, the Schools Commission, teacher training institutions, community arts bodies, state and Catholic schools all gave strong support so that a small professional group could tour New South Wales and interstate. This group was the heart of a plan to foster development of folk dance and to make it widely accessible in Australia. Professional teachers from Australia and overseas, including tribal Aborigines from the Northern Territory, worked with Margaret Walker and many children in cities. The group was also involved with ethnic communities, teacher training and the distribution of a newsletter promoting folk and character dance activities.

In 1977 two significant events took place. One was the formation of the Australian Association for Australian Dance Education at a large national conference of people working in all forms of dance. The other was the sacking of Margaret Walker from Dance Concert by four of the board of directors. (She was unable to agree to their proposal for a budget of $1 million for 1978 — in 1977 it had been $145 000). With the help of the Australia Council, she established the Margaret Walker Folk Dance Centre and carried on the work (Dance Concert collapsed after a few months). In addition she produced a series of displays by large numbers of children. These included the Blacktown Annual Festival for twelve years, the Blacktown City Games in 1981 and annual displays for Children's Week in Martin Place. In 1982 she choreographed the folkloric section of the Brisbane Commonwealth Games and in 1988 the Bicentennial Pageant of Nations for the Wollongong City Council.

Recognition for this work has included the receipt of the Medal of the Order of Australia in 1982 and the Ros Bower Grant from the Australia Council. In 1991 she was honoured in the Story of Sydney at the Rocks for her work with the people of the city. She was one of the first life members of the AADE.

In 1988 the Margaret Walker Folk Dance Centre moved to Canberra where it functions as a national resource for production, teaching and promotion services in folk dance.

'Waltzing Matilda' is probably the most continuously and heatedly debated single item of Australian folklore; there is virtually no aspect of the song's composition, provenance, arrangement and ownership that is not contentious in some way. Such disagreement and the amount of attention and interest generated by it is a reflection of the importance of the song in folklore and also in the broader arena of national consciousness. The orthodox account is that A. B. 'Banjo' Paterson (q.v.) wrote the poem 'Waltzing Matilda' in 1895 to a tune selected by a Miss Christina MacPherson of Dagworth station, Queensland, whom Paterson was courting at the time. This has been challenged by the claim that Paterson tidied up the less professional work of another hand, or even an existing folk song or poem, to produce his own text. This 'original' version was not published by Paterson until 1917, by which time the song had taken on a life of its own. A number of versions emanating from the Paterson–MacPherson collaboration were reported in oral circulation in Queensland in the late nineteenth and early twentieth centuries. In 1903 the song was also used as a promotional gimmick, effectively an advertising jingle, for 'Billy Tea'; copies of the words and music of 'Waltzing Matilda' were given away with packets of tea. The music for this version was 'arranged' by Marie Cowan and the lyrics altered somewhat from Paterson's original poem so that the word 'billy' appeared more frequently than previously.

The emergence of 'Waltzing Matilda' into popular culture, subsequent to its folkloric status in north Queensland, and before it passed into the official literary canon of Paterson's works through publication, is perhaps a clue to the peculiar appeal of the song. This could otherwise be put down to the tune being particularly catchy, but for the fact that there are at least two, and possibly more, distinct tunes to which 'Waltzing Matilda' was and is sung — the most familiar one and that usually known as 'the Queensland version' (see **Manifold, John**). The composer(s) and the provenance of the various tunes to which various versions of Paterson's words are sung have also been strenuously contested, and no doubt will continue to be. At the time of writing the main contenders are numerous. First, there was Miss Christina MacPherson who 'set' Paterson's words to a tune usually said to be 'Thou Bonnie Wood o' Craigielea' (music by James Barr, lyrics by Robert Tannahill, published in 1818) virtually as soon as the text was completed. The second 'composer' or 'arranger' was Marie Cowan who, according to musicological evidence, reworked the MacPherson setting into what most Australians and others recognize as the tune to 'Waltzing Matilda'. The version most people now sing is this 1903 'Billy Tea' adaptation.

However, the story is more complex than this. A third possibility is that an eighteenth-century English folk song related to Marlborough's military campaigns and called 'The Bold Fusilier' was the basis for the Paterson–MacPherson 'Waltzing Matilda', both musically and, to some extent, textually. The most recent complication seems to be evidence suggesting that an Irish song, 'The Bonny Green Tree', bears a close relationship to the Queensland version of the song (also called the 'Buderim version' and the 'O'Neill version'), which has a distinctive tune quite unlike the usual melody, and that both text and tune of this version predate the 1895 Paterson–MacPherson collaboration.

Apart from the debate over the musical aspects of 'Waltzing Matilda', there are also differing interpretations of the text. One view, stemming from Paterson's own account, is that just before his visit to Dagworth a swagman had drowned in the dam or billabong on the property. Paterson took this local tragedy and wove a story around it. This straightforward view of the narrative is undoubtedly that of the majority of listeners with knowledge of the lyrics beyond the chorus. The view has also been put, mainly by Richard Magoffin, that the song is really a thinly veiled allegory or allusion to the rural tensions between workers and squatters during the 1890s. While the lyrics can certainly be interpreted in this way, it seems unlikely that many Australians hear 'Waltzing Matilda' as a proto-revolutionary anthem.

From a folkloric point of view, there are considerable problems in most of these speculations. The description of songs and tunes as 'Irish' or 'Scots' or anything else is untenable, as tunes acknowledge no national boundaries. 'The Bold Fusilier' is of dubious provenance and needs to be further researched in British archives and libraries before any links can be made — or broken — with it and 'Waltzing Matilda'. Above all, the origins and even the variations of 'Waltzing Matilda' are relatively unimportant. What matters is the circulation of the song and the meanings that the song has for those who sing it and hear it.

The essential works in this saga are as follows: S. May, *The Story of Waltzing Matilda* (1955); J. Manifold, *Who Wrote the Ballads?* (1964); O. Mendelsohn, *A Waltz With Matilda* (1966); H. H. Pearce, *On the Origins of Waltzing Matilda* (1971); R. Magoffin, *Fair Dinkum Matilda* (1973); C. Semmler, *The Banjo of the Bush* (2nd edn 1974); H. Pearce, *The Waltzing Matilda Debate* (1974); R. Magoffin, *Waltzing Matilda: The Story Behind the Legend* (1983, rev. edn 1987). There has also been considerable debate on this topic in literary, folkloric and historical journals, notably *Overland, Meanjin, Quadrant* and *Australian Tradition*.

Graham Seal

WANNAN, BILL (William Fielding Fearn-Wannan) (1915–), was born in Victoria and wrote his first articles for the Melbourne *Age* newspaper in 1935 at the age of twenty. In 1938 he spent some months in Japan, and wrote a number of articles for the Melbourne *Argus* and Melbourne *Herald* on the changing culture of Japan and the encroachment of fascism. Back in Australia he was unable to find regular work and 'humped bluey' to the Shepparton–Tatura fruit-growing district, where his experiences provided a wonderful introduction to ordinary Australians in their desperate search for work — and to their idiom, culture and lore. In 1942 Bill Wannan joined the AIF and served in the Pacific during the Second World War. For fifteen months in 1942 and 1943 he was stationed in the Northern Territory, and once again his experiences gave him a valuable insight into various aspects of Australian folklore.

In 1954 Wannan's first collection of Australian folklore was published by the Australasian Book Society with a foreword by Alan Marshall. This book was titled *The Australian: Yarns Ballads Legends Traditions of the Australian People* and was very well received. As a result of the book's success

Bill Wannan was engaged to write regular columns on folklore and Australiana in the Melbourne *Argus* and the *Australasian Post*; the latter continuing for twenty-five years.

Bill Wannan was encouraged by the publisher Lloyd O'Neil to produce numerous books which were published by Currey, O'Neil and by Lansdowne Press. Wannan's fifty-five titles on folklore and Australiana include *A Treasury of Australian Humour* (1960), *Very Strange Tales* 1962) (a history of the turbulent times of Samuel Marsden), *Australian Folklore: A Dictionary of Lore, Legends and Popular Allusions* (1970) and *Hay, Hell and Booligal* (1961). He has also produced numerous articles and essays in scholarly and popular journals and the Grolier *Australian Encyclopaedia*. In 1991 Bill Wannan was awarded membership of the Order of Australia for services to literature and especially folklore.

WARD, RUSSEL (1914–), born in Adelaide, is one of Australia's best-known and most widely read historians. He was Professor in History at the University of New England between 1966 and 1979. He is best known for his book *The Australian Legend* (1958), which not only provided an analysis of Australian mystique about national identity, but also contributed the phrase 'the Australian Legend' (q.v.) which is still widely debated in historical and folkloric analyses. *The Australian Legend* was also the first historical work to emanate from Australian academia which made substantial use of colonial and bush ballads. A number of these were first collected by Russel Ward, and in his Foreword to the book he acknowledges the aid of 'three old folk-singers, Mrs Mary Byrnes, Mr Joseph Cashmere and the late Mr John Henry Lee'. In 1964 he edited *The Penguin Book of Australian Ballads*. Russel Ward has produced numerous other works in Australian history, including *Australia: A Short History* (1965), *The Restless Years* (1968) (with Peter O'Shaughnessy and Graeme Inson), *A Nation for a Continent* (1977), *Australia Since the Coming of Man* (1982) and *Finding Australia* (1987). Russel Ward's socialist views have been a key influence in his work and are discussed in his autobiography *A Radical Life* (1988).

Wartime folklore in Australia can be usefully examined under the following headings: soldier lore (professional/occupational soldiers); digger-lore (volunteer infantry); home front folklore; folklore about the digger, its relationship to Anzac Day (q.v.) and its place in national consciousness.

The folklore of the professional soldier — army, navy or air force — is best understood as occupational lore (q.v.). It consists of considerable traditions of in-group knowledge and activity: custom (initiations and related forms of 'bastardization'); speech (jargon, slang); narrative (legends, yarns, tall tales); song and verse, much of which is of a bawdy nature as might be expected amongst such predominantly male groups with a professional warrior ethic. Military group traditions of this sort may be of considerable age, deriving from similar traditions within the British army and navy. Such traditions are not restricted to times of war, but form a part of the occupational culture of the professional soldier or sailor or aviator.

Of wider interest, perhaps, and certainly of greater national significance, is the folklore of the volunteer Australian infantryman or 'digger'. The 'digger' was brought into existence by the mass mobilization of populations that have been a feature of twentieth-century global conflict. A distinctive wartime culture of the digger was rapidly created during 1914–18 and persisted into peacetime, becoming an important defining element of national consciousness, as well as providing a body of lore and associated values and attitudes that have re-emerged in subsequent conflicts, particularly those of 1939–45 and the Vietnam war. The folklore of the digger was formed in Egypt and Palestine during the Gallipoli campaign and in subsequent campaigns on the Western Front. Although digger lore drew on some pre-existing elements of rural and urban Australian folklore, particularly that surrounding the bushman and the larrikin, a quite distinctive and aggressively nationalistic digger culture began to form on the ships transporting the first AIF to Egypt in early 1915, in training camps in Egypt and on the Gallipoli Peninsula. By the time of the evacuation of Gallipoli in December 1915, the folklore and culture of the digger were well established, undergoing further refinement in France, where the term 'digger' became established as the accepted descriptor of the Australian foot-soldier around early 1917. The essential folkloric elements of digger culture were speech, 'furphies' or rumours, humorous anecdotes or 'yarns', song and verse and behavioural forms such as gambling, and folk belief or 'superstition'.

Embedded in these forms were the fundamental attitudes of the diggers towards all other groups with whom they came into contact — officers, the British, French, Americans, the enemy, and 'Gyppos' (Egyptians). Also twined into these attitudes was the self-image which the digger wished to project to the world at large. In many respects, both these sides of the coin were derived from the attitudes prevailing in Australian society before the First World War: fear of 'others' and a pugnacious Australianism (often equal to, and perhaps caused by, a deep loyalty to the British Empire) being the underlying themes of nineteenth- and early twentieth-century Australian life. Added to this was a military image of toughness and casual effectiveness that began to evolve from Australia's Boer war soldiering, 1899–1901. Perhaps the most common theme involves the digger's relationship to authority within the military hierarchy. This is expressed in an antagonism towards British officers and their perceived insistence on military etiquette, something the Australians found irksome, and conversely in situations in which Australian officers are shown to be more egalitarian in their relationships with their subordinates. The following anecdotes illustrate these themes:

> Two diggers on leave in London pass by a British officer without saluting. Outraged, the British officer upbraids the Australians for their lack of military etiquette. The digger's reply is to the point and includes calling the British officer a 'bastard'. Spotting an Australian officer nearby, the insulted Britisher complains of this insolence, to which the Australian officer replies: 'You're not a bastard, are you? Well, trot over there and tell them they're bloody liars'.
>
> After the war General Birdwood was much in demand at upper-crust functions where he would often recount his experiences on Gallipoli. One of these concerned the time that Birdwood was swimming in the sea as a Turkish bombardment opened up. A digger on the shore yelled at him: 'Duck, Birdie, duck'. 'What was your reply?', asked an astonished British officer. 'No reply', said Birdwood. 'I ducked.'

These are both First World War yarns, but the same problems with rank and Britishness persist in the song 'Horseferry Road'/ 'Dinky-Di'/'The Digger's London Leave', as it is variously known. Originating in the 1914–18 war, this well-known song reached its mature form in the 1939–45 conflict:

Well, he got back to London and straight away strode
To army headquarters in the Horseferry Road,
To see all the Provos who dodge all the strafe,
And see all the bludgers on headquarters staff.

Chorus
Dinky-di, dinky di,
For I am a digger who won't tell a lie.

A hulking great sergeant said 'Pardon me, please,
There's blood on your tunic and dirt on your sleeve,
If you walk round like that all the people will laugh;
Don't hang about here near the headquarters staff'.

The digger just shot him a murderous glance,
Saying, 'I've just come from the shambles in France,
Where whizzbangs are flying and comforts are few,
And brave men are dying for bastards like you.

We're bombed on the left and we're bombed on the right,
Shelled by the day and shelled by the night,
If something don't happen and that pretty soon,
There'll be no bastard left in the bloody platoon'.

The matter soon came to the ears of Lord Gort,
Who gave the whole matter a good deal of thought,
He awarded the digger VC with two bars
Giving the Provo a kick in the arse.

Now when this war's over and we're out of here,
We'll see him in Sydney town begging for beer.
He'll ask for a deener to buy a small glass,
But all he'll get is a kick in the arse.

Another popular theme of digger folklore is the nonchalance of the Australian soldier under fire, as in this yarn:

> A group of diggers are playing cards when a messenger arrives and breathlessly announces that a large enemy attack is almost upon them. No one moves, until one digger throws his hand down, saying 'Aw, I'll go, I'm out anyway'.

Complaints — about everything — are another theme of digger folklore and, indeed, are a venerable tradition among all soldiers. Expressions of contempt for war, food, drink, absence of these, availability of female companionship, the unsatisfactory or restricted nature of such companionship if it is available, living conditions, etc. are legion in digger lore. Probably the best-known yarn is that usually known as 'The World's Greatest Whinger', an often-reprinted tale that is said to have originated in the First World War, though is only known in versions from the Second World War, and later. An example from the Vietnam war, 'Christmas in Vietnam', continues this tradition of complaining through a bitter-sweet parody of 'Jingle Bells':

Driving through the mud in a jeep that should be junk
Over roads we go, half of us are drunk.
Wheels on dirt roads bounce, making arses sore,
Lord, I'd sooner go to hell than finish out this war.

Chorus:
Jingle bells, mortar shells, VC in the grass,
We'll get no merry Christmas cheer until this year has passed.
Jingle bells, mortar shells, VC in the grass,
Take your merry Christmas cheer and shove it up your arse.

Christmas time is here, as everybody knows,
People think it's dear, Aussies think it blows.
All at home are gay, children are at play,
While the boys are stuck out here so bloody far away.

Chorus:
The moral of our song. It's plain as it can be,
Please, no more midnight carols sing, and screw your Christmas tree.
There's one more thing to say, before we have to leave,
Vietnam is not the place to be on Christmas Eve.

An important element of the self-image of the digger was the development of a distinctive form of folk speech (q.v.), or 'diggerese'. The term 'digger' itself is one example of this, and there are many other terms originating in the First World War that have remained part of digger, and even wider, vocabulary ever since. These include 'gutser' (a bad experience), as in to 'come a gutser' and 'possie', a contraction of position. Other terms that were an integral part of First World War diggerese have fallen away, including 'reinstoushment' (reinforcement), 'imshee' (hurry up — from Arabic) and many more.

The other important half of any war effort is the civilian population back home. Relatively little is known of 1914–18 home front folklore, and what has been collected revolves around the conscription debates and generally opposes Prime Minister Hughes's campaign to introduce conscription. The Second World War is a rather different story. There was little opposition to the war and research undertaken to date indicates the circulation of a good deal of material related to relations between the Australians and the American troops stationed here from 1942. Much of this material revolves around sexual relationships between Australian women and American servicemen, as in the following example, a parody of the popular song 'When They Sound the Last All-Clear':

> When they send the last Yank home
> How lonely some women will be.
> When they turn out the lights,
> There'll be long, lonely nights,
> All those good times a memory.
> Evermore they'll be alone,
> Those women no Aussie would own.
> All they'll have are their clothes,
> And a kid who talks through its nose —
> When they send the last Yank home.

It is from this period that the Australian folk-term for Americans originated — 'Septic tank', rhyming slang for the older folk-term 'Yank'.

The Second World War also differed from the experience of the earlier conflict in that women played a much broader role. In addition to the more traditional tasks of nursing, morale maintenance, comforts provision and so on, women took on such tasks as agricultural production, armaments and other manufacturing production. They also moved into areas much more closely associated with active duty in the various women's auxiliary organizations of each service. Such groups speedily developed cultures and folklores of their own, and although research remains to be undertaken in these areas, it is clear that significant folk traditions of serving women were generated and have continued to evolve, in all likelihood underlying the change of attitudes towards women and work that have been a feature of Australian and other Western societies since the 1960s and 1970s.

Finally, an important consideration in discussing wartime folklore is the impact of the essentially private or in-group tradition of the digger on the broader commemorative complex of Anzac Day (q.v.) and its relationship to national identity. The image of the digger and many of his attitudes to authority, to others and his world view in general are closely related to the archetype of the tough, resourceful, anti-authoritarian bushman. The digger, through his celebration on the only non-religious public holiday observed by all states on the same day, has continued to provide a defining image of desirable and typical 'Australianness'. That this image effectively excluded large segments of the population, including non-combatants, women, Aborigines, migrants and children, has only come to be generally lamented in relatively recent years. The convenience of fusing such a centrally important image of nationhood and cultural identity with the volunteer foot soldier for the armed forces, politicians and such institutions as the RSL can hardly be ignored.

REFERENCES: G. Seal, 'Two Traditions: The Folklore of the Digger and the Myth of Anzac', *Australian Folklore* 5, 1990, and *A Broken Song: The Voice of Australians at War*, ABC cassette set, 1985; W. Wannan, *Bill Wannan's Great Book of Australiana*, 1977.

Graham Seal

WATERS, EDGAR (1925–), was born in Sydney. During the 1950s and 1960s he edited folksong recordings for Wattle (q.v.), and wrote notes on songs published in *Australian Tradition*. He completed a doctoral thesis on Australian folk song and popular verse at the Australian National University in 1962 and lectured in history at the University of Papua New Guinea for many years. He is presently a consultant to the National Library of Australia on folklore and oral history.

Wattle Recordings In the early 1950s Peter Hamilton, then an architect working in Sydney, decided that he wished to make films; amongst others, a series of short films using Australian folk songs. He could not find recordings of Australian folk songs that he could use as film soundtracks (the only recordings of Australian folk songs available at the time were on a long-play record of songs from the musical *Reedy River* (q.v.), and not all the songs on this recording were folk songs; some of those that could be considered folk songs had been 'reconstructed' in text or melody or both). Hamilton set up Wattle Recordings to produce both films and sound recordings. He asked Edgar Waters to join him as record editor.

The first Wattle records were produced in the middle of the 1950s, at a time of considerable technical change in the production of gramophone records. Wattle published records in the old 78 rpm format, and in the newer 45 rpm and 33 rpm formats.

The first recordings published were of performances by the Bushwhackers, a group of Sydney collectors and performers including John Meredith (q.v.) and Alan Scott (q.v.). These were 78 rpm recordings. Each record was accompanied by a printed sheet, giving the words of the song and brief notes. Some of these recordings were surprisingly popular, being much played on country radio stations and the 'juke boxes' of country pubs. The Bushwhackers were the most important of the inventors of the 'bush band' tradition, and their Wattle recordings made them much the best known of the early bands of this kind. The popularity of these Wattle recordings thus had much to do with the spread of this new way of performing Australian folk songs, and helped establish the usual bush band repertoire: 'Click Go the Shears', 'The Drover's Dream', 'Nine Miles from Gundagai', and so on.

Wattle published recordings (mostly extended play or long play) of other performers of the 'folk song revival' of the 1950s: the Rambleers (most of whom were also members of the Bushwhackers), the Moreton Bay Bushwhackers, the Bandicoots (led by John Manifold (q.v.)), Bill Scott (q.v.) and Stan Arthur. They also published two long-playing records of performances of Australian songs by the distinguished English folksong scholar and singer A. L. Lloyd (q.v.). Lloyd had spent some time as a station hand in Australia, and had learnt some of his songs during that time.

Many of the songs of these performers were versions put together from various sources: a tune from one singer, most of the text from another, a verse or two from a third. Some of these versions, originally cobbled together for performance on stage or in the recording studio, have also passed into the repertoire of the bush bands and other 'revival' singers. Some have even appeared in the printed collections as versions learned from uncontaminated oral tradition.

In 1957 Wattle published the first collection of songs and dance tunes really recorded from bush oral tradition, as *Australian Traditional Singers and Musicians*. These were performances originally recorded in the field by John Meredith (most of the material used was rerecorded for Wattle in Sydney). The performers included Sally Sloane (q.v.) and 'Duke' Tritton. The record was accompanied by a booklet giving extensive notes on the songs and the singers. In 1960 Wattle published another collection of field recordings, this time of songs and dance tunes from Victoria. The performers included Simon MacDonald (q.v.), who was perhaps — always excepting Sally Sloane — the finest singer in the bush tradition ever recorded in Australia. The recordings were made by members of the Folk Lore Society of Victoria, including Norm and Pat O'Connor, Bob Michel and Maryjean Officer. Very full notes, with transcriptions of the song tunes, were bound up with the record cover.

Wattle published two of the earliest long-play records of Aboriginal tribal music. One was called *Arnhem Land Popular Classics*, at the insistence of the American anthropologist Lamont West, who made the field recordings. The other was called *The Art of the Didgeridoo*. One side of this record consisted of field recordings made in Arnhem Land, the other of demonstrations of didgeridoo technique by the white musicologist Trevor Jones. Wattle also published a collection of field recordings made in what was then the Trust Territory of New Guinea (now the northern half of the independent state of Papua New Guinea). Most of these recordings were made by Ray Sheridan, a Territory health worker and amateur musicologist. All these recordings, in the usual Wattle style, were accompanied by booklets of notes.

Wattle has not published any record since *The Land Where the Crow Flies Backwards* in 1963. This was a collection of songs by a notable Aboriginal singer and song writer, Dougie Young. The field recordings had been made under difficult conditions by the anthropologist Jeremy Beckett. Beckett's original recordings of Dougie Young are held by the Australian Institute of Aboriginal and Torres Strait Islander Studies. Most of the other Wattle material, including master tapes of published records, have been deposited by Peter Hamilton with the National Library of Australia.

Few of the recordings published by Wattle did more than cover their costs of production. The production of Wattle records, nevertheless, took up most of Peter Hamilton's time for several years. He did, during these years, manage to produce a number of short films based around Australian folk songs which are now lodged with the National Film and Sound Archive. No Wattle material is at present available for sale.

Edgar Waters

Western Australian Folklore Archive, The, was established in 1985 within the Centre for Australian Studies at Curtin University by Graham Seal (q.v.), and has extensive holdings of mainly, though not exclusively, Western Australian folklore. The core of the Archive consists of student fieldwork projects and self-administered questionnaires undertaken as a component of study in the folklore units offered as part of the University's BA in Australian Studies. As well, the archive contains a good deal of material from other sources, including the Bicentennial oral history project 'The Other West Australians', tapes and transcripts of the 'Foothills Connection' community history and folklore project and material relating to the 'Chinese-Australian Folklore Project'. Emphasizing contemporary folklore, the archive also covers many aspects of folklore and folklife. Included are written reports of various kinds, questionnaires, audio and video cassettes, photographs and other graphic materials.

'Wild Colonial Boy, The', is a famous Australian folk song, also sung in Ireland, England, Canada, New Zealand and the United States. Probably deriving from the earlier ballads about the activities of bushranger Jack Donohoe (see **Bushrangers**), 'The Wild Colonial Boy' appears around the 1860s. The song deals with the heroic bushranging activities of a Jack Doolan, Dowling, Duggan or other variation who, like Donohoe, was born in Ireland, defies the forces of authority and prefers death to 'slavery, bound down in iron chains'.

See also: **Folk song**.

Women and folklore Much women's culture can be considered as folk culture, since a great deal of it is transmitted informally through observation, imitation and direct instruction in face-to-face settings such as the family. As with children's folklore, the study of women and folklore has more than one dimension, namely folklore *about* women and folklore which belongs to women themselves (folklore *of* women).

Folklore about women is frequently, although not always, transmitted by males. Much of it has been shown by contemporary research to be patently incorrect, such as the beliefs that women are 'bad drivers' or excessively talkative (see, for example, Dale Spender, *Man Made Language*, 2nd edn, 1985, concerning male domination of conversation). Of considerable significance in today's society are a number of male beliefs about women's sexuality and about male 'rights' to women's bodies, irrespective of consent. Some of these attitudes are encoded in commonly used language; for example, a sexually active woman may be described by the derogatory and clinically inaccurate term 'nymphomaniac', or more simply in Australian folk speech as a 'nympho'. There is no similar derogatory term commonly used in Australian English to describe a sexually active man.

Women's own folklore, folklore of women, is normally transmitted from female to female, both between and within generations. Some of its major forms are handcrafts, foodways, folk beliefs, interpersonal relationships and verbal lore. In 1982 Nancy Keesing published *Lily on the Dustbin: Slang of Australian Women and Families*. Keesing's book is an entertaining and insightful account of the special colloquialisms used in a number of female and family contexts, such as the woman who 'puts on her face' before going to work in the morning, using cosmetics resting on the 'duchess set' on the dressing table, anxious to look her best even if she has 'the curse'. The 'Sheilaspeak' and 'Familyspeak' which Keesing describes has a variety of functions, including instructing children ('Were you born in a tent?'), commenting on male behaviour ('the wandering hands society') or on other women ('mutton dressed up as lamb'). Much of it is to enhance female solidarity and for the pleasure of using colourful idioms or euphemisms such as 'a lick and a promise' or 'a canary's bath' to describe hasty ablutions.

Women's traditional handcrafts are important not only for their aesthetic and skilful qualities, but also because they frequently express the values and beliefs of their makers. These values may range from family affection (as in a gift of a hand-knitted jumper) to deep-seated ideological and religious beliefs. In the 1980s, for example, women parishioners from St John's Anglican Cathedral in Parramatta (NSW) made a notable series of tapestry covers for seats and kneeling pads. These tapestries featured the church itself and other historic buildings in Parramatta. In the same decade the Hamilton Quilters (Vic) produced the Bandicoot Quilt to draw attention to this local endangered mammal. Women's handcrafts have been well documented in Australia; for example Jennifer Isaacs's book *The Gentle Arts: 200 Years of Australian Women's Domestic and Decorative Arts* (1987) is not only well illustrated but also includes useful discussions of the meaning and contexts of a number of different activities (see **Craft**; **Patchwork quilts**). The handcrafts made by both women and men from non-English-speaking backgrounds, whether recent immigrants or from old-established communities, have greatly enriched Australian life and have been the subject of a number of exhibitions, such as the *Arti e Mestieri* organized in Melbourne by the Italian Assistance Association (CO.AS.IT).

One of the most interesting aspects of women's folklore and one of no small social and medical significance concerns beliefs about pregnancy, childbirth and infancy. In 1980 Lumley's and Astbury's book *Birth Rites, Birth Rights* described two substantial studies of public hospital patients carried out in the 1970s. The Melbourne study, carried out in 1976 and 1977, found that significant numbers of women believed in 'old wives' tales' about pregnancy, such as that eating certain foods could cause birthmarks, that words or behaviour during pregnancy could cause birth defects and that pregnant women should not lift their arms above their heads. More alarming was the earlier study (Brisbane, 1971), where almost half the 500 women interviewed believed that nothing could affect the foetus during pregnancy, neither drugs, nor rubella, nor smoking. Similar folk beliefs (both of omission and commission) are still common. The Australian Children's Folklore Collection (q.v.) at Melbourne University includes the following 'old wives' tales', collected from persons of diverse ethnic origins in the 1980s:

> A baby shouldn't be allowed to look at itself in the mirror as it may not develop properly. (Anglo-Australian)
>
> Don't give male children chicken feet to eat as this produces shaky hands which are bad for a man's writing. (First generation Chinese-Australian)
>
> If you drive a car while pregnant the baby will be born with the cord around its neck. (Yugoslav-Australian)

Some aspects of women's folk beliefs and practices in contemporary Australian multicultural society have been the subject of controversy. The public debate about genital mutilation of females as practised in some African/Islamic communities has been particularly agitated, although many senior medical personnel regard this practice as insignificant in Australia. Debate has also raged in Australia about the head-scarf worn by traditional Moslem women and female children, although public acceptance of this practice appears to be growing.

A number of women's folk beliefs are beliefs about themselves. Contemporary debate frequently focuses on the lack of women in high positions in Australian life, whether in academic, business or political life. Undoubtedly some of the reasons for this situation are practical (e.g. child-bearing and rearing) and some are the results of derogatory male beliefs about women's competence. Undoubtedly too some reasons are to do with women's lack of self-belief or assertiveness. The corollary to this situation is more positive; traditional women's culture is marked by co-operative and conciliatory attitudes and good communication skills.

Some changes are occurring in aspects of women's folklore. Traditionally, women have taken the main responsibilities for cooking in the domestic arena (see **Foodways**), while professional cooks have almost all been males, but both areas are moving towards greater gender equality. Marriage and childbirth have changed dramatically; birth out of wedlock is no longer a stigma in Australia, and a number of women, both with and without children, are declining to enter into legal marriage with their partners. Despite these changes, the elaborate 'white wedding', largely a female-instigated custom, still flourishes in Australia.

See also: **Folk heroes and heroines**.

REFERENCE: Diane Bell, *Generations: Grandmothers, Mothers and Daughters*, with photographs by Ponch Hawkes, 1987.

Gwenda Beed Davey

Xerox lore; see **Reprographic folklore**.

Z

ZOUGANELIS, KOSTAS (1935–), is an exponent of Karagiozis (q.v.), Greek shadow theatre, in Sydney. Born in Piraeus, he emigrated to Sydney in 1962, working as a boiler attendant. His schooling was limited by poverty and the German Occupation, but at the age of twelve he became assistant to the neighbourhood Karagiozis puppeteer Dimitris Moros, whose father, Yiannis, is credited with creating Stavrakas the Spiv, one of the standard cast of Greek shadow characters. He gave his first performance in Piraeus at the age of sixteen or seventeen using some of Moros's puppets. The sequel to this early effort came twelve years later in 1964 in Sydney at the Migrant Centre in Castlereagh Street and using Theo Poulos's puppets, which were made by a Yugoslav painter from traditional Greek designs. He subsequently assisted Karagiozis puppeteers visiting from Greece, including Vangos Korfiatis in 1967 and Avraam Andonakos and 'Yiannaros' Mourelatos in 1978. In March 1978 he made his own shadow puppets with the help of Takis Constantopedos and started performing Karagiozis at the Sydney *Boîte* (a showcase for the ethnic arts), and during the period 1978–88 gave numerous Karagiozis performances all over the Sydney metropolitan area and several New South Wales country towns, setting up his screen in various Greek community clubs, schools, etc. From 1979 to 1985 he had a permanent venue in Marrickville Uniting Church hall, performing regularly at weekends with the help of his wife and his two children. He claims to have performed not for profit but for love of playing Karagiozis, and to have donated the proceeds from his performances to various Greek community causes.

He has participated in various multicultural festivals such as the September Carnivale. In 1979 he gave performances and demonstrations at the Sydney Opera House and at the exhibition of Folk Arts and Crafts of the World at Roselands; in 1982 at *L'internationale* Village Fair; in 1983 at the International Puppet Festival and the Adelaide Glendi Festival; and in 1985 at the Rocks Theatre, home of the Marionette Theatre of Australia. In 1983 he took up an invitation from Canadian Patchwork Puppets to perform Karagiozis at Summer Fest '83 in Edmonton, Alberta, and in March 1986 he performed at the New Zealand Puppet Festival in Wellington and Auckland. He also performs with marionettes, which he makes himself, playing at schools, shopping malls and private functions. He is a member of the Australian Puppetry Guild and the UNESCO International Marionette Theatre. He owns several hundred puppets and has exhibited them at various venues, including the Sydney Opera House in 1979. In about 1981 he changed from using leather shadow-puppets to coloured acrylic ones, which he makes himself.

Kostas Zouganelis's repertoire includes traditional plays such as *Karagiozis the Servant/the Baker/the Cook/the Bride/the Scribe*, *The Wedding of Karagiozis/Uncle George*, *Karagiozis on the Moon*, *The Two Lords*, *The Princess's Three Riddles and the Suitors*, *The Orphan Girl of Chios*, *Alexander the Great and the Accursed Serpent*. He prefers to avoid the heroic historical plays because they are too chauvinistic and contain violence, which he tries to minimize on his screen, especially the beating of children. Although he claims to adhere to tradition quite conservatively, Lukas observed in the *Sydney Morning Herald*, 31 March 1978, that he 'has introduced features of migrant life in Australia to his repertoire — the conflict between migrant parents and their children about retaining their mother tongue, traditions, and misunderstandings. Sometimes his characters speak in Greek, sometimes in the "pidgin English" used by some of his countrymen'. Kostas Zouganelis has also tried performing shadow theatre in English, using two short sketches about two Australian characters, Charlie and Fred, translated by Angelo Loukakis.

In 1986 he took part in the Sydney-based Hellenic Art Theatre of Australia's production of Miltos Moutafidis's play 'Karagiozis in Australia', which satirizes aspects of the life of Greeks in Australia with traditional shadow-figures being played by live actors on stage. Zouganelis performed the opening of the play manipulating lifesize Karagiozis figures behind a huge white screen, and he also played the role of Uncle George.

He has performed Karagiozis several times at university open days, has lectured on Karagiozis at the Drama School of the University of New South Wales, and in 1981 was awarded an Australia Council grant. He has appeared in SBS TV documentaries and also represented the Greek tradition in the Australian Film & Television School production *A History of Shadow Puppets* in 1978.

Following the departure of Dimitris Katsoulis (q.v.), Kostas Zouganelis seems to be the last exponent of Karagiozis resident in Australia. His son has been his assistant for several years, but does not intend to carry on the family tradition.

Anna Chatzinikolaou

APPENDIX: A BIBLIOGRAPHIC GUIDE TO AUSTRALIAN FOLKLORE

THE BIBLIOGRAPHIC MATERIALS in the field of folklore necessarily reflect the historical development of the subject area within Australia. It is fair to say that, until recently, folklore in Australia has demonstrated a heavy emphasis on collection/fieldwork and much less emphasis on theoretical questions, the analysis of collected material or bibliographic work. The predominant focus of folklore research upon Anglo-Celtic or English-speaking traditions is clear from a survey of bibliographic materials, although a growing interest in ethnic folklore, partially stemming from the input of musicologists such as Peter Parkhill in Greek music and Linda Barwick in Italian musical traditions, is now evident. A widened view of Australia's folklore has also emerged through works such as Ian Turner's *Cinderella Dressed in Yella* (1969), Heather Russell's *Play and Friendship in a Multicultural Society* (1986) and June Factor's *Captain Cook Chased a Chook* (1988). Similar trends are not evident in terms of folklore and Aboriginal culture. Very few folklorists have delved deeply into Aboriginal culture, leaving it largely the domain of cultural anthropologists. This may be the result of folklorists having difficulties in drawing their usual intellectual boundaries within traditional Aboriginal culture with its almost *total* reliance on oral transmission, performance and/or imitation. An exception to the above generalization is the fieldwork conducted by Chris Sullivan and Mark Rummery into European musical traditions among Aboriginal people. Scholars of Aboriginal culture are fortunate in comparison to scholars of folklore, since information is organized within the intellectual framework of one discipline (i.e. anthropology), as opposed to the situation in folklore with its multi-disciplined approach. It appears that the poor coverage and quality of bibliographic materials in Australian folklore stems largely from the general lack, until recently, of a clear theoretical framework.

The Australian national bibliography

SEVERAL PUBLICATIONS PROVIDE bibliographic coverage of Australian materials — i.e. those works which were written by Australians, or published in Australia, or which have Australian content if published overseas — from 1784 to the present. Most state libraries provide access to these publications, as do many university libraries. Printed materials published between 1784 and 1900 are covered by the seven volumes of *Bibliography of Australia* published between 1941 and 1969 by John Alexander Ferguson. The volumes covering 1784 to 1850 included publications with Australian content, whether published in Australia or outside, such as books, pamphlets, broadsides and periodicals/newspapers. Articles which were published at a later date in separate form are also included. Excluded are maps, charts, prints, articles in periodicals, and music. Ferguson's volumes for 1851 to 1900 include printed books, pamphlets and broadsides, but exclude *belles-lettres*; statutes; parliamentary papers; newspapers, periodicals and annual reports; law, medicine, science, technology publications (unless directed to the public, rather than to professionals); reprints of papers in scientific proceedings and journals; elementary school books; programmes; sales catalogues; and memoranda of associations, etc. Overseas materials contain a minimum of 10 per cent Australian content. Information provided by Ferguson includes a bibliographic description of the work (sometimes rather confusing), and detailed notes covering aspects such as physical description, contents, historical background and locations of copies. Folklore materials in the form of actual collections or works with an analytical content are obviously scarce for historical reasons, but researchers have uncovered some valuable primary material in the form of broadsides. In 1975 the National Library assumed responsibility for the Ferguson works when it acquired the copyright from Angus and Robertson. Based upon additional materials, a volume of *Addenda* to Ferguson's volumes was published in 1986 covering the four volumes 1784 to 1850 with listings in chronological order.

The Commonwealth National Library (which preceded the National Library of Australia) took responsibility for the national bibliography in 1936 with the publication of the *Annual Catalogue of Australian Publications* (1936–60). The work was compiled from the books deposited at the National Library under the provisions of the Copyright Act of 1912, and included an incomplete list of annuals/serials and, after 1936, included music. Deposit requirements and frequent lack of compliance made this work rather incomplete. Access to information varies from volume to volume. Early volumes allowed access by author, or by title in the absence of an author. Later volumes allowed access by author, title or subject. Description again varies from volume to volume, but later volumes provided bibliographic description similar to the present *Australian National Bibliography*, including Dewey classification number.

The *Australian National Bibliography, 1901–1950* was published in 1988 by the National Library of Australia. The work overcame the lack of 'accurate and comprehensive coverage' in the volumes dating to 1951 of the *Annual Catalogue of Australian Publications* and bridged what had become known as 'the Ferguson gap', i.e. the years between 1900 when Ferguson's work finished and 1936 when the National Library commenced its responsibility for national bibliography. Coverage extended to all books and pamphlets (of five pages or more) of the period published in Australia, or with Australian authorship, or at least 10 per cent Australian subject content or an Australian setting (in the case of fiction). Excluded are serial publications such as newspapers, periodicals and proceedings of consecutively numbered conferences, directories, manuals, tourist guides where they were updated frequently; all non-text materials such as maps; photographs; sheet music; non-published materials such as theses and manuscripts; Commonwealth and state parliamentary papers; government legislation such as acts, bills, regulations; standards and patents; reprints unless there was a significant bibliographic difference from the original; and ephemeral material such as sales catalogues and programmes unless of particular interest or importance. Two volumes provide detailed information with alphabetic listing by author (or by title in the event of no known authorship); a third volume provided listing by author, title and series, while the final volume was a listing by subject. Bibliographic citation includes information comparable to the present *Australian National Bibliography* citation format, excepting Dewey number and International Standard Book Number (ISBN).

Most researchers will be familiar with the *Australian National Bibliography* (*ANB*), which commenced publication in 1961 on an annual basis under the auspices of the National Library of Australia and is available in major libraries. There is a necessary time lag between publication of a work and its appearance in the *ANB*, but cataloguing-in-publication (CIP) information (found on the title verso page in many publications) has diminished that time lag. Coverage includes books; pamphlets of five pages or more; maps; prints; movies; sheet music; government publications; and first issues of new annuals, periodicals and newspapers, but excludes materials of limited access or availability (e.g. for internal use or limited circulation) or not intended for permanent retention (e.g. colouring books). Items are published in Australia, or have content wholly or substantially Australian, or authors resident in Australia. Access in earlier volumes is via author or index to subjects, secondary authors, titles and form (in the case of films, maps, music and prints). The entries in the editions of *ANB* since 1972 may be accessed by the author, title and series index; subject index or the classified sequence based upon the Dewey classification number which contains the full description of the work. Full description as per 1988 features Dewey number, author, title, publisher, publication place and date, series, physical description, notes, ISBN, price and tracing notes prepared in accordance with Anglo-American cataloguing rules and in-house regulations.

Researchers in folklore are likely to encounter occasional difficulties in using the *Australian National Bibliography* since it does not contain all materials that might be of interest to a researcher, such as videos and sound recordings.

Another area of difficulty concerns what is included and excluded within the *Australian National Bibliography* under the heading of folklore. Folklore appears as a heading within both the Dewey Classification and the Library of Congress subject headings, but with very different meanings.

The Dewey Decimal Classification (edition 19) treats folklore under the more general heading of Customs, etiquette, folklore (390) and provides a restricted, almost literal meaning within that context. The subject matter under the heading of folklore (398) contains several main sub-topics. The first deals with the theoretical, and includes: general philosophy and theory (398.01); sociology and criticism, such as origin, role, function of folklore as a cultural and social phenomenon (398.042); historical and geographical treatment of sociology of folklore (398.09); theories of folklore, such as philological, anthropological, etiological, psychoanalytical, symbolic, historical (398.1); origin and diffusion of folklore, e.g. Pan-Babylonian and Pan-Indian (398.19).

Folk literature (398.2) provides another main topic under the folklore heading. It essentially covers 'short works from oral tradition', but excludes religious mythology. Sub-classification of folk literature is by way of language (398.204); tales and lore of para-natural and legendary beings of human and semi-human form (398.21); tales and lore of historical and quasi-historical persons and events, such as heroes, kings, witches (398.22); tales and lore of places and times, real, legendary and special (398.23); plants, real and imaginary (398.242); animals, real and legendary (398.245); ghost stories (398.25); tales and lore involving physical phenomena, such as heavenly bodies, weather, real and imaginary minerals, etc. (398.26); and tales and lore of everyday human life, e.g. birth, love, marriage, death, occupation, recreation, dwellings, and food (398.27).

A third main topic under the folklore heading is the sociology of specific subjects of folklore (398.3–398.4). One part of this is the natural and physical as subjects (398.3), which deals with places, times, humanity and human existence, and physical phenomena. The other sub-section, the paranatural and legendary as subjects of folklore (398.4), gives coverage to folk beliefs (398.41), including superstitions; legendary places (398.42), such as Atlantis; beings of human and semi-human form (398.45), such as ogres, fairies, giants, vampires; legendary minerals, plants and animals (398.46); and ghosts (398.47).

The final main topic under the folklore heading is minor forms of folk literature, such as chapbooks (398.5); riddles (398.6); rhymes and games (398.8), such as nursery rhymes, counting-out rhymes, street cries and songs; and proverbs (398.9).

If folklore and folklife are considered as synonymous terms, and deemed to extend beyond the realm of oral literatures, then one's search has only just begun. This is partially remedied by looking up headings beginning with 'folk', such as folk dance, folk medicine, folk songs, folk art, etc., but other materials which may satisfy generally agreed-on criteria as to items of folklore, such as jokes, limericks, games, foodways, domestic crafts, etc., remain hidden.

The Dewey approach is in clear contrast to the notion of folklore as provided by the other essential tool of bibliographic work, the Library of Congress subject headings as evidenced by the following statement from the ABN (Australian Bibliographic Network) subject cataloguing manual:

> Subject Headings, as used in the *Australian Bibliographic Network Subject Cataloguing Manual*, generally follow Subject Heading of the Library of Congress. Folklore is defined as . . . those expressive items of culture that are learned orally, by imitation or by observation, including such things as traditional belief, narratives (tales, legends, proverbs, etc.), folk medicine, and other aspects of the expressive performance and communication involved in oral tradition. This is defined in contrast to the concepts designated by the more inclusive subdivision Social life and customs which may include folklore, manners, customs, ceremonies, popular traditions, etc.

While it excludes the area of folk music (providing a separate section), the above definition clearly offers a broader definition of folklore than the Dewey system of classification. This difference may reflect the extensive work of the Library of Congress as a pioneer in the field of folklore scholarship, analysis and collection within the United States.

The *Australian National Bibliography*'s use of Library of Congress subject headings occasionally produces a clash of Australian and American idiom as in the case of 'railroad' and 'railway', where both terms appear to be in use. In other instances the catalogue still reflects the use of Americanisms, as in the case of 'hobo' or 'tramp' for 'swagman'; 'cowboy' for 'stockman', etc.

Researchers working in the area of Australian folklore require vast amounts of dedication and perseverance. Apart from the obvious keywords reflecting specific genres or forms, researchers may have to extend their search, depending on their topic, into less obvious corners of indexes in the effort to second-guess the indexer/cataloguer. Total reliance on keywords to locate titles may not be fully satisfactory in that many folklore titles tend towards the unusual or at least the apparently unrelated. An example is Bill Scott's work on Australian yarns, tales and urban legends, *The Long and the Short and the Tall*. Researchers may find that manual searching of the *Australian National Bibliography* will allow the development of sufficient confidence with keywords/subject headings within a specific and limited topic to allow an on-line computerized search to be conducted by state or academic library staff.

Other important bibliographic tools

TO DATE, TWO versions of a subject bibliography of Australian folklore have been published by the Australian Folk Trust, Inc. The original *Australian Folk Resources: A Preliminary Guide and Select Bibliography*, was compiled by Graham Seal and published in 1983. A revised version (with Ron Edwards) was published in 1988. Essentially, the works are a listing of collections of folklore materials (discussed below) and a subject bibliography covering published collections, discursive works, articles, journals, general works, research aids and unpublished materials. Coverage largely focuses on Anglo-Celtic Australian materials. Both versions proclaim their weakness in the title. The work originated as 'the more or less random collection of references gleaned from, or for, my own research', according to Seal. Nevertheless, the work remains important in the absence of a complete or systematic subject bibliography.

Several other publications provide useful bibliographic information. One is *200 Years of Australian Folk Song: Index 1788–1988*, compiled by Ron Edwards and bringing together much of his extensive research (some of which has been published elsewhere) and building upon previous research of Hugh Anderson published in 1957. The work provides an index as to the location of 1200 Australian folk songs, and cites 86 publications with anecdotal bibliographic information, though in Edwards's rather unique style. In 1992 Edwards produced for the Australian Folklife Centre *Australian Folk Song Publications in Principal Australian Libraries*.

Important for its breadth of coverage is the bibliography provided in Graham Seal's *The Hidden Culture: Folklore in Australian Society* (1989). References are cited for each chapter. In addition, useful specific bibliographies by topic or theme are provided on folklore generally and Australian folklore specifically. The section on Australian Folklore lists approximately 125 works. Seal provides an overview of the theoretical/practical aspects of folklore within Australia, past and present.

The field of children's folklore is well served bibliographically. June Factor compiled *Children's Folklore in Australia: An Annotated Bibliography* published in 1986. The work focuses on children's folklore of and by children as opposed to folklore for or about children, and includes both published and unpublished works, as well as non-book materials. Factor's most recent publication, *Captain Cook Chased a Chook* (1989), provides further bibliographic coverage of the subject, as well as an account of children's folklore in Australia. Further coverage is provided by the *Index to Australian Children's Folklore Newsletter* in issue Nos. 1–21 (December 1991).

The bibliographic limitations existing within folklore are large. With no comprehensive subject bibliography having been produced to date, the present writer hopes to shortly complete such a work with the assistance of the Australian Folk Trust.

Serials and articles

MUCH OF THE valuable work in folklore does not find its way into published books, but rather is published in journals or magazines. In some cases the publication may be little more than a newsletter. All such publications pose additional difficulties for researchers. Some of these publications are not even listed in bibliographies, since many of the publishers/organizations are unaware of or fail to follow the statutory requirement of notifying the National Library and their State Library of the publication's existence. Moreover, folklore serials are sometimes ephemeral, or metamorphize through regular changes of title, publication policy and so on, which poses problems for researchers attempting to follow the trail of a publication through various changes. The *Australian National Bibliography* draws attention to such changes, when known, in the 'notes' section of the bibliographic entry.

Access to periodicals and their contents is difficult in that many of the publications of obvious interest to folklorists are not indexed, which means that while it is known the publication exists, there is little chance of knowing the various contents over the years without consulting each issue.

Several journals which have fairly broadly based interest areas, such as *Meanjin* and *Overland*, have included articles pertaining to folklore over the years. The specialist folklore/folk music publications provided more specific and numerous opportunities for publication. These publications include newsletters of societies, journals of organizations, magazines and academic journals. Most state/territory 'folk' organizations produce some sort of publication which is legally required to be housed in state/territory libraries and the National Library. Examples include *Town Crier*, published by the Western Australian Folk Federation, and *Cornstalk Gazette*, published by the New South Wales Folk Federation. These publications tend to serve organizational interests but have occasional theoretical content (see **Magazines and other publications**).

Children's folklore is again well served by the School of Early Childhood Studies, University of Melbourne (formerly the Institute of Early Childhood Development) through its publication of the *Australian Children's Folklore*

Newsletter, which is indexed 1981–91. More academic concerns are covered in *Australian Folklore: A Yearly Journal of Folklore Studies* (and now sub-titled: *A Journal of Folklore Studies*) first published by the Centre for Australian Studies, Curtin University, and now published by the Australian Folklore Association. This journal provides coverage of fieldwork and analysis of Australian folklore and wide-ranging theoretical considerations pertaining to the study of folklore.

Most notable from a historical perspective are the publications *Mulga Wire/ Singabout* produced by the Bush Music Club, Sydney; *Australian Tradition* and *Gumsucker's Gazette* produced by the Folklore Society of Victoria/Victorian Folk Music Club, and more recently *Stringybark & Greenhide* (June 1979–85). Access to *Australian Tradition* (1963–74) is assisted through an index in the final edition and *Singabout* included an index to the first five volumes in Volume 5, No. 4 (1965).

Many not-so-obvious publications also carry material of interest to folklorists. A listing of articles provided by the National Library of Australia relating to folklore selected from *APAIS* (1973–84) (*Australian Public Affairs Information Service: A Subject Index to Current Literature*) listed publications such as *Daily Telegraph, Daily Mirror, Australian Women's Weekly, Quadrant, Mankind, Hemisphere, Australia Now, Australian Journal of Early Childhood, Orana, Dance Australia, Southern Cross, Reading Time, This Australia, Forum of Education* and even *People with Pix*. Curiously, publications with more obvious folklore content were not listed. *APAIS* is published by the National Library of Australia. The publication, which commenced in 1945, provides comprehensive coverage to approximately 200 publications of scholarly periodical literature in the Social Sciences and Humanities. It provides 'selective' coverage of many other periodical articles, conference papers, book and newspaper articles on Australian economic, social, cultural and political affairs based upon perusal by National Library staff. While the work aims to cover 'current' literature, an item may be included up to two years after its publication. While providing the only real index to materials on folklore contained in periodicals/serials, *APAIS* is very 'unfriendly' in its treatment of the subject area of folklore and related topics. The subject index unfortunately refers users from obvious broad subject headings to even broader subject headings. For example, the subject index under the heading of 'Folk Music' refers users to the much broader heading 'Music'. 'Folklife' is referred to 'Culture' and 'Folklore' is referred to 'Mythology'. In many cases the desired items would become lost in such broad subject areas. In other cases articles of obvious folklore content appear under unexpected subject headings. Hugh Anderson's 'The Making of Australian Traditions: A Survey of "Bush" Literature Since Russel Ward' in *Australian Folkore* 1, 1987 is not listed under mythology or culture, but appears under the subject heading 'national characteristics'. Users of *APAIS* are advised to consult *APAIS Thesaurus: A List of Subject Terms Used in the Australian Public Affairs Information Service* (3rd edn) and published by the National Library of Australia to acquaint themselves with the prescribed subject terms used in the index. Easier access is provided by author listing. On-line computer searching of *APAIS* is available.

Collections

COLLECTIONS ARISING OUT of the fieldwork conducted by folklorists might take the form of field recordings (audio and/or video), notes, transcripts, contextual notes, photographs and so on. The first problem is determining which institutions hold collections. (Incidentally, when folklorists talk about a 'collection', most libraries assume that reference is being made to its stock of published materials rather than to the unpublished results of fieldwork.) It is advisable to contact state libraries and archives to enquire about folklore collections. Both editions of *Australian Folk Resources, A Select Guide and Preliminary Bibliography* contain a listing of institutions with folklore holdings. The listings are now out of date, being based on an Australian Folk Trust survey which began in 1978. The project was plagued with on-going problems, not the least of which was a very poor response rate. The two editions unfortunately differ. The first edition, edited by Seal, lists sixty-five collections of institutions and individuals arranged by state/territory, providing address, telephone, subject areas, physical format, collectors, dates, size of holdings and access. Edwards's revised edition lists only forty-six institutions from the first edition, having removed entries referring to individuals. He has inserted a section devoted to individual collectors dating from 1900 to the present with seventy-three entries. In doing so Edwards had included some and deleted others. His criteria for such severe editing is unstated as is the basis for the information he himself provides. Researchers should obviously consult both editions. In 1992 Edwards produced for the Australian Folklife Centre an Occasional Paper, *Collectors of Australian Folklore*.

The most important single repository for folklore collections is the National Library of Australia, which at the time of writing housed a number of folklore collections within the oral history section. They are the result of work by John Meredith, Alan Scott, Chris Sullivan, Norm O'Connor, Peter Parkhill, Gwenda Davey, Kenneth Goldstein, Dave De Hugard, Wendy Lowenstein, John Marshall, Helen O'Shea, Mark and Cathy Rummery, Bill Scott, Hugh Anderson and Warren Fahey. Also included is the Wattle Collection based upon the Wattle Recording Company series. The National Library manuscript section also contains important papers of people like Russel Ward, Hugh Anderson, Peter Ellis and John Meredith. Documentation, however, is highly variable — being sometimes via indexes, lists or summaries as provided to the library by the collectors, or in some cases as provided by the Library. A good indication of the extent of the above collections can be gained through the National Library of Australia Catalogue, which provides details of informant, collector, date, nature of recording, listing of contents and degree of access. Some collections have access restrictions imposed by collectors and/or informants (as in the case of Meredith's papers). Moreover, immediate access is only available to materials which have been adequately preserved for archival purposes. Unfortunately, less than half of holdings have been preserved. Finally, since many collections contain materials other than field recordings (e.g. photographs, musical notation, notes), searchers may require multiple access points to survey an entire collection.

Other institutions with collections include the La Trobe Library, housing the Australian collections of the State Library of Victoria. This institution houses 111 tapes covering folklore and social history research by Wendy Lowenstein. Peter Parkhill has lodged numerous tapes of the traditional music of the Arabic, Cretan, Italian, Greek, Turkish, Ukrainian, Anglo-Celtic, Macedonian, Czech and Lebanese communities of Australia. The Hmong-Australian Society also houses its tapes (from an on-going study of the Hmong community in Melbourne) in the La Trobe Library.

The National Film and Sound Archive is developing an important collection of materials pertaining to Australian folklore which complements and augments the work of the National Library, and as such is worth researching.

Two other institutional collections are the Australian Children's Folklore Collection at the Australian Centre within the University of Melbourne. This collection is well catalogued. The folklore of Western Australia is housed within the West Australian Folklore Archive which is part of the Centre for Australian Studies, Curtin University of Technology. This collection has a comprehensive computerized catalogue.

Many libraries maintain oral history programmes which obviously collect items of folklore in the overall process of recording oral history. Such materials only infrequently distinguish folkloric content, but should be consulted where possible. Likewise, many state libraries maintain extensive and important music collections (such as the published music collection in the Mitchell Library in Sydney).

Easier access is provided to a variety of collections which have been presented in some published form. Reflecting the predominant interests within Australian folklore, their emphasis is clearly upon folk music and song, folk dance and, to a lesser extent, craft. To name but a few, notable authors/collectors include: Patsy Adam-Smith, Shirley Andrews, Hugh Anderson, W. A. Beatty, 'Bill Bowyang' (A. V. Vennard), Ron Edwards, Warren Fahey, William Fearn-Wannan (Bill Wannan), Nancy Keesing and Douglas Stewart, Wendy Lowenstein, John Manifold, John Meredith, Vance Palmer, A. B. (Banjo) Paterson, Alan Scott, William (Bill) Scott, Charles Thatcher, Ian Turner.

David Hults